ELECTRON CRYSTALLOGRAPHY
OF BIOLOGICAL
MACROMOLECULES

ELECTRON CRYSTALLOGRAPHY OF BIOLOGICAL MACROMOLECULES

Robert M. Glaeser
Kenneth Downing
David DeRosier
Wah Chiu
Joachim Frank

UNIVERSITY PRESS

2007

UNIVERSITY PRESS

Oxford University Press, Inc., publishes works that further
Oxford University's objective of excellence
in research, scholarship, and education.

Oxford   New York
Auckland   Cape Town   Dar es Salaam   Delhi   Hong Kong   Karachi
Kuala Lumpur   Madrid   Melbourne   Mexico City   Nairobi
New Delhi   Shanghai   Taipei   Toronto

With offices in
Argentina   Austria   Brazil   Chile   Czech Republic   France   Greece
Guatemala   Hungary   Italy   Japan   Poland   Portugal   Singapore
South Korea   Switzerland   Thailand   Turkey   Ukraine   Vietnam

Copyright © 2007 by Oxford University Press, Inc.

Published by Oxford University Press, Inc.
198 Madison Avenue, New York, New York 10016

www.oup.com

Oxford is a registered trademark of Oxford University Press

All rights reserved. No part of this publication may be reproduced,
stored in a retrieval system, or transmitted, in any form or by any means,
electronic, mechanical, photocopying, recording, or otherwise,
without the prior permission of Oxford University Press.

Library of Congress Cataloging-in-Publication Data
Electron crystallography of biological macromolecules / Robert M. Glaeser ... [et al.].
   p. cm.
ISBN 978-0-19-508871-7
1. Crystallography. 2. Electron microscopy. 3. Macromolecules. I. Glaeser, Robert M.
QD906.7.E37E39 2007
547'.704425—dc22   2006017707

9 8 7 6 5 4 3 2 1

Printed in the United States of America
on acid-free paper

To all who wish to learn more about the capability offered by electron crystallography within the realm of Structural Biology.

"...It therefore seems that a working knowledge of the phenomena and laws of diffraction might well form a part of the equipment of everyone who uses the microscope and attempts to interpret its indications."
—*A.B. Porter, 1906*

Preface

Electron microscopy of purified macromolecular complexes, viruses, and similar "particles" has long been an important technique in cell biology and biochemistry. What distinguishes *electron crystallography* of biological macromolecules from these more traditional investigations is the use of well-defined computational tools to obtain three-dimensional representations of a structure, rather than relying on more intuitive interpretations of one or more views (projections) of the structure. The strong relationship that these computational tools have with those used in x-ray crystallography is only a partial justification for extending the definition of "crystallography" to include data collection and analysis by electron microscopy, however. The common goal of modeling molecular structures in terms of intra- and intermolecular interactions, extending even to the atomic structure of the macromolecules, establishes a further, deep connection between the newer field of electron crystallography and the long-established field of x-ray crystallography.

Work in structural biology now is currently expanding its emphasis on large complexes and macromolecular assemblies. Electron crystallography has thus come to the fore as a valuable, and in some cases essential, method within the context of biochemistry and cell biology. Compared to the field of x-ray crystallography, however, there are still relatively few review articles and even fewer books that scientists and students new to the field can consult in order to learn what they must know in order to begin work in electron crystallography. The time thus seems appropriate to assemble a unified source of didactic material related to electron crystallography. Our goal in doing so is to make essential background material easily available to those who might otherwise find it difficult to know where to start, and who may never be quite sure — at least at the beginning — what else they might be missing. This is not to say that a single book can come close to covering everything that is relevant and useful! We hope, however, that the current book will provide a useful start.

The planning of the Table of Contents was largely a team effort on the part of all authors of this volume. The drafting of individual chapters, on the other hand, has been done primarily by one or sometimes two members of the team. Each chapter was then circulated for review to the rest of the team, in most cases more than once. Although considerable improvement resulted from the critiques that were exchanged, we have not insisted on unanimity among us regarding the selection of every topic that is covered, which points are emphasized, or even what opinions are expressed. While we hope, of course, that no material included in the book is misleading or wrong, the principle author(s) of each respective chapter want to take individual responsibility — just as they would do in an edited volume that involved less of a team-effort — for their individual viewpoints and certainly for any mistakes that may have survived our necessarily limited process of collaboration. The list of authors responsible for drafting and completing individual chapters is broken down as follows: chapters 1–8, Glaeser; chapters 9–11, Downing and Glaeser; chapter 12, DeRosier; chapter 13, Chiu; chapter 14, Frank; and chapter 15, Glaeser.

The drafting of this book has been a project that extended over quite a few years, long enough, in fact, that it is now difficult to acknowledge the large number of students, postdoctoral scientists, and colleagues who contributed significantly with their comments and advice regarding preliminary drafts of various chapters. A very special "thank you" is owed to Wolfgang Baumeister, who — although not listed as a coauthor — participated in the original planning of the outline, provided extended hospitality while Glaeser wrote early drafts of much of the early chapters in the book, and organized substantial input for a portion of chapter 8. With sincere apologies to those who are not mentioned below but equally deserve to be, we also want to acknowledge the input of Umesh Adiga, Felicia Betancourt, Luis Comolli, Richard Hall, Wen Jiang, Alison Killilea, Huilin Li, Eva Nogales, Raj Pai, Robert Tampe, Mike Schmid, Chuck Sindelar, Henning Stahlberg, Dieter Typke, Janet Vonck, Sharon Wolf, and Z. Hong Zhou.

Finally, we are grateful to many individuals who have provided materials for various figures in the text. In addition to those whose contributions are acknowledged in individual figure legends, we express special thanks to Chuck Sindelar, who invested a considerable amount of time to convert starting materials for many of the figures into electronic form, and even created a number of figures according to rough sketches that were given to him.

Contents

1 INTRODUCTION 3
   1.1   Electron crystallography provides access to a unique class of problems in structural molecular biology 3
   1.2   High-resolution crystallography requires averaging of structures that are present in multiple copies 5
   1.3   Electron crystallography can produce three-dimensional density maps that are interpretable in terms of an atomic model of the structure 7
   1.4   Electron crystallography has developed from rich intellectual origins in optics, electron microscopy, and x-ray crystallography 11
   1.5   Objectives of this book 15

2 STRUCTURE DETERMINATION AS IT HAS BEEN DEVELOPED THROUGH X-RAY CRYSTALLOGRAPHY 17
   2.1   Introduction 17
   2.2   Structure analysis by x-ray crystallography requires well-ordered, three-dimensional crystals 18
   2.3   The practical steps of data collection and data analysis have become very efficient 19
   2.4   The Fourier transform plays a central role in understanding the analysis of diffraction data 19
   2.5   The Fourier transform of a crystal represents discrete, regular samples of the continuous Fourier transform of the molecule 26

x   Contents

- 2.6 The disorder that exists in real crystals can result in easily observed changes in the Fourier transform 32
- 2.7 The Ewald sphere: a powerful mental picture that shows what part of the Fourier transform can be measured for every orientation of the specimen 34
- 2.8 Bragg's law relates the measured scattering angle to the size of the repeat-distance for each sinusoidal term in the Fourier transform of the object 36
- 2.9 Information about the relative phase of each sinusoidal term is lost in diffraction patterns 38
- 2.10 The crystallographic phase problem is usually solved by using additional data obtained from heavy-atom derivatives of the original molecular crystals 39
- 2.11 The three-dimensional electron density of the molecule can be calculated from the experimentally measured amplitudes and phases of the Fourier transform 43
- 2.12 The 3-D density map must be interpreted in terms of other available information, to provide a model of the structure 44
- 2.13 A more accurate estimate of the structure can be obtained by further *refinement* of the model 46
- 2.14 Published structures are made available through a public-domain database 48

3 FOURIER OPTICS AND THE ROLE OF DIFFRACTION IN IMAGE FORMATION 49

- 3.1 Introduction 49
- 3.2 *Abbe's diffraction theory of images*: image formation is the two-dimensional equivalent of the crystallographer's "inverse Fourier transform" 50
- 3.3 Zernike and the invention of phase contrast microscopy 52
- 3.4 The rigorous diffraction theory of image formation describes images in terms of the inverse Fourier transform 54
- 3.5 The lens as a linear system: transfer functions play an important role in Fourier optics 59
- 3.6 The most common applications of Fourier optics in electron crystallography require that the specimen behaves like a weak phase object 63
- 3.7 The image intensity for a weak phase object remains linear in the projected Coulomb potential 64
- 3.8 The concept of a "phase contrast transfer function" is of central importance in the interpretation of high-resolution images 67
- 3.9 Partial coherence imposes an envelope on the phase contrast transfer function 69
- 3.10 Amplitude contrast can also contribute in an important way to images of thin, biological specimens 72
- 3.11 Single side band images: blocking half of the diffraction pattern produces images whose transfer function has unit gain at all spatial frequencies 74

- 3.12 Tilted illumination produces images for which the transfer function includes both phase errors and amplitude modulations 75
- 3.13 Summary: Fourier optics is an important part of the conceptual foundation of electron crystallography 76

4 THEORETICAL FOUNDATIONS SPECIFIC TO ELECTRON CRYSTALLOGRAPHY 77
- 4.1 Introduction 77
- 4.2 The single-scattering (kinematic scattering) approximation and the weak phase object approximation are mathematically similar but not identical 78
- 4.3 Proof of the projection theorem 81
- 4.4 Two important simplifications of crystallographic structure analysis occur when the specimen is approximated as a weak phase object 82
- 4.5 Three-dimensional Fourier space is sampled by collecting data at many different tilt angles 83
- 4.6 The resolution of a 3-D reconstruction is determined by the spatial frequency limit of the measurements and by the completeness of 3-D data collection 85
- 4.7 *Radiation damage* represents a much more important experimental constraint in electron crystallography than in x-ray crystallography 93
- 4.8 Images become very noisy at high resolution due to the finite, "low" exposures which are permitted within acceptable limits of radiation damage 101
- 4.9 Spatial averaging must be used in order to overcome the limited statistical definition that is possible when images are recorded with "safe" levels of electron exposure 102
- 4.10 The amount of averaging required is determined by the number of scattered electrons and by the image quality 104

5 INSTRUMENTATION AND EXPERIMENTAL TECHNIQUES 106
- 5.1 Introduction 106
- 5.2 The basic design of an electron microscope is much like that of a light microscope 107
- 5.3 Technical features that are specific to electron optics 108
- 5.4 Specimen stages 123
- 5.5 Detectors that are suitable for observing and recording images and diffraction patterns 126
- 5.6 Low-dose techniques make it possible to record high-resolution images and diffraction patterns even from easily damaged specimens 131
- 5.7 Spot-scan imaging can minimize beam-induced movement 134
- 5.8 Samples prepared as self-supported specimens within (or over) holes require additional precautions in order to minimize specimen charging 137

6 SPECIMEN PREPARATION   139
- 6.1 Introduction   139
- 6.2 Negative staining provides high contrast as well as excellent stability in the electron beam   140
- 6.3 Metal shadowing produces stable samples which reveal surface topography   142
- 6.4 Glucose and other "sustains" can preserve macromolecular structures to high resolution   145
- 6.5 Contrast matching can be manipulated by using embedding media with different densities   147
- 6.6 Embedding in vitreous ice is the preferred alternative for the preparation of unstained, hydrated specimens   150
- 6.7 Charging and mechanical stability vary with details of the specimen preparation method   159
- 6.8 Preparing extremely flat specimens continues to be one of the most important challenges when working with 2-D crystals   161

7 SYMMETRY AND ORDER IN TWO DIMENSIONS   167
- 7.1 Introduction   167
- 7.2 Classes of symmetry in projection   168
- 7.3 Three-dimensional symmetry classes for monolayer crystals   175
- 7.4 The Fourier transform of a 2-D crystal is sampled at discrete points in two dimensions, but it is continuous in the third dimension   182
- 7.5 Disorder and crystalline defects are an important fact of life   187

8 TWO-DIMENSIONAL CRYSTALLIZATION TECHNIQUES   194
- 8.1 Introduction   194
- 8.2 Integral membrane proteins represent a natural target for 2-D crystallization   195
- 8.3 Many soluble proteins also form very thin crystals   201
- 8.4 Crystallization at interfaces has potential for wide generality   203

9 DATA PROCESSING: DIFFRACTION PATTERNS OF 2-D CRYSTALS   211
- 9.1 Introduction   211
- 9.2 Diffraction intensities are used in a variety of ways in electron crystallography   212
- 9.3 Data that have been recorded on photographic film must be converted to digital form with a scanning microdensitometer   213
- 9.4 Density versus exposure characteristics can be used to convert the film density to the corresponding value of electron intensity   215
- 9.5 Data can also be collected by direct electronic readout rather than on photographic film   217
- 9.6 The digitized diffraction patterns are then indexed and reduced to the final diffraction intensities   219

9.7 Intensities from individual diffraction patterns are merged to form a 3-D data set 225
9.8 Factors that affect data quality 230

10 DATA PROCESSING: IMAGES OF 2-D CRYSTALS 234
10.1 Introduction 234
10.2 Optical diffraction is an effective tool for the preliminary evaluation of image quality 235
10.3 Conversion of the image to a digital form is necessary for computer processing 237
10.4 The fast Fourier transform is an efficient algorithm for numerical computation 244
10.5 Images of crystals: indexing the Fourier transform is similar to indexing the electron diffraction pattern 246
10.6 Extraction of amplitudes and phases from the indexed Fourier transform 247
10.7 Establishing a common phase origin allows data from separate crystals to be merged into a 3-D data set 253
10.8 Evaluation of data quality is based on the signal-to-noise ratio 257
10.9 Quasi-optical filtering reduces the noise in the image 259
10.10 Correction for distortions in the image increases the signal quality 263
10.11 Corrections are also required for other systematic image defects 270

11 HIGH-RESOLUTION DENSITY MAPS AND THEIR STRUCTURAL INTERPRETATION 277
11.1 Introduction 277
11.2 Three-dimensional density maps are computed from discrete samples of the complex structure factors 278
11.3 Options for the display of 3-D density maps 279
11.4 The missing cone of data results in poorer resolution in the direction perpendicular to the plane of the 2-D crystal 282
11.5 Interpretation of the high-resolution map involves building the known chemical structure into the 3-D density 288
11.6 Accurate atomic-resolution models can also be obtained by docking atomic models of individual components into the 3-D density map of a macromolecular complex 291
11.7 Refinement of an atomic-resolution model may proceed in a different way for electron crystallography than is traditionally done in x-ray crystallography 293
11.8 Difference Fourier maps 300

12 ELECTRON CRYSTALLOGRAPHY OF HELICAL STRUCTURES 304
12.1 Introduction 304
12.2 Ideal helices and their diffraction patterns 307
12.3 Real helices and their diffraction patterns 318

xiv Contents

12.4 The hardest step: indexing the diffraction pattern 325
12.5 Gathering amplitudes and phases is the next step in the reconstruction process 330
12.6 Calculating and interpreting three-dimensional maps 336
12.7 Helical particles with a seam can be analyzed by extending the method for helical particles 339
12.8 Helical structures can be analyzed using single-particle methods 340
12.9 The future looks bright 342

13 ICOSAHEDRAL PARTICLES 343
13.1 Introduction 343
13.2 Description of an icosahedron 344
13.3 Local symmetries can be present within an asymmetric unit 347
13.4 Theory of icosahedral reconstruction 347
13.5 Experimental considerations 349
13.6 Data evaluation 351
13.7 Image restoration 352
13.8 Initial model building and structure refinement 354
13.9 Resolution evaluation 360
13.10 Poststructure analysis 362
13.11 Atomic model determination 363

14 SINGLE PARTICLES 365
14.1 Introduction 365
14.2 A certain minimum dose is required to align images of single molecules 368
14.3 Due to the lack of symmetries, 3-D imaging requires coverage of the entire angular space 369
14.4 Conformational variability increases the total number of images needed to achieve higher resolution 370
14.5 Alignment of particles is required for averaging and image reconstruction, and its principal tool is the cross-correlation function 371
14.6 Classification may be used to divide the projection set according to viewing directions, conformations, and ligand-binding states 374
14.7 Variational patterns among images of macromolecules can be found by using multivariate data analysis or self-organized maps 375
14.8 Two useful methods of classification in single particle analysis are hierarchical ascendant classification and K-means clustering 385
14.9 Real-space reconstruction techniques can deal with the general 3-D projection geometries encountered in single-particle reconstruction 388
14.10 Random-conical and common-lines methods can provide angular relationships among the molecule projections, as a way to jump-start a reconstruction project 395
14.11 Angular refinement methods are used to proceed from the initial reconstruction to the final reconstruction 399
14.12 Single-particle reconstruction in practice 401
14.13 What are the prospects of achieving atomic resolution? 413

15 SPECIAL CONSIDERATIONS ENCOUNTERED WITH THICK SPECIMENS 415

- 15.1 Introduction 415
- 15.2 Dynamical diffraction can be described by a number of different, but equivalent mathematical formalisms 416
- 15.3 Conditions when kinematic diffraction theory fails 419
- 15.4 Strong dynamical diffraction effects need not interfere with subsequent refinement of an atomic-resolution model of the structure 424
- 15.5 *Fresnel diffraction* alone can become significant in thick specimens 426
- 15.6 Curvature of the Ewald sphere destroys the appearance of Friedel symmetry at high resolution and at high tilt angles 428
- 15.7 Inelastic scattering becomes an important consideration in thick specimens 430
- 15.8 A final caution: failure of Friedel symmetry for thick specimens can be due to curvature of the Ewald sphere, dynamical diffraction, or inelastic scattering 437

References 441

Index 469

ELECTRON CRYSTALLOGRAPHY
OF BIOLOGICAL
MACROMOLECULES

1

Introduction

1.1 Electron crystallography provides access to a unique class of problems in structural molecular biology

In the language of molecular biology, the term "x-ray crystallography" has become synonymous with using x-ray diffraction data to determine molecular structure. We will use the related term *electron crystallography* to describe the use of electron diffraction data and electron microscope images to determine the structure of biological macromolecules. Our use of the word "crystallography" involves a considerable broadening of terminology from what has been customary, however, because direct images are introduced into crystallographic structure analysis for the first time when electrons are employed. The use of images of single molecules, which is a possibility with electrons, strains even more the definition of what one means by "crystallography," because crystals can even be, in effect, "grown" computationally by the coherent alignment of images of many thousands of such molecules.

Significant progress has been made in the development of electron crystallography, such that it is now possible to get three-dimensional structures at a resolution similar to that obtained by x-ray crystallography, neutron diffraction and multidimensional NMR spectroscopy. As a consequence, high-resolution electron crystallography has become an important, new research tool that further expands the scope of structural research into unique areas of cell biology, biochemistry, and molecular biology.

The uniqueness of electron crystallography as a research method is based on two physical properties of electrons. The first of these is the high scattering power of electrons, which is about 100,000 times greater than that of x-rays or neutrons. As a result, specimens must be very thin; indeed, samples that are only one molecule thick are often ideal. Figure 1.1 shows an example of the type of high-resolution electron

Figure 1.1 Electron diffraction pattern obtained by Dr. Janet Vonck (unpublished) from a small glucose-embedded crystal of bacteriorhodopsin, prepared by the fusion of native patches of purple membrane that were isolated from *Halobacterium salinarum*. The area of this membrane fragment was only about 15 μm^2, with a thickness of only 4.5 nm. The hexagonal unit-cell constant in purple membrane is 6.24 nm. This pattern shows the very high sensitivity available by electron diffraction, due to the much stronger scattering of electrons in comparison to x-rays. The diffraction pattern, recorded initially on film, has been digitized, and the radially symmetric background has been subtracted.

diffraction pattern that can be obtained from thin, small crystals. This specimen was a glucose-embedded crystal of bacteriorhodopsin with a thickness of 4.5 nm and a diameter of about 4 μm, obtained by the fusion of even smaller two-dimensional (2-D) crystals that were isolated from the native cell membrane of *Halobacterium salinarum*. The second important property is the fact that electron waves can be focused by magnetic or electric fields. The ability to focus electrons into a high-resolution image has the consequence that the crystallographic phase information is fully retained in the primary data. The fact that the phase information is not lost in images, as it is in diffraction patterns, is clearly a great advantage.

The diffraction-amplitude information for thin crystals is also preserved in electron microscope images. Diffraction amplitudes can nevertheless be recorded with greater accuracy in electron diffraction patterns than is possible in images. For this reason, it is better to combine amplitude data from electron diffraction patterns with phase data obtained from images, whenever that is possible. When the sample is either too small or not sufficiently well ordered to record electron diffraction patterns, however, there is no choice but to extract the diffraction amplitudes as well as the phases from the image.

The development of electron crystallography makes it possible to tackle new types of structural problems involving very thin specimens, for which there previously was little hope that high-resolution structures might be obtained. X-ray crystallography is an excellent method for suitably thick crystals, but it can give only a limited amount of information about the structure of very thin samples. Examples of the latter include *two-dimensional crystals* of integral membrane proteins; helical and other filamentous assemblies; and even isolated, free-standing macromolecular assemblies. Much more is said in chapter 8 about the kinds of specimens that form 2-D crystals, and in chapters 12, 13, and 14 we present detailed information about methods to study helices, icosahedra, and single particles, respectively.

When samples are made up of *single particles*, far more structural information can usually be obtained by electron microscopy than by any other method. Experimental methods such as x-ray or neutron solution scattering generally yield the spherically averaged (or in some cases cylindrically averaged) values of the diffraction intensities. The averaging process, in turn, results in a major loss of available information. Having direct images of each particle, on the other hand, makes it possible to merge data from many such individual particles, without losing track of their relative orientations.

Finally, it is worthwhile to point out that the mathematical theory of electron scattering is identical to that of x-ray scattering (which is reviewed in chapter 2). The only difference between the two is that the physical object is represented by the electron charge density, $\rho(r)$, in x-ray scattering theory, while it is represented by the Coulomb potential of the atomic nuclei, shielded by their surrounding electrons, in the case of electron scattering theory. We have therefore chosen to represent the *shielded Coulomb potential* in this book by the related mathematical symbol, $\hat{\rho}(r)$, in order to emphasize the formal identity between electron scattering theory and x-ray scattering theory.

Electron crystallography nevertheless differs from x-ray crystallography in a number of significant ways. Among these differences is the use of Fourier optics to describe image theory and the recovery of phase information from images, for which there is no counterpart in x-ray crystallography. Conversely, the use of heavy atom derivatives and anomalous dispersion effects to recover phase information, essential to x-ray crystallography, have not yet had a practical role in the development of electron crystallography of biological macromolecules.

1.2 High-resolution crystallography requires averaging of structures that are present in multiple copies

The familiar diffraction patterns that we know from x-ray crystallography represent the coherent sum of scattering from all of the molecules within the beam. Crystals serve the important role of ensuring that the information in the wave that is scattered by each of the molecules is superimposed, that is, averaged together, only from molecules that are in identical orientations.

The idea that real crystals are needed to obtain high-resolution structural details, although inescapable for x-ray diffraction experiments, no longer applies when electrons are used. If the molecules in the sample are in many different orientations to begin with, as they would be in an aqueous solution, it is only necessary to separate their images into different classes on the basis of the respective views that they present. The indiscriminate averaging of everything within the field of illumination without regard to its orientation, which occurs in the case of solution scattering, can thereby be avoided. Whether the molecules are in a crystal or in random positions and orientations, all that is needed is first to combine data from many molecules that are in identical orientations, and next to merge such data from a large number of different orientations.

If it is possible to get images of individual molecules with electrons, why is it still necessary to average a large number of these images to get high-resolution data? The answer is that one must record a rather large number of electrons at each point of the image in order to produce good statistical definition of the image intensity. At the same time, however, *radiation damage* sets a fundamental limitation on the maximum electron exposure that can be used. The largest electron exposure that can safely be used is, unfortunately, too low to produce a well-defined, high-resolution image of a single molecule (Breedlove and Trammell, 1970; Glaeser, 1971). This dilemma is not unique to electrons; radiation damage is also a limitation for x-rays and all other forms of short-wavelength radiation that might be considered for high-resolution structural studies (Breedlove and Trammell, 1970; Henderson, 1995). The only way around this dilemma is to build up the statistical definition of the data by averaging a large number of images, each of which has been taken at a low, safe exposure.

If one must average images from a large number of individual molecules, then it is easier to do so with crystals than with single molecules. Because the molecules in a crystal form a regular, repeating pattern, the required spatial averaging over the array is carried out automatically when one uses a computer to calculate the Fourier transform of the image. For this and other reasons, numerical calculation of the Fourier transform is one of the most common ways to start the data processing when working with images of thin crystals. The large number of molecules that can be included when calculating the Fourier transform of a crystal ensures that the signal-to-noise ratio in the average will be high.

It is also possible to superimpose images of individual molecules directly, however, rather than doing so indirectly by means of the Fourier transform. Although it may seem surprising at first, *cross-correlation* can be used to align and superimpose two images with an accuracy that is much greater than the resolution at which individual image features are statistically well defined. The larger the areas that are used for the alignment of two images, the more accurate the alignment will be (Saxton and Frank, 1977). This is true because the correlation between two images is much higher when all of the image points are considered than it is if just a single, fixed point is compared.

In the future it is likely that atomic-resolution data will be obtained by aligning and averaging both small crystalline patches and large macromolecular structures. In current practice a resolution better than 1 nm has been reached by merging data from images of over 50,000 individual ribosomes (Halic et al., 2005), which are large particles with no internal symmetry. It has also been shown by Henderson (1995) that a resolution of $\sim$0.3 nm should be possible for particles with molecular weight $\sim$4 MDa, using images of one million particles with the image quality that is currently available. Further improvement in image quality, which is currently still below the limit set by atomic scattering cross sections, would lower this molecular weight limit substantially, and at the same time the required number of particles would also be greatly reduced. The practical results that are currently achievable are therefore far short of what would be physically possible if the image quality were perfect (Glaeser, 1999), and as a result it is still necessary to use relatively large, ordered arrays in order to get data at ~0.35 nm resolution.

Whether molecules are lined up and positioned for spatial averaging by biochemical means (i.e., they really are crystallized), or whether this is done mathematically (i.e., images of individual molecules are, in effect, "crystallized" in the computer), the effect is the same. The signal steadily accumulates as more molecules are included in the average, while the random noise in the image of one molecule tends to cancel out with the random noise in the image of the next molecule. The signal-to-noise ratio therefore increases as the square root of the number of molecules that are included in the average. This law is what makes it possible to use either real crystals or computer-generated "crystals" to overcome the limitations of low statistical significance of the images of single molecules.

Practical considerations nevertheless result in other major differences between the use of single particles and 2-D crystals. On the one hand, for example, it is possible to immediately tell from an electron diffraction pattern of a 2-D crystal (and from the computed diffraction pattern of an image of a 2-D crystal) whether the data can, in principle, be processed to high resolution. On the other hand, collection of high-resolution data from highly tilted samples continues to suffer from a low rate of success. Since it is possible for single-particle specimens to present a near-random distribution of views, at least in favorable cases, it may not be necessary to tilt such specimens in order to generate a 3-D data set. As a result, the yield of acceptable micrographs can be much higher when collecting data from untilted, single-particle specimens than from highly tilted, 2-D crystals. The final resolution that can be achieved with single-particle specimens can only be determined after averaging a large number of images, however, and this involves much more computational work than is the case for a 2-D crystal.

1.3 Electron crystallography can produce three-dimensional density maps that are interpretable in terms of an atomic model of the structure

Electron crystallography has advanced to the point that *three-dimensional (3-D) density maps* can be obtained at a resolution better than 0.4 nm. Density maps of proteins can be interpreted at this resolution in terms of an atomic-resolution model of the specimen if the chemical structure (the amino acid sequence) is known. Thus, electron crystallography is currently able to provide structural information about biological macromolecules at a level of detail that is close to what can be obtained by x-ray crystallography.

Integral membrane proteins represent a particularly large class of structures for which one expects that electron crystallography will play a major role. To begin with, a few integral membrane proteins already exist as 2-D crystals within the native cell membrane. Bacteriorhodopsin is by far the most favorable specimen of this type that is known. The development of an atomic-resolution model of bacteriorhodopsin (Henderson et al., 1990) represents the culmination of pioneering work that first demonstrated that protein structures could be obtained by the method of electron crystallography. Since then, the methodology of crystallographic refinement, a standard part of the structure-determination process in x-ray crystallography, has also been adapted to the unique circumstances that prevail in electron crystallography (Grigorieff et al., 1996; Mitsuoka et al., 1999; Lowe et al., 2001; Gonen et al., 2005).

High-resolution studies on membrane proteins are in no way limited to naturally occurring 2-D crystalline arrays, however. An atomic-resolution model of a photosynthetic light-harvesting complex from green plants, LHC II, has been obtained with 2-D crystals that were prepared from biochemically purified protein (Kuhlbrandt et al., 1994). Figure 1.2 shows one of the sections of the 3-D density map, with a portion of the polypeptide model built into the density; a cartoon of the complete protein structure is also shown as a ribbon drawing in figure 1.2. In this case, 2-D crystals were obtained unexpectedly from the detergent-solubilized, purified protein in the course of attempts to obtain large, 3-D crystals that would be suitable for x-ray crystallography. The formation of thin crystals of integral membrane proteins in the presence of solubilizing detergent appears to have considerable generality. Thin crystals of Ca-ATPase (Stokes and Green, 1990; Taylor and Varga, 1994; Shi et al., 1995) and of Na/K-ATPase (Taylor and Varga, 1994) have been obtained in this way, and the H^+-ATPase of *Neurospora crassa* has been crystallized in detergent by a novel interaction with the air–water interface (Cyrklaff et al., 1995). More is said about these and other methods of producing 2-D crystals in chapter 8.

Integral membrane proteins can also be crystallized as monolayer crystals, following detergent solubilization and purification, simply by reconstituting the protein with limiting quantities of lipid (Leonard et al., 1987; Engel et al., 1992; Jap et al., 1992;

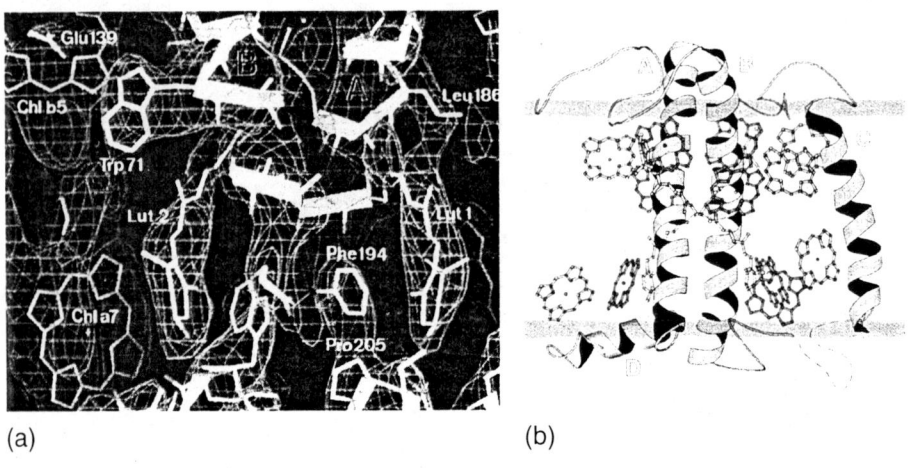

Figure 1.2 Molecular graphics representations of the light-harvesting protein LHC II from green plant photosynthetic membranes. (a) A portion of the 3-D electron crystallographic density map is shown, with the corresponding pieces of the atomic-resolution models of the protein, carotenoids, and chlorophyll molecules built into the density. (b) A ribbon diagram is used to represent the overall fold of the protein part of the structure. Illustrations are from Kuhlbrandt et al. (1994), with permission.

Kuhlbrandt, 1992). An early example that demonstrated that high-resolution diffraction patterns could be obtained after crystallization of a membrane protein by reconstitution with phospholipid is represented by the work done on PhoE porin, a pore-forming protein from the outer membrane of *E. coli* (Jap, 1988; Walian and Jap, 1990). More recent examples of membrane proteins that have been purified and then reconstituted with lipid as well-ordered 2-D crystals include the water-channel protein isolated both from red blood cell membranes (Fujiyoshi et al., 2002; de Groot et al., 2003) and from lens tissue (Gonen et al., 2004, 2005); a glutathione transferase from microsomal membranes (Hebert et al., 1997; Schmidt-Krey et al., 1999); and two bacterial secondary transporters (Williams, 2000; Hirai et al., 2003).

Yet another method to obtain crystals of integral membrane proteins is based upon the spontaneous or induced ordering of the target protein within isolated membranes that are highly enriched in a single protein component. In practice this approach has resulted mainly in small patches of 2-D crystals or in tubular, helical cylinders, as is described further in chapter 8. Work on helical tubes of the nicotinic acetylcholine receptor has led to a chain-trace model at a resolution of ~0.4 nm (Miyazawa et al., 2003), using images of frozen-hydrated tubes such as the one shown in figure 1.3. It is worth noting that the chain-trace structure of a bacterial flagellum has also been solved by electron crystallography (Yonekura et al., 2003), again at a resolution of ~0.4 nm, although this is not a membrane protein, of course. Even higher-resolution

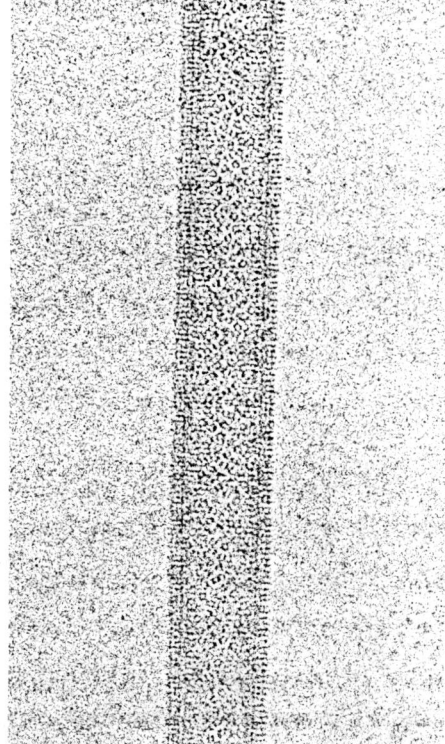

Figure 1.3 High-resolution electron micrograph of a helical tube of the nicotinic acetylcholine receptor, which is formed from membrane vesicles isolated from *Torpedo marmorata* (Unwin, 1993). This is an unstained, fully hydrated specimen, prepared by the frozen-hydrated specimen technique. Figure provided courtesy of Dr. Unwin.

structural information will become available from "crystals" of this sort, as the technical development of electron crystallographic methods continues to advance.

The fortuitous formation of thin 2-D crystals of soluble proteins, produced in the course of efforts to obtain 3-D crystals or as the result of other biochemical studies, can also lead to significant opportunities for electron crystallography. A particularly good example is the formation of 2-D monolayer crystals of tubulin that occurs when Zn^{2+} is included in the polymerization buffer (Larsson et al., 1976). With refinement of the crystallization conditions, it proved to be possible to obtain electron diffraction patterns like the one shown in figure 1.4. A high-resolution, 3-D density map obtained with these specimens has, in turn, been used to solve the structure of the α, β tubulin dimer at atomic resolution (Nogales et al., 1998).

A variety of other methods are able to provide thin 2-D crystals that are suitable for electron crystallography. One general-purpose method involves the adsorption (or binding) of the protein of interest onto a lipid monolayer or a lipid bilayer. It has been shown that monolayer crystals of streptavidin can be grown by binding the protein to biotinylated lipids at the air–water interface, and these crystals diffract to at least 0.3 nm resolution (Kubalek et al., 1991, Avila-Sakar and Chiu, 1996). Even structures as large and complex as RNA polymerase (Darst et al., 1988, 1991b; Hemming et al., 1995)

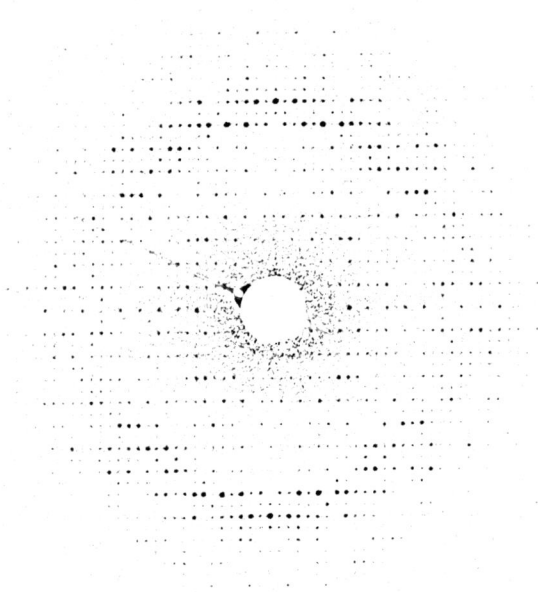

Figure 1.4 Electron diffraction pattern from a 2-D crystal of tubulin (Downing and Li, 2001). Well-diffracting, two-dimensional crystals of tubulin can be grown by including Zn^{2+} ions in the tubulin-polymerization buffer (Downing and Jontes, 1992). The diffraction pattern, which extends to a resolution of 0.28 nm, was recorded on a CCD camera, and both the radially symmetric background and the locally determined background have been subtracted.

or the 50S ribosomal subunit (Avila-Sakar et al., 1994) can be induced to form 2-D crystals by simple adsorption to a positively charged lipid monolayer.

Another important application of electron crystallography involves the production of *pseudo-atomic models* that combine (1) relatively low-resolution Coulomb potential maps of a complex macromolecular assembly with (2) atomic-resolution models of the component molecules. Experience with a number of systems supports the idea that atomic models can be docked together accurately when the overall envelope of the assembled complex is known to a resolution of only ~2 nm (Volkmann and Hanein, 1999; Wriggers et al., 1999; Rossmann and Tao, 1999; Wriggers and Birmanns, 2001; Volkmann and Hanein, 2003). In an exciting example of this approach, the high-resolution x-ray structures of myosin S1 ATPase and actin, determined separately, were fitted together into the lower-resolution envelope of the actin–S1 complex, determined by helical reconstruction from images of actin–S1 filaments (Rayment et al., 1993). As can be seen in figure 1.5, the fit is unambiguous. Apart from a piece of S1 ATPase that was present in the x-ray crystal structure, but which was intentionally removed from the protein used for the electron microscopy studies, there is little volume within the envelope that is not occupied by the model, and there is little of the model structure that lies outside of the envelope. In a second example, the molecular envelopes obtained for the 70S *E. coli* ribosome with transfer RNA, bound as a complex, have revealed alternative positions at which the transfer RNA is located under different biochemically controlled buffer conditions (Agrawal et al., 1999b). Thus, having an atomic-resolution structure for just one member of a multicomponent complex can be enough to allow interpretation of a medium-resolution density map in a way that gives significant biochemical insight.

1.4 Electron crystallography has developed from rich intellectual origins in optics, electron microscopy, and x-ray crystallography

The use of an essentially crystallographic approach to obtain 3-D density maps from electron microscope images of biological macromolecules began with the work of DeRosier and Klug (1968). Digital *image processing* and Fourier transform mathematics were combined to reconstruct the structure of the T4 bacteriophage tail, shown in figure 1.6. While this 3-D model of the negatively stained specimen gives only a low-resolution picture of the structure of the phage tail, the completion of this work launched a major new research method, for which Aaron Klug was recognized through the award of the 1982 Nobel Prize in Chemistry. The DeRosier and Klug paper was soon followed by *reconstructions* of other helical structures (Moore et al., 1970), and the method was then extended to icosahedral viruses (Crowther et al., 1970a; Crowther, 1971). The idea that the Fourier transforms of electron microscope images effectively constituted diffraction data that could be used for a crystallographic structure analysis was recognized independently by Walter Hoppe in this period (Hoppe et al., 1968; Hoppe, 1970), and his emphasis was directed strongly toward large, single particles such as fatty acid synthetase (Hoppe et al., 1974) rather than symmetrical macromolecular assemblies.

The early development of electron crystallographic structure analysis in Cambridge began with negatively stained specimens, and thus the image intensity was initially

Figure 1.5 Atomic models of multiprotein complexes can be obtained by fitting previously known atomic models into a 3-D density map obtained by electron microscopy. In this example, atomic models of myosin S1 ATPase and actin have been docked together within the density map (of the complex), which was obtained by electron crystallography at a resolution of ~3 nm (Rayment et al., 1993). The portion of the S1 light-chain crystal structure that sticks out of density was, in fact, removed from the S1 fragment by chymotryptic cleavage before decorating the actin filaments, thus accounting for the absence of density for this portion of the molecule. Figure provided courtesy of Dr. Milligan.

analyzed in terms of the "mass thickness" model of image contrast. The wave-optical diffraction theory of *electron microscope image formation* that had been developed extensively by German theorists (Lenz, 1971; Hanssen, 1971; Frank, 1973) and by the materials science community (Hirsch et al., 1965) was soon incorporated into the analysis of negatively stained samples, however (Erickson and Klug, 1971). *Electron diffraction* experiments, already well established for structural studies with other specimen materials (Vainshtein, 1964; Hirsch et al., 1965), were also beginning to be applied to structural studies with biological specimens at about this time (Parsons and Martius, 1964; Ferrier and Murray, 1966; Murray and Ferrier, 1967; Glaeser and Thomas, 1969).

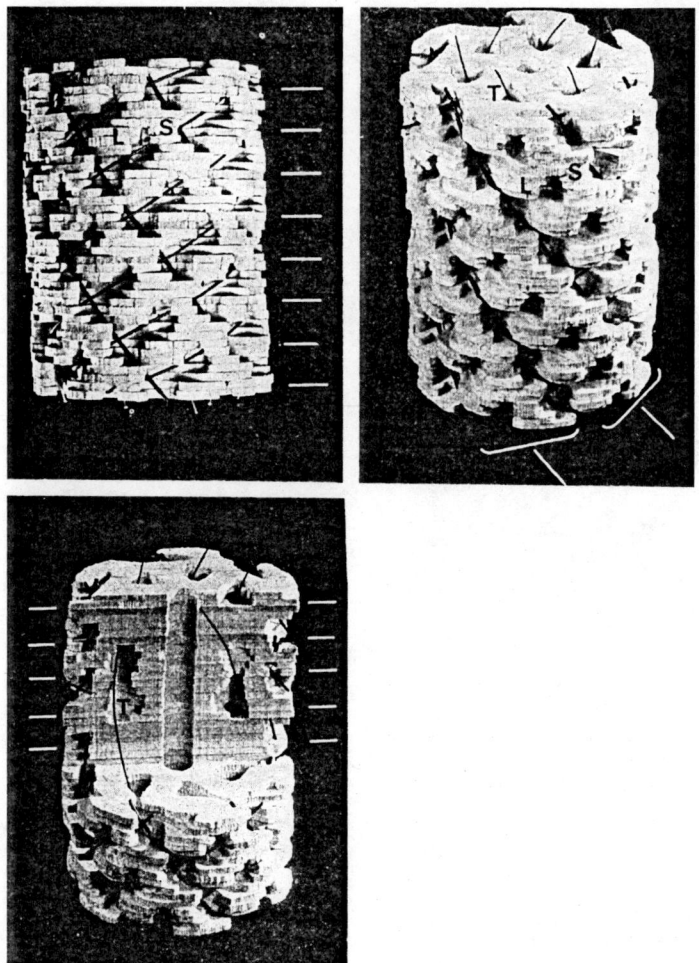

Figure 1.6 A solid model of the 3-D density of T4 bacteriophage tail, reconstructed from crystallographic amplitude and phase data contained in electron micrographs of negatively stained virus particles (DeRosier and Klug, 1968).

Two significant contributions were made by the early electron diffraction work. First of all, electron diffraction patterns were used to demonstrate the need to preserve a hydrated state of the specimen, within the vacuum of the electron microscope, if high-resolution detail was to be retained. High-resolution electron diffraction patterns of protein crystals, such as the one shown in figure 1.7, were first obtained with a differentially pumped hydration stage in which one could examine specimens that were immersed in a thin film of water or mother liquor (Matricardi et al., 1972). Similar results were obtained shortly thereafter with *frozen-hydrated* specimens (Taylor and Glaeser, 1974) and with *glucose-embedded* specimens (Unwin and Henderson, 1975). Secondly, the rapid fading of high-resolution electron diffraction patterns of biological

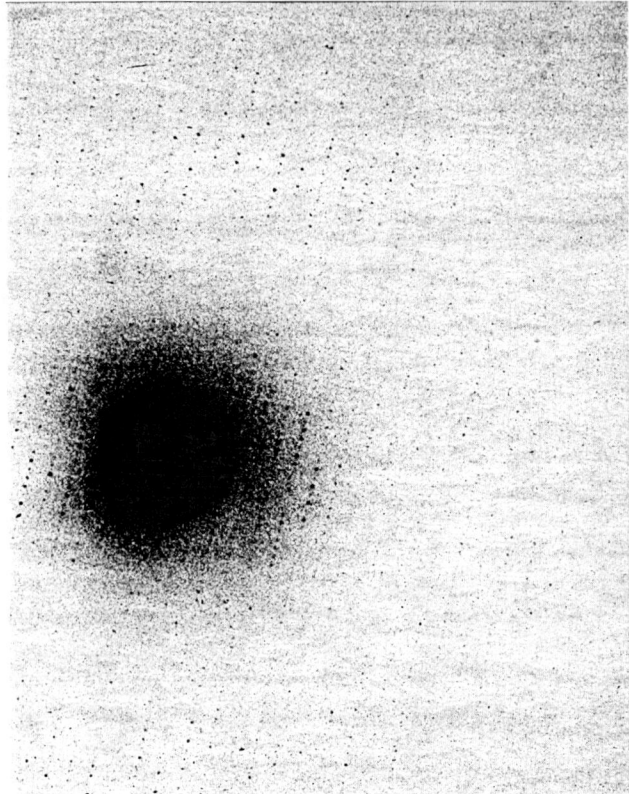

Figure 1.7 High-resolution electron diffraction pattern of a thin crystal of catalase, obtained by Dr. Douglas Dorset (unpublished) in the wet, hydrated state through the use of a differentially pumped hydration stage (Parsons et al., 1974). The unit-cell parameters are approximately 17.0 nm by 6.8 nm, and such diffraction patterns extend to a resolution close to 0.3 nm.

molecules made it clear that image resolution would be limited in a fundamental way by radiation damage. It thus became apparent that spatial averaging would have to be employed in order to overcome that limitation (Glaeser, 1971; Glaeser et al., 1971). In addition, low specimen temperatures were shown to allow biological macromolecules to tolerate higher radiation exposures, as is illustrated in figure 1.8. These radiation damage studies established the point that very stable cold stages are an important asset for high-resolution electron crystallography, whether or not frozen-hydrated specimens are used.

A major landmark in the development of electron crystallography of biological macromolecules was set when Unwin and Henderson (1975) combined (1) the use of electron diffraction patterns of thin, 2-D crystals, (2) glucose embedding, a novel method for maintaining the hydrated state of the specimen without any specialized cold stage or hydration chamber, (3) spatial averaging, by Fourier transform techniques, of images that had been recorded at very low electron exposures, and (4) the wave optical theory of image contrast in the electron microscope. Immediately thereafter,

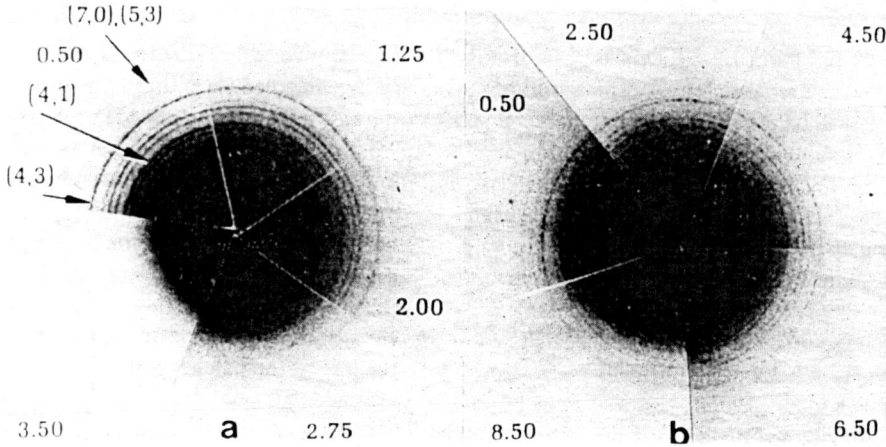

Figure 1.8 Comparison of the rate of fading of electron diffraction intensities at room temperature and at low temperature. A series of electron diffraction powder patterns of glucose-embedded purple membrane were recorded after specified periods of previously accumulated electron exposure (Hayward and Glaeser, 1979). The results show that about 5 to 7 times greater electron exposure can be tolerated at low temperature than at room temperature, for the same extent of specimen damage.

Henderson and Unwin (1975) applied these tools to produce a 3-D density map of bacteriorhodopsin at 0.7 nm resolution, giving the first crystallographic view of the structure of an integral membrane protein. Although the complete development of electron crystallography required many further advances, extending over a period of 15 years before the first atomic resolution model would finally be published (Henderson et al., 1990), the basic outline of how this work would be done was already clearly laid out in the two seminal papers that were published in 1975.

1.5 Objectives of this book

The objective in writing this book has been to provide a comprehensive treatment of the experimental and theoretical topics that are involved in electron crystallography of biological macromolecules. We have not included any material on the use of electron diffraction and electron microscopy with small molecules or "simple" organic polymers. Readers who want to know more about applications in these areas are referred to an earlier book by Dorset (1995).

With only a few exceptions, the material presented here is drawn from the primary literature, review articles, and other pertinent books. Although these sources may be familiar to those who are active in the field of electron crystallography, the information that we have sought to bring together would only be assembled with considerable effort when attempted by anyone new to the subject. By bringing this material together in one place, therefore, our goal is to make the subject more easily accessible to everyone who would like to take advantage of this valuable, new tool for determining the structure of biological macromolecules.

It is not necessary that we develop all of the standard aspects of the subject of crystallographic structure analysis. We assume that those who wish to use this book will already be familiar with the subject of x-ray crystallography of biological macromolecules, or will get the necessary background by reference to books on x-ray crystallography. As a start, however, chapter 2 briefly reviews some of the major highlights of x-ray crystallography. The purpose of including the material in chapter 2 has been to identify, and give a short introduction to, those topics in macromolecular crystallography with which the reader must be familiar in order to understand the subject of electron crystallography.

A large part of the material in this book is concerned with crystalline specimens, because of the advantages that crystals offer for very high-resolution work. At the present time it is possible to obtain density maps at ~0.35 nm resolution only if the specimen is a well-ordered 2-D crystal or helix, although we emphasize that the situation is likely to change in the years ahead. Nevertheless, the intentional emphasis on 2-D crystals in the content of this book reflects the fact that a molecular structure analysis becomes of interest to a much wider community when an atomic-resolution model of the structure can be obtained. Most of the topics that we develop in the context of thin 2-D crystals are directly applicable to all types of specimen, however, and to electron crystallography at any level of resolution. Reinforcing this latter point, separate chapters are devoted specifically to electron crystallographic structure analysis of helical macromolecular assemblies, icosahedra, and true individual macromolecules, respectively. In addition, we draw attention to the book by Frank (2006), which gives a more complete treatment of the type of three-dimensional structural studies that are possible for individual macromolecular assemblies.

2

Structure Determination as it has been Developed through X-ray Crystallography

2.1 Introduction

This chapter gives an overview of how *x-ray crystallography* is used to determine the structure of proteins and other biological macromolecules at high resolution. Our brief review of x-ray crystallography is designed to identify the most important mathematical concepts that electron crystallography has in common with x-ray crystallography. Thus, this chapter is meant to be a summary of concepts that the reader should be familiar with in order to work independently and proficiently in the field of electron crystallography. Our review is not intended to teach these concepts in full detail; that would be the subject of an entire book of its own.

Readers interested in more background than they will find in this chapter should consult classic texts in protein crystallography such as Blundell and Johnson (1976) and Stout and Jensen (1989), or a number of more recent books, such as the ones by Drenth (1994), Glusker et al. (1994), McRee (1993), or Rhodes (1993). In subsequent chapters it will also become apparent that many of the practical details of electron crystallography are quite different from their counterparts in x-ray crystallography. Nevertheless, x-ray crystallography represents one of the key foundations upon which electron crystallography has been built, and the present chapter provides a broad outline of topics that are an important part of the background that one should have in order to carry out similar work with electron crystallography.

2.2 Structure analysis by x-ray crystallography requires well-ordered, three-dimensional crystals

Specimen crystallization is the first, and arguably the most critical step for any structural study by x-ray crystallography. The conditions that produce good crystals for any given protein (or other macromolecule) are likely to be specific, in some way, to that particular specimen. Nevertheless, there are general strategies that are used for the *crystallization of proteins*, and numerous articles have been written on these techniques (McPherson, 1990; Jancarik and Kim, 1991; Weber, 1991; Samudzi et al., 1992; Weber, 1997). In brief, the idea is to slowly change the solution conditions from those in which the protein is highly soluble to those in which the protein is less soluble. In doing so there is always the danger that the protein will flocculate or precipitate as an amorphous, noncrystalline mass. Fortunately, in many cases the protein comes out of solution in the crystalline state.

Because the initial step of *crystal nucleation* is thermodynamically unfavorable, nucleation events will be rare if the protein concentration is only slightly above the limit of solubility. Once nucleation has occurred, protein molecules in the supersaturated solution will tend to add to already nucleated crystals, causing them to grow in size rather than to create new nuclei. If the limit of solubility is exceeded too rapidly, however, there is the danger of "showering", that is, the formation of a large number of small crystals, or even the danger of noncrystalline precipitation. Interestingly, some proteins crystallize well by growing out of, and in near equilibrium with, an amorphous precipitate.

A number of parameters can be used to decrease the solubility of a protein. Ionic strength, pH, polyethylene glycol (PEG), specific divalent ions such as Zn^{2+} or Mg^{2+} and exotic solutes such as spermine or imidazole, the temperature, and many other factors can all influence the crystallization process. Initial use of a coarse matrix of proven conditions provides an efficient scheme for exploring a variety of conditions that have the greatest promise for protein crystallization (Jancarik and Kim, 1991). Even when one is as economical as possible in screening potential conditions for crystallization, however, it is necessary to plan on using as much as 10 mg of protein or more to initiate the crystal search.

The purity of the protein solution is usually an important factor in the success of crystallization, especially when one is trying to increase crystal size or crystalline perfection. Purity must be taken in this context in its strictest chemical meaning. More than one form of post-translational modification, more than one (functional) conformational state, and partially denatured species may all have to be rigorously excluded in order to have hope of a successful crystallization. In this connection, it should be mentioned that the inclusion of a natural ligand or cofactor, or even a mechanistic inhibitor, can stabilize a unique functional conformation or prevent denaturation of the protein, thereby improving the chances for good crystallization. All of these rules are also expected to govern the likelihood of success in forming good two-dimensional crystals.

2.3 The practical steps of data collection and data analysis have become very efficient

As in electron crystallography, the crystals that are used for x-ray diffraction experiments must be kept fully hydrated so that their native, well-ordered structure is preserved. Historically, protein crystals were drawn into a thin-walled glass capillary tube, whose inner diameter was not much larger than the size of the protein crystal itself. As much of the surrounding mother liquor as possible would be "wicked away" to minimize the background scattering, and the capillary tube was then sealed at both ends. More recently it has been found that many protein crystals can be frozen in a small, open loop, and that superior quality data can be obtained from such *frozen hydrated* crystals. The original paper of Hope (1988) — see also the reviews by Hope (1990) and by Rodgers (1997) — can be consulted for more of the practical details on the way in which "large" protein crystals are frozen for x-ray diffraction experiments, and the reasons why this can be a very advantageous method of specimen preparation.

Data collection involves obtaining diffraction intensities from as complete a set of Bragg reflections as possible. This means that it is necessary to illuminate the crystal from many different angles, each time satisfying one or more new Bragg conditions, as will be described in more detail in sections 2.5 and 2.7. The requirement to view the object from many different directions also exists in 3-D imaging of noncrystalline objects, as in tomographic medical imaging, radioastronomy, or — as we will see later — electron microscopy. If possible, diffraction intensities should therefore be collected for every view that provides data that are independent of all previously collected data. As more independent measurements are added, one begins to build up the full set of data that are needed to determine the 3-D distribution of matter within the crystal.

The method used most frequently to collect diffraction intensities in modern work is simple rotation (oscillation) through a small angle around an axis perpendicular to the x-ray beam. As the crystal slowly rotates relative to the incident x-ray beam, different sets of lattice planes come briefly into an orientation that satisfies their respective Bragg conditions. The geometrical conditions required to satisfy Bragg's law are reviewed further in sections 2.7 and 2.8. Each diffraction spot is recorded on a two-dimensional "area detector," and more and more spots are added as the crystal continues to rotate. Data on the detector are read off periodically to avoid confusion when indexing the data, and even overlap of separate diffraction spots, if too many spots are accumulated in one frame.

2.4 The Fourier transform plays a central role in understanding the analysis of diffraction data

Because *Fourier transformation*, a mathematical operation, plays a central role in crystallographic structure analysis, it is important to acquire a familiar feeling for what a *Fourier transform* is, and how it relates to crystallography.

Without going into a formal proof, we can start with the mathematical fact that an arbitrary function, such as the *electron density* everywhere in a protein crystal, can be represented as a sum of cosine functions. Equation 2.1 is a one-dimensional

Fourier series, which illustrates the point.

$$\rho(x) = \sum_{n=0}^{N} F(n) \cos\left\{-2\pi \frac{n}{L} x + \alpha(n)\right\}. \tag{2.1}$$

In order for this sum to correctly represent a particular electron density, $\rho(x)$, each cosine term must be allowed to have its own *amplitude*, $F(n)$, the value of which is appropriate for that particular structure. At a fixed point in the unit cell of the crystal, chosen to be the origin of the coordinate system, each such cosine term must also be free to start at the correct point in its cycle. In other words each cosine term must also be allowed to have its own *phase*, $\alpha(n)$, the value of which is appropriate for that particular structure. One last requirement is that the periodic repeat-distance of the cosine terms in this mathematical sum must be allowed to be different for each type of protein crystal, because they are determined by the *unit cell periodicity*, L, of the crystal itself. The three fundamental concepts of *amplitude*, *phase*, and *period* of a cosine function are summarized in box 2.1.

BOX 2.1 Cosine functions

Each cosine term in the summation that is shown on the right hand side of equation 2.1 is characterized by three parameters: its amplitude, its phase, and the distance over which it periodically returns to the same value. The reciprocal of the periodic repeat interval is called the *spatial frequency*, and the repeat period itself is often referred to as the wavelength of the cosine function. The meaning of each of these parameters is illustrated in figure B2.1.

Two cosine functions are shown in panel (a); they have the same wavelength (the same spatial frequency), and they are perfectly in phase with one another, but the function represented by the dashed line has an amplitude that is half the amplitude of the function shown by the solid line.

The two cosine functions in panel (b) also have the same wavelength, and in this case they have the same amplitude as well, but they are no longer in phase with each other. We might say that the function shown by the dashed line has been shifted to the left relative to the function shown by the solid line. In other words, the origin — that is, the point where the cosine function is equal to 1 — is no longer the same for the two functions in panel (b). In crystallographic terminology, we would say that "there is a difference in the phase, α, of the two cosine functions."

In panel (c), we show two cosine functions that have the same amplitude, and which are perfectly in phase with each other at the origin, but which have different spatial frequencies. In fact, the wavelength of the function shown by the dashed line is only 0.8 times as long as the wavelength of the function shown by the solid line. As a result, the two functions quickly get out of phase with one another, the farther one goes from the chosen origin of the coordinate system.

(continued)

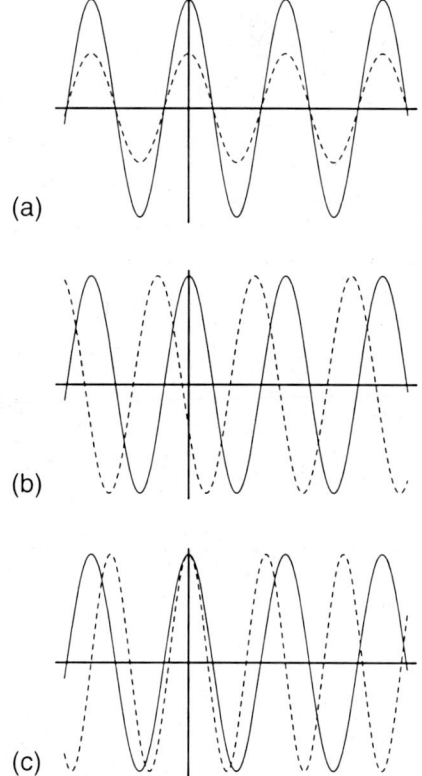

Figure B2.1 Three examples of cosine functions that differ from one another in (a) amplitude, (b) phase, or (c) spatial frequency. Each of the three panels is explained further in the text of this box.

In a *Fourier series*, such as the one written in equation 2.1, the values of the amplitude and phase are likely to be different for each wavelength, that is, for each term in the sum. This point is illustrated by the drawing in panel (a) of figure B2.2, where three different cosine functions are shown; as can be seen, they each have different amplitude, phase, and spatial frequency. When these three cosine functions are added together, as they would be in equation 2.1, they form a new function, which has an irregular shape. This irregular shape, or *motif*, repeats with the same frequency as does the longest-period, or "fundamental" cosine function, as long as the spatial frequencies of successive cosine functions are integer multiples of the fundamental, as is called for in equation 2.1. The repeating motif therefore forms the *unit cell* of a one-dimensional crystal. The actual motif that one obtains as the result of adding up a series of different cosine functions will look very different, depending upon the values of the amplitude and phase for each cosine term. By implication, then, a function of any arbitrary shape can be built up — or "represented" — by an appropriate Fourier summation, as long as we choose the right values of amplitude and phase for each term in the series.

(continued)

BOX 2.1 *(continued)*

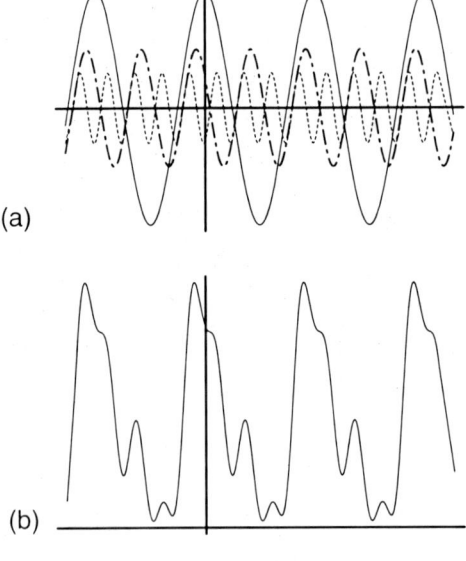

Figure B2.2 Three cosine functions are shown in panel (a), each of which has a different amplitude, phase, and period from that of the others. Their sum, shown in panel (b), has an irregular shape, which is made up of a repeating motif, as is explained in the text of the box. This illustration is modified from figure 1 of Chiu et al. (1993).

It is important to point out that the actual value of the phase is not important; rather, it is the relative values of the phases for each cosine term in equation 2.1 that affect the appearance of the final motif, or unit cell. To understand the distinction between actual phase and relative phase, we could define the actual phase as being 2π times the distance between the origin of the coordinate system and the maximum of the cosine function that is nearest to the origin, divided by the period (the wavelength) of the cosine function. In one dimension, therefore, moving the origin to a new position, a distance a from the previous origin, will change the phase of each cosine function by

$$\Delta\phi = 2\pi \frac{a}{\text{(period of the cosine term)}} = 2\pi \frac{n}{L} a, \qquad (B2.1)$$

or in three dimensions,

$$\Delta\phi = 2\pi \mathbf{a} \cdot \mathbf{S}. \qquad (B2.2)$$

In moving the origin of the coordinate system, however, every cosine term in equation 2.1 (or equation 2.2) remains fixed in its position relative to every other cosine term — that is, the relative phases do not change, and thus the motif itself does not change.

Since the phase changes from ϕ to $\phi + 2\pi \mathbf{a} \cdot \mathbf{S}$, when the origin shifts by an amount $\mathbf{a}$, each structure factor will change from $\mathbf{F}(\mathbf{S})$ to $F(\mathbf{S})e^{i2\pi \mathbf{a} \cdot \mathbf{S}}$. This analysis represents a proof of the *Fourier shift theorem*, and when applied to the

(continued)

Fourier transform of the electron charge density it states that

$$\mathcal{F}\{\rho(\mathbf{R}-\mathbf{a})\} = \mathcal{F}\{\rho(\mathbf{R})\}e^{i2\pi \mathbf{a}\cdot\mathbf{S}}. \tag{B2.3}$$

In other words, if the position of a function is moved by an amount **a**, the effect on its Fourier transform is represented by a *linear phase ramp*. We will use this result shortly, in box 2.3.

A final caveat is necessary regarding the sign of the linear phase ramp in equation B2.3. In optics and many other areas of physics, it is conventional to define the Fourier transform with the opposite sign convention compared to that which is used in crystallography, as we will explain in box 2.2. In effect the inverse Fourier transformation operation in crystallography is defined to be the Fourier transform and vice versa. When one adopts the convention used in crystallography, the Fourier shift theorem becomes:

$$\mathcal{F}\{\rho(\mathbf{R}-\mathbf{a})\} = \mathcal{F}\{\rho(\mathbf{R})\}e^{-i2\pi \mathbf{a}\cdot\mathbf{S}}. \tag{B2.4}$$

Equation B2.4 is often the form in which one finds the Fourier shift theorem stated in reference books; the reader should be aware that the correct form depends only on the convention of what one identifies as being the Fourier transform and the inverse Fourier transform.

Since $\frac{L}{n}$ is the repeat-period of the cosine term, its reciprocal, $\frac{n}{L}$ is called the *spatial frequency* of the cosine function. The use of a negative sign in the argument of the cosine function may seem curious, but it is used here to conform to the standard sign convention that is used in x-ray crystallography to designate the Fourier transform and its inverse, as is explained in box 2.2. Each term of the form $F(n)\cos\{-2\pi \frac{n}{L}x + \alpha(n)\}$ can be considered to be a *Fourier component* of the electron density, $\rho(x)$, which is made up of N such terms, or components.

In three dimensions the sum in equation 2.2 takes the form

$$\rho(\mathbf{R}) = \sum_{h,k,l=0} F(hkl)\cos\{-2\pi\,\mathbf{g}_{hkl}\cdot\mathbf{R} + \alpha(hkl)\}, \tag{2.2}$$

where the bold-faced variable **R** designates a 3-D vector in real space, the capital letter R being used to designate the region of space within which the object is confined, while $\mathbf{g}_{hkl}$ is a 3-D vector in Fourier space, also called *reciprocal space*. The integers h, k, and l are known in crystallography as the *Miller indices*; their function is to keep track of each separate cosine term in the sum, above. More will be said about the *spatial frequency vector*, $\mathbf{g}_{hkl}$, in section 2.5. Note, also, that it is conventional to use the "shorthand" notation $F(hkl)$ in place of $F(\mathbf{g}_{hkl})$.

A more general mathematical formulation of the sum in equation 2.2 is given by

$$\rho(\mathbf{R}) = \sum_{h,k,l} F(hkl)e^{i\alpha(hkl)}e^{-i\,2\pi\,\mathbf{g}_{hkl}\cdot\mathbf{R}}, \tag{2.3}$$

where it should be understood that the sum now goes equally over positive and negative values of the Miller indices. As a result, the amplitudes in equation 2.2 and those in equation 2.3 differ from one another by a factor of two.

BOX 2.2 Sign convention for the Fourier transform and its inverse

The Fourier transform operation and its inverse differ only in the sign of the argument in the complex exponential within the defining integral. This statement is expressed mathematically by the pair of equations:

$$\mathbf{F(S)} = \int \rho(\mathbf{R}) e^{i 2\pi \mathbf{S \cdot R}} d\mathbf{R} \tag{B2.5}$$

$$\rho(\mathbf{R}) = \int \mathbf{F(S)} e^{-i 2\pi \mathbf{S \cdot R}} d\mathbf{S}, \tag{B2.6}$$

where $\rho(\mathbf{R})$ can be any physically well behaved function. There is no absolute criterion, however, that one can use to say which of the two integrals in equation B2.5 should be identified as the Fourier transform, and which is to be identified as the inverse Fourier transform.

In crystallography it is customary to adopt a sign convention in which it is said that the structure factor, $\mathbf{F(S)}$, is the Fourier transform of the electron charge density, $\rho(\mathbf{R})$. Once this choice has been made for the definition of the Fourier transform, we must say that the electron charge density, $\rho(\mathbf{R})$, is the inverse Fourier transform of the structure factor, $\mathbf{F(S)}$. In many other fields the opposite sign convention is used, however. Thus, in Fourier optics it would be conventional to say that the function $\rho(\mathbf{R})$ in equation B2.5 is the Fourier transform of the function $\mathbf{F(S)}$, and $\mathbf{F(S)}$ is the inverse Fourier transform of $\rho(\mathbf{R})$.

In casual scientific conversation, little attention is paid to which sign convention is being used when speaking about the Fourier transform. All that is crucial to remember is that the argument of the complex exponential must change sign when speaking of the Fourier transform and its inverse. Although it would be simpler if we were able to say that the Fourier transform is its own inverse, and that is almost the case, it is easy to see that this is not quite true. Simple inspection of the defining integrals shows, instead, that

$$(\mathcal{F}^{-1}[\mathcal{F}\{\rho(\mathbf{R})\}]) = \rho(\mathbf{R}) \tag{B2.6a}$$

but

$$(\mathcal{F}[\mathcal{F}\{\rho(\mathbf{R})\}]) = \rho(-\mathbf{R}). \tag{B2.6b}$$

Thus, use of the Fourier transform in place of the inverse Fourier transform does return the original function, $\rho(\mathbf{R})$, but the original function is now inverted through the origin, as is indicated by the change in sign of the argument of the original function.

In equation 2.3 we have introduced, for the first time, the concept of *complex-valued* functions, that is, functions that contain both real and imaginary parts. As usual, the imaginary-valued part is preceded by the mathematical symbol, i, which stands for the square root of -1. In making the change in notation from equation 2.2 to equation 2.3, we have used the *Euler identity*,

$$e^{i\theta} = \cos\theta + i\sin\theta, \tag{2.4}$$

to replace the cosine function with the complex exponential. Complex numbers provide a mathematically convenient way to represent both the amplitude and the phase of each term in a Fourier series. The more general notation used in equation 2.3 is the standard way of representing Fourier transforms in crystallography.

Finally, the most general Fourier representation is obtained by replacing the summation with a continuous integral

$$\rho(\mathbf{R}) = \int \mathbf{F}(\mathbf{S})e^{-i2\pi \mathbf{S}\cdot\mathbf{R}}d\mathbf{S}, \tag{2.5}$$

where the continuous, three-dimensional spatial frequency vector $\mathbf{S}$ has replaced the discrete vector $\mathbf{g}_{hkl}$. The complex-valued function $\mathbf{F}(\mathbf{S})$ represents the product of the amplitude, $F(\mathbf{S})$, and the complex exponential of the phase, $e^{i\alpha(\mathbf{S})}$, as is written in equation 2.3. $\mathbf{F}(\mathbf{S})$ is called the (complex-valued) *structure factor*; it is written in bold face because it can be represented as a vector in a two-dimensional space for which one axis is the real part of $\mathbf{F}(\mathbf{S})$ and the other axis is the imaginary part of $\mathbf{F}(\mathbf{S})$.

Using words rather than mathematical symbols, equation 2.5 is read "the electron density, $\rho(\mathbf{R})$, is the (inverse) *Fourier transform* of the complex structure factors, $\mathbf{F}(\mathbf{S})$." The reason for inserting the qualifying word "inverse" in describing equation 2.5 is explained in box 2.2. At the moment we only want to emphasize that equation 2.5 is a completely general mathematical expression of what we mean when we speak of the "Fourier transform." The integral notation, while valid for all situations, is necessary only for functions that are not periodic. The simpler cosine sum, equation 2.2, is all that is needed to represent a real-valued periodic function, such as the electron density within a crystal.

The values of amplitude and phase that give rise to the correct representation of the electron density function can themselves be calculated by the Fourier transformation integral, if the electron density function is already known. The product of the amplitude and the complex exponential of the phase can be calculated by taking the Fourier transform of $\rho(\mathbf{R})$:

$$F(\mathbf{S})e^{i\alpha(\mathbf{S})} = \int \rho(\mathbf{R})e^{i2\pi \mathbf{S}\cdot\mathbf{R}}d\mathbf{R}. \tag{2.6}$$

Note that the sign of the argument in the complex exponential in equation 2.5 is negative, while that in equation 2.6 is positive. Fourier transformation thus has the surprising property that it is its own inverse operation, that is, one can reverse the action of a Fourier transform by doing another Fourier transform in which the sign of the argument, $2\pi \mathbf{S}\cdot\mathbf{R}$, is reversed, as we have mentioned in box 2.2.

In order to determine a crystal structure, the amplitude and phase of the structure factor must first be derived from experimental data, rather than from equation 2.6.

The electron charge density function of the crystal is then calculated as a discrete sum, according to the following equation:

$$\rho(\mathbf{R}) = \sum_{h,k,l} F(hkl) e^{i\alpha(hkl)} e^{-i2\pi \mathbf{g}_{hkl} \cdot \mathbf{R}}. \tag{2.7}$$

Equation 2.7 is identical to equation 2.3, but it is repeated here for convenience. As we will show in section 2.5, the integral in equation 2.5 reduces to the simpler summation, equation 2.7, when $\rho(\mathbf{R})$ itself is a periodic function.

A remarkable mathematical result in the theory of diffraction is the fact that the *diffraction intensity* is proportional to the square of the structure factor, defined in equation 2.6. Proof of this important result is postponed until chapter 4, however. For the moment, the point of interest is that $F(\mathbf{S})$, *the modulus of the structure factor*, is simply the square root of the diffraction intensity.

Furthermore, there is a unique correspondence between the scattering angle for a given diffraction spot and the spatial frequency (or the periodicity) of a particular cosine term, or Fourier component of the structure. This relationship is explained in section 2.8. If, then, we have both the amplitude and the spatial frequency for every cosine term, all that is missing in order to calculate the electron charge density itself, by equation 2.7, is the phase of the structure factor.

Since only the intensity of the diffracted wave can be measured, and not the wave function itself, the phase information is lost in the diffraction measurement (see section 2.9). This loss of information gives rise to the famous *phase problem* in x-ray crystallography. As we will see in section 2.10, however, the structure factor phases can be recovered by measuring the changes in diffraction intensity that occur after heavy atoms have been bound to the protein in a specific, reproducible way. Thus, diffraction experiments ultimately do give the complete information that is needed to calculate the electron density function.

2.5 The Fourier transform of a crystal represents discrete, regular samples of the continuous Fourier transform of the molecule

The electron density function of a crystal, $\rho_c(\mathbf{R})$, can be described in terms of two separate functions. The first of these is the electron density of the unit cell, $\rho_u(\mathbf{R})$, which is repeated over and over again to make up the crystal. The second is a function which marks the position of every unit cell. More specifically, this *lattice function*, $L(\mathbf{R})$, is a sum of *Dirac delta functions*, one at every lattice point. The lattice function can be represented mathematically as

$$L(\mathbf{R}) = \sum_j \delta(\mathbf{R} - \mathbf{R}_j), \tag{2.8}$$

where $\mathbf{R}$ is a continuous (vector) variable and $\mathbf{R}_j$ is the vector that specifies the jth point in the lattice. The electron density of the crystal, $\rho_c(\mathbf{R})$, can then be written in terms of these two functions by means of the convolution operation (see box 2.3).

BOX 2.3 Convolution

The mathematical definition of a *convolution product* (and the convolution operation) is stated in equation 2.9, but in the context where one of the functions of the convolution product is a lattice function. It is worthwhile to repeat this mathematical definition in a more general way, making no assumptions at all about the two functions that are to be convoluted.

The convolution of $f(x)$ with $h(x)$ makes a new function, $g(x)$, defined by the integral,

$$g(x) = f(x) * h(x) = \int f(x')h(x - x')dx'. \qquad (B2.7)$$

In words, the right-hand side of equation B2.7 prescribes that the function $h(x')$ must first be inverted to give $h(-x')$. Next, the inverted function must be shifted relative to $f(x')$ by a given amount, x. The two functions are then multiplied together, and the product function is integrated over all space. This prescription is repeated over and over again, each time for a different value of the amount by which $h(-x')$ is displaced relative to $f(x')$. Each integration results in a new value of the function, $g(x)$. Repeated performance of this operation eventually results in a new continuous function.

Although the convolution product may at first seem to be a rather complicated operation, a few simple examples will give a very good, intuitive feeling for what happens when two functions are convoluted with each other. The convolution of a *rectangle function* (shown in panel (a) of figure B2.3) with itself represents a good place to begin. The rectangle function, rect(x), is chosen to be symmetric about the origin, to have a value of 1 over a length L, and to be zero every where else. This is an easy example to start with, because inversion of rect(x)

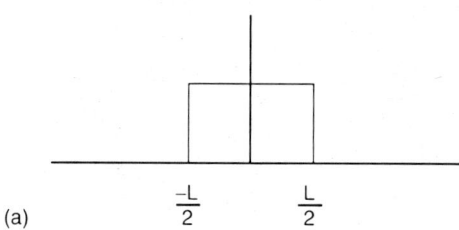

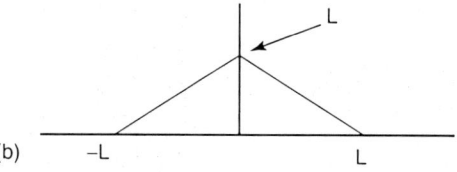

Figure B2.3 The convolution product of a rectangle function, shown in panel (a), with itself results in a "triangle" function, shown in panel (b). The explanation of this result is discussed in the text of the box.

(continued)

BOX 2.3 *(continued)*

[i.e., creation of rect($-x$)] just gives back rect(x), making the next steps easier to visualize. If $x = 0$, that is, if there is no displacement between the two copies of rect(x'), the product is again a rectangle function of width L and height 1. The integral therefore has a value L. If, on the other hand, one copy of rect(x') is displaced by an amount $\pm x$ from the second copy, the product will be a narrower rectangle function, of width $(L - x)$. In this case the integral has a value of $(L - x)$. The reader can now easily verify that the convolution product, rect(x) * rect(x), is a "triangle" function, as is shown in panel (b) of figure B2.3.

This simple example already illustrates a very important property of the convolution operation, which is the fact that the convolution product is always a function that is "broader" than either of the "parent" functions that are being convoluted together, except when one of the functions is a Dirac delta function, in which case the convolution product is identical to the companion function. In the previous illustration, for example, rect(x) has non-zero values only from $-\frac{L}{2}$ to $\frac{L}{2}$, while rect(x) * rect(x) is non-zero between $-L$ and L.

In order to further consolidate one's understanding of the convolution operation, the reader should show that the convolution of $\text{rect}_1(x)$, of width L_1, and $\text{rect}_2(x)$, of width L_2, is a truncated "triangle" function, or "mesa," as illustrated in figure B2.4. The mathematically inclined reader will also be able to show, without difficulty, that the convolution of two Gaussian functions, e^{-ax^2} and e^{-bx^2}, is a new Gaussian function, $e^{-(a+b)x^2}$.

What happens when a given motif, or unit cell (represented by the function $\rho_u(\mathbf{r})$), is convoluted with a lattice function? Let us begin by considering first the case where a one-dimensional motif, $\rho_u(x)$, is convoluted with a single *Dirac delta function*, which is located at the origin. The formal answer is

$$\rho_u(x) * \delta(x) = \int \rho_u(x')\delta(x - x')dx' = \rho_u(x). \tag{B2.8}$$

The right-hand side of equation B2.8 is nothing more than a *definition* of what is meant by a Dirac delta function; that is, the right-hand of equation B2.8 serves to define $\delta(x - x')$ as that function for which the indicated equality is true. At the same time, the right-hand side of equation B2.8 also shows that an integral involving a Dirac delta function is the easiest of all

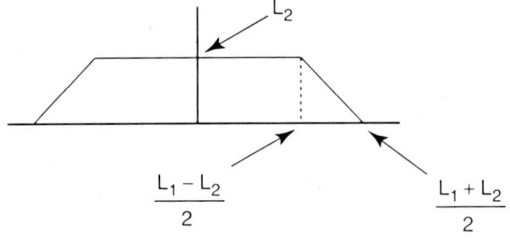

Figure B2.4 The convolution of two rectangle functions of different widths, L_1 and L_2, respectively, results in a truncated triangle, or "mesa," as shown here. The derivation of this result is left up to the reader.

(continued)

possible integrals to carry out: the answer is always the value of the remaining function within the integrand, evaluated at the point where the argument of the delta function is zero. Generalizing equation B2.8, then, we see that

$$\rho_u(x) * \delta(x-a) = \int \rho_u(x')\delta(x-x'+a)dx' = \rho_u(x+a). \quad (B2.9)$$

In other words, the function $\rho_u(x)$ gets displaced to the position $x = -a$ when it is convoluted with a Dirac delta function that is located at $x = a$.

The concept of convolution of a unit cell with a lattice function is now readily visualized by starting with a one-dimensional lattice function, which is shown in figure B2.5 as an evenly spaced row of lines, schematically representing Dirac delta functions. This one-dimensional lattice function is often referred to as a *comb function*, because of its resemblance to the one-dimensional, evenly spaced teeth on a comb. The unit cell in figure B2.5 is a line-drawing "cartoon," meant to represent a cup of hot coffee. As the figure shows, the convolution of the cup with the comb function results in a one-dimensional crystal, where every unit cell in the crystal contains an identical cup.

The same concept can now be extended to illustrate the convolution of a unit cell with a two-dimensional lattice function. In this case the lattice function is represented in figure B2.6 as a two-dimensional array of evenly spaced black dots, each spot again representing a Dirac delta function. The unit cell is now taken to be a four-helix bundle, and this time it is shown as a shaded surface, to emphasize that a single molecule is really a three-dimensional object. Convolution of the 2-D lattice function with the four-helix bundle results in a 2-D crystal, as is shown by the illustration.

The reader can appreciate that the concept illustrated in figures B2.5 and B2.6 could now be extended to three dimensions. In this case, however, it is easier to imagine a three-dimensional array of Dirac delta functions than it is to show it as a printed illustration. The 3-D crystal that would result from convolution of the contents of one unit cell with the 3-D lattice function would then consist of

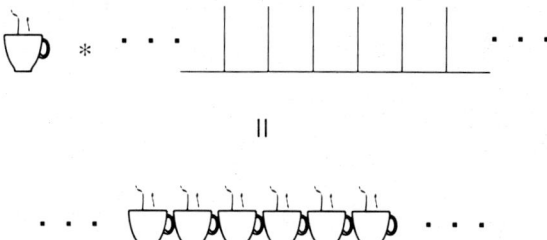

Figure B2.5 A one-dimensional lattice function, $L(x)$, is a regularly spaced sequence of Dirac delta functions, and this is represented by the uniform array of lines at the top right section of this figure. This one-dimensional lattice function is also called a "comb" function, $comb(x)$ because of its resemblance to the uniformly spaced teeth of a hair comb. This figure illustrates the concept, described in the text of the box, that the convolution of a motif with a comb function results in a one-dimensional crystal.

(continued)

BOX 2.3 *(continued)*

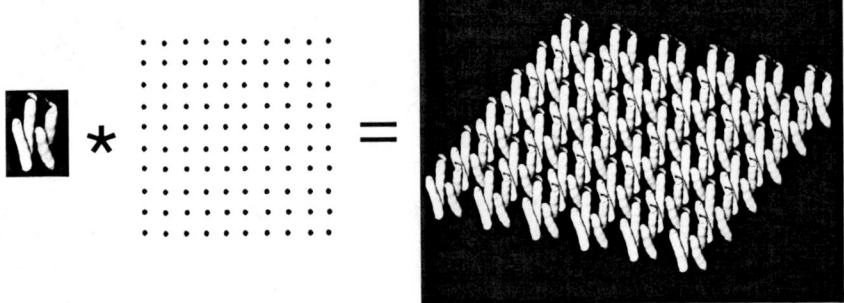

Figure B2.6 A single unit cell, taken here to be a cartoon representation of a four-helix bundle, is shown being convoluted with a 2-D lattice function to give a monolayer, 2-D crystal. The 2-D lattice function is represented by the periodic array of black dots, each of which should be thought of as being a Dirac delta function. This figure is modeled after a similar illustration in Chiu et al. (1993), and it was prepared with the assistance of Dr. Michael Schmid.

successive layers, one stacked on the next, each of which would be identical to the monolayer crystal shown in figure B2.6.

As we have indicated in the text, the *convolution theorem* tells us that the rather difficult, or at least unfamiliar, concept of the convolution product is converted by Fourier transformation into the simple concept of direct multiplication. In mathematical notation,

$$\mathcal{F}\{f(x) * h(x)\} = F(s) \cdot H(s), \tag{B2.10}$$

where F and H are the Fourier transforms of f and h, respectively.

The formal proof of this theorem follows by elementary manipulations of integral calculus. We begin by writing the Fourier transform of the convolution product explicitly:

$$\mathcal{F}\{f(x) * h(x)\} = \int \left\{ \int f(x')h(x-x')dx' \right\} e^{i2\pi sx} dx. \tag{B2.11}$$

The integration over x can be moved into the integral over x', to give the result

$$\mathcal{F}\{f(x) * h(x)\} = \int f(x')[\mathcal{F}\{h(x-x')\}]dx' \tag{B2.12}$$

$$= \int f(x')H(s)e^{i2\pi sx'} dx'$$

$$= H(s)\mathcal{F}\{f(x')\} = F(s) \cdot H(s).$$

Note that the Fourier shift theorem, described in box 2.1, has been used in the middle line of equation B2.12.

(continued)

> The *convolution operation* described above, and the equally important *convolution theorem*, represent core mathematical concepts that are essential to both a practical and a theoretical understanding of crystallographic structure analysis. Those readers who are not already well familiar with these two mathematical concepts should be prepared to return frequently to this box, or to other sources of mathematical explanation, as they proceed through the following chapters.

This gives the result

$$\rho_c(\mathbf{R}) = \rho_u(\mathbf{R}) * L(\mathbf{R}) = \int \rho_u(\mathbf{R}')L(\mathbf{R}-\mathbf{R}')d\mathbf{R}'. \tag{2.9}$$

The asterisk notation used within the middle of equation 2.9 is commonly used to symbolize the *convolution product* (of the two functions), and the right-hand side defines the actual integral that is involved in the formation of the convolution product.

While the convolution product may seem at first to be a bit complicated, our work in crystallography is made simple by the fact that the convolution product is converted into ordinary multiplication (i.e., a scalar product) when one carries out the Fourier transformation operation. As is explained more fully in box 2.3, the *convolution theorem* states that the Fourier transform of the convolution product of two functions is the scalar product of their respective Fourier transforms. The reverse is also true, that is, the Fourier transform of the scalar product of two functions is the convolution product of their respective Fourier transforms. Restating just the first part of the convolution theorem, one writes

$$\mathcal{F}\{\rho_u(\mathbf{R}) * L(\mathbf{R})\} = \mathcal{F}\{\rho_u(\mathbf{R})\} \cdot \mathcal{F}\{L(\mathbf{R})\}, \tag{2.10}$$

where we have introduced the symbolic notation $\mathcal{F}$ to represent the Fourier transform operation.

The second topic that we must now discuss is the relationship between the lattice function, $L(\mathbf{R})$, and its Fourier transform, $\mathcal{L}(\mathbf{S})$, which we state here without proof:

$$\mathcal{L}(\mathbf{S}) = \mathcal{F}\{L(\mathbf{R})\} = \mathcal{F}\left\{\sum_j \delta(\mathbf{R}-\mathbf{R}_j)\right\} = \sum_{h,k,l} \delta(\mathbf{S}-\mathbf{g}_{hkl}). \tag{2.11}$$

In other words, $\mathcal{L}(\mathbf{S})$ is itself a lattice function, which we call the *reciprocal lattice function*.

The points of the reciprocal lattice are determined by *reciprocal lattice basis vectors*, $\mathbf{a}^*$, $\mathbf{b}^*$, and $\mathbf{c}^*$, which themselves are determined by the crystal lattice basis vectors (*unit cell vectors*), $\mathbf{a}$, $\mathbf{b}$, and $\mathbf{c}$, and a combination of vector cross-products, indicated by "×", and vector scalar-products, indicated by "·":

$$\begin{aligned}\mathbf{a}^* &= \frac{\mathbf{b}\times\mathbf{c}}{\mathbf{a}\cdot\mathbf{b}\times\mathbf{c}} \\ \mathbf{b}^* &= \frac{\mathbf{c}\times\mathbf{a}}{\mathbf{a}\cdot\mathbf{b}\times\mathbf{c}} \\ \mathbf{c}^* &= \frac{\mathbf{a}\times\mathbf{b}}{\mathbf{a}\cdot\mathbf{b}\times\mathbf{c}}.\end{aligned} \tag{2.12}$$

The vector to an arbitrary reciprocal lattice point in Fourier space can finally be written in terms of $\mathbf{a}^*$, $\mathbf{b}^*$, $\mathbf{c}^*$ and the Miller indices as

$$\mathbf{g}_{hkl} = h\mathbf{a}^* + k\mathbf{b}^* + l\mathbf{c}^*. \tag{2.13}$$

The *Miller indices* used in equation 2.13, often written as the triplet (*hkl*), provide a unique identification for every point in the reciprocal lattice. As we will soon see, the Miller indices therefore serve to uniquely identify every individual diffraction spot in a diffraction pattern.

Combining equations 2.10 and 2.11 we get the result

$$\mathcal{F}\{\rho_c(\mathbf{R})\} = \mathcal{F}\{\rho_u(\mathbf{R}) * L(\mathbf{R})\} = \sum_{h,k,l} \mathbf{F}(\mathbf{g}_{hkl})\delta(\mathbf{S} - \mathbf{g}_{hkl}). \tag{2.14}$$

In other words, the Fourier transform of a crystal is zero everywhere except at the reciprocal lattice points, $\mathbf{g}_{hkl}$, where it has the same value that the Fourier transform of the unit cell, $\mathbf{F}(\mathbf{S})$, would have at that spatial frequency.

The fact that the Fourier transform of a crystal is just a regularly sampled version of the (otherwise) continuous Fourier transform of the unit cell tells us two very important things. The first point is that there are no new "complications" that are added to the Fourier transform of a crystal, beyond those that are already present in the Fourier transform of the unit cell. The second point is that the continuous Fourier transform of the unit cell represents "redundant information" about the structure of the unit cell, in the sense that the discrete set of samples that are obtained in the Fourier transform of the crystal are all that one needs to have in order to recover the structure of the unit cell.

2.6 The disorder that exists in real crystals can result in easily observed changes in the Fourier transform

Two distinguishable types of disorder are found in crystalline materials, the effects of which are usually quite apparent in the diffraction pattern. For reasons that will become clear in the following paragraphs, the two types of disorder are called short-range disorder and long-range disorder, respectively.

In crystals with *short-range disorder*, the diffraction spots are less intense than they would be for a perfect crystal, the effect becoming increasingly severe at higher resolution, that is, at higher spatial frequency, $\mathbf{S}$. It is found, empirically, that the fall-off of intensities is modeled quite well by multiplying the diffraction intensities for the ideal crystal by a Gaussian function:

$$I_{real}(\mathbf{S}) = I_{ideal}(\mathbf{S})e^{-\frac{B}{2}S^2} \tag{2.15}$$

where the parameter B is known as the *Debye–Waller thermal parameter*. The Gaussian function $\left(e^{-\frac{B}{2}S^2}\right)$ is known as the *Debye–Waller temperature factor*, or simply the *temperature factor*. The temperature factor is written here in the form that is conventionally used in x-ray crystallography. Equation 2.15 implies that the diffraction amplitudes are also modified by a Gaussian function, because the amplitudes are nothing but the square root of the intensities. As a result,

$$F_{real}(\mathbf{S}) = F_{ideal}(\mathbf{S})e^{-\frac{B}{4}S^2}. \tag{2.16}$$

The seemingly arbitrary factor of 4 by which the parameter B is divided in equation 2.16 has historical origins that are related to the fact that earlier versions of equation 2.16 had been written with $\left(\frac{2}{\lambda}\sin\frac{\theta}{2}\right)^2$ rather than s^2.

The inverse Fourier transform that is obtained with the real amplitudes will tell us how our averaged structure is affected by short-range disorder. The Fourier transform of $\mathbf{F}_{ideal}(\mathbf{S})e^{-\frac{B}{4}S^2}$ is easily calculated if we employ the convolution theorem and if we use the additional result that the Fourier transform of a Gaussian function is itself a Gaussian function. Using these two tools, we see that the real electron charge density is the convolution product of the ideal electron charge density and the real-space Gaussian function obtained from the Fourier transform of $e^{-\frac{B}{4}S^2}$. Fourier transformation of equation 2.16 therefore shows that

$$\rho_{real}(\mathbf{R}) = \rho_{ideal}(\mathbf{R}) * e^{-\frac{2\pi^2}{B}R^2}. \tag{2.17}$$

The effect caused by short-range disorder can thus be modeled mathematically by the convolution of the ideal unit cell structure with a Gaussian function.

The idea that short range disorder leads to a "smearing" of high-resolution features in the average electron density map is quite a plausible one. As shown by Debye and Waller, thermal vibrations are expected to generate exactly the type of short-range disorder that we have been discussing, in which the diffraction intensities decrease exponentially with the square of the resolution. The constant, B, properly referred to as the *Debye–Waller thermal parameter*, is also often informally called the "temperature factor." In the model of Debye, the parameter B is determined by the square of the mean displacement of the atoms due to thermal vibrations, $\langle u^2 \rangle$, according to the equation:

$$B = 8\pi^2 \langle u^2 \rangle. \tag{2.18}$$

Note that $\langle u^2 \rangle$ represents the mean square atomic displacement in the direction perpendicular to the diffracting planes.

While thermal vibrations are one factor that causes short-range disorder, there are many additional reasons why the structure of a crystal does not repeat perfectly from one unit cell position to another. In the case of large, complex molecular structures such as proteins, for example, small differences in internal structure or in the packing of one molecule against another will exist. Essentially static short-range disorder of this type will normally contribute a bigger effect to the temperature factor of a protein crystal than that which is due to thermal vibrations.

Referring to equation 2.15, we can see that the natural logarithm of the ratio between the real diffraction intensities and the ideal diffraction intensities, when plotted against the square of the resolution, should be a straight line with a slope of $-\frac{B}{2}$. This simple relationship cannot be used directly to determine the temperature factor, of course, because the ideal intensities cannot be known in advance. By a statistical argument, however, Wilson (1942) showed that it is sufficient to take the average diffraction intensity within a narrow, given interval of scattering angles and normalize the average by the average value of the atomic scattering cross sections for all atoms in the unit cell. This sum must be weighted according to the relative abundance of the different types of atoms in the structure. The resulting logarithmic plot of the normalized intensity versus the square of the resolution is called a *Wilson plot*. In practice, only data at a resolution higher than about 0.3 nm show the expected linear relationship on a Wilson plot (Drenth, 1994), making this a relatively unimportant tool for electron crystallography. A *relative Wilson plot*, which is a plot of the ratio of Fourier amplitudes computed from images to electron diffraction amplitudes, is insensitive to the structural correlations (in chemically bonded systems) which make the Wilson plot itself less useful at lower resolution. As a result, the relative Wilson plot has proven to be an important tool in electron crystallography.

Crystals that have *long-range disorder* produce diffraction spots that are more broad than what is expected for a perfect crystal. Two examples of long-range disorder are worth describing. The first example envisions a continuous distortion of the lattice function, $L(\mathbf{R})$. The picture here is that the crystal is an elastic block that can be stretched and twisted irregularly and by small amounts. The second example is the idealized case in which small, perfect blocks of the lattice are displaced by random amounts from one another; these displacements can be both translational and rotational. Such a crystal can be thought of as a three-dimensional mosaic, built up from a large number of not-quite-perfectly positioned tiles.

The *mosaic model* of long-range disorder lends itself to a particularly direct understanding of the broadening of diffraction spots. Each mosaic block can be represented by the product of an infinite, perfect crystal and a finite "shape function," that is, a rectangle function which defines the boundaries of the block in real space. The concept of a *rectangle function*, previously defined in box 2.3 in the one-dimensional case, must here be generalized to mean a *shape function* that is one (unity) everywhere within the boundary of the three-dimensional crystal, and zero elsewhere. The Fourier transform of this product is the Fourier transform of the rectangle function convoluted with the Fourier transform of the infinite, perfect crystal. The perfectly sharp diffraction spots of the latter therefore become broadened when they are convoluted with the Fourier transform of the real-space shape function. The Fourier transform of a rectangle function has a width that is inversely proportional to the size of the mosaic block. The diffraction spots, broadened now because of the finite size of the mosaic domains, can be broadened even further because of angular misalignment of the separate mosaic domains.

While the actual effect of any one or another form of long-range disorder may be quite complicated to describe mathematically (Vainshtein, 1966), the hallmark of all forms of long-range disorder is a blurring — or even streaking — of diffraction spots, so that they are no longer as sharp as they would be for a perfect crystal.

2.7 The Ewald sphere: a powerful mental picture that shows what part of the Fourier transform can be measured for every orientation of the specimen

The Fourier transform of a three-dimensional crystal is non-zero on a three-dimensional lattice of points, the reciprocal lattice (see section 2.5). The x-ray diffraction pattern that is recorded on an area detector is only a two-dimensional distribution of spots. What then is the relationship between the measured distribution of spots on the two-dimensional detector, and the three-dimensional array of reciprocal lattice "spots"? The *Ewald sphere construction*, which we will explain in the following paragraphs, is a kind of bookkeeping device that allows us to answer that question. The Ewald sphere concept also helps us to better understand that the experimental data collection must involve a large number of diffraction experiments, in each of which different parts of the Fourier transform of the crystal are measured by using a new orientation of the specimen.

We begin by stating the important fact, without derivation, that the *spatial frequency vector*, **S**, which occurs in the three-dimensional Fourier transform (see equation 2.6), is related to the direction of both the incident wave and the scattered wave by the vector equation

$$2\pi \mathbf{S} = \frac{2\pi}{\lambda}(\hat{\mathbf{k}} - \hat{\mathbf{k}}_0), \tag{2.19}$$

where $\hat{\mathbf{k}}_0 = $ a unit vector in the direction of the incident wave and $\hat{\mathbf{k}} = $ a unit vector in the direction of the scattered wave. We also note that $\mathbf{k} = \frac{2\pi}{\lambda}\hat{\mathbf{k}}$ is termed the *momentum vector of the wave*. The physical connection between the scattered wave and the Fourier transform, and thus the spatial frequency vector defined in equation 2.19, is discussed more extensively in chapter 4.

The locus of all vectors, **S**, at which the Fourier transform can be measured is specified by equation 2.19. For a fixed orientation of the specimen, the direction of the incident beam, $\hat{\mathbf{k}}_0$, has a fixed value, but the scattered beam can occur at any angle. Thus the locus of all vectors, **S**, is a spherical surface that passes through the origin, $S = 0$, as is illustrated in figure 2.1. This spherical surface is called the *Ewald sphere*. Although we are introducing the Ewald sphere in the context of a crystalline specimen and its reciprocal-space lattice, it is important to emphasize that the concept of the Ewald sphere is extremely useful for any type of specimen, even for a single molecule. In all cases, regardless of the nature of the specimen, the Ewald sphere is the locus of all vectors, **S**, at which the Fourier transform can be measured if $\hat{\mathbf{k}}_0$ is held constant.

It is important to keep in mind that the Ewald sphere is only a geometrical *construction*, and that it exists only in *reciprocal space*, that is, in the mathematical, three-dimensional space of Fourier transforms. The Ewald sphere does not exist in the real-space laboratory frame of reference, even though the directions of the incident and scattered beams certainly do exist in the laboratory frame. It must be remembered that the Ewald sphere is only a construction that is intended to be a bookkeeping device. The value of the Ewald sphere concept is that it helps us to keep track of how our physical measurements can be mapped onto the Fourier transform representation of the data.

As can be seen in figure 2.1 (and later, more explicitly in figure 2.3), the intersection of the Ewald sphere and the reciprocal lattice defines completely all spatial frequencies, **S**, at which non-zero values of the Fourier transform of the crystal can be measured. These are the only values of **S** at which the crystal structure factors can be measured. By changing the direction of the incident beam relative to the unit cell of the crystal, however, the Ewald sphere can be made to intersect new positions of

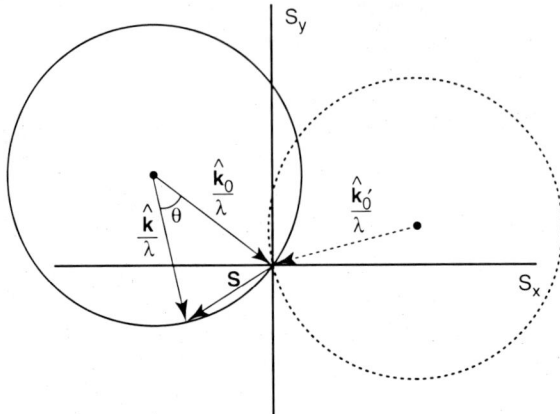

Figure 2.1 A simplified, two-dimensional diagram that illustrates the basic concept of the Ewald sphere. As shown here, the Ewald sphere is constructed by first placing the incident wave-vector $\frac{1}{\lambda}\hat{\mathbf{k}}_0$ such that it points to the origin of the coordinate system in reciprocal space. Next, the scattered wave-vector $\frac{1}{\lambda}\hat{\mathbf{k}}$ is added, with its origin coincident with that of the incident wave-vector $-\frac{1}{\lambda}\hat{\mathbf{k}}_0$. This construction is designed to display the locus of all vectors, $\mathbf{S}$, that satisfy the relationship $\mathbf{S} = \frac{1}{\lambda}(\hat{\mathbf{k}} - \hat{\mathbf{k}}_0)$. As is evident by inspection, the circle of radius $\frac{1}{\lambda}$, centered at $-\frac{1}{\lambda}\hat{\mathbf{k}}_0$ and drawn with a solid line, represents the position of all possible spatial frequency vectors, $\mathbf{S}$, for which one can measure the structure factors. In order to measure more of the Fourier transform of the object, it is necessary to change the orientation of the incident wave-vector relative to the specimen, as indicated by $\hat{\mathbf{k}}'_0$ in the figure, so that data will be collected on the surface of a new Ewald sphere, shown here with a dotted line.

the reciprocal lattice. A change in direction of the incident beam relative to the crystal is easily accomplished by rotating the crystal relative to the stationary, incident x-ray beam. Thus it is clear that data must be collected over a continuously varied range of orientations between the crystal and the incident beam; every time the Ewald sphere passes through a reciprocal lattice point a new diffraction spot gets recorded on the detector at the angle formed between $\hat{\mathbf{k}}$ and $\hat{\mathbf{k}}_0$. In practice, the range of angles over which a crystal is rotated during the course of recording a single x-ray photograph should be kept to approximately one or two degrees. This is done, as is mentioned in section 2.3, in order to avoid having too confusing an array of spots, and especially to prevent two spots from overlapping on the detector.

2.8 Bragg's law relates the measured scattering angle to the size of the repeat-distance for each sinusoidal term in the Fourier transform of the object

Bragg's law expresses a simple relationship that exists between the x-ray wavelength, the scattering angle for diffracted beams, and the distance between parallel planes of scattering density within the crystal. Bragg's law is usually written as

$$\lambda = 2d \sin \frac{\theta}{2}, \tag{2.20}$$

where λ is the *x-ray wavelength*, d is the *Bragg spacing*, and θ is the angle between the unscattered beam and the scattered beam. The Bragg spacing is often described as being the separation between parallel crystal planes, but it is better described as being the repeat-distance of one of the possible sinusoidal components of the electron density function (see equation 2.2). From equation 2.20 it is clear that beams scattered at very small angles will correspond to large Bragg spacings, while those scattered at large angles will correspond to small Bragg spacings. Thus, there is a *reciprocal relationship* between the size of the Bragg spacing and the size of the scattering angle, which in more precise terms is specified by the sine of half the scattering angle, as in equation 2.20.

Bragg's law is often derived by showing that beams that are reflected from successive planes will interfere constructively if the difference in path length is precisely equal to the x-ray wavelength, as is shown in figure 2.2. Although this derivation of Bragg's law gives an extremely important insight into the constructive interference that is the basis of x-ray diffraction by crystals, it is even more important to understand the origin of Bragg's law from the perspective of the Ewald sphere construction. Referring to figure 2.3, we can see immediately that a diffraction spot will be recorded whenever the angle between $\hat{\mathbf{k}}$ and $\hat{\mathbf{k}}_0$ satisfies the relationship

$$\sin\frac{\theta}{2} = \lambda \frac{g_{hkl}}{2}. \tag{2.21}$$

If we then take the further step of equating the spatial period, $\frac{1}{g_{hkl}}$, with the spacing between successive "Bragg planes," d_{hkl}, then equation 2.21 becomes identical to equation 2.20. In this way the somewhat artificial idea of "Bragg planes" is given a deeper physical meaning, in which a particular sinusoidal component of the continuous electron density function is regarded as a kind of diffraction grating.

The concept that sinusoidal components of the electron density serve as diffraction gratings for x-rays is not restricted to crystals, of course. If we think of a noncrystalline

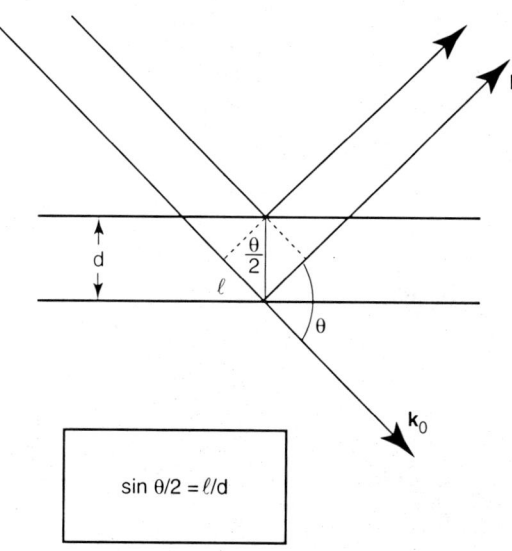

Figure 2.2 Diagram to explain the traditional derivation of Bragg's law, which pictures the interference of waves that have undergone reflection from successive planes of a crystal. By inspection of the drawing shown here, $\sin\frac{\theta}{2}$ is equal to $\frac{l}{d}$. Constructive interference will occur when the difference in path length, $2l$, for rays reflected from successive planes is equal to the wavelength, λ, ensuring that equation 2.20 will be satisfied.

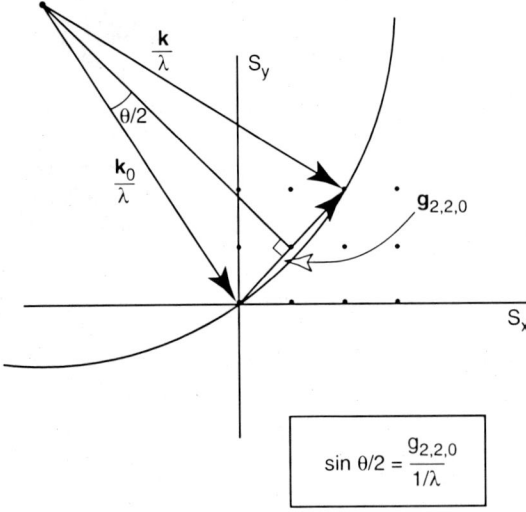

Figure 2.3 The central part of the Ewald sphere construction is shown on a magnified scale, and a two-dimensional representation of a small portion of the reciprocal lattice has also been added. The drawing has purposely been constructed to show the Ewald sphere intersecting the reciprocal lattice at the point $\mathbf{S} = \mathbf{g}_{2,2,0}$. By inspection we can see that the point of intersection satisfies equation 2.21, where θ is the angle between $\hat{\mathbf{k}}_0$, the direction of the incident beam, and $\hat{\mathbf{k}}$, the direction of the scattered beam.

object, for example a single molecule, its Fourier transform will no longer be confined to a discrete set of reciprocal lattice points. Instead, its Fourier transform will be a continuous function in reciprocal space. Thus, we understand that the electron density function of a noncrystalline object must be made up of a continuous set rather than a discrete set of sinusoidal components. What this means, in turn, is that the Ewald sphere will now intersect a function that has finite values at essentially all values of $\mathbf{S}$. Each sinusoidal component of the density again acts like a diffraction grating, which has an effective Bragg spacing of $\frac{1}{S}$, where the continuous spatial frequency, $\mathbf{S}$, now takes the place of the discrete spatial frequency, $\mathbf{g}_{hkl}$, in equation 2.21. Therefore a scattering experiment that uses a noncrystalline object will measure something about the Fourier transform of the object at every angle, rather than just at a discrete set of Bragg angles.

2.9 Information about the relative phase of each sinusoidal term is lost in diffraction patterns

The *phases of the structure factors*, equation 2.6, are lost when the diffraction intensities are recorded, for the simple reason that it is the scattered intensity, not the scattered wave function, that can be measured experimentally. The phases must thus be recovered in a separate experiment in order to calculate the electron density function.

As we will explain more fully in chapter 4, the scattered wave function is proportional to the Fourier transform of the electron charge density:

$$\Psi_{scattered} \propto F(\mathbf{S})e^{i\alpha(\mathbf{S})}, \qquad (2.22)$$

where $F(\mathbf{S})$ is the amplitude of a given sinusoidal Fourier component of the electron charge density, and $\alpha(\mathbf{S})$ is its phase. The intensity is the absolute square of the scattered wave function, that is,

$$I(\mathbf{S}) = \Psi \cdot \Psi^*, \qquad (2.23)$$

where Ψ^* designates the complex conjugate of Ψ. If we substitute equation 2.22 into equation 2.23, we have

$$I(\mathbf{S}) \propto F(\mathbf{S})e^{i\alpha(\mathbf{S})} \cdot F(\mathbf{S})e^{-i\alpha(\mathbf{S})} = F^2(\mathbf{S}), \qquad (2.24)$$

and it then becomes clear why $I(\mathbf{S})$ no longer contains any information about the phases, $\alpha(\mathbf{S})$.

2.10 The crystallographic phase problem is usually solved by using additional data obtained from heavy-atom derivatives of the original molecular crystals

The recovery of phase information can be accomplished by preparing crystals in which the protein has been modified by the addition of one or more heavy atoms at specific locations. The positions of the added heavy atoms must be the same from one molecule to the next within the crystal. Historically, the addition of heavy atoms was accomplished by soaking crystals of the native protein in dilute solutions of an appropriate heavy-atom reagent (Blundell and Johnson, 1976). This method requires that the structure of the native protein crystal and the heavy-atom derivative are *isomorphous*, that is, that there is no structural change between them other than the addition of a heavy atom where there previously was none. Mathematically, isomorphism is represented by the following equation:

$$\rho_D(\mathbf{R}) = \rho_P(\mathbf{R}) + \rho_{HA}(\mathbf{R}), \qquad (2.25)$$

where $\rho_P(\mathbf{R})$ is the electron density in the crystal of the native protein, $\rho_D(\mathbf{R})$ is the electron density in the crystal of the *heavy-atom derivative*, and $\rho_{HA}(\mathbf{R})$ is the electron density of the heavy atoms alone.

The process of phasing then begins by recording, spot-by-spot, the diffraction intensities for the native and the derivative crystals. The presence of the heavy atoms in the derivative crystals gives rise to additional scattering, which will, in general, cause a change in the diffraction intensities. Crick and Magdoff (1956) have derived the following equation to estimate the statistically expected amount of the intensity change, assuming perfect isomorphism and 100 per cent occupancy of the heavy atom binding sites:

$$\langle \Delta I \rangle = \sqrt{\frac{2N_{HA}}{N_P} \cdot \frac{f_{HA}}{f_P}}, \qquad (2.26)$$

where $\langle \Delta I \rangle$ is the statistically expected change in diffraction intensity when N_{HA} heavy atoms, which have an atomic scattering amplitude of f_{HA}, are added to a protein that is made up of N_P atoms, each with a scattering amplitude of f_P. The structure factor amplitudes of the native protein crystal and its heavy atom derivatives, that is, the modulus of the native structure factor and the modulus of the heavy atom derivative structure factor, are then obtained as the square root of the respective diffraction intensities:

$$F(\mathbf{S}) = \sqrt{I(\mathbf{S})}. \tag{2.27}$$

The location of the heavy atoms within the unit cell is next deduced by using the measured differences between the native and derivative structure factor amplitudes. The process usually begins with the calculation of a *difference Patterson* map. The difference Patterson function is defined as the Fourier transform of the square of the difference between the respective structure factor amplitudes:

$$P(\mathbf{R}) = \mathcal{F}\{[F_D(\mathbf{S}) - F_P(\mathbf{S})]^2\}. \tag{2.28}$$

We state here without proof — but see books such as Blundell and Johnson (1976) or Drenth (1994) for a derivation — that the difference Patterson function is a close approximation to the autocorrelation function of the electron density of the heavy atoms alone. If the heavy atom density map itself has N well-resolved peaks within the unit cell, then the difference Patterson map will have $\{N(N-1)+1\}$ peaks, and the vectors from the origin to each peak will constitute a map of all possible interatomic vectors between the heavy atoms. If there are not too many heavy atom binding sites within the unit cell, one can — by experience and a little bit of trial-and-error — deduce the arrangement of heavy atoms that is responsible for the experimentally obtained difference Patterson map.

Solving the difference Patterson means that we now have a model of the heavy atom density function, $\rho_{HA}(\mathbf{R})$. The structure factor of the heavy atom electron density function alone can then be calculated from this model. Estimating the heavy atom structure factor is a major step in recovering the phases of the native structure factors. To see why this is such an important step we need to look again at equation 2.25, which relates the electron density of the native protein crystal to that of the heavy atom derivative crystal.

Calculating the Fourier transform of both sides of equation 2.25, we get the result that the structure factor of the derivative is the sum of the structure factor of the native crystal and the structure factor of the heavy atom(s),

$$\mathbf{F}_D(\mathbf{S}) = \mathbf{F}_P(\mathbf{S}) + \mathbf{F}_{HA}(\mathbf{S}). \tag{2.29}$$

Since an estimate of both the amplitude and the phase of the heavy atom structure factor are available, it is possible to solve equation 2.29 to estimate the native protein phases, using only the modulus of the native and derivative structure factors. The way in which this can be done is best illustrated by a graphical method of solution, shown in figure 2.4. The graphs shown in this figure are called *Argand diagrams*; these are vector representations of complex-valued quantities in which the horizontal axis represents the real part and the vertical axis represents the imaginary part of the complex-valued quantity.

The first step in the graphical solution is to draw a circle whose radius is equal to the modulus of the native structure factor (for a particular diffraction spot), as is illustrated in figure 2.4. This circle is obviously the locus of all possible structure factors that would be consistent with the measured value of the native-crystal diffraction intensity. Similarly, a circle of radius equal to the modulus of the derivative structure factor,

Structure Determination as it has been Developed through X-ray Crystallography

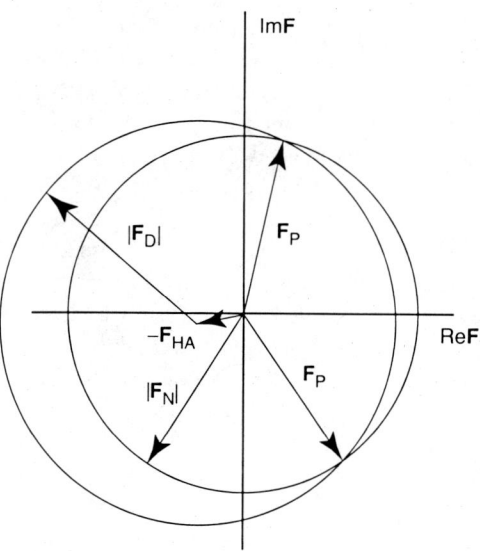

Figure 2.4 Argand diagrams can be used to show how to find the phase of the native protein structure factors by a graphical method of solution. The circle centered at the origin of the coordinate system represents the locus of all vectors, $\mathbf{F}_P$, that would be consistent with the measured amplitude, F_P. The circle centered at $-\mathbf{F}_{HA}$ represents the locus of all vectors $\mathbf{F}_P$ that would be consistent with the experimentally inferred location of the heavy atoms in the derivative crystal, and with the measured amplitude, F_D. The correct value of $\mathbf{F}_P$ must be the same in both cases, and therefore it must lie at the intersection of these two circles. There are two such points of intersection, however, only one of which can represent the correct solution. This remaining ambiguity in the phase of $\mathbf{F}_P$ can be resolved by repeating the same experiment for one or more additional heavy atom derivatives. The experimental errors described in the text cause errors in where these circles intersect, however. Even so, it will usually be clear which set of points of intersection is more tightly clustered than the other. The center of the best cluster, in turn, gives the best experimental estimate of $\mathbf{F}_P$. Estimates of the experimental error in the measured diffraction intensities can be used to generate a *phase probability function*, that is, a bell-shaped distribution lying on a circle of unit radius. This distribution has a center of mass that is located at the vector position, **m**. If the probability distribution is very narrow, m will be close to 1, whereas m will be close to 0 if the distribution is spread uniformly over the circle. Thus the length of the vector **m** is a quantitative measure of how well the phase is known, and m can be used as a weighting factor in computing a Fourier map.

centered at the vector location $-\mathbf{F}_{HA}$, represents the locus of all vectors

$$\mathbf{F}_P(\mathbf{S}) = \mathbf{F}_D(\mathbf{S}) - \mathbf{F}_{HA}(\mathbf{S}). \tag{2.30}$$

In other words, this second circle represents the locus of all possible native structure factors that are consistent with the measured value of the derivative-crystal diffraction intensity plus the model of the heavy atom positions within the unit cell. The two points of intersection of these two circles represent estimates of the alternative values of the native structure factor that are consistent with all of the data.

A word of caution is appropriate here, because the model of the heavy atom density function, $\rho_{HA}(\mathbf{R})$, will necessarily have a certain degree of experimental error in it. This error will be carried over in the calculation of $\mathbf{F}_{HA}(\mathbf{S})$, which has to be used in equation 2.29. The presence of error in $\mathbf{F}_{HA}(\mathbf{S})$, along with the inevitable error in the amplitude, $\mathbf{F}_D(\mathbf{S})$, means that there will be a gap between the tip of the vector $\mathbf{F}_P(\mathbf{S})$, which we do not yet know, and the modulus of the vector difference $\mathbf{F}_D(\mathbf{S}) - \mathbf{F}_{HA}(\mathbf{S})$, which we will see shortly is something that we can measure. The fact that the experimental vector difference does not perfectly close the triangle with the vector $\mathbf{F}_P(\mathbf{S})$ means that the *closure error* will propagate into errors in estimating the phase of $\mathbf{F}_P(\mathbf{S})$. How these errors are handled will be discussed later in this section.

The phase problem has now been reduced to the job of deciding which of two vectors represents the correct estimate of the structure factor for the native crystal. The traditional way of making this decision is to prepare two or more heavy atom derivatives. The Argand diagrams for all derivatives should share one point of intersection with the circle that represents the native structure factor, that is, the correct choice must always be included in the two choices offered by each derivative. Because of the closure error that was discussed in connection with equation 2.29, the circles for two different heavy atom derivatives will not intersect perfectly at a single point, but one set of intersection points should at least be close to one of the intersections shown in figure 2.4. When this is the case, we can be confident that the vector $\mathbf{F}_P(\mathbf{S})$ must itself lie close to this particular set of points.

If one were to construct Argand diagrams for a large number of independent heavy atom derivatives, one would begin to build up a distribution of estimated positions for the structure factor, $\mathbf{F}_P(\mathbf{S})$. These estimated positions would, in turn, give estimates of the phase that are distributed randomly about the correct value. While it is not uncommon to use more than two heavy atom derivatives to solve the phase problem, it is impractical to actually generate such a phase probability function by using a large number of different derivatives. Instead, one can use estimates for the experimental error involved in measuring the diffraction intensities in order to estimate the closure error. Thus, a well-founded estimate of the probability distribution for $\mathbf{F}_P(\mathbf{S})$ can be derived from the data. This probability distribution has a center of mass at the vector position $\mathbf{m}$. The length of the vector to the center of mass, $\mathbf{m}$, is known as the *figure of merit* of the phase of the structure factor, $\mathbf{F}_P(\mathbf{S})$. It has been shown that a Fourier synthesis that uses the values of the figure of merit as weights for each diffraction spot will produce a map with the smallest mean square error (see, for example, Drenth, 1994).

After the twofold ambiguity shown in figure 2.4 has been solved, the phase of the native structure factor is given simply as the angle between the positive real axis (horizontal axis) of the Argand diagram and the vector which represents the correct choice for $\mathbf{F}_P(\mathbf{S})$.

Other methods of solving the phase problem also exist. Widespread access to synchrotron radiation sources has recently made it practical to solve the phase problem by using *anomalous intensity differences*, which are intensity differences that occur between diffraction spots located at $\mathbf{g}$ and $-\mathbf{g}$, when the x-ray wavelength is close to an absorption edge for one of the atoms in the structure. Exceptionally good phases can be obtained, for example, by incorporating selenomethionine in place of methionine during bacterial expression of a protein. This biosynthetic approach to incorporating one or more heavy atoms directly into the protein structure can make *MAD* (*multiple anomalous dispersion*) *phasing* the preferred method. We will not digress here to explain how MAD phasing works, however, but refer the reader instead to books on protein crystallography (Blundell and Johnson, 1976; Drenth, 1994) or the review by Hendrickson and Ogata (1997).

2.11 The three-dimensional electron density of the molecule can be calculated from the experimentally measured amplitudes and phases of the Fourier transform

The 3-D density map in real space is calculated as a Fourier series, using all available structure factors for the native protein crystal:

$$\rho(\mathbf{R}_j) = \sum_{h,k,l} m(hkl) F(hkl) e^{i\alpha(hkl)} e^{-i2\pi \mathbf{g}_{hkl} \cdot \mathbf{R}_j}. \quad (2.31)$$

The numerical calculation, done using a computer, takes a given position, $\mathbf{R}_j$, in real space and does the called-for summation. The result is stored at the position $\mathbf{R}_j$, a new position is selected, and the summation is repeated. This operation continues until the density function has been calculated on a full 3-D grid of points. In reality, the computer code which is used to do these calculations would employ the *fast Fourier transform* (FFT) algorithm. The *FFT algorithm* involves algebraic manipulations that are mathematically equivalent to those represented by equation 2.31, but which are far more efficient than a computer algorithm that would literally carry out the complex multiplications and additions as they are written here. The mathematical basis for this speedup is explained in section 10.4, and more information about the FFT algorithm can be found in the book by Drenth (1994).

The *resolution* available in the density map is specified as the Bragg spacing of the highest resolution diffraction spot (at a spatial frequency $\mathbf{g}_{max}$) for which phase information has been obtained. This *Fourier transform definition of resolution* is different from, but nevertheless closely related to, *Rayleigh's criterion of resolution*, which is commonly used in optics. A more detailed description of how the crystallographic definition of resolution is related to the Rayleigh criterion is found in box 4.1. For the moment we only need to say that a spatial frequency limit of $\mathbf{g}_{max}$ in incoherent optics would resolve two points that are separated by a distance of $0.61\left(\frac{1}{g_{max}}\right)$. Two points that are separated by only 0.61 of the smallest Bragg spacing will not necessarily be resolved from each other in a Fourier map, however, as the Fourier map is equivalent to a coherent optical image, for which a more conservative estimate of the resolution is appropriate.

Different features of protein structure begin to emerge in a 3-D map as the resolution is improved. The overall size and shape of a large macromolecule are already well defined at a Fourier resolution of ~ 2.5 nm. Elements of internal, secondary structure begin to be resolved from one another at a Fourier resolution of ~ 1 nm. It may not be possible to distinguish β-sheet from α-helix yet at this resolution, however, especially if the density is still a bit fragmented in the map. At a Fourier resolution of ~ 0.45 nm, on the other hand, the individual strands of β-sheet can become well resolved from one another. Finally, bulky side chains begin to take on sufficient mass at a resolution

44   ELECTRON CRYSTALLOGRAPHY OF BIOLOGICAL MACROMOLECULES

Table 2.1 New levels of structural information become available for interpretation as the resolution of a density map increases. While the values of the resolution that are required to visualize specific structural information cannot be specified in precise steps, the following estimates represent a practical, approximate guide

Resolution range (nm)	Interpretable structural features
~2.5	The size and shape of large macromolecules Subunit organization of a particle Position and orientation for docking known, atomic-resolution structures into portions of the density map
~1.5 <0.8–0.9	Improved definition of shapes for subunits Helices become recognizable as irregular rods or "sausages"
~0.45	Separate strands of density for individual polypeptide chains in beta sheet and favorable loops or "random coil" structures Screw of helices becomes apparent
0.35–0.4	Bulky side chain residues become apparent Chain trace is normally possible
0.3–0.35 or better	Refinement of an atomic model (with limited use of B-factors)
0.25 or better	Refinement begins to provide accurate side chain rotamer conformations, positions of water molecules, ligand occupancies, etc.

of ~0.35 nm that it is possible to make a *chain tracing interpretation* of the map by fitting the known primary structure into the density. More is said in section 2.12 about the process of interpreting a high-resolution density map.

The precise Fourier resolution at which tracing the peptide chain becomes possible will vary with the quality of the phases. The ability to make such an interpretation of the map may also depend upon how much additional information is known about the structure in advance. Indeed, all of the estimates given in this paragraph, which are summarized again in table 2.1, should only be regarded as rough guidelines.

2.12 The 3-D density map must be interpreted in terms of other available information, to provide a model of the structure

The most common method used to display the 3-D density function shows surface contours of the density in "wire-basket" graphics, like the example shown in figure 2.5(a). By itself, this 3-D density map tells us surprisingly little about the molecular structure. The next stage in a high-resolution structure analysis therefore requires that one should make a detailed *chemical interpretation of the density map*. Because the resolution of the initial Fourier synthesis may not be better than 0.3 nm, and the high-resolution features of the map are usually quite noisy as well, it is not normally possible to interpret the map without first knowing the amino acid sequence of the protein.

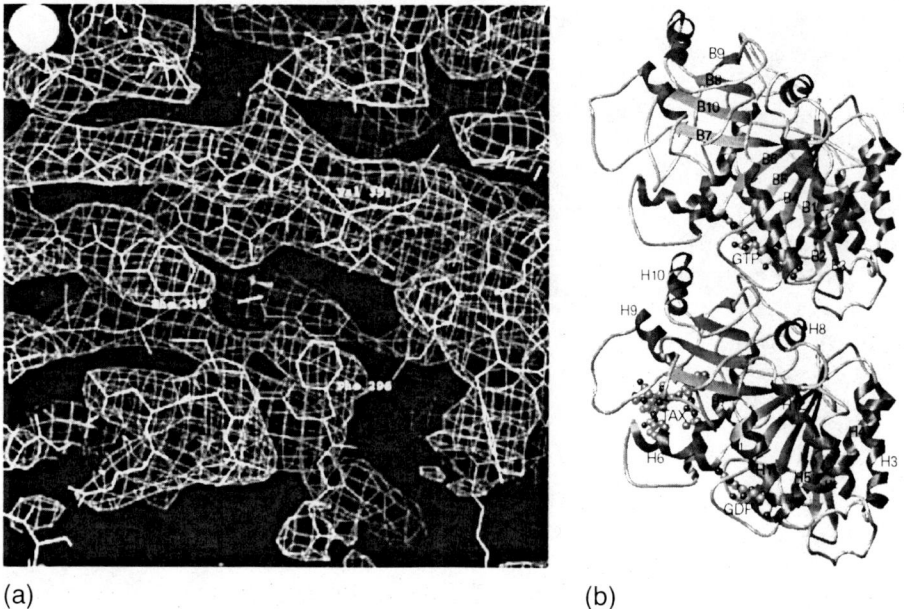

Figure 2.5 Example of molecular graphics representations that are commonly used to visualize protein structures at atomic resolution. (a) A region of the density map of tubulin, obtained by electron crystallography (Nogales et al., 1998), is shown at a resolution ~ 0.35 nm. The three-dimensional density is contoured at 1.0σ, and the resulting closed surface is shown as a "wire basket" representation. The positions of bulky amino acid side chains can be identified, but their chemical identity cannot generally be recognized at this resolution. (b) A "ribbon drawing" of the backbone-trace for the entire α, β tubulin dimer that has been drawn with the PYMOLE software package (http://pymol.sourceforge.net/). Qualitative representations like this one give a clear idea of the "fold" of a protein, once the structure has been determined at atomic resolution.

For example, it is not uncommon that one can see bulky side chains coming off of the peptide backbone, as in figure 2.5(a), but it may be impossible to say whether the side-chain density corresponds to that of amino acids as different from one another as tryptophan and arginine.

Prior knowledge of the full amino acid sequence is usually sufficient to lead to a correct interpretation of a high-resolution density map, however. At the initial stage, major features of secondary structure such as helix or beta-sheet need to be identified. Patterns of bulky versus small side-chain residues are then identified in the primary sequence, which match up with corresponding patterns of bulky side-chain features in the density map. At the beginning there may be multiple, alternative interpretations for any one part of the map. Constraints such as the requirement of continuity of the peptide chain, or the fact that a small amino acid (glycine or alanine) should not fall where there is a large side-chain density coming off the backbone, usually impose such strict limitations that an incorrect interpretation of the density map occurs only rarely. Already existing spectroscopic, mutational, biochemical, and functional information can also be very helpful in guiding and constraining the early stages of interpretation of the density map. The end-product of this stage of interpretation of the density map is a molecular model of the structure of the protein. The model is "built" in a graphics computer, using the known stereochemical constraints of the planar peptide bond to help fit the backbone, and keeping all of the molecular model "within density" as well as possible.

When such a model has been completed, the overall structure of the protein can be very effectively communicated by graphics programs, which display a so-called ribbon diagram, such as the example shown in figure 2.5(b). The molecular model of the structure at this point can be extremely informative. Nevertheless, it is important to realize that the atomic coordinates of such a model, before refinement, can easily have an r.m.s (root mean square) error as large as 0.1 or 0.2 nm. It is normal that chemically and functionally important features, such as the position and orientation of critical side-chain residues, will not yet be precisely correct at this stage.

2.13 A more accurate estimate of the structure can be obtained by further *refinement* of the model

The molecular model that is first built into the density map can be refined very substantially by using the higher resolution, native-crystal diffraction amplitudes that could not be used to compute the density map, due to the lack of experimental phases. The *crystallographic refinement* procedure seeks to find the best possible match between (1) the square root of the experimental diffraction intensity and (2) the modulus of the structure factor that can be calculated from the atomic coordinates of the model. The latter is given by the modulus of the Fourier series,

$$F_{calc}(hkl) = |\sum_j f_j(g_{hkl}) e^{i 2\pi g_{hkl} \cdot R_j}|, \qquad (2.32)$$

where the atomic scattering factors, $f_j(g_{hkl})$, are assumed to be real and spherically symmetrical. The refinement process allows the atomic coordinates of the molecular model to move, subject to stereochemical restraints on acceptable bond angles and bond lengths as well as requiring unfavorable van der Waals contacts to be minimized. At each point during the refinement, the discrepancy between measured and calculated modulus of the structure factor is expressed as a residual, called the *R-factor*, which is defined by

$$R = \frac{\sum |F_{obs} - F_{calc}|}{\sum F_{obs}}, \qquad (2.33)$$

where F_{obs} represents the square root of the observed diffraction intensity and F_{calc} is the modulus of the structure factor calculated for the model. The summations include all diffraction spots that are to be used in the refinement. Molecular energy calculations (Brunger et al., 1987; Karplus and Petsko, 1990) can also be used to assist the refinement by letting the model structure "settle into" the energetically most favorable conformation that is close to the starting model structure.

At an intermediate stage of refinement, the structural model can be improved by assigning individual temperature factors to separate atoms or atom clusters. These individual temperature factors enter into the calculated structure factors in the form of products with the individual atomic scattering factors, so that equation 2.32 is rewritten as

$$F_{calc}(hkl) = |\sum_j f_j(g_{hkl})e^{-\frac{B}{4}g_{hkl}^2}e^{i2\pi g_{hkl}\cdot R_j}|. \tag{2.34}$$

An important consideration at this and later stages of the refinement is that it is necessary to keep the number of adjustable parameters much smaller than the number of observed intensities. As a result, a refinement in which individual atomic B-factors are included can only occur if the resolution is high enough to give the number of diffraction spots needed to keep the ratio of observations to parameters sufficiently high.

If the native data set extends to very high resolution, and there are enough observed intensities, the model structure can be further refined by adding tightly bound water molecules to the model of the protein structure. Likely positions of the water molecules can be seen when one calculates the difference between the density map obtained with experimental diffraction amplitudes, F_{obs}, and the density map obtained with F_{calc}, the amplitudes calculated for the current version of the atomic-resolution model of the structure. Well-ordered features of the structure that have not yet been built into the atomic-resolution model of the structure will contribute to the experimental measurement, that is, F_{obs}, but not to F_{calc}, and thus will show up as density peaks in the *difference Fourier map*.

Deciding just how far one can go in refining a model structure is a matter that requires some experience and care. The residual difference between the calculated and experimental diffraction intensities will always continue to decrease, especially as new parameters are added to the refinement. As a result, improvement between the measured and the calculated diffraction intensities is not itself a reliable indication that one is getting closer to the truth. A very effective technique to deal with this problem is to hold back a certain fraction of the measured data, for example 10 per cent of the diffraction spots, and not use those data in refining the model. The data that have been set aside can be used to calculate an independent estimate of the R-factor, which is called the *free R-factor* (Brunger, 1992, 1997) because the data involved have not been used to determine the values of the adjustable parameters. If the free R-factor continues to decrease during refinement, one can be confident that the model of the structure has improved.

A few examples of substantial errors in the model structure have been reported, even though the refinement operation had indicated in each case that all was going well (Branden and Jones, 1990; Colovos and Yeates, 1993; Jones and Kjeldgaard, 1997). The report of a more recent search for likely errors by Hooft et al. (1996), examining a collection of over 3000 solved structures, has itself been strongly criticized (Jones et al., 1996), since the automated identification of "likely structural errors" used by Hooft was found to greatly exaggerate the true number of errors. Nevertheless, the point remains that the danger of error is as real in crystallographic structure determinations as it is in any other experimental method. It is therefore necessary to regard the model of the structure obtained at any one stage as being a hypothesis, rather than a fact. As is true of any other scientific hypothesis, a model structure may do a good job of explaining current knowledge, but it must also stand the test of being useful in designing new experiments. We should always be prepared for the eventuality that new information may require a major revision of the "hypothesis."

2.14 Published structures are made available through a public-domain database

X-ray crystallographic structure analysis of biological macromolecules has become a fast and efficient process. As a result, there are thousands of new protein (and other macromolecule) structures published each year. In the midst of this explosion of x-ray crystallographic structural data there has also emerged the capability to determine the structure of macromolecules by NMR spectroscopy, and, of course, by electron crystallography.

The description of a single protein structure entails an enormous amount of information, corresponding to the atomic positions and (usually) the thermal parameters for most — if not all — of the thousands of atoms in the structure. Some of this information is traditionally summarized in the initial publication of the structure in the form of a ribbon drawing or other, equivalent representations of the chain trace. Other details, showing atomic representations of specific side chains, the binding of a substrate or other ligand, etc., may also be included in the initial publication. Even so, hard-copy journal publications capture only a fraction of all the information about the structure that really exists as the outcome of the structure determination.

Since 1989, in accord with the policy of the International Union of Crystallographers (Acta Cryst. A45: 658, 1989), scientific journals have required that authors deposit coordinates of new structures shortly after publication. The resulting ready access to solved structures makes it possible for other investigators to search for rules that may determine protein folding; to compare proteins that have homologous structures even though there is no recognizable sequence homology; and to pursue a host of other theoretical investigations where access to a wide range of solved structures is essential. As recounted by Bernstein et al. (1977), such a database began to operate at the Brookhaven National Laboratory in 1971 following formal recognition of the need for such a service. The *Protein Data Bank* (PDB) now provides a central point of distribution for all atomic-resolution structures that have been placed into the public domain (Berman et al., 2000). Further information about the database can be obtained directly from the PDB website: www.rcsb.org/pdb/.

3

Fourier Optics and the Role of Diffraction in Image Formation

3.1 Introduction

This chapter provides a brief review of the basic concepts of image formation, described from a wave-optical point of view. Chapter 2 has already given an outline of important concepts that emerge from x-ray crystallography. The concepts of x-ray crystallography contribute only one part of the foundations of electron crystallography, however. The mathematical understanding of image formation contributes a second part that is equally basic to the development of electron crystallography.

There is, in fact, a very close relationship between the wave theory of diffraction, covered in chapter 2, and the wave theory of images, covered in this chapter. As we will discuss in sections 3.2, 3.3, and 3.4, the wave function produced when the scattered wave is combined with the unscattered wave (during formation of a so-called bright-field image) is actually the inverse Fourier transform of the diffracted wave. The inverse Fourier transformation that is produced by wave propagation and the focusing properties of the lens is, speaking mathematically, a linear operation. As we explain in section 3.5, it is often quite useful to describe a linear system by a *transfer function*, which in this case is simply the Fourier transform of the image wave function divided by the Fourier transform of the wave function transmitted through the object. Generating the inverse Fourier transform of the scattered wave function is also the ultimate goal of diffraction experiments in x-ray crystallography. In the case of diffraction experiments, however, the inverse Fourier transform must be calculated numerically, in the way that was described in chapter 2.

Much of the discussion in this book focuses its attention on specimens that produce only a very weakly scattered wave. In most cases, in fact, we make the further approximation that scattering is primarily due to a weak perturbation of the phase of the transmitted wave, and not its amplitude. The formal definition of what is meant by a *weak phase object* is thus introduced in section 3.6. In the case of a weakly scattering object, not only the image wave function, but even the (bright-field) image intensity remains a linear function of the scattered wave, as we show in section 3.7.

The fact that the image intensity is linear in the scattered wave amplitude means that the crystallographic phases are directly accessible in the measured image intensity. The weak phase object approximation is thus an essential part of the theory on which electron crystallography is founded. Among other things, the property of linearity makes it possible to describe image formation in terms of a *phase contrast transfer function*, as we will show in section 3.8. The transfer function provides an elegant formalism that shows how to correct, at least partially, for systematic defects in the measured data that arise during image formation. These defects include the effects of lens aberrations and imperfect wave coherence. As we will see in section 3.8, however, spherical aberration and defocus are not so much a "defect" but rather an essential asset, in that they are necessary for the formation of phase contrast in images of weakly scattering objects.

The mathematical theory that is described in section 3.8 applies rigorously to weak phase objects, and only in the limit of perfect wave coherence. The experimental sources that are practical to use in electron microscopy are, however, only partially coherent. As a result, the high-resolution components of the Fourier transform of the image intensity are attenuated by an overall *envelope function*, which is described in section 3.9 for the specific case of a weak phase object. Section 3.10 then goes on to describe how one can also extend the theory of image formation for weakly scattering objects so as to include (weak) amplitude contrast.

The mathematical theory presented up to that point also makes the assumption that the objective lens collects electrons that have been scattered in all directions, and that the diffracted wave is symmetrical (shows inversion symmetry) across the center of the objective aperture. Since it is a formal possibility that one could use an objective aperture to selectively block part of the scattered wave before forming an image, section 3.11 makes a digression to describe image formation in such a case. This formal treatment is useful in providing an alternative view of image formation, and, as we will see in chapter 15, it is even of practical assistance in understanding image formation when the diffracted wave itself is not fully symmetrical on both sides of the diffraction pattern. Another example of a practical situation in which the wave function in the back focal plane of the objective lens is not symmetrical, described in section 3.12, occurs when the incident electron wave is not parallel to the optical axis of the electron microscope.

3.2 Abbe's diffraction theory of images: image formation is the two-dimensional equivalent of the crystallographer's "inverse Fourier transform"

Image formation by a lens can be thought of as occurring by a two-step process. In the first step the lens collects the radiation that has been scattered by the object and produces a diffraction pattern in the back focal plane (figure 3.1). As we have discussed in chapter 2, the diffraction intensity represents the square of the Fourier transform

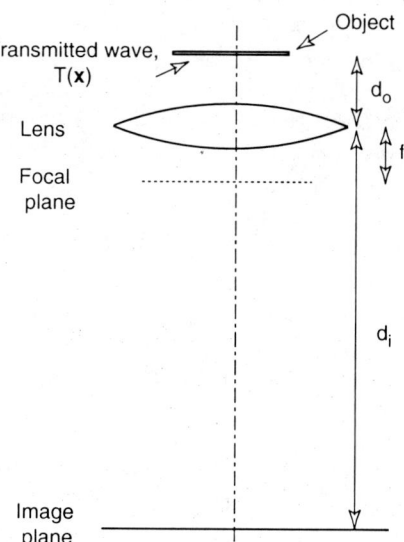

Figure 3.1 A schematic diagram that illustrates some of the key terminology and notation used in discussing image formation. The object is located a distance d_o in front of the lens, and the in-focus image is formed at a distance d_i behind the lens. In the theoretical treatment of coherent optics, the specimen is illuminated by a monochromatic plane wave. The wave function in the focal plane is then the Fourier transform of the transmitted wave. The wave function in the image plane is the transmitted wave, convoluted with the impulse response of the optical system, as is described in section 3.4.

of the object, but with the important qualifying remark that only terms whose spatial frequency vectors lie on the Ewald sphere contribute to the diffraction pattern. If the incident wave is parallel to the optical axis, the unscattered radiation will be focused to a point at the intersection of the optical axis and the back focal plane, which then represents the origin of the Fourier-space coordinate system. In the second step of image formation, the lens acts like an analog computer and carries out an inverse Fourier transformation of the wave amplitude in the back focal plane. This inverse Fourier transformation (or Fourier synthesis) returns us back to a real-space representation, which is the image of the object.

According to this simple, two-step description of image formation, the lens can be expected to modify the diffracted wave to some extent before the inverse Fourier transform is carried out. At the very least, the finite aperture of the lens will accept only part of the scattered wave. The fact that some of the diffracted wave falls outside the limiting aperture, and thus is not included in the Fourier synthesis of an image, generally means that the resolution at which there is information about an object is less in an image than it is in a diffraction pattern. In light microscopy the finite aperture of the optics is often the limiting factor in determining the resolution, but in electron microscopy it is generally the case that other factors, not the objective lens aperture, limit the resolution.

Lens *aberrations* and *defocus* are among the other factors that modify the diffracted wave during image formation, and both have a significant effect in electron microscopy. These enter the theory of image formation in the form of a *phase distortion*, $\gamma(s)$, causing the diffracted wave to be multiplied by the function $e^{i\gamma(s)}$ before the Fourier synthesis is carried out.

Even if the lens is perfect, that is, free of aberrations, and precisely in focus, the Fourier synthesis that generates the image of the object will still be truncated by the aperture function, as we have just mentioned. As a result, the image will always be a *diffraction-limited*, imperfect representation of the object.

The theoretical understanding of image formation in terms of the two-step process described above was put forward by the German physicist Ernst Abbe, who first proposed the double Fourier transform character of image formation in 1873. Abbe did not present a formal, mathematical proof of the theory (Porter, 1906), but he did design a number of convincing optical demonstrations of its validity. Perhaps the most famous of these was the use of a linear *aperture* (slit), placed in the back focal plane of a lens. When correctly oriented with respect to the diffraction pattern of the object, such an aperture caused the image of a *two-dimensional grating* formed by intersecting, parallel lines to be reduced to only one set of parallel lines. Exactly the same result is obtained when the mathematically calculated Fourier transform of a two-dimensional grating is multiplied by a slit-like rectangle function before calculating the inverse Fourier transform. In spite of Abbe's ingenuity in providing experimental proofs of the double Fourier transform character of image formation, his great theoretical insight was not readily accepted by all. Even 30 years after the date of the first cited publication of Abbe's major contribution (Abbe, 1873), the American physicist A. B. Porter, who had elaborated on Abbe's work, remarked that many were still skeptical about, or even unfamiliar with the "diffraction theory" of image formation (Porter, 1906).

3.3 Zernike and the invention of phase contrast microscopy

The Fourier transform representation of image theory provided the basis for a practical advance of tremendous importance when Zernike showed that it is possible to produce stunning contrast in otherwise invisible specimens by simply imposing a quarter-wave phase shift on the transmitted (unscattered) wave (Zernike, 1955). Unstained biological specimen materials could thereby be seen in microscopic, subcellular detail, going about their business without any notice of being observed. The *Zernike phase contrast microscope* is a routinely used research tool in any cell biology laboratory, being surpassed in only the most demanding applications by the Nomarski differential interference contrast microscope, particularly when the latter is aided by modern facilities for video image capture and digital image manipulation.

The genius of the Zernike phase contrast microscope was inspired from the realization that the image of a pure phase object — that is, an object which modifies only the phase of the transmitted wave and not its intensity — must be almost completely devoid of contrast, the more so the better the quality of the lens! Why is this the case?

A pure *phase object* will transform an incident plane wave into a transmitted wave that still has constant amplitude but variable phase. Recall that the incident plane wave has constant amplitude and constant phase on a plane that is perpendicular to the direction of propagation (identified here as the z-axis direction). The mathematical expression for a plane wave therefore is

$$\psi_0(\mathbf{x}) = e^{i\ \mathbf{k}_0 \cdot \mathbf{z}}, \qquad (3.1)$$

where $\mathbf{k}_0$ is the momentum vector of the incident wave (defined previously in connection with equation 2.19). The two-dimensional vector notation, $\mathbf{x}$, is used to designate a point in the (x, y) plane, perpendicular to the direction of the incident wave, while the vector **z** specifies the position of such a plane along the direction of the incident wave. The transmitted wave can be represented mathematically as

$$\psi_{trans}(\mathbf{x}) = e^{i\,\eta(\mathbf{x})} e^{i\,\mathbf{k}_0 \cdot \mathbf{z}}, \qquad (3.2)$$

where $\eta(\mathbf{x})$ is the amount by which the phase of the wave differs from that of a plane wave after passing through the specimen.

A perfect lens, free of aberrations, will simply produce a magnified version of the transmitted wave, ignoring for the present the effects of the finite aperture of the lens. Thus,

$$\psi_{image}(\mathbf{x}) \propto \psi_{trans}\left(\frac{\mathbf{x}}{M}\right), \qquad (3.3)$$

where M is the magnification. After accounting for the difference in magnification, the image intensity — which is the square of the image wave amplitude — is given by

$$I(\mathbf{x}) \propto (e^{i\,\eta(\mathbf{x})} e^{i\,\mathbf{k}_0 \cdot \mathbf{z}}) \cdot (e^{-i\,\eta(\mathbf{x})} e^{-i\,\mathbf{k}_0 \cdot \mathbf{z}}) = 1. \qquad (3.4)$$

Equation 3.4 shows that all information about the phase of the image wave function is lost. As a result, there is no contrast in the image intensity function when a phase object is imaged by a perfect lens.

Referring to Abbe's diffraction theory of image formation, however, it is easy to see that images of exactly the same object can be formed with high contrast just by using an optical device to shift the phase of the scattered wave relative to the unscattered wave. The principle is illustrated in figure 3.2 for the case of a *weak phase object*.

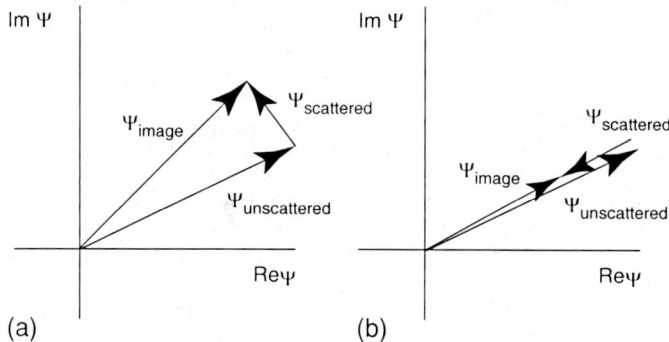

Figure 3.2 The unscattered and the scattered components of the wave transmitted through the specimen are represented by two vectors in the complex plane, that is, by an *Argand diagram*. In panel (a), the scattered wave is rotated by 90° from the unscattered wave, because of the factor of i in equation 3.5. When the scattered wave is weak relative to the unscattered wave, the length of the vector sum of the two, in the image, is almost the same as the length of the unscattered wave. The vector sum is therefore insensitive to the size of the scattered wave. As a result, the image intensity has almost no contrast. However, if an additional phase shift of ±90° is applied, as is shown in panel (b), the vector representing the scattered wave becomes either antiparallel or parallel (not shown) to the unscattered wave. The length of the vector sum in the image now becomes very sensitive to the size of the scattered wave. As a result, applying an additional 90° phase shift between the scattered and the unscattered waves results in a large amount of *phase contrast* in the image intensity.

Because the phase modulation, $\eta(\mathbf{x})$, is small for a weak phase object, the complex exponential that represents the phase modulation of $\psi_{trans}(\mathbf{x})$ can be approximated by a linear expansion, thereby giving the result

$$\psi_{trans}(\mathbf{x}) = e^{i\,\eta(\mathbf{x})} e^{i\,\mathbf{k}_0 \cdot \mathbf{z}} \simeq e^{i\,\mathbf{k}_0 \cdot \mathbf{z}} + i\,\eta(\mathbf{x}) e^{i\,\mathbf{k}_0 \cdot \mathbf{z}}. \tag{3.5}$$

$\psi_{trans}(\mathbf{x})$ is therefore represented in equation 3.5 by two terms. The first term, $e^{i\,\mathbf{k}_0 \cdot \mathbf{z}}$, is identical to the incident plane wave, and it is therefore referred to as the *unscattered wave*. The second term, $i\,\eta(\mathbf{x}) e^{i\,\mathbf{k}_0 \cdot \mathbf{z}}$, corresponds to the *scattered wave*. Noting the identity

$$i = e^{i\,\pi/2}, \tag{3.6}$$

we can see that the scattered wave has a 90° phase shift relative to the unscattered wave. The two components of ψ_{trans} therefore add in quadrature, as is shown in figure 3.2(a), and the resultant intensity remains essentially constant over the field of the image, although the relative length of the two vectors, $\psi_{unscattered}$ and $\psi_{scattered}$, may vary. By using an optical device that applies a 90° phase shift to either $\psi_{unscattered}$ or to $\psi_{scattered}$, however, there can be either constructive or destructive interference when the two components are recombined, as is shown in figure 3.2(b). As the figure shows, the 90° phase shift causes the vector representing the scattered wave to become parallel (or antiparallel) to the unscattered wave. As a result, there will be corresponding variations in the image intensity that reflect the variations in phase, $\eta(\mathbf{x})$, of the wave transmitted at each point of the specimen.

In light microscopy many objects also absorb some of the incident radiation. These specimens are called *amplitude objects* because they produce a spatial modulation of the amplitude of the transmitted wave. In this case a perfect lens will simply produce a magnified version of the transmitted wave, and therefore the image intensity will replicate the pattern of varying intensity — the contrast — that is already present in the transmitted wave.

Thus, to summarize, a key lesson obtained from the diffraction theory of image formation is that the best images of amplitude objects are obtained with a perfect (albeit diffraction limited) lens, while the best images of a phase object are those obtained with a lens that has a built-in phase shift of 90° between the scattered and unscattered wave.

3.4 The rigorous diffraction theory of image formation describes images in terms of the inverse Fourier transform

Compelling though Abbe's experiments were in demonstrating the relationship between diffraction and image formation, and recognizing the triumph of Abbe's theory of image formation when Zernike invented the phase contrast microscope, even further insight is gained by discussing the mathematical proof that lenses really do carry out the two Fourier transformations that have been described in section 3.2.

Arguing first from physical plausibility rather than rigorous mathematics, we can draw on the fact that a *Fraunhofer diffraction* pattern, which is formed in the "far field" limit, that is, infinitely far from the specimen, is simply the Fourier transform of the wave that is transmitted through the specimen. At the same time, parallel rays leaving the specimen arrive at a single point in the focal plane of an objective lens (of a microscope), just as they do, in effect, at infinity (in the absence of the lens). A *prima facie* case exists, then, that the wave amplitude at the back focal plane will be the same Fourier transform as that which is formed at infinity.

A formal proof is obtained by the straightforward application of scalar diffraction theory (Goodman, 1968). In brief, the derivation first assumes that the wave incident on the specimen is a monochromatic plane wave, $e^{i\,\mathbf{k_0 \cdot z}}$. This assumption means that the theoretical derivation is only valid for conditions of illumination that have perfect *spatial and temporal coherence*. As we will see in section 3.9, however, the theory appropriate to any condition of *partial coherence* can be developed directly from the results obtained for perfect coherence.

The formal derivation next assumes that the interaction of this plane wave with the specimen can be represented by a *transmittance function*, $T(\mathbf{x})$. The transmittance function simply describes the change, if any, in both the phase and the amplitude of the incident wave, after it has passed through the specimen. What is important, for the purpose of the derivations, is that the transmitted wave can be represented as the product of $T(\mathbf{x})$ and the incident plane wave:

$$\psi_{object}(x, z) = T(\mathbf{x}) e^{i\,\mathbf{k_0 \cdot z}}. \tag{3.7}$$

Note, as before, that $\mathbf{x}$ is a vector in the (x, y) plane. Unless $T(\mathbf{x})$ is a constant, the transmitted wave is no longer a plane wave, but it does remain monochromatic, with the same wavelength as the incident wave.

After the wave has passed through the specimen, it next propagates through space by *Fresnel diffraction* from the specimen to the lens, as described in box 3.1. Propagation through the lens itself is again described by an appropriate transmittance function for the lens (Goodman, 1968). After passing through the lens, the wave next propagates

---

BOX 3.1 The wave function in the focal plane is proportional to the Fourier transform of the transmitted wave

In the formal derivation by Goodman (1968), the transmitted wave is first allowed to evolve by Fresnel propagation as it travels from the specimen to the lens. This means simply that the wave entering the lens is represented as the convolution of $\psi_{object}(\mathbf{x})$, in equation 3.7, and the parabolic approximation of a spherical wave,

$$P_{Fresnel}(x, y) = \frac{-i}{\lambda z} e^{i \frac{k}{2z}(x^2 + y^2)}. \tag{B3.1}$$

The symbols λ and k are the wavelength and the magnitude of the momentum vector, respectively, as in previous usage, while z is the distance from the object to the plane where the wave function is being evaluated.

Next, the lens itself is represented by a *transmittance function* of the form

$$T_{lens}(x, y) = e^{i\,k\,n\,\tau} e^{-i \frac{k}{2f}(x^2 + y^2)}, \tag{B3.2}$$

(continued)

BOX 3.1 *(continued)*

where τ is the thickness of the lens, n is its refractive index, and f is the focal length of the lens. If the lens is ground with a spherical shape, representation of the transmittance function as having a parabolic phase curvature, as in equation B3.2, requires that the rays being considered lie close to the optical axis. *Spherical aberration* is the optical defect that occurs in (glass lens) optics when the face of the lens is actually a sphere rather than a parabola, and thus equation B3.2 becomes increasingly inaccurate as x and y become large. The equivalent aberration also occurs in electron optics, where it plays a useful role in determining the phase contrast transfer function.

In the final step, the wave that is transmitted through the lens again evolves by Fresnel propagation over the distance f, the focal length of the lens. The formal result that is obtained for the wave function at the focal plane is

$$\Psi_f(x_f, y_f) = e^{i\frac{k}{2f}\left(1-\frac{d_0}{f}\right)\left(x_f^2+y_f^2\right)} \int T(x, y) e^{-i2\pi\left(\frac{xx_f}{\lambda f}+\frac{yy_f}{\lambda f}\right)} dx\,dy, \quad (B3.3)$$

where x_f and y_f are Cartesian coordinates in the focal plane, and d_0 is the distance between the object and the lens. The integral in equation B3.3 is easily seen to be a Fourier transform when one makes the substitutions:

$$s_x = \frac{x_f}{\lambda f}$$

$$s_y = \frac{y_f}{\lambda f}. \quad (B3.4)$$

It is not precisely correct, therefore, to say that Ψ_f is the Fourier transform of $T(\mathbf{x})$, because of the additional, quadratic phase term, represented by the factor $e^{i\frac{k}{2f}\left(1-\frac{d_0}{f}\right)\left(x_f^2+y_f^2\right)}$ in equation B3.3, that is present in the wave function at the back focal plane. This quadratic phase curvature does vanish in the special case when d_0 is made equal to f, that is, the object is placed in the front focal plane. Putting the specimen in the front focal plane of the lens is not useful for producing images, however, as the image is then formed at infinity. On the other hand, this phase curvature in the wave function is also lost, and thus has no effect, when one records the intensity of the wave at the back focal plane, as one does when recording diffraction patterns.

---

through space to the back focal plane, once again by Fresnel diffraction. A schematic outline is given in box 3.1 for each of these steps, describing the way in which they together produce a Fourier transform of the transmitted wave function at the position of the back focal plane. The essence of this derivation is based on two major approximations: (1) the incident wave is a plane wave, which means that the illumination is both parallel and monochromatic; and (2) the wave front of interest is close to the optical axis of the lens (thus allowing one to represent the spherical wave that is involved in Fresnel propagation by a parabolic wave). The derivation of the wave function in the back focal plane does not require the additional approximation that one is in the far-field region, however, as is the case for the formation of a Fraunhofer diffraction pattern (without a lens). A more detailed exposition of the mathematics can be found in the book by Goodman (1968).

The fact that a second (inverse) Fourier transform then generates the image wave function from the wave function in the back focal plane is derived in a similar way. Arguing first from physical plausibility, the inverse Fourier transform is the only mathematical step that could convert the wave function in the back focal plane into a wave function that is identical to the one we started with just below the specimen. Perfect replication of the wave function transmitted through the specimen is, of course, the purpose of a perfect lens, albeit with a stage of magnification as well.

The formal derivation is again based upon Fresnel propagation of the wave between the lens and the image plane. The approach used here is to derive, according to the Huygens principle, what happens to the wave originating from a single point in the transmitted wave. The full image is then assembled by superimposing all of the resulting wave functions that originate from different points in the object. The steps of Fresnel propagation from the object to the lens, transmission through the lens, and Fresnel propagation from the lens to the plane of observation are the same as those that were used in box 3.1. When considering the image wave, however, we need to take account of the greater distance of Fresnel propagation that is required in order to reach the image plane. As is explained schematically in box 3.2, and in more detail by Goodman (1968), the increased distance of wave propagation results in a wave function that is similar to the one that had been transmitted through the object. This image wave is, however, convoluted by the *impulse response* (point spread function) of the optical system. In the case of *diffraction limited optics*, this point spread function is the Fourier transform of the lens aperture.

The derivation of image formation described above is interesting in that it does not formally use Fresnel propagation between the focal plane and the image plane to determine what the image wave must be. Instead, Fresnel propagation, in a single step from the lens to the image plane, is used to show that the object wave becomes convoluted by the point spread function of the optical system. The convolution theorem then tells us that the image wave function must be the Fourier transform of the product of two

---

BOX 3.2 The image wave function is given by the Fourier transform of the wave function at the plane of the objective aperture

Goodman's approach to the mathematical description of images begins with the assumption that the image wave is the convolution of the wave transmitted through the object and the *impulse response* of the imaging system. With this approach, the mathematical problem reduces to the task of finding the image of a mathematical point, that is, the impulse response of the imaging system (Goodman, 1968).

Goodman uses the same mathematical tools of Fresnel propagation and multiplication by a lens transmittance function, which are described in box 3.1,

(continued)

BOX 3.2 *(continued)*

to show that a point-object will produce an image wave function

$$h(x_i, y_i; x_o, y_o) = \frac{1}{\lambda^2 d_o d_i} e^{i\frac{k}{2d_i}(x_i^2+y_i^2)} e^{i\frac{k}{2d_o}(x_o^2+y_o^2)}$$

$$\cdot \int \left[A(x,y) e^{i\frac{k}{2}\left(\frac{1}{d_o}+\frac{1}{d_i}-\frac{1}{f}\right)(x^2+y^2)} \right.$$

$$\left. \cdot e^{-ik\left[\left(\frac{x_o}{d_o}+\frac{x_i}{d_i}\right)x + \left(\frac{y_o}{d_o}+\frac{y_i}{d_i}\right)y\right]} \right] dx dy. \quad \text{(B3.5)}$$

The following notation is used in equation B3.5: (x_o, y_o) are the coordinates (position) of the point-like object (modeled as a Dirac delta function) in a plane that is perpendicular to the optical axis, located a distance d_o in front of the lens; (x_i, y_i) are coordinates in a plane that is perpendicular to the optical axis, located a distance d_i behind the lens; f is the focal length of the lens; $A(x, y)$ is a rectangle function that describes the action of the lens aperture, normally a circular disc; k is the modulus, $2\pi/\lambda$, of the momentum vector of the wave; and λ is the wavelength.

The quadratic phase factor, $e^{i\frac{k}{2d_i}(x_i^2+y_i^2)}$, can be ignored on the grounds that it will be lost when the image wave function is squared to get the image intensity. The quadratic phase factor, $e^{i\frac{k}{2d_o}(x_o^2+y_o^2)}$, can be ignored by a similar argument, even though the coordinates (x_o, y_o) will later serve as variables of integration in the convolution integral mentioned above (Goodman, 1968). In addition to the algebraic simplifications that are obtained by the elimination of these two quadratic phase factors, the term $e^{i\frac{k}{2}\left(\frac{1}{d_o}+\frac{1}{d_i}-\frac{1}{f}\right)(x^2+y^2)}$ becomes equal to 1 when the image is in focus. This last simplification follows from the *lens law* in geometrical optics, which states that the distance to the object, the distance to the in-focus image, and the focal length satisfy

$$\frac{1}{d_o} + \frac{1}{d_i} = \frac{1}{f}. \quad \text{(B3.6)}$$

The impulse response for image formation therefore simplifies to

$$h(x_i, y_i; x_o, y_o) = \frac{1}{\lambda^2 d_o d_i} \int A(x,y) e^{-ik\left[\left(\frac{x_o}{d_o}+\frac{x_i}{d_i}\right)x + \left(\frac{y_o}{d_o}+\frac{y_i}{d_i}\right)y\right]} dx dy. \quad \text{(B3.7)}$$

The fact that the impulse response in equation B3.7 is the Fourier transform of the aperture function can be made more evident by the algebraic substitutions

$$s_x = \frac{x}{\lambda d_i}$$

$$s_y = \frac{y}{\lambda d_i}, \quad \text{(B3.8)}$$

(continued)

and by using the fact that the image magnification is given by the relationship

$$M = \frac{d_i}{d_o}. \quad (B3.9)$$

When these substitutions are made in equation B3.7, we obtain

$$h(x_i, y_i; x_o, y_o) = M \int \left[A(\lambda d_i s_x, \lambda d_i s_y) \right.$$

$$\left. \times e^{-i \, 2\pi[(x_i + Mx_o)s_x + (y_i + My_o)s_y]} \right] ds_x \, ds_y. \quad (B3.10)$$

The impulse response is seen to be the Fourier transform of the aperture function, and it is centered at the image coordinates $x_i = -Mx_o$ and $y_i = -My_o$.

functions; the first of these is the wave function for what is effectively a Fraunhofer diffraction pattern of the specimen, and the second is the (coherent) transfer function of the optical system (see section 3.5). This transfer function will normally have contributions from phase distortions due to lens aberrations and errors in focus as well as the effect due to the lens aperture.

It is worthwhile to emphasize that the product (i.e., multiplication) of two functions in Fourier space is a mathematical fiction; it does not occur at a specific point during propagation from the lens to the detector, say at the back focal plane. That it is fictitious is made obvious by the fact that there must be a different point spread function — and thus a different transfer function — for each setting of the defocus, even though the lens "does not know" where the specimen or the detector will be placed. Nevertheless, this fiction is so helpful in understanding image formation that we will treat it as being physical reality.

3.5 The lens as a linear system: transfer functions play an important role in Fourier optics

Linear systems theory is a well developed branch of mathematical analysis with rich applications in fields such as signal transmission, electrical engineering, and many other areas of science. Optical systems in general, and image-forming lenses in particular are also describable as linear systems under appropriate circumstances. One key condition is that the optical system must be *isoplanitic*, which means that the image of a point-object is independent of where that point is in the field. In other words, there should be no "off axis" aberrations in the optical system. When the appropriate conditions apply, then all of the formal results of linear systems theory can immediately be taken over and applied to image theory (Goodman, 1968).

A universal result, which we will not derive here, states that the performance of a *linear system* can be fully characterized by a transfer function. The *transfer function* of the system has a defined effect on any given sinusoidal input, multiplying the amplitude of the input sine wave by a specified scale factor, called the *gain* of the transfer function (which is commonly less than one!), and applying a specified *phase shift*. Both the gain and the phase shift may vary with the frequency of the sine wave. The frequency-dependent gain (or attenuation) and phase shift are properties of the particular linear system, however, and they are independent of the initial amplitude and phase of the sinusoidal input function.

When a more complicated input function is used, made up of the sum of several different sinusoidal functions, the property of linearity implies that the system applies its amplitude and phase modulations to each Fourier component separately, without any influence or "cross-talk" from the other Fourier components. In other words, the transfer-function characteristics of a linear system are independent of the input function. In optics, the equivalent statement says that the transfer function is specimen-independent.

The properties of a particular linear system are therefore completely defined by a *complex-valued transfer function*,

$$\mathbf{H}(\mathbf{s}) = |\mathbf{H}(\mathbf{s})| \cdot e^{i\,\gamma(\mathbf{s})}, \tag{3.8}$$

where $\mathbf{s}$ is the frequency of the sinusoidal input. Note that we are using vector notation with the lower-case argument, $\mathbf{s}$, to indicate a two-dimensional vector in Fourier space that is conjugate to the two-dimensional vector, $\mathbf{x}$, in real space. The modulus of the transfer function, $|\mathbf{H}(\mathbf{s})|$, is the gain of the system at each frequency, $\mathbf{s}$, and $\gamma(\mathbf{s})$ is the phase shift.

For a given input function, $\psi_{in}(\mathbf{x})$, it is possible to calculate the output of the linear system by a three-step process. The first step is to compute the Fourier spectrum (i.e., the Fourier transform) of the input, $\Psi_{in}(\mathbf{s})$ – note that we are using lower-case ψ to designate a complex-valued wave function in real space, and upper-case Ψ to designate a complex-valued wave function in reciprocal space. The second step is to multiply $\Psi_{in}(\mathbf{s})$ by the transfer function, $\mathbf{H}(\mathbf{s})$. This product applies a frequency-dependent gain, $|\mathbf{H}(\mathbf{s})|$, and a frequency-dependent phase shift, $\gamma(\mathbf{s})$, to each term in the Fourier transform of the input function. The third and final step is to calculate the inverse Fourier transform of the product, $\Psi_{in}(\mathbf{s}) \cdot \mathbf{H}(\mathbf{s})$, in order to obtain the output, $\psi_{out}(\mathbf{x})$. The behavior of a linear system is therefore commonly represented as a "black box," as shown in figure 3.3, in which the mathematical form of the transfer function is known but the detailed implementation of $\mathbf{H}(\mathbf{s})$ (for example, the specific hardware that is involved) does not need to be specified.

The reader may have already noticed that application of linear systems theory to the process of image formation in a coherent-optical system results in nothing more than a restatement of *Abbe's theory of image formation*. To summarize, image formation in a coherent optical system proceeds first by the formation of the diffracted wave function, that is, the Fourier transform of the object wave function. The diffracted wave is then modified by the aperture function (which corresponds to a gain of one within the aperture and zero outside the aperture) and by phase changes that are due to lens aberrations and defocus. The inverse Fourier transform of the modified diffracted wave function then produces the image wave function, that is, the output of the coherent optical system.

Application of the convolution theorem to the second, that is, inverse Fourier transform gives the result

$$\psi_{out}(\mathbf{x}) = \mathcal{F}^{-1}\{\Psi_{in}(\mathbf{s}) \cdot \mathbf{H}(\mathbf{s})\} = \psi_{in}(\mathbf{x}) * h(\mathbf{x}), \tag{3.9}$$

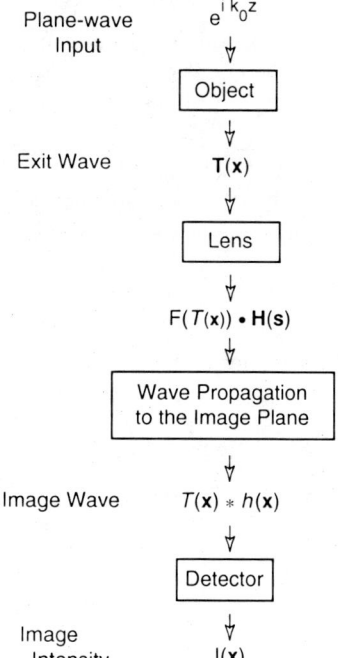

Figure 3.3 A flow diagram that represents image formation as a series of "black box" operations. In *linear systems theory*, one only needs to know the mathematical form of each operation, for example the *transfer function*, **H**(s), and one does not need to have any information about the specific, physical process by which that operation is implemented. The abstract diagram representing the linear systems concept of image formation complements the physical detail that is specified in a conventional optical diagram, such as figure 3.1.

where the convolution product is indicated symbolically as ∗, and the notation $\mathcal{F}^{-1}$ indicates the inverse Fourier transform operation. If we consider the special case when the input function is a Dirac delta function, then we know that the diffracted wave, $\Psi_{in}(\mathbf{s})$, is unity. Equation 3.9 thus tells us that the Fourier transform of the transfer function, **H**(s), is the image wave that is produced by the optical system when the object wave function is a single point. This image wave, $h(\mathbf{x})$, is necessarily broader than a Dirac delta function, and thus it is called the *point-spread-function*, or *impulse response* of the system. For all other objects, equation 3.9 tells us that the image wave function, $\psi_{out}(\mathbf{x})$, is equal to the wave function transmitted through the object, $\psi_{in}(\mathbf{x})$, convoluted by the point spread function of the system.

If the only defect in an optical system is the finite limit of the lens aperture, and image formation is otherwise perfect, then the transfer function is simply the rectangle function, $A(\mathbf{s})$, which describes the aperture:

$$\mathbf{H}(\mathbf{s}) = A(\mathbf{s}). \tag{3.10}$$

In describing $A(\mathbf{s})$ as a rectangle function, we mean that the aperture, usually a circular disc, is either completely transparent ($A(\mathbf{s}) = 1$) or it is completely opaque ($A(\mathbf{s}) = 0$), depending upon the spatial frequency. More complicated apertures that have regions of partial transmission would also be possible, of course. Optical systems that are limited only by their finite aperture are referred to as *diffraction limited systems*; they produce images that are identical to the transmitted wave except for convolution by the Fourier transform of the aperture function, $A(\mathbf{s})$.

A circular aperture, as we have just said, is described by a (circular) rectangle function, which is unity everywhere within the aperture and zero everywhere outside the aperture. A circular aperture therefore produces an impulse response

$$h(\mathbf{x}) \propto \frac{J_1(2\pi s_{max} \cdot \mathbf{x})}{2\pi s_{max} \cdot \mathbf{x}}, \qquad (3.11)$$

where J_1 is a Bessel function of the first kind, of order 1. The function $\frac{J_1(u)}{u}$, which occurs in the case of a circular aperture, has characteristics very similar to those of the function $\frac{\sin(u)}{u}$, which occurs in the case of an aperture with a square (or rectangular) shape (Goodman, 1968). In the optics literature the square of the impulse response in equation 3.11 is known as the *Airy pattern* (or Airy disc), and it represents the intensity distribution in the Fourier transform of a circular aperture.

From a linear systems perspective, the effect of Fourier truncation by the aperture is to convolve the input wave by the impulse response in equation 3.11. The resulting image wave function is said to be *diffraction limited* because the point spread function, which is the wave function corresponding to the Fraunhofer diffraction pattern of the aperture, limits the image resolution through the broadening effect of the convolution.

The mathematical viewpoint of linear systems theory shows us that many systematic distortions or defects of an optical system, if they are known in advance, can be corrected by *image processing*. It is therefore possible to restore the original object wave from what may have seemed to be badly corrupted image data. Recall first of all that linear systems theory (or Abbe image theory) says that the image wave is the inverse Fourier transform of $\Psi_{object}(\mathbf{s}) \cdot \mathbf{H}(\mathbf{s})$, which in turn means that

$$\mathcal{F}\{\psi_{image}(\mathbf{x})\} = \Psi_{image}(\mathbf{s}) = \Psi_{object}(\mathbf{s}) \cdot \mathbf{H}(\mathbf{s}). \qquad (3.12)$$

Equation 3.12 has very great conceptual importance, which is readily seen by the algebraic rearrangement

$$\Psi_{object}(\mathbf{s}) = \frac{\Psi_{image}(\mathbf{s})}{\mathbf{H}(\mathbf{s})}. \qquad (3.13)$$

The formal use of equation 3.13 thus allows us to *deconvolute* the system response from the corrupted, real-space image wave. Equation 3.13 tells us that information about the specimen, in the form of $\Psi_{object}(\mathbf{s})$, can be fully recovered from noise-free images, provided that one has an accurate characterization of the transfer function of the system.

The use of equation 3.13 (and associated techniques) to recover an improved estimate of the Fourier spectrum of the object wave is called *image restoration*. It is important to keep in mind that image restoration by means of equation 3.13 is not possible for those spatial frequencies where the modulus of the transfer function, $|\mathbf{H}(\mathbf{s})|$, approaches zero. Division by a small number will lead to nonsense through the amplification of even the smallest amount of noise in $\Psi_{image}(\mathbf{s})$. There nevertheless are other restoration schemes, including the use of multiple images produced with different transfer functions as mentioned in section 3.8, that are effective in overcoming the problem of *noise amplification*. These restoration schemes are described at length in sections 10.6 and 11.2, and in connection with numerous examples that are presented in part IV of this book.

In order to calculate the Fourier transform of the image wave function, as is prescribed in equation 3.12, one would first have to retrieve the image wave function, $\psi_{image}(\mathbf{x})$, from the experimentally measured image intensity. Retrieval of the image wave function, for example by electron holography, is not the approach that is taken for image restoration of weakly scattering objects, however. Instead, as we will show in sections 3.6, 3.7, and 3.8, it is much easier to take advantage of the approximation that the image intensity of a weakly scattering object is linear in the projected coulomb potential of the specimen. As we will show below, (1) the transfer function that then describes the linear relationship between the object and its image intensity is a real-valued function rather than a complex-valued function, and (2) rather different transfer functions apply to images of weak phase objects and weak amplitude objects, respectively.

3.6 The most common applications of Fourier optics in electron crystallography require that the specimen behaves like a weak phase object

In electron crystallography, the phases of the structure factors can be obtained from the Fourier transform of the recorded image intensity, provided that the specimen is a weak phase object. As we will show below, recovery of the *structural* ("crystallographic") *phases* depends upon the fact that the image intensity of a weak phase object is linear in the (shielded) Coulomb potential.

A *weak phase object*, in electron microscopy, is any specimen for which the transmittance function can be written in the form:

$$T(\mathbf{x}) = 1 - i\, 2\pi \frac{e}{hv} \hat{\rho}'(\mathbf{x}), \tag{3.14}$$

where

h = Planck's constant

v = the velocity of the incident electrons

e = the elementary electron charge, used here as a positive number, and

$\hat{\rho}'(\mathbf{x}) = \int \hat{\rho}(x, y, z)dz$, the projection of the Coulomb potential of the atomic nuclei of the specimen, shielded by their respective electron clouds.

Our use of the symbol $\hat{\rho}(\mathbf{r})$ to designate the shielded Coulomb potential, in analogy to the use of $\rho(\mathbf{r})$ to designate the electron charge density function in x-ray crystallography, has been mentioned previously in section 1.1. We draw attention, however, to the new feature of notation, that is, the use of the prime to designate the 2-D *projection* of a three-dimensional function.

The concept of a weak phase object is an idealization, of course, but even so the theory that can be developed for this ideal case is of great practical value. In section 4.2 we will discuss more completely the relationships between the weak phase object approximation, the kinematic approximation, and more general connections to scattering theory. At this point our interest will be confined to the task of deriving the image intensity for an object that is simply defined, according to equation 3.14, to be a weak phase object.

3.7 The image intensity for a weak phase object remains linear in the projected Coulomb potential

According to Fourier optics, the image wave function is obtained by first multiplying the wave function for the Fraunhofer diffraction pattern — which is the Fourier transform of the transmittance function — by the aperture function, $A(\mathbf{s})$, and by the wave aberration function, $e^{i\,\gamma(\mathbf{s})}$, to get the (fictitious) wave function "at the back focal plane":

$$\Psi_{backfocalplane}(\mathbf{s}) = \mathcal{F}\{T(\mathbf{x})\} \cdot A(\mathbf{s}) \cdot e^{i\,\gamma(\mathbf{s})}, \tag{3.15}$$

where

$A(\mathbf{s})$ = the mathematical representation of the aperture

$\gamma(\mathbf{s})$ = the phase distortion due to the lens.

The Fourier transform of a projection of the structure is a central section through the 3-D Fourier transform of the object. The mapping of a projection into a central section, which we will refer to as the *projection theorem*, is a well-known result in crystallography. A formal proof of this result is given in section 4.3, and for the present we will just use the projection theorem without proof. Since the transmittance function of a weak phase object, $T(\mathbf{x})$, defined in equation 3.14, involves the projection of the Coulomb potential, we obtain

$$\Psi_{backfocalplane}(\mathbf{s}) = [\delta(\mathbf{s}) - i\,\sigma\,\mathbf{F}(\mathbf{s})] \cdot A(\mathbf{s}) \cdot e^{i\,\gamma(\mathbf{s})}, \tag{3.16}$$

where

$\delta(\mathbf{s})$ = the Dirac delta function

$\sigma = 2\pi \frac{e}{hv}$

$\mathbf{F}(\mathbf{s})$ = the Fourier transform of $\hat{\rho}'(\mathbf{x})$, that is, the structure factor of the specimen evaluated on a central section in Fourier space.

We next take the inverse Fourier transform of the wave function at the back focal plane in order to obtain the image wave function:

$$\Psi_{image}(\mathbf{x}) = 1 - i\,\sigma\hat{\rho}'(\mathbf{x}) * \mathcal{F}^{-1}\{A(\mathbf{s}) \cdot e^{i\,\gamma(\mathbf{s})}\}. \tag{3.17}$$

The image intensity, in turn, is the square of the image wave function:

$$I(\mathbf{x}) = \Psi_{image}(\mathbf{x}) \cdot \Psi^{*}_{image}(\mathbf{x})$$

$$= 1 - 2\mathcal{R}e[i\,\sigma\hat{\rho}'(\mathbf{x}) * \mathcal{F}^{-1}\{A(\mathbf{s}) \cdot e^{i\,\gamma(\mathbf{s})}\}]$$

$$+ \text{a term in } \{\sigma\hat{\rho}'(\mathbf{x})\}^2. \tag{3.18}$$

The symbol $\mathcal{R}e$ in equation 3.18 denotes the real part of the quantity that is in the square brackets. The term that is quadratic rather than linear in $\{\sigma\hat{\rho}'(\mathbf{x})\}$ is neglected in the subsequent analysis on the grounds that $\sigma\hat{\rho}'(\mathbf{x}) \ll 1$ for a weak phase object.

Equation 3.18 can be carried one step further by using the *Euler relationship*:

$$e^{i\,\gamma(\mathbf{s})} = \cos\gamma(\mathbf{s}) + i\,\sin\gamma(\mathbf{s}). \tag{3.19}$$

Dropping the quadratic term in equation 3.18, as mentioned above, we can write the image intensity:

$$I(\mathbf{x}) = 1 + 2\sigma \hat{\rho}'(\mathbf{x}) * \mathcal{F}^{-1}\{A(\mathbf{s}) \cdot \sin \gamma(\mathbf{s})\}. \tag{3.20}$$

The real significance of all of these mathematical steps is made apparent if we now look at the Fourier transform of the image intensity:

$$\mathcal{F}\{I(\mathbf{x})\} = \mathcal{I}(\mathbf{s}) = \delta(\mathbf{s}) + 2\sigma \, \mathbf{F}(\mathbf{s}) \cdot A(\mathbf{s}) \cdot \sin \gamma(\mathbf{s}). \tag{3.21}$$

The possibility of retrieving the crystallographic phase information from the image intensity becomes clear after making a simple algebraic rearrangement of equation 3.21, which gives us

$$\mathbf{F}(\mathbf{s}) = \frac{\mathcal{I}(\mathbf{s})}{2\sigma \, \sin \gamma(\mathbf{s})}. \tag{3.22}$$

Equation 3.22 simply says that the phase of $\mathbf{F}(\mathbf{s})$ is the same as the phase of the Fourier transform of $I(\mathbf{x})$, taking into account the 180° phase change that occurs whenever $\sin \gamma(\mathbf{s})$ is negative. Dividing $\mathcal{I}(\mathbf{s})$ by $\sin \gamma(\mathbf{s})$, as in equation 3.22, restores the correct sign, and thus the correct phase, as well as making the appropriate corrections to the amplitudes.

The remarks made following equation 3.13 regarding the amplification of noise still apply, of course. Surprisingly (in the context of equation 3.22), the restoration scheme that gives the smallest mean square error at spatial frequencies where the noise is a problem involves multiplying (rather than dividing) $\mathcal{I}(\mathbf{s})$ by $\sin \gamma(\mathbf{s})$. This improvement in image restoration is discussed further, in connection with equation 10.13.

Two important assumptions have been made in the derivation of equation 3.20. First of all it was assumed that the incident illumination was a plane wave, with wave vector $\mathbf{k}_0$ parallel to the optical axis. The derivation of the transfer function discussed here requires that the phase distortion function, $\gamma(\mathbf{s})$, must be symmetrical on either side of the diffracted wave function, $F(\mathbf{s})$, and this is possible only when the illumination is parallel to the optical axis of the objective lens. Additional corrections must be applied to the data when the illumination is not well aligned, and these are described in section 3.12.

Secondly, it was assumed that $\mathbf{F}(\mathbf{s})$, the Fourier transform of $\hat{\rho}'(\mathbf{x})$, obeys the special symmetry relationship,

$$\mathbf{F}(-\mathbf{s}) = \mathbf{F}^*(\mathbf{s}), \tag{3.23}$$

which is called *Friedel symmetry* in crystallography. Friedel symmetry is a mathematical consequence of the fact that the projected Coulomb potential, $\hat{\rho}'(\mathbf{x})$, is a real-valued function; indeed, Friedel symmetry would not be obeyed if $\hat{\rho}'(\mathbf{x})$ were a complex-valued function. The reason why these two conditions must be met in order for equation 3.20 to be true is made evident in an alternative, step-by-step derivation, which is developed in box 3.3.

BOX 3.3 Abbe's diffraction theory of images can be applied to each diffraction spot, one at a time

An alternative derivation of equation 3.20 for a weak phase object is based upon a formalism that can be used even when the incident wave is not parallel to the optical axis, or when the structure factors, **F(s)**, fail to obey *Friedel's Law* (defined in equation B3.15 below). Significant departures from Friedel symmetry can occur either due to curvature of the Ewald sphere or due to multiple elastic scattering (failure of the first Born approximation); these unfavorable conditions are given further consideration in chapter 15.

This derivation again starts with equation 3.16, where the wave function at the back focal plane is represented as the product of the structure factor, **F(s)**, and the wave aberration function, $e^{i\,\gamma(s)}$. In the following we will ignore the aperture function, $A(s)$. We begin by considering a single Fourier component at $s = g$ along with the Dirac delta function at $s = 0$. The wave function at the back focal plane for this highly simplified case is:

$$\Psi_{backfocalplane}(s) = \delta(s) - i\,\sigma\,\mathbf{F(g)} \cdot e^{i\,\gamma(\mathbf{g})} \cdot \delta(s - g)$$

$$= \delta(s) - \sigma\,\mathbf{F(g)} \cdot e^{i\left(\gamma(\mathbf{g}) + \frac{\pi}{2}\right)} \cdot \delta(s - g). \quad (B3.11)$$

The Fourier transform of equation B3.11 gives the image wave function:

$$\psi_{image}(\mathbf{x}) = 1 - \sigma\,F(\mathbf{g}) \cdot e^{i\left(2\pi\,\mathbf{g}\cdot\mathbf{x} + \alpha(\mathbf{g}) + \gamma(\mathbf{g}) + \frac{\pi}{2}\right)}. \quad (B3.12)$$

Squaring the wave function in equation B3.12, and ignoring the term in $F^2(\mathbf{g})$, gives us the image intensity:

$$I(\mathbf{x}) = 1 - 2\sigma\,F(\mathbf{g})\cos\left(2\pi\,\mathbf{g}\cdot\mathbf{x} + \alpha(\mathbf{g}) + \gamma(\mathbf{g}) + \frac{\pi}{2}\right). \quad (B3.13)$$

Equation B3.13 demonstrates that the phase distortion, $\gamma(\mathbf{g})$, shifts the phase origin of the sinusoidal component of the image; in other words, $\gamma(\mathbf{g})$ produces a phase error but no modulation of the amplitude of the gth Fourier component of the image.

The next step in the derivation is to include a Friedel-symmetric term at $s = -g$ in the expression for the diffracted wave, equation B3.11. The reader can easily verify that this additional term produces a second sinusoidal term in the image intensity, so that equation B3.13 becomes:

$$I(\mathbf{x}) = 1 - 2\sigma\,F(\mathbf{g})\cos\left(2\pi\,\mathbf{g}\cdot\mathbf{x} + \alpha(\mathbf{g}) + \gamma(\mathbf{g}) + \frac{\pi}{2}\right)$$

$$- 2\sigma\,F(\mathbf{g})\cos\left(-2\pi\,\mathbf{g}\cdot\mathbf{x} - \alpha(\mathbf{g}) + \gamma(\mathbf{g}) + \frac{\pi}{2}\right). \quad (B3.14)$$

In deriving equation B3.14, we have used the *Friedel-symmetry* relationships:

$$F(-\mathbf{g}) = F(\mathbf{g})$$

$$\alpha(-\mathbf{g}) = -\alpha(\mathbf{g}). \quad (B3.15)$$

(continued)

Friedel's law states that the Fourier transform of a real-valued function will possess the symmetry-relationships given in equation B3.15. In addition, we have assumed that

$$\gamma(-\mathbf{g}) = \gamma(\mathbf{g}). \tag{B3.16}$$

We will see later (in section 3.12) that equation B3.16 is true only when the illumination is parallel to the optical axis.

The use of trigonometric identities, applied to equation B3.14, results in the further simplification:

$$I(\mathbf{x}) = 1 + 4\sigma \ F(\mathbf{g})\cos(2\pi \ \mathbf{g} \cdot \mathbf{x} + \alpha(\mathbf{g})) \cdot \sin \ \gamma(\mathbf{g}). \tag{B3.17}$$

Equation B3.17 shows that the phase distortion, $\gamma(\mathbf{g})$, now causes a modulation of the amplitude of the gth Fourier component of the image, that is, the coefficient, $F(\mathbf{g})$, is replaced by the "weighted" coefficient, $F(\mathbf{g}) \cdot \sin \gamma(\mathbf{g})$. We want to emphasize that the phase distortion in equation B3.13 (or equation B3.14) gets converted into an amplitude modulation only if the diffraction pattern possesses Friedel symmetry, and the plane-wave illumination is parallel to the optical axis. As mentioned earlier, the reasons why these restrictions exist for the validity of the phase contrast transfer function are made quite transparent through this derivation.

The final step in the derivation proceeds by induction, adding first another Friedel pair, with frequency $\mathbf{g}'$, and then another, eventually generalizing to a sum and then to an integral. The result is:

$$I(\mathbf{x}) = 1 + 2\sigma \int [2F(\mathbf{s}) \cdot \sin \ \gamma(\mathbf{s})]\cos(2\pi \ \mathbf{s} \cdot \mathbf{x} + \alpha(\mathbf{s}))d\mathbf{s}. \tag{B3.18}$$

Note that in equation B3.18 we have substituted the symbol **s** for **g**, when generalizing the discrete summation over **g** into the continuous integral over **s**. The integral in equation B3.18 can be recognized as being the Fourier (cosine) transform (see, for example, equation 2.1) of the product of $2F(\mathbf{s})$ and $\sin \gamma(\mathbf{s})$. The Fourier transform of the product of these two functions will, of course, be the convolution product of their respective Fourier transforms. The Fourier transform of $2F(\mathbf{s})$ is simply the projected potential, $\hat{\rho}'(\mathbf{x})$. The end result, then, is that equation B3.18 becomes:

$$I(\mathbf{x}) = 1 + 2\sigma \hat{\rho}'(\mathbf{x}) * \mathcal{F}^{-1}\{\sin \ \gamma(\mathbf{s})\}, \tag{B3.19}$$

repeating the result given in equation 3.20.

3.8 The concept of a "phase contrast transfer function" is of central importance in the interpretation of high-resolution images

The function $\sin \gamma(\mathbf{s})$ is known as the *phase contrast transfer function*, because it determines how well each Fourier coefficient, $\mathbf{F}(\mathbf{s})$, of the transmitted wave, $T(\mathbf{x})$, is retained in the image of a weak phase object. Referring to equation 3.21, written again here without the aperture function, $A(\mathbf{s})$, the Fourier transform of the image of a weak phase object is

$$\mathcal{I}(\mathbf{s}) = \delta(\mathbf{s}) + 2\sigma\, \mathbf{F}(\mathbf{s}) \cdot \sin \gamma(\mathbf{s}). \qquad (3.24)$$

It is evident that the image intensity, $I(\mathbf{x})$, will have maximum contrast for those spatial frequency components where $|\sin \gamma(\mathbf{s})| = 1$. This means, in turn, that the image will have maximum contrast for those spatial frequencies for which $\gamma(s) = \pm\pi/2, \pm 3\pi/2, \pm 5\pi/2, \ldots$. By the same token, the image intensity will have no contrast at all for those spatial frequencies for which $\gamma(s) = 0, \pm\pi, \pm 2\pi, \pm 3\pi, \ldots$. In addition, we emphasize that the phase of $\mathcal{I}(\mathbf{s})$ can be 180° from the phase of $\mathbf{F}(\mathbf{s})$, depending upon the sign of $\sin \gamma(\mathbf{s})$.

The *wave aberration function* of the electron microscope objective lens, $\gamma(\mathbf{s})$, can normally be represented by just two terms, one for *spherical aberration* and one for *defocus*. The analytic form of the wave aberration function is:

$$\gamma(\mathbf{s}) = 2\pi \left\{ \frac{C_s}{4} \lambda^3 s^4 - \frac{\Delta Z}{2} \lambda s^2 \right\} \qquad (3.25)$$

where C_s is the coefficient of spherical aberration, ΔZ is the amount of defocus, and λ is the electron wavelength. An additional term, $-\frac{\Delta Z_{ast}}{2} \sin 2(\phi - \phi_0) \lambda s^2$, can be added within the brackets in equation 3.25 in order to account for *objective lens (axial) astigmatism*; note that ΔZ_{ast} represents the astigmatic defocus value, ϕ_0 is the angle at which the effect of astigmatism is null, and ϕ is the azimuthal angle within the plane of the image.

The wave aberration function in electron microscopy, $\gamma(\mathbf{s})$, acts in a way that is analogous to the "quarter wave plate" in the Zernike phase contrast (light) microscope. Phase contrast in the electron microscope is of poorer quality in comparison to the light microscope, however, in that maximal phase contrast (in the electron microscope) is realized over only a rather limited range of spatial frequencies. In addition, reversal of contrast is inevitable at higher spatial frequencies, when $\sin \gamma(\mathbf{s})$ changes sign. These drawbacks are substantially reduced by the ready ease of changing focus (and thus $\sin \gamma(\mathbf{s})$) in successive images, so as to cover different frequency bands with high contrast. In addition, image restoration can be used to correct for reversals in the sign of $\sin \gamma(\mathbf{s})$, as described above.

Figure 3.4 shows some representative examples of the wave aberration function, $\gamma(\mathbf{s})$, calculated for a range of defocus values and a realistic estimate of C_s. The corresponding phase contrast transfer functions are shown in figure 3.5. One of the curves has a particularly wide band of spatial frequencies over which $|\sin \gamma(\mathbf{s})| \approx 1$. This curve corresponds to a defocus value that compensates the effect of spherical aberration to an optimal extent, giving a net phase shift of nearly $\pi/2$ for the largest possible band of spatial frequencies. This value of defocus, known as the *Scherzer defocus* according to the analysis of Scherzer (1949), is given by

$$\Delta Z = 2.5 \left(\frac{\lambda C_s}{2\pi} \right)^{\frac{1}{2}}. \qquad (3.26)$$

The curves shown in figure 3.5 not only illustrate the reversals in contrast that must occur at higher spatial frequency, they also illustrate that oscillations in the

Figure 3.4 Six representative examples of the wave aberration function, $\gamma(\mathbf{s})$, are shown, corresponding to the wide range of defocus values that is likely to be used in the course of collecting images. The examples shown are for (a) Scherzer focus (see equation 3.26); (b) "perfect (Gaussian) focus," $\Delta Z = 0$ (rarely used); (c) underfocus by 264 nm; (d) overfocus by 264 nm; (e) underfocus by 1 μm; and (f) overfocus by 1 μm. The calculations assumed an accelerating voltage of 100 kV ($\lambda = 3.7$ pm) and $C_s = 2$ mm.

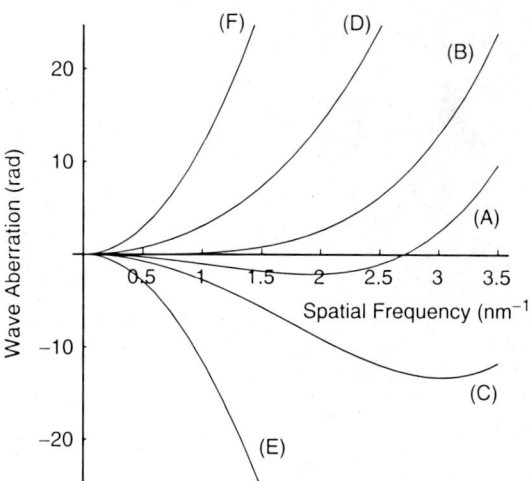

transfer function set in at progressively lower resolution, and become increasingly rapid, as the amount of defocus increases. Large defocus values are therefore not suitable for producing good phase contrast at high resolution. On the other hand, large defocus values are precisely what is wanted in order to get good phase contrast at low resolution. Conversely, the contrast is extremely weak if the image is taken close to focus.

3.9 Partial coherence imposes an envelope on the phase contrast transfer function

Up to this point our discussion of phase contrast in electron microscopy has been based upon the assumption that the incident wave is perfectly coherent, that is, it is a monochromatic plane wave. The assumption of perfect coherence is implicit in the mathematical representation of a plane wave, $e^{i\,\mathbf{k}_0 \cdot \mathbf{r}}$, which is used to approximate the wave function for the incident electron. The fact that the incident wave vector, $\mathbf{k}_0$, has a unique direction implies that the incident illumination is perfectly parallel, a condition that is described as perfect *spatial coherence*. In addition, the fact that the incident wave vector also has a unique wavelength implies that the incident illumination is perfectly monochromatic, a condition that is described as perfect *temporal coherence*. The spatial coherence and the temporal coherence of the incident electron beam are never truly perfect, however.

The degree of collimation of the incident beam (i.e., the spatial coherence) is, to a large extent, something that is under the control of the operator, as is described in section 5.3. Under conditions used for low-dose electron microscopy, the beam can be easily collimated to approximately 10^{-4} radians. While even more stringent collimation is certainly possible, there is a trade-off involved in that the beam intensity necessarily decreases as the collimation is improved.

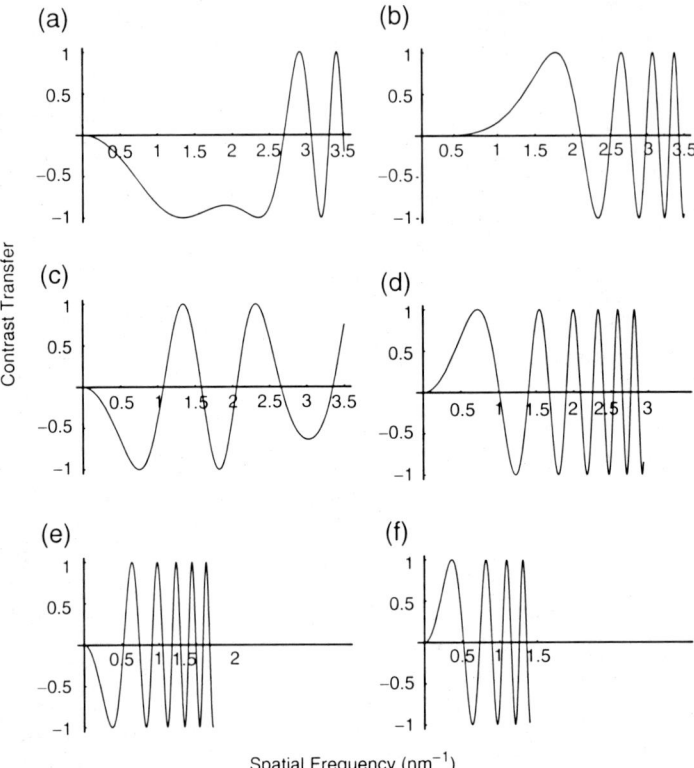

Figure 3.5 The *phase contrast transfer function*, sin $\gamma(\mathbf{s})$, is shown for the six representative defocus values used in figure 3.4. The labeling of each panel corresponds to the labeling of individual curves in figure 3.4. When the phase contrast transfer function oscillates rapidly, as it does at high spatial frequency, it is important to remember that image contrast actually becomes strongly damped by the effects of imperfect coherence, as is discussed in section 3.9. The curves shown here are meant to be an idealized representation, which must be multiplied by an appropriate envelope function in order to give a correct impression of experimentally achievable information transfer at high resolution.

The monochromaticity (i.e., the degree of temporal coherence) of the incident beam, on the other hand, is something that is determined largely by the type of electron gun that is provided on the electron microscope. Unless a cold field-emission gun is used, the spread in wavelengths will be roughly 1 part in 10^5, the exact value depending upon the type of gun, how it is operated, and the value of the accelerating voltage. Further details are, again, discussed in section 5.3.

Although the degree of spatial and temporal coherence that is available in electron microscopy is impressive, the residual imperfection in coherence nevertheless does limit the microscope performance at very high resolution. The theoretical analysis of how resolution depends on both spatial and temporal coherence treats the incident illumination as a distribution of beams with slightly different energies and angles, each of which, however, has perfect spatial and temporal coherence. The image intensity is then described as the sum of the image intensities that are produced by each of the members in the incident distribution of beams. Since both the direction of the incident wave vector and the value of the incident wavelength change from one member to the next, the resulting contrast transfer function (CTF) is slightly different for the contribution that each beam makes to the total image intensity. The way in which the wavelength affects the CTF has been described in equations 3.24 and 3.25. The effect of variation in the direction of the wave vector relative to the optical axis is still to be discussed in section 3.12. The net result is that the positions of zeros in the contrast transfer function are dependent on the direction of the incident beam as well as its wavelength.

When the zeros in the CTF are quite close to one another, which is necessarily more the case at higher values of defocus, even small shifts in their positions will lead to a partial cancelation between the positive and negative contributions to the contrast transfer that come from different beams in the distribution. The effect of partial coherence is naturally greatest at high resolution, where the oscillations in the CTF become increasingly rapid, as is shown in figure 3.4. As can be appreciated from the curves shown in figure 3.4, the effect of partial spatial coherence is also greater for larger values of the image defocus. The effect of partial temporal coherence, on the other hand, is independent of the image defocus.

When the contributions from beams with slightly different wavelengths and slightly different directions are added together, the overall oscillations of the CTF are still retained, but the maxima and minima are limited to progressively smaller values of their amplitudes. The amplitude of the CTF is thus confined within what is called the *envelope function* of the CTF, designated $E(s)$.

Approximate, but analytical expressions have been derived for the envelope function that results from various distributions of beam angle and electron wavelength (Hansen and Trepte, 1971; Frank, 1973; Wade and Frank, 1977). As an example, a Gaussian distribution of beam directions produces the envelope function

$$E_{SpatialCoherence} = \exp\left[-\pi^2 s_0^2 \left(C_s \lambda^3 s^3 - \Delta Z \lambda s\right)^2\right], \tag{3.27}$$

where s_0 is the spatial frequency that corresponds to the incident beam direction at which the Gaussian angular distribution falls to half of its maximum value. If the angular width of the distribution of incident beams is θ_0, for example, then

$$s_0 = \frac{\theta_0}{\lambda}. \tag{3.28}$$

A distribution of beam wavelengths, on the other hand, produces an envelope function that can be characterized in terms of the corresponding energy spread of the beam, δE. This energy spread produces an effective spread in the defocus, $\delta Z = C_c \frac{\delta E}{E}$, where C_c is the coefficient of chromatic aberration of the objective lens, and E is the average energy of the distribution. When the wavelength distribution is Gaussian, the resulting temporal coherence envelope function is

$$E_{TemporalCoherence} = \exp\left[-\frac{\pi^2}{4}(\delta Z)^2 \lambda^2 s^4\right]. \tag{3.29}$$

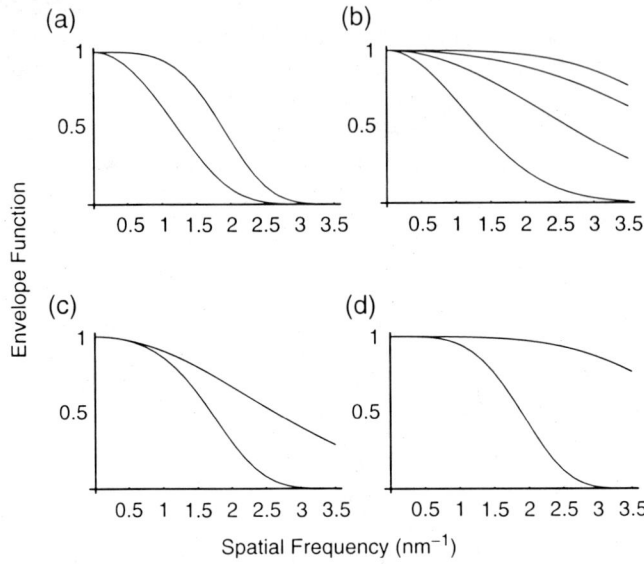

Figure 3.6 Curves that represent the envelope function due to partial spatial and temporal coherence expected under good experimental conditions. The product of an envelope function, such as the ones shown here, and the corresponding phase contrast transfer function, similar to those shown in figure 3.5, gives a more accurate impression of the limitations in information transfer that are determined just by the electron optics. Additional limitations, such as radiation damage, specimen charging, stage drift, etc., will generally cause a further dampening of the signal at increasing resolution. (a) The calculations shown here assume an accelerating voltage of 100 kV (λ = 3.7 pm); spherical aberration coefficient, 2 mm; chromatic aberration coefficient, 2 mm; illumination angle of 10^{-4} radians; and energy spread of 2 eV. The two curves in this panel compare the envelope functions for an underfocus of 100 nm and an underfocus of 2 μm. (b) Similar comparison of envelope functions at 300 keV, for a finer range of underfocus (100 nm, 500 nm, 1 μm, and 2 μm, respectively). Other parameters (except the wavelength) are as in panel (a). (c) Comparison of the envelope functions at 100 keV and 300 keV under otherwise identical conditions, in this case assuming an underfocus of 1 μm. (d) Comparison of the envelope functions at 100 keV and 300 keV under otherwise identical conditions, in this case assuming an underfocus of only 100 nm.

Representative examples are shown in figure 3.6 for the envelope functions that limit the transfer of information at high resolution under practical conditions, due to imperfect coherence in the incident electron beam.

3.10 Amplitude contrast can also contribute in an important way to images of thin, biological specimens

A brief digression from our discussion of phase contrast is appropriate at this point, to describe the use of contrast transfer function theory in the case of a *weak amplitude object*. Phenomenologically, some of the incident radiation will appear to be absorbed in the sample, or — more correctly stated — it will be removed from the incident, coherent plane wave in a spatially varying way. Without describing inelastic scattering or other physical processes that can cause this to happen, the spatially varying depletion of the coherent illumination can be represented by an effective *mass absorption function*, $\mu(\mathbf{x})$, so that the intensity of the transmitted wave becomes:

$$I(\mathbf{x}) = e^{-\mu(\mathbf{x})}. \tag{3.30}$$

As we did for the pure phase object, we can now consider the weak amplitude object, in which the transmitted intensity is approximated by the linear term in the series expansion of the exponential in equation 3.30:

$$I(\mathbf{x}) \cong 1 - \mu(\mathbf{x}). \tag{3.31}$$

The transmitted wave function, in turn, is therefore approximated as

$$\psi(\mathbf{x}) = \sqrt{I(\mathbf{x})} \cong 1 - \frac{1}{2}\mu(\mathbf{x}). \tag{3.32}$$

Using the same type of derivation that we used previously for the image intensity of a weak phase object, we find that the image intensity for a weak amplitude object is given by:

$$I(\mathbf{x}) = 1 - \mu(\mathbf{x}) * \mathcal{F}^{-1}\{\cos \gamma(\mathbf{g})\} \tag{3.33}$$

The result for a weak amplitude object is therefore similar to the result for a weak phase object, in that information about different Fourier components of the object is transferred in zones. However, these zones now correspond to regions where $|\cos \gamma(\mathbf{g})| \cong 1$ rather than to zones where $|\sin \gamma(\mathbf{g})| \cong 1$, as is the case for a weak phase object.

The image intensity for an amplitude object thus differs from that for a phase object in that the contrast transfer is nearly perfect at low resolution for the amplitude object, while contrast transfer for a phase object always goes to zero at low resolution. Unlike the case of a phase object, the image contrast of an amplitude object is only improved as the optics become more perfect; a phase object, on the other hand, requires a wave aberration of some type in order to generate contrast in the image. This point has been made previously, at the end of section 3.3.

In general, the transmittance function of a weak object will have both a phase-object term and an amplitude-object term. Since the image intensity of either a pure phase object or a pure amplitude object is a linear function of the appropriate term in the transmittance function, the image intensity for a mixed phase and amplitude object will just be the sum of the two pure-object functions, that is, if

$$T(\mathbf{x}) = 1 - i\,\sigma\hat{\rho}'(\mathbf{x}) - \frac{1}{2}\mu(\mathbf{x}), \tag{3.34}$$

then

$$I(\mathbf{x}) = 1 + 2\,\sigma\hat{\rho}'(\mathbf{x}) * \mathcal{F}^{-1}\{\sin \gamma(\mathbf{g})\} - \mu(\mathbf{x}) * \mathcal{F}^{-1}\{\cos \gamma(\mathbf{g})\}. \tag{3.35}$$

3.11 Single side band images: blocking half of the diffraction pattern produces images whose transfer function has unit gain at all spatial frequencies

It is often felt that the zeros in the phase contrast transfer function, $\sin \gamma(\mathbf{g})$, represent a serious defect of electron microscope images. It is therefore worth pointing out that it is mathematically possible to produce perfect contrast transfer at all spatial frequencies by using an aperture that blocks one half of the diffraction pattern but lets the other half through. Such an aperture is called a *single side band aperture*, and the image produced with such an aperture is called a *singe side band image* (Downing, 1979). Referring to equation B3.13, the image intensity produced by the superposition of one diffracted beam and the unscattered beam is

$$I(\mathbf{x}) = 1 - 2\sigma \, F(\mathbf{g}) \cos\left(2\pi \, \mathbf{g} \cdot \mathbf{x} + \alpha(\mathbf{g}) + \gamma(\mathbf{g}) + \frac{\pi}{2}\right). \quad (3.36)$$

This result can be generalized to a discrete sum, and then to the integral over a continuous distribution of diffracted beams, just as was done for "double side band" images in box 3.3. In the case of single side band images, however, each term included in the sum (or integral) no longer has a Friedel mate.

The Fourier transform of the image intensity then becomes

$$\mathcal{I}(\mathbf{s}) = \delta(\mathbf{s}) - \sigma \, F(\mathbf{s}) e^{i[\alpha(\mathbf{s}) + \gamma(\mathbf{s}) + \frac{\pi}{2}]}. \quad (3.37)$$

Although the contrast in the single side band image is perfect for every spatial frequency, regardless of defocus or the effect of spherical aberration, the phase of each Fourier component of the image is now distorted by the wave aberration, $\gamma(\mathbf{s})$.

Experimental methods to produce single side band images have not become well developed, in spite of the fact that they have the attractive property of giving uniform contrast transfer at all spatial frequencies. One of the technical problems that has never been solved experimentally is the question of measuring the wave aberration, $\gamma(\mathbf{s})$, from the image data. There is currently no scheme to do this that would be equivalent to measuring the zeros in the phase contrast transfer function. An even more important, practical difficulty comes from the fact that the aperture that is used to block half of the back focal plane is prone to becoming electrically charged when it is hit by the scattered electrons (Downing and Siegel, 1973, 1975). The amount of this charge fluctuates with time, causing a fluctuating wave aberration, and no correction can be made to restore the information that is lost in the time-average of these fluctuations.

There is good reason to work towards overcoming the experimental difficulties described above. If single side band imaging could be used in a practical way, it would not be necessary to use extremely large defocus values in order to get good contrast at low resolution, for example with frozen hydrated specimens. In addition, severe complications in data processing due to curvature of the Ewald sphere, which are important at high resolution for thicker samples or for samples tilted to very high angles, disappear when a single side band aperture is used. We will not discuss the problem of curvature of the Ewald sphere any further here, but the topic is discussed at length in sections 15.6 and 15.8.

3.12 Tilted illumination produces images for which the transfer function includes both phase errors and amplitude modulations

When the illumination is a plane wave, but the direction of propagation is no longer parallel to the optical axis, then the unscattered wave will be focused to a point in the back focal plane that is shifted away from the optical axis. In other words, the origin of reciprocal space, $s = 0$, will no longer coincide with the "origin" of the wave aberration function, $\gamma(s)$. Because the centers of the two functions are displaced from one another, the value of $e^{i\,\gamma(s)}$ will no longer be the same at a given diffraction spot and at its Friedel mate. The amount by which the center of the diffraction pattern is displaced from the center of the lens aberration function, $\gamma(s)$, is given by $\Delta s = \frac{\theta}{\lambda}$, where θ is the angle that the incident wave vector, $\mathbf{k}_0$, makes with the optical axis.

The image wave function can still be obtained, as before, by taking the Fourier transform of the product of the diffracted wave function and $e^{i\,\gamma(s)}$, but taking account of the fact that the wave aberration function is off-set from the center of the diffraction pattern. On the other hand, the values of $\gamma(s)$ must be symmetrical for the derivation to lead to the result that $\sin \gamma(s)$ represents the contrast transfer function for a weak phase object. Clearly, that derivation is no longer valid when the illumination is tilted so strongly that $\sin \gamma(s)$ and $\sin \gamma(-s)$ differ by an appreciable fraction of $\pi/2$.

When the *beam tilt* is well corrected, such that the residual misalignment is less than 10^{-3} radians, the contrast transfer function will still be written as $\sin \gamma(s)$, where $\gamma(s)$ is defined in equation 3.25. However, the introduction of a residual amount of beam tilt changes the apparent defocus by a small amount, and a small amount of astigmatism is introduced into the defocus, as well. The additional defocus and astigmatism are made apparent by a shift in the radial positions of the zeros and the maxima of the contrast transfer function, and by the introduction of an "elliptical" shape rather than a circular shape in the contrast transfer rings. Not only are these focus (and astigmatism) changes small, they can easily be compensated during image processing in the same way that is done for genuine defocus and astigmatism. Thus, unless the beam tilt varies across the field of view — which can well be the case — one might think that small amounts of beam tilt have no practical consequences.

Unfortunately, however, the introduction of beam tilt also changes the apparent crystallographic phase, that is, the apparent phase of the structure factor of the specimen, $\mathbf{F}(s)$. The phase errors introduced by beam tilt represent a serious effect, because they are not automatically apparent as are the tilt-related changes in focus and astigmatism. The analytical form of the phase error is

$$\Delta\alpha(\mathbf{g}) = -2\pi\,\theta\,C_s\,\lambda^2 g^3 \hat{\mathbf{g}} \cdot \Delta\hat{\mathbf{s}}, \tag{3.38}$$

where

- $\mathbf{g}$ = the spatial frequency of a given diffraction spot
- θ = the tilt angle of the illumination
- C_s = the coefficient of spherical aberration
- λ = the electron wavelength

$\hat{g}$ = a unit vector in the direction of g

$\Delta\hat{s}$ = a unit vector in the direction of the beam tilt.

The phases of diffraction spots that lie in a direction perpendicular to the direction of beam tilt (i.e., in the direction parallel to the tilt axis) will be unaffected because of the vector dot-product in equation 3.38. The effect for the spots that lie parallel to the direction of beam tilt might be small at lower resolution, but the phase error can rise suddenly to unacceptable values at high resolution because of the third-power dependence on g. To illustrate this, the phase error due to a beam tilt of 10^{-3} radians will be only about 20° at a resolution of 0.75 nm, but it is over 180° at a resolution of 0.35 nm.

It is clear from this example that it would be good if the beam could be aligned to an accuracy of 10^{-4} radians or better, so that the phase error would be negligible even at high resolution. This degree of accuracy is difficult to guarantee experimentally, however. As a result, it is helpful to correct the image phases as part of the overall data processing (Henderson et al., 1986), and the details for making this correction are discussed in section 10.11. Unfortunately, it is only practical to apply such a correction for images of 2-D crystals, and no corresponding correction has been attempted for images of single particles.

If the beam tilt is 10^{-3} radians or more, both the amplitudes and the phases of the Fourier transform of the image are affected in a much more complicated way than we have been discussing up until now (Smith et al., 1983). Fortunately, alignment of the illumination to better than 10^{-3} radians is easily achieved by normal procedures, which are described in section 5.6.

3.13 Summary: Fourier optics is an important part of the conceptual foundation of electron crystallography

The formal background of Fourier optics and the associated concepts of transfer functions and linear-systems theory covered in this chapter are used very extensively in everyday work in electron crystallography. Electron crystallography differs from x-ray crystallography in a fundamental way, in that it is possible in electron crystallography to recover the phases of the structure factors from high-resolution electron micrographs. In order to understand how it is possible to "solve" the crystallographic phase problem by using images rather than heavy atom derivatives of the structure, it is first necessary to understand the wave-optical theory of image formation. As we have seen, the theory of image formation is as strongly founded in the mathematics of Fourier transforms as is the theory of diffraction. Thus, the additional burden of mathematical theory that arises in electron crystallography is actually very easy to master, once one already understands the mathematics that underlies x-ray crystallography.

4

Theoretical Foundations Specific to Electron Crystallography

4.1 Introduction

The extremely short wavelength that is normally encountered in electron crystallography, about 4 pm (0.04 Å) or less, results in such a small curvature of the Ewald sphere that the sphere can be closely approximated by a plane in the region close to the origin of reciprocal space. Approximation of the surface of the Ewald sphere as a plane results in very useful simplifications of crystallographic theory, some of which have already been introduced in our preliminary discussion of the phase contrast transfer function in chapter 3. We will now elaborate further on the role that these simplifications play (1) in justifying the use of the weak phase object approximation, covered in section 4.2; (2) in recovering phase information from images, covered in the proof of the projection theorem in section 4.3, and in a discussion of the consequences of the projection theorem that are discussed in section 4.4; and (3) in the way that we approach the task of collecting and indexing three-dimensional data in reciprocal space, covered in section 4.5.

Some of the work that is done in electron crystallography uses specimens that are prepared in the form of thin, sheet-like crystals. The 2-D crystalline specimen is actually the most favorable one for high-resolution structure analysis because of the possibility that it gives for recording electron diffraction intensities and because averaging hundreds of thousands of images (or more) of individual molecules is relatively easy to perform with such specimens. On the other hand, the geometry of the 2-D crystal makes it very difficult to obtain images (or diffraction patterns) at tilt angles much greater than ±60°. As a result, the resolution in a 3-D reconstruction of the density map is degraded in the direction of the missing data, an effect that is not normally encountered in x-ray crystallography. The cause of this anisotropic resolution and the potential that it introduces for creating artifactual features are discussed in section 4.6.

Finally, the subject of radiation damage has to be given much more attention in electron crystallography than is the case in x-ray crystallography. Radiation damage is certainly well known in x-ray crystallography, where it is made evident by the progressive fading of diffraction intensities, especially at high resolution, with successive x-ray exposures. In x-ray crystallography, however, it is rare that radiation damage makes it impossible to collect at least some data at 0.35 nm resolution or higher before it is necessary to discard the specimen and continue data collection with a new crystal. In electron crystallography, on the other hand, the crystals are very much smaller than those that are commonly available for x-ray diffraction experiments. The relatively small number of molecules that are present in a typical 2-D crystal must therefore be exposed to much higher radiation doses per molecule than is required when the much larger 3-D crystals are irradiated in x-ray crystallography. Consequently, radiation damage is an issue that must be confronted with greater urgency in electron crystallography, as will be discussed in sections 4.7 and 4.8.

4.2 The single-scattering (kinematic scattering) approximation and the weak phase object approximation are mathematically similar but not identical

There is a very close relationship between the kinematic scattering approximation, which is the standard approximation used in the theory of x-ray crystallography (see chapter 2), and the weak phase object approximation, which has been introduced in chapter 3. Both approximations consider only single-scattering events, and thus they are linear in the scattering potential, and both are limited to elastic scattering events. In contrast to these single-scattering approximations, theoretical descriptions of multiple elastic scattering are referred to as *dynamical scattering* approximations. More is said about dynamical diffraction in chapter 15.

The *kinematic scattering* approximation, also known as the *first Born approximation*, says that the scattered wave function, $\Psi_{kin}(\mathbf{S})$, is proportional to the Fourier transform of the scattering potential, $\hat{\rho}(\mathbf{R})$, evaluated at those spatial frequencies that lie on the surface of the Ewald sphere:

$$\Psi_{kin}(\mathbf{S}) = \mathcal{F}\{\hat{\rho}(\mathbf{R})\} \mid_{EwaldSphere}. \tag{4.1}$$

We want to draw special attention to the fact that we use upper-case $\mathbf{R}$ to designate a three-dimensional vector for points lying within an object, and we will use lower-case $\mathbf{r}$ to designate points anywhere in space, not just those within the object. Later, we will also use the two-dimensional, real-space vector, $\mathbf{x}$, when describing points lying within 2-D projections of the object.

A key concept that is implicit within the notation used in equation 4.1 is that the Ewald sphere is a curved surface. This curvature can become important at the highest spatial frequencies that are of interest to us, but in most cases we will assume that the Ewald sphere is well approximated by a plane.

A simple graphical illustration, shown in figure 4.1, can be used to derive the result stated in equation 4.1. The derivation begins by considering what happens if the

Theoretical Foundations Specific to Electron Crystallography

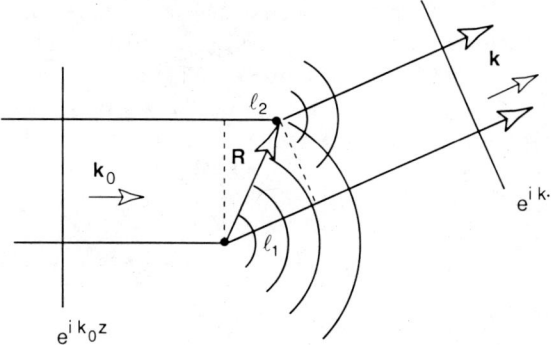

Figure 4.1 A schematic diagram that can help to understand that the scattered wave is the Fourier transform of the scattering potential. As shown here, the derivation begins with just two scattering points, separated by the vector **R**. A plane wave is incident in the direction of the unit vector, $\hat{\mathbf{k}}_0$, and the scattered wave is observed in the direction of the unit vector, $\hat{\mathbf{k}}$. Although each scattering point initially generates a spherical wave, these asymptotically become equal to the plane wave, $e^{i\,\mathbf{k}\cdot\mathbf{r}}$, at long distances from the object. The path difference for the two rays shown in this figure can be seen, by inspection, to be $l_1 - l_2$. The two scattered waves must then be superimposed with the correct relative phase which is $(\mathbf{k} - \mathbf{k}_0) \cdot \mathbf{R}$, when one wishes to calculate the scattered wave function.

specimen is made up of only two scattering points, one at the origin of a real-space coordinate system and one at the vector position **R**. When our incident plane wave, $e^{i\,\mathbf{k}_0\cdot\mathbf{r}}$, is scattered from each point, spherical waves will be created whose amplitudes are proportional to the strength of the scattering potential at that point. When we then go a far distance from the object, in any direction specified by the vector **k**, the spherical wave asymptotically approaches being a plane wave of the form $e^{i\,\mathbf{k}\cdot\mathbf{r}}$. Furthermore, the path length difference, $l_1 - l_2$, for a ray scattered from the point at the origin and for a ray scattered from the point at vector position **R** satisfies the relationship

$$2\pi \frac{(l_1 - l_2)}{\lambda} = (\mathbf{k} - \mathbf{k}_0) \cdot \mathbf{R} \tag{4.2}$$

The left-hand side of equation 4.2 is easily recognized as being the phase difference between the two different scattered plane waves.

The scattered wave function for two scattering points is given by the superposition of the two separate scattered waves:

$$\Psi_{kin} = \hat{\rho}(0)e^{i\,\mathbf{k}\cdot\mathbf{r}} + \hat{\rho}(R)e^{i(\mathbf{k}-\mathbf{k}_0)\cdot\mathbf{R}}e^{i\,\mathbf{k}\cdot\mathbf{r}}, \tag{4.3}$$

where $\hat{\rho}(0)$ is the scattering strength of the point located at the origin, and $\hat{\rho}(R)$ is the scattering strength of the point located at the vector position **R**. In writing equation 4.3 we have weighted each of the two scattered plane waves by the strength of the shielded Coulomb potential at each point, and we have accounted for the difference in their respective phases, corresponding to the difference in their respective path lengths.

Before continuing this derivation, we first make a substitution of variables, defining a vector **S** such that

$$2\pi\,\mathbf{S} = \mathbf{k} - \mathbf{k}_0. \tag{4.4}$$

As we already know, **S** will emerge as the spatial frequency variable in the Fourier transform relationship between the scattered wave and the scattering potential. We also drop the common term, $e^{i\,\mathbf{k}\cdot\mathbf{r}}$, in equation 4.3. Doing this is justified by noting that only the intensity of the scattered wave is physically observable, and when the wave function in equation 4.3 is squared, the common factor $e^{i\,\mathbf{k}\cdot\mathbf{r}}$ will drop out. With both of these points in mind, we now rewrite equation 4.3 as

$$\Psi_{kin}(\mathbf{S}) = \hat{\rho}(0) + \hat{\rho}(\mathbf{R})e^{i\,2\pi\mathbf{S}\cdot\mathbf{R}}. \tag{4.5}$$

It is clear that we can equally well draw a diagram for three scattering points, just like the one shown in figure 4.1 for two scattering points. When we then add up the three scattered plane waves, we get the result:

$$\Psi_{kin}(\mathbf{S}) = \hat{\rho}(\mathbf{R}_0)e^{i\,2\pi\mathbf{S}\cdot\mathbf{R}_0} + \hat{\rho}(\mathbf{R}_1)e^{i\,2\pi\mathbf{S}\cdot\mathbf{R}_1} + \hat{\rho}(\mathbf{R}_2)e^{i\,2\pi\mathbf{S}\cdot\mathbf{R}_2}, \tag{4.6}$$

where we have introduced the artificial vector, $\mathbf{R}_0$, of zero length, to create a self-evident progression of the notation. By induction, then, we see that the scattered wave for N discrete scattering points becomes:

$$\Psi_{kin}(\mathbf{S}) = \sum_{j=0}^{N} \hat{\rho}(\mathbf{R}_j)e^{i\,2\pi\mathbf{S}\cdot\mathbf{R}_j}. \tag{4.7}$$

Generalizing to a continuous distribution of scattering points (i.e., to the case where $\hat{\rho}(\mathbf{R})$ is a continuous function) gives the result:

$$\Psi_{kin}(\mathbf{S}) = \int \hat{\rho}(\mathbf{R})e^{i\,2\pi\mathbf{S}\cdot\mathbf{R}}d\mathbf{R}. \tag{4.8}$$

We have shown, therefore, that the far-field scattered wave function (i.e., the wave function in the Fraunhofer diffraction pattern) is the Fourier transform of the scattering potential. This result was used earlier in section 3.4, but no proof was given at that time.

Why do we say that the scattered wave is the Fourier transform "evaluated on the surface of the Ewald sphere"? The shielded Coulomb potential, $\hat{\rho}(\mathbf{R})$, is a 3-D function in real space, and the right hand side of equation 4.8 formally maps this function into another 3-D function in frequency space (reciprocal space). Recalling the definition of **S** that is written in equation 4.4, the values of the vectors **S** that appear in the Fourier transform relationship, equation 4.8, are not free to lie just anywhere in reciprocal space. Instead, they are constrained to lie on the locus given by $(\mathbf{k} - \mathbf{k}_0)/2\pi$, which is the surface of the Ewald sphere.

The weak phase object approximation says that the transmitted wave, $T(x, y)$, is linear in the two-dimensional projection, $\hat{\rho}'(x, y)$, of the scattering potential:

$$T(x, y) = 1 - i\sigma\hat{\rho}'(x, y), \tag{4.9}$$

where $\sigma = e\,2\pi/hv$ has been defined previously, in connection with equation 3.16. The transmittance function written in equation 4.9 is the linear approximation to the more accurate phase object approximation, for which the transmittance function is given as:

$$T(x, y) = e^{-i\sigma\hat{\rho}'(x,y)}. \tag{4.10}$$

The phase object approximation can itself be derived as an approximate representation within multiple elastic scattering theory. We will put off further discussion of its derivation (and its limitations) until chapter 15.

It follows from equation 4.9 that the scattered wave in the weak phase object approximation, Ψ_{wpo}, is proportional to the Fourier transform of the scattering potential evaluated on a central section within Fourier space:

$$\Psi_{wpo}(\mathbf{S}) \propto \mathcal{F}\{1 - i\sigma\hat{\rho}'(x, y)\} = \delta(\mathbf{S}) - i\sigma \,\mathcal{F}\{\hat{\rho}(\mathbf{R})\} \,|_{CentralSection}. \qquad (4.11)$$

The relationship between projections and central sections that is invoked in equation 4.11 is based upon the *projection theorem* of the Fourier transform, for which a proof is given in section 4.3. The key insight given by comparing equation 4.1 and equation 4.11 is that the more convenient weak-phase approximation resembles the more accurate kinematic approximation only to the extent that the Ewald sphere is closely approximated by a plane passing through the origin of reciprocal space.

The kinematic approximation and the weak phase object approximation are in fact identical only in the asymptotic limit of very high energy. This is because the wavelength of the incident electron goes asymptotically to zero at high energy. In other words, the surface of the Ewald sphere asymptotically approaches a plane in the high-energy limit, and therefore the discrepancy between the kinematic approximation and the weak phase object approximation decreases as the energy increases. Between the two approximations, of course, the kinematic approximation is always the more accurate one.

Regardless of the energy, there will always be a range of small scattering angles, that is, small **S**, over which the deviation of the curved Ewald sphere from a central section is so small that the discrepancy can be neglected. This simple insight permits us to describe the weak phase object approximation as being a "small angle" approximation. The effect of increasing the energy can therefore be seen to be nothing more than to increase the range of spatial frequencies over which the finite amount of curvature of the Ewald sphere can be ignored.

As the specimen thickness increases, dynamical diffraction as well as curvature of the Ewald sphere (or Fresnel propagation within the object itself) may cause the weak phase object approximation to fail. These issues will not be addressed until chapter 15, however. For the present, it is sufficient to say that the weak phase object approximation provides an accurate basis for electron crystallographic structure analysis of objects that are thinner than 10 to 20 nm, and for resolutions up to 0.35 nm. A large fraction of the work in which structure analysis of biological macromolecules can be done at high resolution will, in fact, involve specimens that are only one molecule thick, and thus the weak phase object approximation will be valid. When thicker specimens are involved, however, more careful attention must be given to the factors discussed in chapter 15.

4.3 Proof of the projection theorem

In equation 4.11 we had stated without proof that the Fourier transform of a real-space 2-D projection is equal to the set of values of the 3-D Fourier transform that lie on a central section in reciprocal space. A *central section* is defined as a 2-D plane that goes through the origin of the 3-D Fourier transform. The direction of the projection in real space is, of course, parallel to the momentum-vector of the incident beam, $\mathbf{k}_0$. Similarly, the vector normal to the central section in reciprocal space is also parallel to $\mathbf{k}_0$. This statement, describing the relationship between a 2-D projection and its Fourier transform, is known as the *projection theorem*.

For convenience of derivation, and without loss of generality, we use a Cartesian coordinate system in reciprocal space, for which the S_x and S_y axes lie in the plane of a given central section. We let $\mathbf{F(S)}$ represent the 3-D Fourier transform of $\hat{\rho}(\mathbf{R})$, and thus $\mathbf{F}(S_x, S_y, 0)$ represents the values of $\mathbf{F(S)}$ that lie in the plane of the chosen central section. Since $\mathbf{F(S)}$ is the Fourier transform of $\hat{\rho}(\mathbf{R})$, we have:

$$\mathbf{F}(S_x, S_y, S_z) = \int \hat{\rho}(x, y, z) e^{i\ 2\pi(S_x \cdot x + S_y \cdot y + S_z \cdot z)} dx\,dy\,dz \tag{4.12}$$

It therefore follows that the values of $\mathbf{F(S)}$ on the central section are obtained by setting $S_z = 0$, to give:

$$\mathbf{F}(S_x, S_y, 0) = \int \hat{\rho}(x, y, z) e^{i\ 2\pi(S_x \cdot x + S_y \cdot y)} dx\,dy\,dz$$

$$= \int \left[\int \hat{\rho}(x, y, z) dz\right] e^{i\ 2\pi(S_x \cdot x + S_y \cdot y)} dx\,dy$$

$$= \int \hat{\rho}'(x, y) e^{i\ 2\pi(S_x \cdot x + S_y \cdot y)} dx\,dy. \tag{4.13}$$

As the third line of equation 4.13 states, the integral within the square bracket (in the second line) is simply the 2-D real-space projection of $\hat{\rho}(\mathbf{R})$ along the z-axis of the coordinate system. Equation 4.13 is therefore identical to equation 4.11.

4.4 Two important simplifications of crystallographic structure analysis occur when the specimen is approximated as a weak phase object

The fact that the Fourier transform of the image represents a central section through the 3-D Fourier transform of the object makes it particularly easy to keep track of precisely which structure factors have been measured. In other words, the diffraction pattern is particularly easy to *index* because the observed values of the Fourier transform of the object, and of its image, lie on a plane (central section) in reciprocal space. This point represents the first simplification that results from the use of the weak phase object approximation.

The second simplification comes from the fact that the values of the Fourier transform, both the real part and the imaginary part, are computed from the measured image. This point is especially important because the ratio of the real and imaginary parts of each Fourier component of an image is what determines the phase of that Fourier component. As a result, the phase information is preserved when one uses a computer to calculate the Fourier transform of an image. This powerful result contrasts with the case in which the square of the Fourier transform is recorded in a Fraunhofer diffraction pattern. In this latter case, the phase of the Fourier transform is lost when the complex-valued wave function is squared. The resulting, so-called *phase problem*, which is so familiar in the theory of x-ray crystallography, does not occur when one can record images that are linearly related to a projection of the object structure.

4.5 Three-dimensional Fourier space is sampled by collecting data at many different tilt angles

In deriving the projection theorem, equation 4.13, we used a coordinate system in which the incident wave-vector, k_0, was parallel to S_z, which is to say that the projection in real space is onto the (x, y) plane. A more general version of the theorem states that the projection in real space is onto a plane that is perpendicular to the direction of the incident beam, and that the associated central section is also perpendicular to k_0. Thus, if projections of an object are obtained in two different directions, their respective Fourier transforms will result in data in reciprocal space that lie on two different central sections. This fact serves as the basis for the strategy of collecting a full 3-D data set by tilting the object over a wide range of angles.

To illustrate further how data obtained on a 2-D central section are related to the 3-D Fourier transform of the object, we will discuss the specific example of a specimen that is a thin, 2-D crystal. The same concepts that we will illustrate in the case of 2-D crystals apply equally well to all types of specimen, whether they consist of single particles, helices, or symmetrical assemblies such as icosahedra. The only details that are different for each such case have to do with the 3-D Fourier transform of the object itself.

The (3-D) Fourier transform of a thin, monolayer crystal (assumed here to be only one molecule thick) is a sampled version of the Fourier transform of the unit cell, just as is true for the Fourier transform of a 3-D crystal (see section 2.5). The way in which the Fourier transform of the unit cell is sampled is quite different for a 2-D crystal, however. Without going into the mathematical justification now, as this is taken up in section 7.4, we simply state that the Fourier transform is sampled along a set of parallel lines in reciprocal space. The parallel lines themselves, which are called *reciprocal lattice lines* (or reciprocal lattice rods) are found to occur in a geometrically regular arrangement, which itself is determined by the 2-D translational symmetry within the plane of the 2-D crystal. A square lattice in real space will produce a square arrangement of the parallel reciprocal lattice lines, for example, while a hexagonal lattice in real space will produce a hexagonal arrangement of the reciprocal lattice lines. The unique result is that the continuous Fourier transform of the unit cell is sampled at discrete, regularly spaced points along two dimensions, corresponding to the 2-D translational symmetry of the crystal, but it is sampled continuously in the third dimension (i.e., along each rod), corresponding to the absence of periodic repetition of the real-space structure in the third dimension.

As is shown schematically in panels (a) and (b) of figure 4.2, the data collected on a given central section, obtained at a particular tilt angle, consist of measurements at a set of points where the central section intersects the lattice of reciprocal lattice rods. As a result, the diffraction pattern will always look like a regular arrangement of spots, even though the underlying (3-D) Fourier transform has non-zero values on a regular arrangement of lines. The data measured on a given central section can be indexed by specifying the z^* positions where the central section intersects each of these reciprocal

84 ELECTRON CRYSTALLOGRAPHY OF BIOLOGICAL MACROMOLECULES

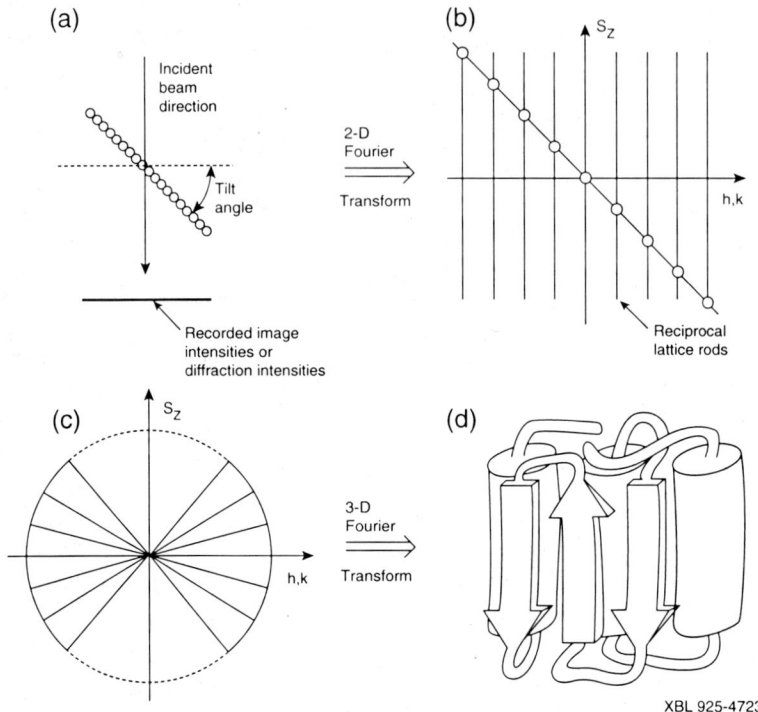

Figure 4.2 The overall process of 3-D data collection and analysis used for thin, 2-D crystals is presented in four steps. In (a), images and electron diffraction patterns are collected from a crystal that is tilted at a given angle relative to the incident electron beam. In (b) the diffraction amplitudes and the image phases are plotted at the correct z^* position, according to the 2-D Miller indices of each diffraction spot, and according to the tilt angle of the crystal. Note that the central section for the projection shown in panel (a) is rotated in the opposite direction to the central section shown as an example in panel (B). In (c), data have been collected for a large number of central sections, indicated by the solid lines, sufficient to give a complete sampling over as much of Fourier space as is possible. Nevertheless, data within a cone surrounding the S_z axis cannot be measured, as is indicated by the dashed extension of the circle (sphere). The 3-D Fourier transform of the measured data then gives a 3-D density map. At sufficiently high resolution, this map can be interpreted, as indicated in (d), by fitting the known amino acid sequence into the density map.

lattice lines. This indexing will consist of two integers (Miller indices), which identify each reciprocal lattice line, and the continuous variable, z^*, which will be different for each lattice line. Data are then collected, one plane (central section) at a time, over as wide a range of tilt angles as possible. Within the angular range for which data can be collected, care must also be taken that the angular separation between different central sections is made fine enough to ensure a complete sampling of all values of the structure factor. This latter point is addressed further in section 4.6, and the same basic consideration will come up again in more specific detail in later chapters.

Whenever possible, it is a real advantage to use electron diffraction intensities, rather than the amplitudes calculated in the Fourier transforms of images, to measure the *modulus of the structure factor*, $F(\mathbf{S})$. The use of the square root of the electron diffraction intensities to measure the *amplitude* (i.e., the modulus) of the structure factor is preferred because the values obtained from diffraction patterns are considerably more accurate than those obtained from images. The larger experimental error in the amplitudes obtained from images is partly due to the focus-dependent value of $\sin \gamma$ as well as the steep fall-off of the envelope function at high resolution (see section 3.9). In addition, the amplitudes obtained from images are much more prone to errors associated with noise in the transform. The systematic errors in amplitude (i.e., the modulus of $F(\mathbf{S})$) can be partially corrected, of course, if the functions $\sin \gamma(S)$ and $E(S)$ are well known. However, such corrections tend to result in noise amplification when the amplitudes are weak. The result is that the best structure factors are obtained when the modulus of $\mathbf{F}$ is obtained from diffraction intensities, while the phase of $\mathbf{F}$ is obtained from images. Diffraction intensities cannot be obtained for all types of specimens, however, as we will describe in subsequent chapters. In that case one can still proceed with the less perfect estimates of the amplitudes that are obtained from the Fourier transform of the image.

When the process of 3-D data collection is as complete as it can be, as is suggested in figure 4.2(c), the measured structure factors, $F(S_x, S_y, S_z)$, are used to carry out an inverse Fourier transform. Inverse Fourier transformation of the experimental data results in a real-space 3-D density map. As is implied in figure 4.2(d), interpretation of the density map then provides an experimental model of the structure. If the density map is produced from data that extend to sufficiently high resolution, the map can be interpreted in terms of a fit of known chemical information, such as the primary amino acid sequence of a protein, as is described in section 2.12. With lower resolution data it may only be possible to interpret the map in terms of major domains, or in terms of distinct elements of secondary structure such as α-helices and regions of β-sheet.

4.6 The resolution of a 3-D reconstruction is determined by the spatial frequency limit of the measurements and by the completeness of 3-D data collection

The amplitudes and phases used in producing a 3-D density map are necessarily limited to some maximum spatial frequency. The limitation in resolution may lie in the specimen itself, due, for example, to imperfect short-range order. Alternatively, the limitation may be due to the experimental difficulties that are encountered in collecting data at very high resolution. Either way, we want to know how limitations in the data will, in turn, limit our ability to visualize different features in the density map.

We can first discuss the rather idealized case in which data have been collected isotropically (i.e., in all directions) to some maximum spatial frequency, S_{max}. By limiting the Fourier synthesis at a maximum radius, S_{max}, we are in effect multiplying the Fourier transform of the ideal electron density function by a *spherical rectangle function*, $\text{rect}(\mathbf{S}; S_{max})$, which is equal to 1 if $S \leq S_{max}$, and equal to 0 if $S > S_{max}$:

$$\mathbf{F}_{limited}(\mathbf{S}) = \mathbf{F}_{ideal}(\mathbf{S}) \cdot \text{rect}(\mathbf{S}; S_{max}). \tag{4.14}$$

The resulting density map is thus convoluted with the inverse Fourier transform of the rectangle function:

$$\hat{\rho}_{limited}(\mathbf{R}) = \hat{\rho}_{ideal}(\mathbf{R}) * \mathcal{F}^{-1}\{\text{rect}(\mathbf{S}; S_{max})\}. \quad (4.15)$$

The (inverse) Fourier transform of a spherical "rectangle function" is itself spherically symmetrical, and it is analogous to the 2-D Fourier transform of a circular aperture discussed in section 3.5:

$$\mathcal{F}^{-1}\{\text{rect}(\mathbf{S}; S_{max})\} = \int_0^{S_{max}} e^{i2\pi \, \mathbf{S} \cdot \mathbf{R}} d\mathbf{S} = \frac{j_1(2\pi \, RS_{max})}{2\pi \, RS_{max}}, \quad (4.16)$$

where the function j_1 is the ordinary first-order, spherical Bessel function.

The inverse Fourier transform (of the spherical rectangle function) described in equation 4.16 is, in effect, a *three-dimensional point-spread-function*, as can be seen from the convolution relationship that is shown in equation 4.15. Convolution of the three-dimensional density function, $\hat{\rho}_{ideal}(\mathbf{R})$, by a three-dimensional point-spread-function must, in turn, blur out and degrade the resolution of the density map. The resolution of the 3-D density function obtained as a result of this convolution is limited by the highest spatial frequency of the rectangle function itself. The 3-D resolution of $\hat{\rho}_{limited}(\mathbf{R})$ is thus "diffraction limited" by the 3-D rectangle function in exactly the same way that the finite lens aperture encountered in coherent image formation results in a diffraction-limited representation of the object. This is an important point, because the calculation of a 3-D density map from experimental data is analogous to the formation of the wave function in a coherent optical image. The relationship between the familiar "Rayleigh criterion" for resolution and Fourier resolution, or, similarly, the diffraction-limited resolution of a coherent optical system, is explained in box 4.1.

BOX 4.1 The relationship between "Fourier resolution" and the Rayleigh criterion

The *resolution* in an image — we here use the term "image" in a very general sense, to include graphical representations of density maps — can be discussed in a clear way in terms of a point spread function, $h(x)$. When invoking the concept of a point spread function, we suppose that the image at hand, $I(x)$, is derived from a higher resolution, ideal image, $I_{ideal}(x)$. In practical circumstances, it will indeed be the case that the resolution of an experimental image or density map is always less than it might be under perfect circumstances. Thus, because of imperfections in the experimental process, the ideal image becomes convoluted with a point spread function, to give

$$I(x) = I_{ideal}(x) * h(x). \quad (B4.1)$$

Convolution by the point spread function broadens the features inherent in the ideal image, and thereby degrades the resolution.

(continued)

The conventional definition of resolution refers to the closest distance to which two points can be brought before their images merge together into a single, unresolved peak. If one point is at x_0, for example, and the other point is at $-x_0$, then the image will be made up of two copies of the point spread function, which are centered at x_0 and $-x_0$ respectively:

$$I(x) = h(x - x_0) + h(x + x_0). \tag{B4.2}$$

From this idealized example, it is clear that the resolution will depend upon how broad the point spread function is.

The *Rayleigh criterion for resolution* defines two identical points as being resolved from one another when the separation between them is equal to the distance between the peak and the first zero in the point spread function. In an incoherent, diffraction-limited optical system the Rayleigh criterion provides a very sensible definition of resolution, but the same criterion is less appropriate for coherent optical systems and for Fourier maps. The explanation for this difference lies in the mathematical details that distinguish incoherent images from coherent images, as we explain in the following paragraphs.

The point spread function of a diffraction-limited image is related to the inverse Fourier transform of a circular rectangle function:

$$\mathcal{F}^{-1}\{\text{rect}(s; s_{max})\} = \frac{J_1(2\pi r s_{max})}{2\pi r s_{max}}, \tag{B4.3}$$

where r is the distance from a central point in the image and s_{max} is the highest spatial frequency (i.e., the diffraction limit) used in the Fourier synthesis. Use of the same notation for the three-dimensional case, in which the spatial frequencies would be represented by upper case symbols, should be self evident.

For a diffraction-limited, coherent image wave function, and for a Fourier map, the point spread function is equal to the inverse Fourier transform of the diffraction-limiting rectangle function, and thus

$$h_{coherent}(r; s_{max}) = \frac{J_1(2\pi r s_{max})}{2\pi r s_{max}}. \tag{B4.4}$$

For a diffraction-limited, incoherent image, however, the image intensity (not the image wave function) is convoluted with the point spread function of the system (Goodman, 1968). In this case the point spread function is the square of the inverse Fourier transform of the diffraction-limiting rectangle function,

$$h_{incoherent}(r; s_{max}) = h_{coherent}^2(r; s_{max}). \tag{B4.5}$$

Recall that the square of $h_{coherent}(r; s_{max})$ is called the *Airy disc*, as we have mentioned in section 3.5.

The distance between the peak and the first zero of the point spread function is the same for $h_{coherent}(r; s_{max})$ and for $h_{coherent}^2(r; s_{max})$, of course. In both cases, this distance is given by

$$d = \frac{0.61}{s_{max}}. \tag{B4.6}$$

(continued)

BOX 4.1 *(continued)*

The full width at half maximum will not be the same for the two functions, however, and therefore there will be a distinct difference between the image that is formed by the superposition of the two different point spread functions.

When two copies of $h_{incoherent}(r; s_{max})$, that is, $h^2_{coherent}(r; s_{max})$, are separated by a distance of $\frac{0.61}{s_{max}}$, a clear trough separates the two peaks, as is shown in figure B4.1. The depth of the trough is about 19% of the respective peak heights. In a Fourier map and in a coherent image, however, the situation is more complicated because the result depends upon the relative phase of the wave at the two points. If the wave transmitted through the specimen has the same phase and the same amplitude at both points, then the superposition

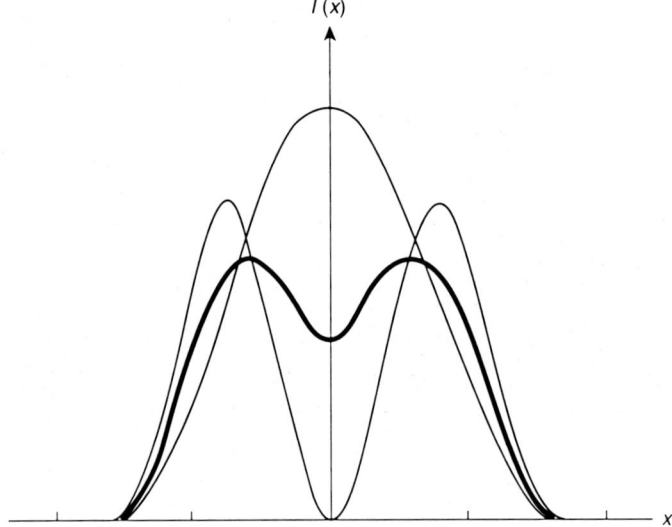

Figure B4.1 The ability to resolve the images of two points that are separated by a distance of $\frac{0.61}{s_{max}}$, the Rayleigh resolution limit, is markedly different for a perfectly incoherent image and a perfectly coherent image. In an incoherent image, the point spread function is actually the square of the coherent point spread function, that is, $h_{incoherent}(r; s_{max}) = h^2_{coherent}(r; s_{max})$. As a result, the point spread function for an incoherent image is narrower than that for a coherent image. The image intensity for the incoherent image of two points, shown in this figure as a heavy line, meets the Rayleigh criterion for resolution, which requires that the intensity half way between the two points is 81% of the intensity at the points themselves. In a coherent image, however, the image intensity is the square of the sum of the two coherent point spread functions. As a result, the two points are no longer resolved, as is shown by the curve with a single peak midway between the two points. Just to emphasize the difference between coherent and incoherent images further, we also show a profile through the coherent image for two points that have opposite sign. In this case the two points are more than adequately resolved when separated by the resolution limit for incoherent imaging. The example of an object-function whose values can change sign is of didactic interest only.

(continued)

(in the image) of two copies of $h_{coherent}(r; s_{max})$ at a separation of $\frac{0.61}{s_{max}}$ results in a single, bell-shaped maximum. The two points are not resolved from one another in this case. As was mentioned above, the reason why the resolution in an incoherent image is better than that in a coherent image (or in a Fourier map) is that the full width of the point spread function at half maximum is narrower for the incoherent optical image, due to the quadratic relationship that is stated in equation B4.5.

If the phase at the two points differs by 180°, however then the transmitted wave will have the opposite sign at the two points — one positive and one negative. In this case the two points are resolved much better in a coherent optical image than in an incoherent optical image (Goodman, 1968). While examples of this type are intriguing and sometimes counterintuitive, we refer the reader to discussions in Goodman (1968) or in Lipson et al. (1995) rather than to go into further detail here.

Although one would ideally like to collect data for tilt angles over the full range from +90 to −90°, a number of practical factors commonly limit the maximum tilt range to ±60° or less. The limitation in tilt angle therefore means that the structure analysis will be carried out with partially incomplete data. The existence of a *cone of missing data* has already been anticipated in the schematic illustration shown in figure 4.2.

A cone of missing observations will always result in a real-space map that has anisotropic resolution, the resolution being degraded more severely in the direction of the missing cone than in the other two directions. The convolution theorem of Fourier transforms provides an elegant proof of this statement. Let us start first with a complete 3-D data set that extends to the same resolution in all directions. Obviously, a real-space density map that is calculated from this complete data set will have equal resolution in all directions. The incomplete data set that would be collected with only a limited range of tilt angles, for example tilting over a range of ±60°, is related to the complete data set by simply multiplying the latter by an appropriately shaped, 3-D rectangle function in reciprocal space. Referring to figure 4.2, for example, the rectangle function would have the value 1.0 within the region where data are measured, and the value zero outside. Using the notation $C_\theta(\mathbf{S})$ to represent this *cone-shaped rectangle function*, where use of the symbol "C" is meant to indicate that there is a cone-shaped region of missing data, and the subscript θ indicates the maximum tilt-angle,

$$F_{obs}(\mathbf{S}) = C_\theta(\mathbf{S}) \cdot F(\mathbf{S}). \tag{4.17}$$

The inverse Fourier transform of $F_{obs}(\mathbf{S})$ is then

$$\hat{\rho}_{obs}(\mathbf{R}) = \hat{\rho}(\mathbf{R}) * c_\theta(\mathbf{R}), \tag{4.18}$$

where

$$c_\theta(\mathbf{R}) = \mathcal{F}^{-1}\{C_\theta(\mathbf{S})\}.$$

Equation 4.18 shows that the resolution in the experimental 3-D map is degraded by convolution of $\hat{\rho}(\mathbf{R})$ with the three-dimensional point spread function, $c_\theta(\mathbf{R})$. Since high-resolution terms have been blocked out by $C_\theta(\mathbf{S})$ in one direction, the three-dimensional point spread function, $c_\theta(\mathbf{R})$, is itself elongated in the direction of the cone of missing data.

When data are collected over the range $\pm 60°$, however, the 3-D point spread function is elongated by less than 30% in the direction parallel to the axis of the missing cone (Glaeser et al., 1989). The shape of the full 3-D point spread function is shown in figure 4.3. Panel (a) shows a contour map of $c_\theta(\mathbf{R})$ in the (x, z) (or, equivalently, the (y, z)) plane, while panel (b) shows quantitative graphs of $c_\theta(\mathbf{R})$ along the x-axis and along the z-axis. Density maps that were calculated at a resolution of 0.36 nm, using a missing cone of 30°, confirmed that the modest loss of resolution in the z-direction does not have a noticeable affect on how the map would be interpreted when building an atomic model into density. Even when the tilt angle covers a range of only $\pm 45°$, distortion of the computed high-resolution density map remains small enough that the natural interpretation of the map remains the same as for a map computed without a missing cone of data. These points are discussed more fully in section 11.4.

The anisotropy of resolution introduced by the 3-D point spread function is a function of the cone-angle only, and it does not depend upon the absolute scale of the resolution. Stating this another way, the resolution in the direction parallel to the cone, relative to the resolution perpendicular to the cone, will be in the same ratio for all values of the limiting (i.e., best) resolution.

Although the degree of anisotropy of resolution depends only on the geometry of collected versus missing data, and not on the absolute value of the resolution, the extent to which the anisotropic resolution can affect the interpretation of a map does depend upon the resolution. At a resolution that is high enough to permit a chain-trace interpretation, the relatively small anisotropy (less than 30%) associated with a

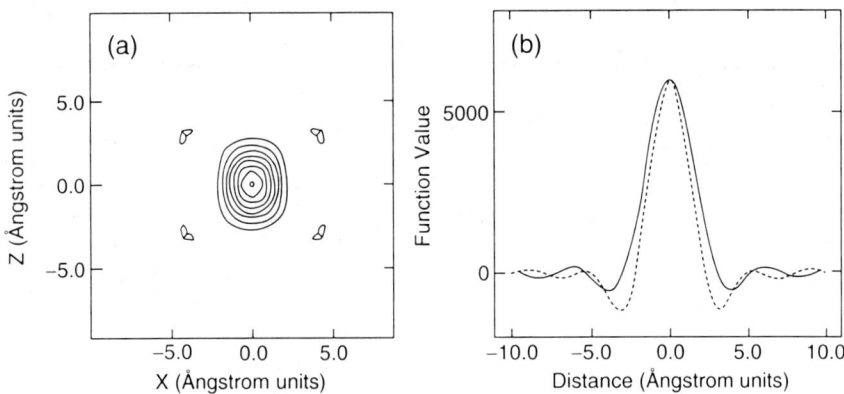

Figure 4.3 The 3-D Fourier transform of a rectangle function, spherical in shape but with a missing cone of $\pm 30°$, gives the 3-D point spread function shown here. In (a) the point spread function is represented with a contour plot in the (x, z) plane; the full, 3-D function is a figure of revolution about the z axis. In (b) the same function is represented as line-trace functions along the x-axis and the z-axis, respectively.

±30° missing cone has no discernible effect on the interpretability of the map (Glaeser et al., 1989). On the other hand, at a resolution of 1.5 nm, which is an exceptionally good value for 3-D reconstructions of negatively stained samples, the anisotropic point spread function is able to eliminate real features in some regions of the map and create entirely artifactual features elsewhere (Baumeister et al., 1986).

From a mathematical point of view, the effect of convolution with the point spread function shown in figure 4.3 will be to broaden somewhat the density features of a map, provided that the central maximum of the point spread function is narrower than the separation between the sought-after features of the map. On the other hand, when the width of the central maximum is similar to the separation of individual features in the initial structure, then convolution with the point spread function will lead to mixing of adjacent densities, which is likely to generate artifacts in the density map. This particular point is taken up again more extensively in section 11.5, where we discuss the question of how the missing cone can limit our ability to correctly interpret a high-resolution map in terms of the atomic-resolution structure of the specimen.

A different form of incompleteness in the data can still occur even when the tilt angles — and their corresponding central sections — are uniformly distributed over the full range of 90°. Incomplete measurement of **F(S)** still arises in this case if there simply are not enough central sections to measure all independent values of **F(S)**. Intuitively, the central sections must be sufficiently close to one another that all required points in reciprocal space are sampled. Adequate sampling is not usually a problem close to the origin, but as one goes out to higher resolution, the central sections fan out from one another, and eventually gaps of missing data begin to open up between them. The result, as is shown in figure 4.4, is that the isotropic data are multiplied by a three-dimensional, *star-shaped rectangle function*. In the absence of a detailed characterization of the real-space point spread function for a particular star-shaped aperture, the conservative estimate is to say that the resolution extends only to the point where gaps begin to appear in the well-sampled data. As we will show next, the number of evenly spaced

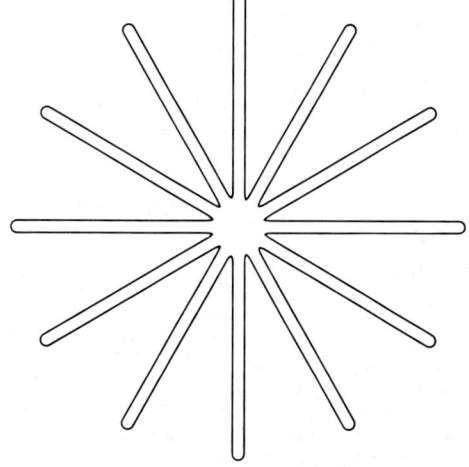

Figure 4.4 If the angular difference between two successive central sections is too large (as in the exaggerated example shown here), there are gaps of missing data that cannot be measured experimentally, lying between each central section. Beyond a certain radius, the Fourier transform of the object is known only in the close vicinity of the central sections, but not in the spaces between them. The effect of having such sparsely sampled data is as if the full, 3-D Fourier transform had been multiplied by a three-dimensional star-shaped mask, or aperture function, as is shown here.

projections that are needed to achieve a particular resolution is determined by the size of the object.

To understand why the required number of projections depends upon the size of the object, we first must mention that the Fourier transform of a bounded (i.e., not infinite) object changes smoothly from one point in reciprocal space to the next. Because of this smooth, continuous change, the values of the Fourier transform at two adjacent points cannot be independent of one another if the points are sufficiently close. The values of the Fourier transform eventually become independent of one another, of course, if the points are far enough apart. Thus, while the points in Fourier space at which experimental measurements must be made can be spaced apart by a certain amount, it is still necessary to use an appropriate number of different projections to measure all of the independent values of the Fourier transform. We will now show that the distance in reciprocal space by which two, independent values of the Fourier transform are separated is inversely related to the size of the object.

In order to develop some intuition about the general case, we can first discuss the simple case of an object that is a cube with an edge-length, L. This object is represented by a *cubic rectangle function* in real space, rect($\mathbf{R}$; L), where the notation now implies that the rectangle function has a value of one within the cube and a value of zero outside the cube. The Fourier transform of such a cube is a three-dimensional sinc function:

$$\mathbf{F}(\mathbf{S}) = \mathcal{F}\{\text{rect}(\mathbf{R}; L)\} = \text{sinc}(2S_x L)\text{sinc}(2S_y L)\,\text{sinc}(2S_z L)$$

$$= \frac{\sin 2\pi S_x L}{2\pi S_x L} \frac{\sin 2\pi S_y L}{2\pi S_y L} \frac{\sin 2\pi S_z L}{2\pi S_z L}. \tag{4.19}$$

$\mathbf{F}(\mathbf{S})$ therefore oscillates sinusoidally in reciprocal space with a "wavelength" of $1/L$. This means that the central sections in Fourier space have to be placed sufficiently close to one another to guarantee that the sampling is able to capture these oscillations. If the size, L, of the cube is small, then $\mathbf{F}(\mathbf{S})$ will oscillate only very slowly, and relatively few central sections will be required to sample $\mathbf{F}(\mathbf{S})$ accurately, whereas $\mathbf{F}(\mathbf{S})$ will oscillate more rapidly if L is large, thus requiring a greater number of evenly spaced central sections to accurately sample the 3-D transform.

According to the *Shannon sampling theorem*, which we now apply in Fourier space, one must make at least two measurements within each interval of length $1/L$. In other words, when the most rapidly oscillating component of any function is sampled at least twice in each cycle, then the continuous variation of this component (along with all other, more slowly oscillating components) can be recovered perfectly from the discrete, evenly spaced measurements. Looking now at the diagram in figure 4.5, the goal is to have two central sections intercepting a short segment of the perimeter of a circle whose radius is S_1, in order to measure $\mathbf{F}(\mathbf{S})$ to a resolution equal to S_1. If the length of that segment of the perimeter is chosen to be $1/L$, where L is the size of the object, one can see that the total number of equally spaced projections needed to get a resolution $d = 1/S_1$ is

$$N = 2\pi S_1 L = 2\pi \frac{L}{d}. \tag{4.20}$$

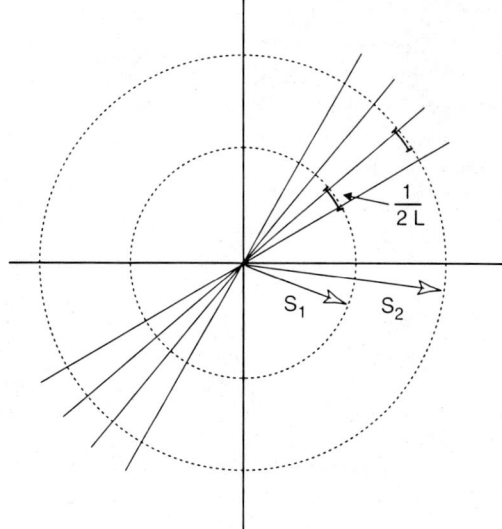

Figure 4.5 This figure illustrates the concept that a large number of evenly spaced rays (central sections) can provide two or more evenly spaced measurements along an arc of length $1/L$, only to a maximum radius, S_1. At higher resolution, the average number of samples per interval of $1/L$ falls below the minimum of two that is required by the Shannon sampling theorem.

A more detailed analysis, made by Klug and Crowther (1972) on the basis of Bessel function interpolation, gives a similar formula, but one that corresponds to sampling at intervals $1/L$ rather than $1/2L$. As a general rule of thumb, however, it is best to have at least three rather than just two samples for each interval, $1/L$, which means that the total number of equally spaced projections ought to be 50 per cent more than the number given in equation 4.20.

4.7 *Radiation damage* represents a much more important experimental constraint in electron crystallography than in x-ray crystallography

Everything that we have described up until now would suggest that three-dimensional density maps could be obtained at high resolution for individual macromolecules simply by recording images of the same macromolecule over a wide range of closely spaced tilt angles. This would indeed be true if the electron beam were only a flux of focusable short-wavelength radiation. Unfortunately, a beam of electrons is also a flux of ionizing radiation (as is a beam of x-rays, of course).

Electron-induced excitation or ionization of the valence electrons within the single bonds (aliphatic bonds) of biological macromolecules inevitably leads to bond rupture, the formation of molecular ions and radicals, and a cascade of secondary chemical reactions. As a result of all of these chemical events, the specimen undergoes a progression of structural changes and eventually winds up in a terminal, "stable" chemical form that has little relationship, if any, to the biologically relevant structure.

The theoretical strategy for data collection and structure analysis that was described in section 4.5 obviously cannot be applied to an object whose structure changes continuously during the course of data collection. In order to proceed with a meaningful structure analysis, it is necessary to reduce the "dose" of electron exposure that is given to a single molecule to such a low level that the damage incurred is too small to cause a detectable change in structure. The highest possible exposure that can be used, without altering the structure more than can be tolerated, is commonly called the *critical dose* in electron crystallography. (Although the technical meaning of "dose" in the context of ionizing radiation refers to the energy absorbed per unit volume, this word is used in electron microscopy as a convenient syllogism for the electron exposure (in units of electrons per unit area) that results in a given amount of energy being deposited per unit volume of the sample.)

As we will explain in section 4.8, the electron exposures that are low enough to be safe, from the point of view of radiation damage, are much too low to provide statistically well-defined images of single molecules at high resolution. The limitations of statistical shot-noise therefore require that we integrate the data from a large number of images of identical, undamaged molecules. The problem of poor statistical definition in images of "undamaged," single molecules, and the consequent requirement for spatial averaging have both been mentioned already in section 1.2. At this point we will be more specific about how one can estimate the maximum "safe" exposure that can be used, and how one can estimate the number of molecular images that must be combined in order to build up the required statistical definition of the data.

Radiation damage causes a progressive fading of the electron diffraction pattern

Observation of what happens to the electron diffraction patterns of protein molecules during continuous irradiation provides an excellent experimental method for determining the maximum electron exposure that can be used before too much radiation damage is incurred. The principle of the measurement is illustrated in figure 4.6, where four electron diffraction patterns are shown for the same catalase crystal, taken at different times during a continuous exposure to 100 kV electrons (Taylor and Glaeser, 1976). The diffraction pattern shown in panel (a) is taken with very low exposure, such that there is almost no radiation damage. Very low exposures can still result in good diffraction intensities if the crystal covers a rather large area, as it did in this example. When the same area of the crystal continues to be exposed, however, the diffraction intensities begin to fade away. Finally, at an exposure greater than 1000 electrons/nm^2, even the low-order reflections have faded completely, which tells us that there is no longer any periodic structure within the irradiated area. On the basis of many studies such as the one illustrated in figure 4.6, using many different types of proteins or other organic materials, we can roughly say that a safe dose for high-resolution work on biological macromolecules, at 100 keV, is no more than 100 electron/nm^2 at room temperature and no more than 5–10 times that much at low temperature (Glaeser, 1975; Isaacson, 1977; Glaeser and Taylor, 1978; Hayward and Glaeser, 1979; Chiu et al., 1986; Henderson, 1990).

Somewhat higher electron exposures can be tolerated in microscopes that operate at higher electron energies, such as 200 or 300 keV, because the *linear energy transfer* (LET) of the incident electron (i.e., the energy deposited in the specimen per unit path-length) decreases somewhat with increasing energy. Representative values of the LET relative to the value of 4.1 MeV-cm^2/g for 100 keV electrons (and the corresponding electron energy) are 2.8 (200 keV), 2.3 (300 keV), 2.1 (400 keV), and 1.8 (1 MeV).

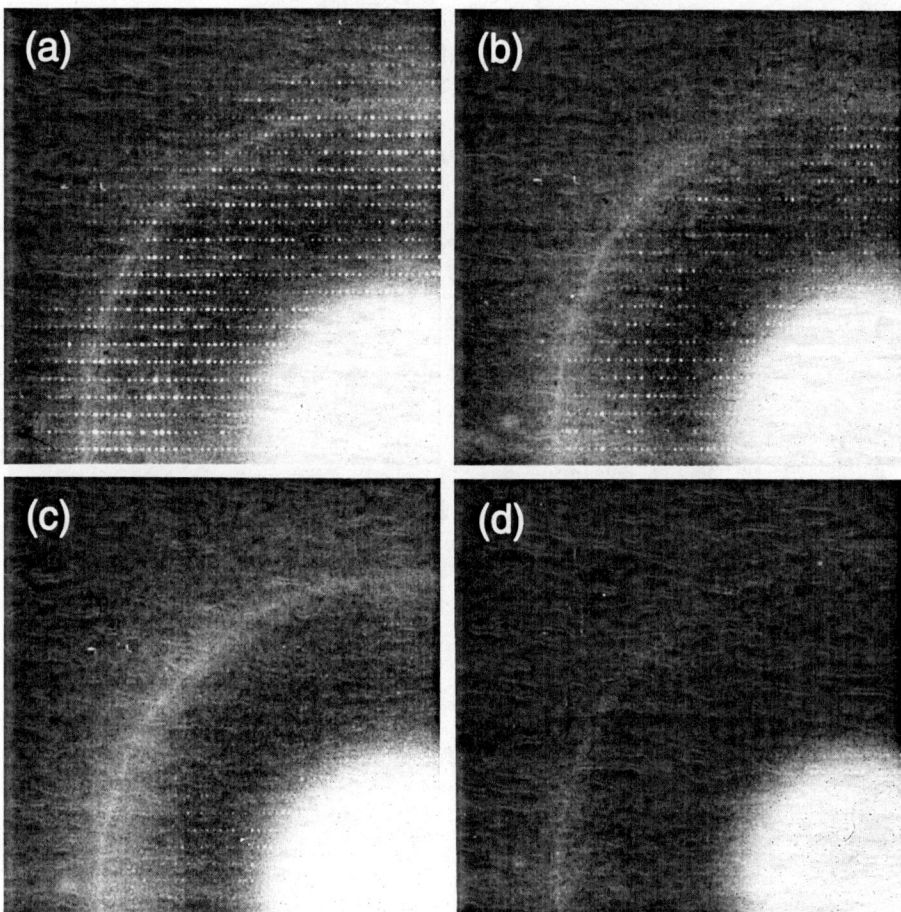

Figure 4.6 Four different electron diffraction patterns that were recorded with 100 keV electrons at different times from the same area of a thin, frozen-hydrated catalase crystal. Panel (a) shows the initial diffraction pattern, recorded with fewer than 100 electrons/nm^2. After a total exposure of 250 electrons/nm^2, the diffraction pattern, shown in (b), extends to a similar resolution as before, but the diffraction intensities have faded significantly. In (c) the intensities, recorded after an exposure of 500 electrons/nm^2, have faded even further, and they are no longer detectable at high resolution. Finally, in (d), after an exposure of 1100 electrons/nm^2, even the low-resolution diffraction spots have faded away almost completely. Reproduced from Taylor and Glaeser (1976).

Although less energy is deposited in a specimen per incident electron, there is no net gain in the signal-to-noise ratio that is obtained with higher voltage electrons. The reason why this is true is that the elastic scattering power also decreases at nearly the same rate as does the LET (or the inelastic scattering) as the electron energy is increased.

The "fading of the electron diffraction pattern" is an especially appropriate criterion to use for judging how much electron exposure can be employed when collecting data for high-resolution structural studies. It is quite obvious that electron diffraction data can no longer be collected after the diffraction pattern has faded away completely, as in figure 4.6. It is also quite plausible to assert that one can safely continue to collect diffraction data as long as the fading of the diffraction intensities has not gone too far. This latter assertion relies on the argument that the structure itself cannot have changed too much if the diffraction intensities have not yet been greatly altered. In practice it is common to collect data with exposures that are high enough to cause the initial diffraction intensities to fade down to $\sim e^{-1}$ of their initial value.

Strictly speaking, it is not necessary for complete fading of the diffraction pattern to correspond to any damage to the structure of the macromolecule itself. In principle, conversion from a well-ordered crystalline arrangement to a randomly aggregated, glass-like packing of exactly the same molecules would also be sufficient to account for the conversion of a crystalline diffraction pattern to an amorphous scattering pattern. The fact that destruction of the molecular structure itself is only a sufficient but not a necessary basis for the observed fading of the diffraction pattern requires that other experimental measurements be used to confirm that destruction of the macromolecule itself does occur at the same dose that leads to fading of the diffraction pattern.

Radiation damage causes molecular disintegration

The majority of radiation damage that occurs in biological macromolecules can be attributed to excitation and ionization of valence electrons. The primary excitation/ionization events occur with an energy loss (to the incident electron) in the range ~ 5 to ~ 100 eV. Two other types of extremely damaging events can occur, but their respective cross-sections are so small that their role is insignificant in the context of biological polymers, and they are only thought to become important for the most radiation-resistant aromatic compounds. Ionization of K-shell electrons is the first of these two processes. K-shell ionization is highly damaging because it results in multiple valence-electron ionization through the so-called Auger process, which occurs in preference to x-ray fluorescence in low-Z atoms. Even highly conjugated π-electron systems are not stable following the loss of two or more valence electrons. The second process involves so-called knock-on collisions between the incident electron and a target atom. Knock-on collisions have a small cross-section because the incident electron must approach the target atom with a very small impact parameter, that is, at a distance very close to the atomic nucleus, in order to transfer enough momentum to the target atom that it is then literally knocked out of its restraining chemical bonds. Both the K-shell ionization process and the knock-on collision process have cross-sections that are about 1000 times smaller than the cross-section for valence-electron ionization, and that is why they are of no practical significance in relation to radiation damage of most biological materials.

Important insight regarding radiation damage is provided by calculating the radiation dose in rads (or the official units of grays) that are deposited in the specimen for an electron exposure of 100 electron/nm^2. The conversion from "exposure" to dose is achieved easily by knowing the linear energy transfer for the specimen. A close estimate for the linear energy transfer in proteins can be obtained from values that are tabulated for a variety of simple materials and a range of electron energies (Berger and

Selzer, 1964). Taking polyethylene, lucite, CO_2, and water as materials whose density, atomic composition, and mean ionization potential will be not too far different from that of biological materials, the linear energy transfer in biological specimens is estimated to be 4.1 ± 0.4 MeV cm^2/g for 100 keV electrons. A flux of 100 electron/nm^2 will therefore result in 4.1×10^{22} eV/g being deposited in the sample. One rad corresponds to 100 erg/g, and one gray is 100 rad. For 100 keV electrons the final answer, then, is that a flux of 100 electron/nm^2 will deliver 6.6×10^8 rad (6.6×10^6 Gy) to the sample. This is an extremely high radiation dose, as can be appreciated when one refers to the extensive literature on radiation chemistry of various organic molecules.

Direct chemical measurements of radiation damage have been carried out on a wide variety of small organic molecules. The results in such experiments are commonly presented as the number of parent molecules that are destroyed per 100 eV of energy deposited in the target by the ionizing radiation, and as the number of product molecules created per 100 eV of energy deposited. The standard notation for these quantities is $G(-$ parent molecule) and G(product molecule), respectively. There usually are many different product molecules formed, each with a different G-value. Typical values for irradiation of solid amino acids are $G(NH_3) = 2$ to 4, $G(CO_2) = 0.2$, and $G(H_2) = 0.2$ (Garrison et al., 1968). As these experimental values reveal, the irradiation of biological materials results in the production of a certain amount of low molecular weight, volatile products, resulting in the preferential loss of heteroatoms.

We can now use a typical estimate for $G(-$ parent amino acid), to compare (1) the rad dose that would convert half the sample to radiolysis products and (2) the rad dose that causes complete fading of the diffraction pattern. For the purpose of calculation, let us take the conservative estimate that $G(-$ parent amino acid) $= 2$, and let us use a molecular weight of 110 g/mol for a typical amino acid. A dose of 2.2×10^9 rad would in this case lead to destruction of at least half of the amino acids. By way of comparison, an exposure of about 500 electrons/nm^2 (100 keV) leads to fading of the high-resolution spots of a protein crystal, at low temperature, and this corresponds to a dose of 3.3×10^9 rad. From these comparisons we can clearly see that the fading of the diffraction pattern occurs on the same scale of electron exposure as that which leads to severe chemical destruction of the biological specimen.

Another informative comparison can be made between the exposure that leads to appreciable mass loss, through the formation of volatile or rapidly diffusing fragments of the type mentioned above, and the exposure that leads to the fading of the diffraction pattern. At room temperature, where small fragments can readily escape, thin films of organic polymers, including proteins, are observed to lose about 20 per cent or more of their mass as a result of exposure to very high amounts of irradiation. The mass loss occurs with an exponential decay, for which the characteristic exposure is about 300 electrons/nm^2 at 100 keV, according to a range of experiments cited in reviews (Glaeser, 1975; Isaacson, 1977). The characteristic exposure for mass loss thus appears to be similar to that which causes fading of diffraction patterns; the important point learned from this comparison is the fact that molecular damage is unquestionably severe at the time that significant loss of diffraction intensity occurs. Mass loss is significantly reduced at low temperature (Egerton, 1982; Lamvik, 1991), as one would expect from the temperature dependence of the volatility and diffusional mobility of small molecular fragments. Although the fundamental radiolytic damage to individual molecules cannot be avoided by going to low temperature, the use of low temperature to prevent the escape of smaller molecular fragments can nevertheless be beneficial in extending the lifetime of high-resolution structural features, at least in crystalline specimens, as we will explain below.

Caging (trapping) of molecular fragments leads to modest protection against radiation damage

As we indicated above, use of low specimen temperatures does result in significant improvement, by a factor of about 5 to 10, in the amount of electron exposure that can be used to record electron diffraction patterns and high-resolution images. While there is also some evidence that the degree of "radiation protection" that is provided at helium temperature may be as much as twice that which is obtained at liquid nitrogen temperature (Chiu et al., 1986), this level of difference is effectively within the uncertainty in the measurement. A 5- to 10-fold increase in specimen lifetime is very significant, and structural studies at atomic resolution would hardly be possible without it. The use of sample temperatures of 110 K or less has thus become a standard part of high-resolution electron microscopy of biological macromolecules.

A number of factors probably contribute to the "radiation protection" effect that is seen at low temperature. A reduction in the extent to which secondary chemical reactions can occur is expected to be one factor. Another factor may be a reduction in the amount of displacement and reorientation of molecular fragments that occurs following bond scission, as may be seen from the following ideas. If the temperature is so low that even small molecular fragments are retained (i.e., there is a reduction in mass loss), the larger molecular fragments tend to be held in their original position by a "cage" of surrounding material. That such a *cage effect* actually does occur at low temperature is demonstrated by the fact that molecular free radicals can be trapped in crystallographically reproducible orientations when organic materials are irradiated at low temperature; this is an effect that has been measured by electron spin resonance spectroscopy (Box, 1975).

Not all problems are solved when the products of radiolysis are caged and trapped at low temperature, however. The radiolytic cleavage of covalent bonds results in the formation of fragments that must move apart from one another as the atom–atom distance increases from that of a covalent bond to that of a van der Waals contact. The result is that radiolysis can generate tremendously large internal pressures. In three-dimensional organic crystals, which would eventually crack rather than merely bend in order to relieve the stress, the measured values of pressure approach half that required to convert graphite to diamond (McBride et al., 1986). In the case of thin crystals of paraffin, these stresses apparently produce a local expansion of the irradiated area, resulting in buckling and bending of the sample (Downing, 1988, 1991; Brink and Chiu, 1991) when the escape of small fragments is prevented at low temperature.

Catastrophic damage ultimately occurs at high electron exposures

If chemical radicals can be trapped at low temperature, in crystallographically precise orientations, why is there still a limited dose that can be tolerated before the electron diffraction pattern fades into the broad rings that are characteristic of an amorphous material? The answer to this question may be that there is a limit to the density (concentration) of radicals that can be built up, because the reaction between two radical species is spontaneous and exothermic. Thus, at high enough densities two radical species are bound to be formed in van der Waals contact with each other, leading immediately to a chemical reaction between them. The heat of that reaction may also permit reorientation of neighboring molecules as well as further chemical reactions among the radiolysis products, such that all of the crystalline order will eventually be lost.

A second stage of radiation damage becomes apparent in frozen-hydrated specimens at electron exposures that are 2 or 3 times higher than the exposure that causes "complete fading" of even the low-order Bragg reflections in electron diffraction patterns. At exposures greater than about 3000 electrons/nm^2 (for 100 keV electrons, and about twice that value for 300 keV electrons), frozen-hydrated samples develop small voids or "bubbles," which grow and coalesce as the dose increases (Glaeser and Taylor, 1978; Dubochet et al., 1982a,b; Cohen et al., 1984). Pure ice does not show this effect at low temperature, although a related effect has been reported at temperatures close to those that lead to measurable sublimation (Unwin and Muguruma, 1971). The high-dose image of microtubules shown in figure 4.7 provides a dramatic illustration of the damage that occurs for ice-embedded biological specimens at very high electron exposures. As this figure illustrates, the overall appearance and morphology of the specimen becomes completely distorted as the bubbles grow in size. The reaction eventually becomes exhausted, leaving a relatively stable product comprised of holes and specimen debris, in which little, if anything, of the original specimen can still be recognized. This figure also illustrates the fact that bubbling does not occur in the surrounding ice itself. In addition, it is worthwhile to mention that unstained air-dried biological samples do not show such an effect.

The overall impression given by this *bubbling effect* is that gaseous products are generated from reactions between water and the organic specimen, initiated by radiolysis, and these gases are trapped and accumulate at the site of reaction. In the case of cryo-sections of a solution of 6-molar glycerol plus 6% bovine serum albumin, as well as liver tissue that was cryo-preserved in the same solution, Leapman and Sun (1995) were able to use electron energy loss spectroscopy to demonstrate that molecular hydrogen gas appears to be the major component of these bubbles. In addition, Leapman and Sun estimated that one molecule of H_2 was trapped in bubbles for every 20 electrons incident on a specimen whose thickness is one mean free path for inelastic scattering. Finally, Leapman and Sun estimate that the pressure in the bubbles increases to about 1000 bar before the bubbles rupture and the signal from molecular hydrogen is lost.

Aromatic molecules can tolerate much higher doses of ionizing radiation than the values discussed above for proteins

For completeness of discussion it should be mentioned that compounds with extensively conjugated double-bond structures are much more resistant to radiation damage than are aliphatic compounds. Even adenosine, which is only partially made up of a π-bonded ring structure, is about eight times more resistant to radiation damage than proteins

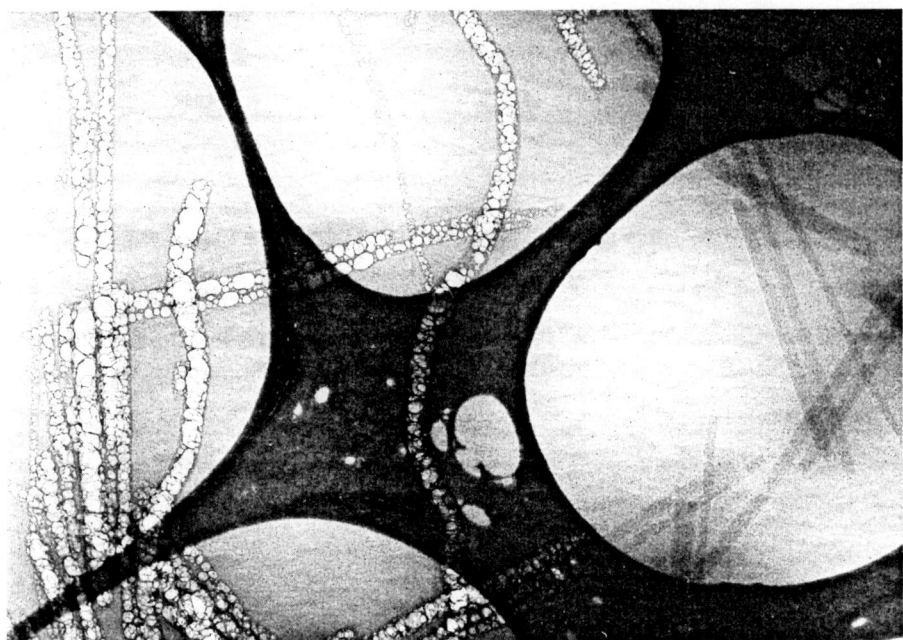

Figure 4.7 Example of the "bubbling" damage that occurs in frozen-hydrated specimens at high electron exposures. The specimen consists of unstained microtubules embedded in vitreous ice (section 6.6). The area on the right-hand side of the figure has received less electron exposure than that over most of the field of view, and the microtubules there still retain a normal appearance for ice-embedded specimens. This micrograph illustrates how the bubbling produced at high exposures is confined to the regions where both ice and protein are located. In addition, the bubbles grow to a much larger size where there are two or more microtubules adhering to one another, indicating that the bubble size is limited by the supply of organic material that can react with water molecules in the ice. Illustration provided courtesy of Dr. Eva Nogales.

or amino acids (Glaeser, 1971), and pure nucleotide bases are even more resistant to radiation damage (Isaacson, 1977).

The study of purine and pyrimidine bases is particularly interesting because radiation-induced changes in their characteristic energy loss spectra, in the region of 10 eV, provide a direct way in which to observe a change in chemical structure as a result of radiation damage (Isaacson, 1977). The relatively high radiation resistance of π-bonded organic molecules is due to the delocalized character of valence electrons in the highest occupied molecular orbitals. If one of the valence electrons is removed by ionization, the delocalized wave functions of the remaining π-electrons may still be able to hold the molecule together by continuing to provide sufficient electron charge density in the region between nuclei to make a stable bond. In aliphatic, single-bond systems, however, ionization removes a full electron from one bond, and the deficiency in electron charge density is not shared, or spread out, among several bonds as it is in conjugated π-electron systems. In this case the Coulomb repulsion between adjacent nuclei is no longer compensated sufficiently by the residual electron charge density in the overlap region between nuclei, and the chemical bond ruptures.

4.8 Images become very noisy at high resolution due to the finite, "low" exposures which are permitted within acceptable limits of radiation damage

In section 1.2 we stated that the very low electron exposures that can be tolerated by biological materials, as little as 100 electron/nm² at room temperature, are far less than what is required in order to get a statistically well-defined image at high resolution. As an extreme example, suppose that we would like to get an image at ~0.3 nm resolution. To compute the Fourier transform of such an image we should sample (digitize) the image in pixels that are about 0.1 nm on edge. Even at the highest exposure level that would be acceptable at low temperature, say 10^3 electron/nm², the statistical fluctuation in the number of electrons per pixel will be about 30%. This figure is far greater than the relative variation in the actual image intensity, that is, the image contrast, from one pixel to the next. At these low exposures the signal will therefore be buried in the shot noise produced by the low counting statistics.

The situation at lower resolution is not as bad as it is at 0.3 nm resolution, simply because the pixel size needed for sampling is larger. For example, there could be as many as 10^3 electrons in a 1.0 nm square pixel, the sample size that would be used for 2–3 nm resolution. In that case the shot noise within one pixel will be only 3%. If the contrast at this resolution is high enough, the signal will no longer be buried in the noise.

In the early days of the development of commercial television, A. Rose created test patterns made up of features with intensity I_1 in a surround of intensity I_2, which he then used to established the limits that shot-noise imposes on the visual detection of features of different size and different contrast (Rose, 1948). Briefly, Rose found that the image contrast, defined as

$$C = \frac{I_1 - I_2}{(I_1 + I_2)/2}, \tag{4.21}$$

must exceed the shot-noise by a factor K in order for the feature to be detectable by eye. Empirical tests show that $K \simeq 5$ for uniform circles on a uniform background, while for parallel lines $K \simeq 1$. The results obtained with Rose's test patterns can be summarized by the inequality

$$Cd \geq \frac{K}{\sqrt{N}} \tag{4.22}$$

where

- d = the smallest feature size that can be detected visually
- C = contrast between the feature and its surround
- K = an empirical constant
- N = number of quanta per unit area used to record the image.

The *Rose equation* (equation 4.22) shows very clearly the dilemma that one faces in trying to get a high-resolution image of a single molecule. If the exposure, N, is higher than the critical dose, then radiation damage destroys the specimen. The image itself may be statistically well defined, even at high resolution, but the structure so obtained has been destroyed in some random fashion and cannot be related to the structure of the parent molecule. If, on the other hand, N is kept below the value at which such damage is significant, then the image is so noisy that high-resolution features cannot be detected. Either way, the original structure of the molecule cannot be seen in the images.

4.9 Spatial averaging must be used in order to overcome the limited statistical definition that is possible when images are recorded with "safe" levels of electron exposure

There is an obvious way to get around the limitations of shot-noise, and that is to increase the statistical definition by averaging the images of a large number of identical molecules. In this way the exposure, N, given to each molecule is kept below the critical dose, but the shot noise is also reduced relative to the accumulated signal.

The easiest way to superimpose the images of many identical molecules is to start with a 2-D crystalline array. Use of a crystalline array guarantees that every molecule (or more correctly, every unit cell) is rotationally aligned with all the others such that they all are facing in the same direction. In addition, there is a fixed translational relationship between all of the molecules in the crystalline array. This predetermined translational and rotational relationship between all molecules makes the superposition of their respective images a trivial operation.

There are two ways in which one can average the individually noisy images of separate molecules within a perfect 2-D crystalline array. The intuitive way is to superimpose the separate images in real space, taking advantage of the translational symmetry that exists between molecules in a crystal. The second way is to calculate the Fourier transform of the entire crystalline array, read off the values of the Fourier transform from the reciprocal lattice points, and use these values to calculate the inverse Fourier transform. The use of just the values found at the reciprocal lattice points to calculate the inverse Fourier transform produces what is called a *Fourier average* of the 2-D array.

It may not be apparent to everyone that the results obtained with these two procedures are rigorously identical. However, it will help to point out that complete information about the perfect, periodic features of the image is contained in the values at the reciprocal lattice points, as we know from our discussion of the Fourier transform of crystals in chapter 2 (and more will be said in chapter 7). Non-zero values of the Fourier transform that lie between reciprocal lattice points can only come from nonperiodic features in the image, for example, from noise. By summing up the separate images in real space, the periodic features add coherently to one another whereas the noise in each image adds incoherently. With enough unit cells, the (normalized) sum approaches the periodic features of the image, and the nonperiodic noise goes to zero. Similarly, application of the Fourier transform method leads to removal of all of the nonperiodic features (noise) except for the inevitable residue that falls under the diffraction peak. Even that bit of noise becomes vanishingly small relative to the height of the diffractions peaks as more unit cells are included in the initial Fourier transform. In practice the Fourier transform method represents the most convenient way to carry out the desired spatial averaging, provided that the image of the 2-D crystalline specimen is sufficiently free of distortions.

The advantage of the Fourier transform is that it immediately gives one the spacing and the orientation of the translational symmetry vectors (the so-called unit-cell vectors) in real space. This information is needed in any case, so even the real-space averaging procedure might start by using the Fourier transform to determine the unit cell vectors. Once the Fourier transform has been calculated, therefore, one may as well extract the values at the diffraction peaks and then compute the inverse Fourier transform.

There is a another way to determine the unit cell vectors, however. The translational vectors that specify the positions of each unit cell can be identified by computing the real-space cross-correlation between a reference image and the statistically noisy image of the 2-D crystal. Unit cell positions found in this way can then be used to generate an average image from the real-space superposition of the images of each molecule. This alternative procedure of averaging the data from many different molecules is called *correlation averaging*. The advantage of correlation averaging is that it does not require perfect translational symmetry, as does Fourier averaging; indeed, correlation averaging can even be used with images of individual molecules, as is described more fully in chapter 14.

Correlation averaging in one form or another has become an essential step in electron crystallography because it can be used to compensate for distortions, microtwins, and a variety of other real-crystal imperfections in ways that the simple Fourier averaging method cannot. Specific versions and implementations of correlation averaging are described in subsequent chapters, according to the particular application. At this stage we will simply point out that the cross-correlation function between a reference motif and a full image will have a series of local maxima when the reference is translated and rotated so as to be superimposed perfectly on each of the unit cells of the crystal. As a result, there will be a cross-correlation peak at each repeating position of the unit cell within the image. If there are distortions within the image of the specimen — and at some level both long-range disorder in the specimen and electron optical distortions are inevitable — the peaks in the cross-correlation function will identify the true unit-cell position. Thus the peaks of the cross-correlation function do not need to lie on a perfect lattice, and their actual positions can be used to follow the distortions that really exist in the image of the 2-D array.

It is quite clear that structurally nonequivalent regions of two different molecules will be superimposed if Fourier averaging is used when there are distortions in the image (or the specimen). If no allowance is made for this fact in the averaging procedure, high-resolution features will be combined in an incoherent fashion, leading to loss of resolution. Thus, the (uncompensated) Fourier averaging method, described above, will only work well if the long-range order is good enough that the true unit-cell positions are identical to the positions that are predicted from a knowledge of the reciprocal lattice vectors. In practice it is almost always the case that a significant improvement can be made by using the cross-correlation function to compensate for defects in the long-range order.

The height of each cross-correlation peak can also be used to reject bad parts of the image, that is, parts where the structure is poorly correlated with the reference motif. Regions of the specimen that are poorly preserved, or regions that differ markedly from the reference motif for any reason, can be included with reduced weight, or can be excluded from the average altogether. Domains where molecules are packed in a particular twin orientation can be identified under conditions where twinning would

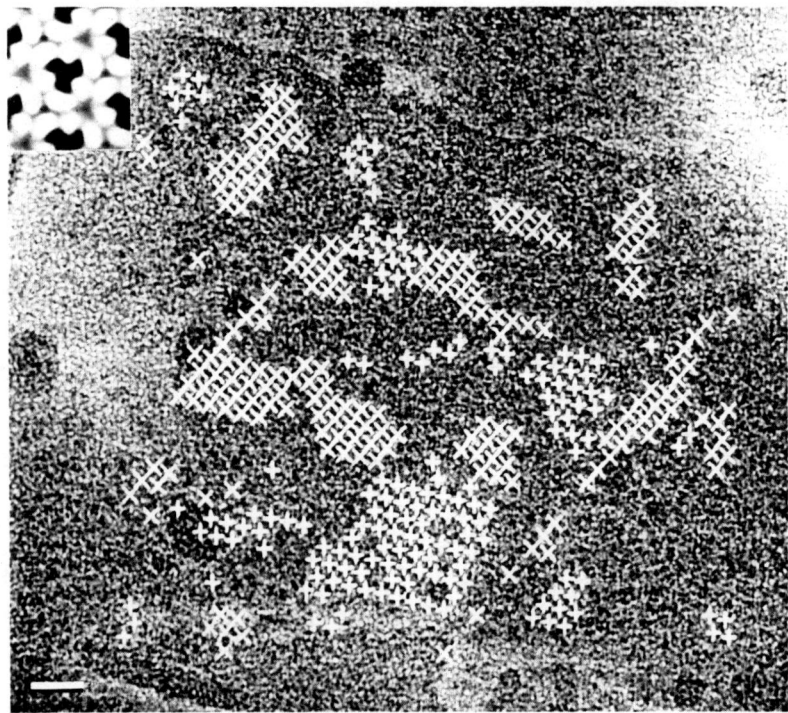

Figure 4.8 An image of negatively stained, 2-D crystals of maltoporin, decorated with maltose binding protein, was cross-correlated with the motif of the maltoporin trimer alone, which is shown in the upper left-hand corner. Cross-correlation with the motif in one orientation yielded strong peaks at the positions marked with the "plus" sign, whereas the same reference rotated by 60° yielded strong peaks at the positions marked by the symbol "x". We thank Dr. Andreas Engel (Basel) for providing this unpublished example of the sensitivity of the cross-correlation function to the molecular orientation of individual trimers. Further details of this structure analysis are given by Stauffer et al. (1992).

otherwise go undetected. The 2-D crystalline surface layer of *Sulfolobus acidocaldarius* represents an example in which the diffraction pattern alone suggests that the protein array is an untwinned, p6 lattice, but correlation averaging reveals it to be a microtwinned mosaic of p3 domains (Lembcke et al., 1991). An even more extreme example of the intergrowth of unit cells of a p3 lattice, rotated by 60° rather than by the required 120°, is shown in figure 4.8.

4.10 The amount of averaging required is determined by the number of scattered electrons and by the image quality

How much *spatial averaging* is required in order to reach atomic resolution? In current practice a rough rule-of-thumb is that one must average $\sim 10^4$ unit cells for each projection-image. Even larger areas are used to collect high-resolution electron diffraction patterns. It is common, for example, to use areas that are 3 μm or more in diameter, containing more than 2×10^5 unit cells, to record diffraction intensities from monolayer crystals.

In theory one could get accurate phases from images of approximately 100 unit cells or fewer. As Henderson (1992) has pointed out, the expected phase error in a particular Fourier component (i.e., structure factor) is only 45° when the electron exposure has reached the point that there would be just a single electron that had been diffracted into that Bragg spot. Most of the diffraction spots in purple membrane have an intensity that is at least 10^{-6} times the intensity of the incident beam, and those which are weaker than this can be neglected, as they will contribute little to the final inverse Fourier transform. If images are recorded at low temperature with an exposure of 500 electrons/nm^2, one must average over an area of 2×10^3 nm^2 in order to reach a total exposure of 10^6 electrons, the exposure needed to have an expected "diffraction intensity" of one single electron in the weaker reflections. The area per unit cell of bR is about 34 nm^2, and therefore the phase errors in perfect images would be 45° or less if one used as few as 60 unit cells for spatial averaging. An even more detailed analysis of the possibilities associated with the alignment of images of single molecules (Henderson, 1995) has shown that two-dimensional arrays would not be needed for proteins with molecular weight larger than $\sim$40 kDa if the images were of perfect quality.

The discrepancy between current practice and the result predicted for theoretically perfect data is largely due to remaining technical problems in the quality of high-resolution image data. The quality of image data is improved somewhat by using electron microscopes that operate at higher voltage, for which the envelope function due to chromatic aberration, $E(S)$, remains high at 0.3 nm resolution (see figure 3.6). Image quality is also improved by using the spot-scan method of illumination, discussed in section 5.8, to reduce the effects of beam-induced specimen movement and the effects of specimen charging (Downing, 1988). Even when using the best currently available imaging techniques, however, the image contrast observed at high resolution is about 10 times less than what would be predicted from the measured ratio of the intensities of the diffracted and the incident beam intensities (Henderson and Glaeser, 1985; Downing, 1991; Mitsuoka et al., 1999). Since the experimentally available signal is 10 times less than that of a theoretically perfect image, we should expect that it will be necessary to improve the statistical definition by averaging over 100 times as many unit cells, and that is indeed what is found to be needed, as is described above.

It is interesting to point out that it will always be necessary to average over about 100 to 1000 times more unit cells when collecting diffraction intensities than is required when collecting diffraction phases. Recall that a very satisfactory phase error of 45° can be obtained with a single diffracted electron (Henderson, 1992). On the other hand, achieving no more than a 10 per cent error in the diffraction intensity requires at least 100 diffracted electrons, and a 3 per cent error requires at least 1000 diffracted electrons. Using bR as an example, we can see that it will be necessary to have about 10^4 unit cells in the crystal in order to reach an accuracy of 10 per cent in the diffraction intensity for weaker reflections (those that are 10^{-6} of the incident beam), even under theoretically perfect conditions, assuming that the incident flux is kept under 500 electrons/nm^2. In practice the background scattering from the support film, additional background from inelastic scattering, and the effect of imperfect detector performance will require that even somewhat larger crystal sizes must be used in order to reach an accuracy of 10 per cent in the weaker diffraction intensities.

5

Instrumentation and Experimental Techniques

5.1 Introduction

In this chapter we first present a general overview (section 5.2) of the instrumental design of an electron microscope. This is followed in section 5.3 by more specific and detailed descriptions of several key components such as the electron gun, the condenser lens system, the objective lens, and the projector lens system. Along with this we develop some of the practical aspects that are related to optical coherence and alignment of the illumination, as well as the effects that these factors have on the contrast transfer function. These are topics that we discussed previously from a more theoretical perspective in section 3.9. In section 5.4 we move on to describe options that are available in the design of specimen stages, emphasizing the need for low specimen temperatures and explaining the relative advantages of top-entry and side-entry stage designs. Section 5.5 deals with the main parameters that are important in the various types of detectors used for recording both images and electron diffraction patterns. Two further sections then discuss some of the experimental details of how high-resolution images are recorded for beam-sensitive specimens. The first section on experimental image collection, section 5.6, addresses the need to record images with very low electron exposures, thus ensuring that the specimen receives only a negligible exposure before the data collection begins. The second section on experimental image collection, section 5.7, discusses the "spot-scan" exposure method, in which only a small area of the whole field of view is illuminated by the electron beam at any one time. Section 5.8 then concludes this chapter with a description of precautions that must be taken in order to avoid image artifacts associated with specimen charging.

5.2 The basic design of an electron microscope is much like that of a light microscope

The design of an electron microscope is very similar to that of an ordinary light microscope. As is shown in figure 5.1, there is an electron source (light source), a condenser lens, a specimen stage, an objective lens, a "projector lens" system (ocular lens) — actually involving two or more lenses in the electron microscope — and a way to view and record the image. There are many variations on this overall design in the electron microscope as well as in the light microscope. Even so, the basic idea of the design is simple, with the fundamental pieces being arranged as they are shown in figure 5.1.

An important constraint in the design of an electron microscope is the fact that the optical path must be in vacuum, in sharp contrast to the light microscope. A fairly high vacuum is required because electrons are scattered very easily, even by a low density of atoms or molecules. As we will see later, in chapter 6, one practical consequence of the need to maintain a good vacuum is that special techniques in specimen preparation have to be used when we wish to preserve the native, hydrated structure of a biological specimen.

The physical principle required to focus electrons is very different from that involved when focusing light. Electrons are focused by magnetic fields (or, in some applications, by electrostatic fields), while glass lenses are used to focus light. By changing the electric current in the lens winding, the focal length of an electron lens can be varied up to a certain limit (determined by the saturation of the magnetic pole pieces of the lens). The continuous control of focal length is used not only to focus the objective lens (rather than by moving either the lens or the specimen up and down, as in the light

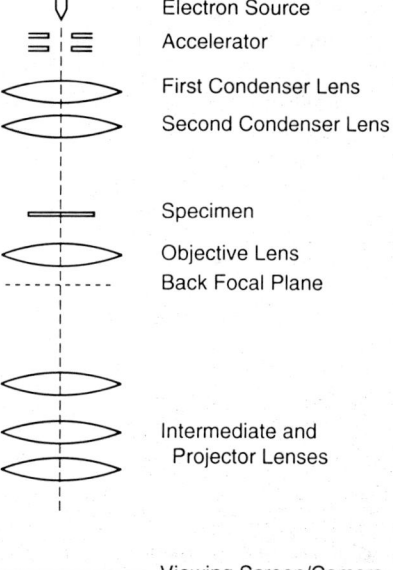

Figure 5.1 Schematic diagram of the major elements of an electron microscope. The principal elements are shown from top to bottom, as they are arranged in nearly all electron microscopes. Electrons are first emitted by an electron gun (which itself has one or more additional electrodes that are not shown here) and accelerated into the condenser lens system. The use of a double condenser lens, standard in all research microscopes, provides valuable flexibility in determining the diameter of the illumination spot and the beam intensity. The specimen is mounted on a stage that is located just above the objective lens. As was discussed in section 3.4, the wave function in the back focal plane of the objective lens corresponds to the Fraunhofer diffraction pattern of the specimen. A series of additional lenses, often three or four of them, finally serve to relay the image (or the diffraction pattern) onto a viewing screen (or camera) at the bottom of the microscope.

microscope), but also to control the collimation and the intensity of illumination, and to vary the magnification in the final image.

The most significant difference between the electron microscope and the light microscope is the higher resolution that can be obtained in the electron microscope. The highest spatial frequency for which information can be collected by a lens is determined by the wavelength of the radiation. If radiation scattered at angles up to 90° is collected by the lens, for example, the highest spatial frequency will be $\sqrt{2}/\lambda$. The light microscope normally uses wavelengths of 400 to 600 nm (in air, but smaller than that by a factor equal to the refractive index in an immersion medium), and thus the resolution (see box 4.1) in the light microscope is limited to about 200 nm. The resolution of the electron microscope, on the other hand, is actually limited by chromatic aberration and spherical aberration in the objective lens, and not by the electron wavelength. Even so, the resolution in an electron microscope can easily be as good as 0.2 nm.

5.3 Technical features that are specific to electron optics

Electron guns

The *electron source* — usually called the *electron gun* — is most commonly a heated tungsten filament or a heated rod of lanthanum hexaboride (LaB_6). In both cases a significant emission of electrons occurs when the temperature is high enough to cause many electrons to be thermally excited to energy levels higher than the work function of the material. The tungsten filament (or LaB_6 rod) is held at a negative potential relative to ground, and as a result the thermally excited electrons that escape from the tip are accelerated towards the anode, which is at ground potential. The principle of emission in such an electron gun is called *thermionic emission*, and the source is called a *thermionic source*.

Field emission sources represent an attractive alternative to the thermionic source. Field emission guns may be operated at ambient temperature ("cold" field emission) or at temperatures as high as 1800°C (thermally assisted field emission). Even in the latter case, the temperature is still not high enough to cause any significant number of electrons to have energies greater than the work function of the tip. Instead, emission is brought about by applying a very strong electric field to the tip. When the electrical potential at a point that is only a few tenths of a nanometer away from the surface of the tip reaches a value that is below the work function, electrons within the tip can *tunnel* through the narrow potential barrier and then be accelerated to ground, as before.

The need to illuminate the specimen with a beam that is both coherent and intense represents an important, practical consideration in choosing the "best" type of electron gun. As is described in box 5.1, the coherence of the illumination is determined by both the energy spread and angular spread of the electron beam. High coherence is needed in order to achieve maximum phase contrast at high resolution, as will be discussed in the next subsection, which deals with the condenser lens. At the same time, however, high intensity is needed to achieve short exposure times, so as to minimize the effects of long-term electrical or mechanical instabilities. The increased brightness that can be obtained with a field emission gun (FEG) makes it possible to have very high spatial

BOX 5.1 The physical distinction between spatial and temporal coherence

The coherence of illumination depends upon two physical parameters, one that specifies how monochromatic the incident illumination is and one that specifies how well collimated it is. An incident plane wave is perfectly coherent, for example. Represented mathematically by the equation

$$\psi_{incident} = e^{i\frac{2\pi}{\lambda}\hat{\mathbf{k}}\cdot\mathbf{r}}, \qquad (B5.1)$$

a plane wave consists of radiation with a single wavelength, λ, propagating in a single direction, $\hat{\mathbf{k}}$. Incident illumination that consists of a single wavelength is said to have perfect *temporal coherence*, while incident illumination that can be made to propagate in a single direction is said to have perfect *spatial coherence*.

Incident radiation that consists of a single wavelength must be perfectly monoenergetic, that is, every particle in the beam must have the same energy. This is so because the energy is a function of the wavelength, and vice versa. For electrons the kinetic energy, eV, is related to the electron wavelength by de Broglie's equation. In the relativistic case (encountered in electron microscopes)

$$eV = \frac{h^2}{2m\lambda^2}, \qquad (B5.2)$$

where

$$m = \frac{m_0}{\sqrt{1-(v/c)^2}} \qquad (B5.3)$$

and e is the magnitude of the electron charge, V is the accelerating voltage, h is Planck's constant, λ is the electron wavelength, m_0 is the rest mass of the electron, v is the velocity of the electron, and c is the velocity of light.

In a real physical situation, the electrons in the illuminating beam will have a range of energies, in part because of fluctuations in the accelerating voltage, but primarily because of the fact that they emerge from the surface of the emitting material with nonzero kinetic energy, due to the Boltzmann distribution of kinetic energies that they have within the material. A real electron beam will therefore have only partial temporal coherence.

Monochromatic radiation that is emitted as a spherical wave from a point source can be converted to a plane wave by an ideal (perfect) lens, if the point source is placed on the optical axis of the lens, at a distance from the lens that is equal to the focal length of the lens. This statement is just the reverse of the usual one, in which one says that a lens will bring an incident plane wave to a point focus at the focal plane of the lens. From this basic idea, however, one will recognize that an extended source can never produce perfectly collimated, parallel illumination. If placed in the (front) focal plane of a lens, each point within the extended source will produce perfectly parallel illumination on the opposite side of the lens, but the angle that the parallel illumination makes with the optical axis of the lens will increase in proportion to the distance that each point is from the optical axis. As a result, the distribution of points within the extended source must always produce a distribution of angles within the illumination.

coherence as well as high intensity. For this reason the FEG is the preferred, though more expensive choice for the electron gun.

A single parameter, the source brightness, combines information about both the spatial coherence and the intensity. *Brightness*, when used as a technical term, is defined as the number of electrons that pass through a unit area per second, per steradian. An important theorem in optics (Born and Wolf, 1997) states that brightness cannot be increased, by any optical device, to a value greater than its value directly at the source, but that it can be permanently decreased by lens aberrations, apertures, etc. In electron optics, however, the brightness achievable with a given type of electron gun increases linearly with the kinetic energy of the accelerated electrons (Hall, 1966; Spence, 1981). Assuming that the accelerating voltage is 100,000 V, the brightness obtained with a tungsten filament is approximately 10^6 ampere/cm^2-steradian, while that for a LaB$_6$ source approaches 10^7, and the value for a heated filament FEG is estimated to be in the range of 10^7 to 10^8 (Spence, 1981). While LaB$_6$ guns have now largely replaced the traditional "tungsten hairpin" as the standard source because of their superior brightness, the higher cost of the even brighter FEGs currently limits their use to the more advanced research instruments.

The standard *accelerating voltage* of commercial high-resolution electron microscopes was fixed at 100 kV for many years, but recent design and engineering advances have resulted in wider use of microscopes that operate at much higher voltage. It is unusual, however, for the accelerating voltage to be higher than 300 kV. Higher accelerating voltages result in several different improvements in the quality of data that can be obtained. In addition to improved temporal coherence (described below), higher voltages result in an increase in the depth of field, a decrease in the curvature of the Ewald sphere, and a reduction in the number of double scattering (or multiple scattering) events. These latter factors become important as the specimen thickness increases (see chapter 15).

The *energy spread* of the emitted electron beam, divided by the accelerating voltage, is the primary factor that determines the *temporal coherence (longitudinal coherence)* of the incident electron wave. Even low levels of instability of the high-voltage supply and the objective lens current could, in principle, also contribute to degradation of the temporal coherence. When all is working properly, however, the effect of such electronic instabilities is small compared to that of the energy spread of the electron gun.

Imperfect temporal coherence causes a high-resolution cutoff on information transfer in the image, as we have described in section 3.9. It is important, therefore, that the energy spread should be as small as possible, and that the accelerating voltage should be as high as possible. The heated tungsten filament has an energy spread that is determined by the Boltzmann energy distribution of electrons in the heated filament. The width of the energy distribution for thermionic sources is generally considered to be about 2.5 eV. The energy spread for the LaB$_6$ filament and for the thermally assisted FEG is somewhat smaller, roughly half this figure. The energy spread for a cold FEG is only about 0.25 eV, giving it superb temporal coherence as well as the best available spatial coherence.

From a practical point of view, the cold field-emission source is tedious to use for routine applications, and it has not been widely adopted in electron crystallography. The primary inconvenience is related to the adsorption of residual gas on the field-emission tip, causing the brightness to drop continuously during use and making it necessary to periodically clean the tip by heating it for a brief period. After prolonged use the cold FEG also becomes unstable due to the sputtering action of positive ions that have been accelerated toward the tip. Flashing (heating) of the tip can again reverse the ion-sputtering damage. The brightness of a cold FEG is so high that it is possible to use only a single condenser lens and the extraction voltage (i.e., first anode voltage) to adjust the beam intensity and the area of illumination. Such a streamlined optical design is to be absolutely avoided, however. The problem with a single condenser lens system is that it does not provide the convenience needed for setting up readily interchangeable illumination conditions that are used during low-dose imaging and electron diffraction. As discussed in section 5.6, as many as five different conditions of beam size and beam intensity may be required in the course of data collection.

Since the energy spread for a heated FEG is only marginally better than that of a LaB_6 gun, one must be aware that the resolution cut-off at optimal defocus, due to imperfect *temporal coherence*, will not be much better for the (thermally assisted) FEG than it is for the LaB_6 gun. Thus, the chief advantage of a heated FEG is in its high level of brightness and *spatial coherence*.

The high brightness of a thermally assisted FEG is useful in two important ways. The first is in applications involving spot-scan imaging (described in section 5.7), where one needs to have a very high beam intensity but a very small illumination angle. The second situation in which high brightness is important is when imaging specimens with an intentionally large amount of defocus of the objective lens, so as to achieve high phase contrast at low resolution. Good spatial coherence helps in this situation because of the strong defocus dependence of the envelope function (see section 3.9). To illustrate this point, the spatial coherence envelope drops to $1/e$ at a resolution of about 1.2 nm when the defocus is set at 2 μm, even if the beam divergence is only 2×10^{-4} radians. In order to keep the contrast high at high resolution, the illumination must be made more parallel than this. The cost of making the beam more parallel, however, is that the illumination intensity drops (as the square of the illumination angle), requiring a corresponding increase in exposure times. The benefit of high brightness is to keep the exposure times reasonably short while still making it possible to collimate the beam to the level required for a good envelope function at large defocus values.

The condenser lens system

After being accelerated, the electrons enter the *condenser lens system*. The function of the condenser lens is to collimate and focus the electron beam onto the specimen. The use of two lenses, one placed immediately after the other, provides for flexibility in adjusting the beam size, the beam intensity, and how parallel the illumination is. The condenser lens system is also fitted with electromagnetic deflection coils. These are used to position the focused beam anywhere that is desired and to align the tilt angle of the beam with respect to the optical axis of the objective lens.

The first condenser lens (known as C1) is normally operated at a rather short focal length, thereby producing a demagnified image of the area from which electrons are emitted in the electron gun. This demagnified image serves, in turn, as the source of electrons for the second condenser lens (known as C2). A small aperture is always

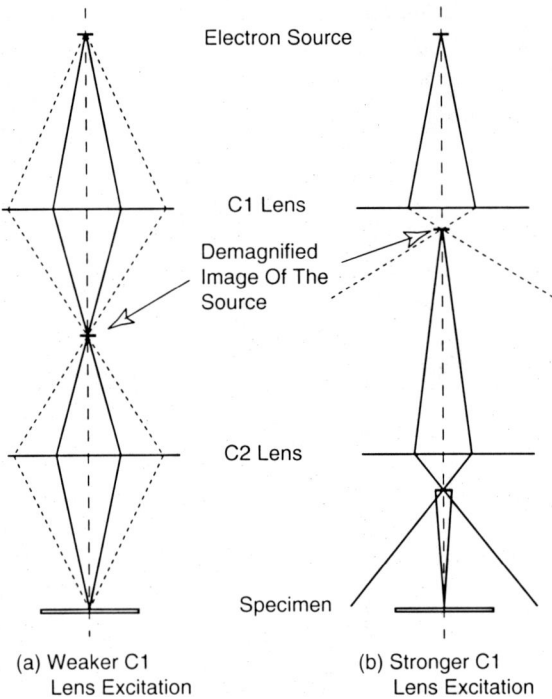

Figure 5.2 Schematic diagram to illustrate the condenser lens configurations that are used to achieve different illumination conditions at the specimen. In each of the configurations shown here, the first and second condenser lenses are represented schematically as lines (in effect, the principal planes) at which the electron rays are bent abruptly to bring them to focus at different positions along the optical axis. The position of the specimen is shown at the bottom, and the illumination may or may not be brought to focus on the specimen, as desired. (a) When the illumination is focused onto the specimen, the beam intensity is at its maximum value, but the illumination is not as parallel as may be desired. The angular range of illumination can be controlled in this case by adjusting the size of the condenser lens aperture. Dashed lines are thus shown to indicate rays that would be transmitted by a larger aperture, but could be eliminated with a smaller aperture in order to achieve more nearly parallel illumination. (b) When the second condenser lens is excited more strongly, the beam "crossover" moves up relative to the specimen. In this case a given point on the specimen is illuminated by a nearly parallel bundle of rays, whose angular range is determined by the size of the crossover and by its distance from the specimen.

inserted into the second condenser lens to limit the angular acceptance of the rays coming from this demagnified image of the original source. The excitation (and thus the focal length) of the second condenser lens is varied in order to control the size of the area that is illuminated. Two common illumination conditions are shown schematically in figure 5.2.

Assuming that the excitation of the first condenser lens and the size of the C2 aperture are both held constant, the highest possible beam intensity is achieved when the electron beam is focused into the smallest possible spot at the specimen, as is shown in figure 5.2(a). This condition is known colloquially as bringing the beam to *crossover*. For beam-sensitive specimens, however, it is only during spot-scan imaging (section 5.7) that one would want to record images with the beam close to crossover. More commonly, the second condenser lens is over-focused, that is, it is run at a higher lens current than that which would focus the source onto the specimen, as is shown in figure 5.2(b). The primary effect of overfocusing C2 is to spread the beam so that it illuminates a larger field of view.

The electron beam is the least parallel, and thus the spatial coherence of the illumination is poorest, when C2 is focused to crossover at the specimen. Figure 5.2(a) shows that the angular range over which electrons are incident on the specimen at crossover is determined by the size of the aperture in the second condenser lens and by the distance between C2 and the specimen. When operating at crossover, therefore, the only way to make the illumination more parallel (i.e., the only way to improve the spatial coherence) is to reduce the size of the C2 aperture. One must be aware, however, that the beam size at crossover will not change significantly when the aperture size is reduced, but the beam intensity will decrease in proportion to the square of the diameter of the C2 aperture. On the other hand, the size of the area illuminated at crossover can be changed with little change in the intensity, by changing the excitation of the first condenser, an adjustment that is labelled "spot size" on some microscopes. Higher excitation of the first condenser decreases the focal length of this lens, which in turn demagnifies the image of the source.

Overfocusing C2, on the other hand, has the effect of making the illumination more parallel at each point of the specimen, thereby improving the spatial coherence of the illumination. The explanation for this is shown in figure 5.2(b), where it is seen that the beam crossover moves above the specimen plane as the excitation of the second condenser is increased. As a result, the angular range of illumination that is incident on any one point of the specimen is no longer determined by the size of the C2 aperture, but instead it soon becomes determined by the size of the beam crossover and by the distance between the crossover and the specimen. By using short focal lengths for both C1 and C2, therefore, one can greatly improve the transverse ("spatial") coherence of illumination. As before, however, achieving increasingly better collimation of the beam is only done at the expense of decreasing the intensity (current density) at the specimen. In addition, a progressively larger area of the specimen is illuminated when C2 is overfocused. It is therefore sensible to use a C2-lens excitation that spreads the beam to an area not much larger than that which covers the detector, at the chosen magnification.

Figure 5.2(b) also shows that different points on the specimen are illuminated at differing angles when C2 is overfocused. Although the illumination at any one point of the specimen may be quite parallel, the range of angles covered at different points across the illuminated area is now determined by the size of the illuminated area, which is determined by the aperture size, as can be seen from figure 5.2(b). It should thus be apparent that the full range of angles for the incident illumination remains the same, whether C2 is overfocused or whether it is focused to crossover at the specimen.

In our discussion of illumination angles we have so far ignored the prefield effects of the objective lens. In normal high-resolution optics, where the specimen is placed well within the magnetic field of the objective lens, the incident electrons begin to be focused by the top part of the magnetic field, even before they reach the specimen.

The *prefield of the objective lens* thus acts as an additional condenser lens element, causing the illumination to be more convergent than it would otherwise be.

Optimal alignment of the illumination

Careful *alignment* of the illumination is essential if one intends to merge data from images at a resolution higher than ~1 nm. What is meant by "alignment" in this case is to adjust the beam tilt so that the illumination is parallel to the axis of the objective lens. We have already said, in the previous section, that the angular spread of illumination angles should ideally be less than 10^{-4} radians. By the same token, the mean angle that the incident electrons make with the axis of the objective lens should also be less than 10^{-4} radians. Alignment to within this very low tolerance is required in order to avoid large phase errors that would otherwise be introduced in the direction of the beam tilt, an effect that was discussed in section 3.12. It is critical that this beam-tilt-related *phase error* be kept as small as possible if data for images of single particles are to be merged at high resolution, since software has not yet been devised to correct the error (by computational means) as can often be done when processing high-resolution images of two-dimensional crystals (section 10.11).

Imperfect alignment causes two other effects in high-resolution images, but these are normally less troublesome than is the phase error discussed above. The first of these effects is to change the objective lens defocus relative to what it would be if the illumination were parallel to the axis of the objective lens. A change in defocus is not a problem, however, unless the tilt angle varies substantially across the field of view. In this latter case, however, the local value of the defocus will also vary across the field of view, just as is the case if the illumination were parallel to the optical axis of the objective lens but the specimen itself were tilted. The second effect is that the tilt angle of the illumination introduces astigmatism, that is, an azimuthal variation in defocus. Once again, however, this effect is not troublesome unless there is a substantial variation in the tilt angle over the field of view.

The first step in alignment might be to set the *pivot points* for *beam deflection* in the condenser lens. Beam deflections are produced by electromagnetic deflection coils that can be used to adjust both the beam tilt angle and the x, y shift position of the illumination. The goal is to be able to change the beam tilt without shifting the illumination, and also to shift the illumination without changing the beam tilt. Commercial software is normally used to guide the user through the process of setting the ratio of currents applied to the deflection coils so that this goal is satisfied. A second step of alignment is needed to keep the illumination centered at the same point when the C1 lens excitation is varied in order to change the *spot size* over the full range provided by the microscope. This alignment step is, again, normally carried out under the guidance of commercial software.

Alignment of the *condenser aperture* is done in either of two ways. One method is to illuminate a relatively large area and then switch the microscope to the diffraction mode. If the illuminated area is large enough, the position of the edge of the condenser lens will then become apparent, because off-axis aberrations prevent the perimeter of the illuminated area from being focused to a spot at the same time that the rays near the center of illumination are focused to a spot. This is a useful effect, since one can now observe directly whether the C2 aperture is centered about the optical axis of the objective lens. The alternative way in which one can center the C2 aperture is to vary the focus of the C2 lens above and below crossover while moving the position of the aperture. The illumination will converge and diverge concentrically if the aperture is properly centered, but it will sweep in from one side and sweep out the other side when the aperture is positioned off-center.

The initial alignment of the illumination is normally done by adjusting the beam tilt while a sinusoidal variation is applied to either the accelerating voltage or the objective lens current. If the illumination is not parallel to the axis of the objective lens, distinctive features in the specimen will be seen to sweep back and forth as the focus is "wobbled." If this is indeed what happens, the beam tilt is simply adjusted in order to minimize the amount of movement of the visible features of the specimen.

This initial step is then followed by a more accurate alignment, which aims to achieve *coma-free imaging* conditions (Zemlin et al., 1978; Zemlin and Zemlin, 2002). The word "coma" refers to a specific aberration that occurs when the illumination is not parallel to the objective lens. The procedure involved in coma-free alignment is to tilt the beam in opposite directions, by equal amounts, and observe whether the focus of the image is the same for both directions. Tilting the illumination causes the image-focus to change, as was mentioned above. If the nominally untilted beam is already parallel to the axis of the objective lens, however, then the beams that are tilted in the two (opposite) directions will be at equal angles to the axis of the objective lens. As a result, the new values of the defocus will be the same in the two cases. If the nominally untilted beam is not parallel to the axis of the objective lens, however, then the net beam tilt will not be equal in the two, opposite directions, and as a result the defocus of the image will also be different. In this latter case, the angle of the nominally untilted beam is changed until the defocus value is the same when an additional, equal but opposite amount of deflection is imposed on the beam.

The objective lens

The first lens that plays a role in forming a magnified image of the specimen is called the *objective lens*, following the terminology used in light microscopy. As we will explain more fully below, the properties of the objective lens are critical in determining the quality of the final image. The key parameter, of course, is the *objective lens focal length*. The values of the *spherical aberration* and the *chromatic aberration* coefficients of the objective lens are also of great importance, since these objective-lens aberrations normally set the ultimate resolution limit of the electron microscope.

As in light optics, the image is said to be focused when the distance between the object and the lens, d_o, the distance between the image and the lens, d_i, and *focal length of the lens*, f, all satisfy the equation

$$\frac{1}{d_o} + \frac{1}{d_i} = \frac{1}{f}. \qquad (5.1)$$

Equation 5.1 is known in geometrical optics as the *lens law*. What the phrase "in focus" means, in fact, is that the objective lens current has been adjusted so that the focal length, f, satisfies the lens law relationship in equation 5.1. Similarly, if the focal length, f, does not satisfy equation 5.1, the objective lens is said to be "defocused." As we mentioned above, the focal length of electromagnetic lenses can be varied continuously over a wide range by varying the lens current, and thus the magnetic field strength or "lens excitation."

In practice there will be a small range of positions above and below d_o for which the image is within an acceptable range of being in focus, even though there is only one position, d_o, that satisfies equation 5.1 precisely. In other words, if a point is misfocused into a disc of diameter d, where d is the desired resolution, one would still consider the image to be "in focus." The range in axial position of the specimen for which the image remains in focus is called the *depth of field*, which we will designate by the symbol δZ.

The depth of field actually varies with the resolution that one wishes to retain. A simple ray-diagram construction shows that the depth of field is obviously related to the objective-aperture angle and the specified diameter of the disc, d, mentioned above. Although the depth of field increases as the aperture angle is decreased, it is not possible to make the aperture angle arbitrarily small, just in order to increase the depth of field. Instead, the smallest aperture angle that is allowed is determined by the angle for which the diffraction-limited resolution (explained previously in section 3.4) is equal to the chosen disc diameter, d. With this additional relationship between the objective aperture angle and d, the depth of field becomes

$$\delta Z = \frac{d^2}{2\lambda}, \tag{5.2}$$

where d is the resolution, λ is the electron wavelength, and δZ is half the axial distance over which features of size d will remain "in focus."

The reader can readily confirm that $\gamma(s)$, the *wave aberration function* defined in equation 3.25, changes by $\pi/2$ when the defocus, ΔZ, is changed by $\frac{d^2}{2\lambda}$. As a result, the depth of field defined in equation 5.2 can also be understood as the greatest distance above and below a given point in the specimen for which a Fourier component of period d (of a weak phase object) could be imaged with the same sign of the *contrast transfer function* (equation 3.24).

From a wave-optical point of view it is important to emphasize that a change in the focus of the objective lens results in a resolution-dependent change in the phase of the wave function. The two situations drawn in figure 5.3 help to explain why it is that the wave function behind the lens is affected by the value of the lens defocus. If the point lies exactly in the front focal plane of the lens, as is shown in panel (a), the diverging spherical wave emitted by the point is converted into a plane wave behind the lens. If the lens is then made stronger, the position of the front focal plane will move closer to the lens, as is shown in panel (b) of figure 5.3. The distance, ΔZ, by which the front focal plane moves is known as *the amount of defocus*. If the point source were moved together with the front focal plane, the spherical wave that it emits — shown by dashed lines to illustrate that hypothetical case — would still be converted into a plane wave on the other side of the lens. If the point source remains in its original position, however, the diverging spherical wave that it emits (still shown by the solid lines) is now converted into a spherical wave that converges to a point on the other side of the lens. The phase of the spherical wave in the back focal plane of the lens can next be compared to that of the plane wave. At small angles, where the spherical wave front can be approximated by a parabola, the phase difference, $\Delta\varphi$, increases as the square

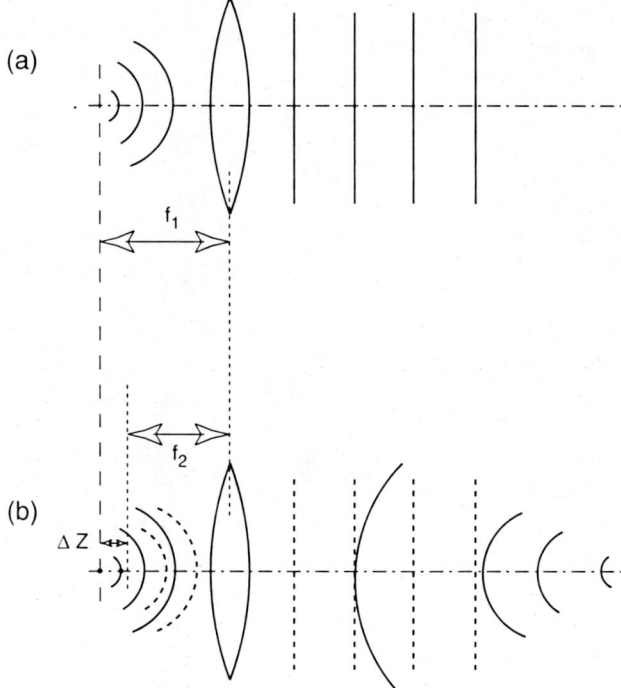

Figure 5.3 "Cartoon" representation of the way in which a change of focus causes a change in the convergence of the beam behind the lens. In (a) the specimen is imagined to be placed in front of the lens by a distance that is equal to the focal length of the lens. A spherical wave that emerges from the point shown on the optical axis is then converted by the lens into a plane wave. In (b), however, the focal length is decreased by the amount ΔZ, so that the distance between the specimen and the lens is larger than the focal length. The spherical wave that emerges from the point on the optical axis is now converted to a converging spherical wave. The difference in phase between the converging spherical wave in (b) and the plane wave in (a), when measured at the back focal plane of the lens, is referred to as the *phase shift due to defocus*, the value of which is given in equation 5.3.

of the spatial frequency, giving the relationship

$$\Delta\varphi = 2\pi \frac{\Delta Z \lambda}{2} s^2, \tag{5.3}$$

where λ is the electron wavelength and s is the spatial frequency. The phase change written in equation 5.3 is the same, of course, as the focus-dependent phase in the wave aberration function that was written in equation 3.25.

When the lens is made stronger, the defocus is said to be negative. We then describe the lens as being *overfocused*. Conversely, when the lens is made weaker, and the defocus is positive, we say that the lens is *underfocused*. The same conventions are carried over, by analogy, when we are talking about a change in focus relative to the specimen, rather than relative to the position of the front focal plane.

If it becomes necessary, under some conditions, to change the objective lens focus by a large amount, one must remember that a change in lens focal length will also produce a change in lens magnification. A simple ray-optical construction shows that the *magnification* is given by

$$M = \frac{d_o}{d_i}. \tag{5.4}$$

Equation 5.4 is often presented as part of the lens law, stated previously in equation 5.1. The two equations are linked together because a change in the focal length, f, will cause d_i to change as well (if d_o is held fixed, for example), and thus the magnification will change. In practice, the magnification of the objective lens is normally about 100 times, and the objective lens focal length is changed by no more than a fraction of a percent when focusing the image. As can then be worked out from equations 5.1 and 5.4, the percentage change in magnification associated with refocusing the image is almost the same as the percentage change in the focal length itself. This change in magnification is normally so small that it can be ignored. It is important to point out, however, that there are some conditions in which significant changes in magnification can result from large changes in focus. One clear example occurs when working with a top-entry stage and a highly tilted specimen. In this case the focus may have to be varied by as much as one millimeter or more from one end of the grid to another. The resulting change in magnification can then be 50 percent or more.

Axial astigmatism is a common image defect that is intimately related to defocus. More precisely, an image is said to be astigmatic if the amount of defocus varies azimuthally around the optical axis of the objective lens. An azimuthal variation in focus can occur due to several different factors. Imperfect machining of the objective lens pole piece, or small variations of the magnetic susceptibility of different parts of the pole piece will certainly lead to astigmatism of the lens, just as imperfect shaping of a glass lens does in light optics. Electrostatic charging of the objective aperture, caused by nonconducting layers of contamination at the edge of the aperture, is also an important cause of astigmatism. Even nonuniform charging of the specimen itself can lead to significant amounts of astigmatism in the image. Fortunately, astigmatism is easily corrected by means of auxiliary magnetic field coils that are placed within the objective lens. Adjustment of the current in these coils allows one to correct the asymmetry so that the azimuthal variation of focus can be made less than 10 nm.

Chromatic aberration arises because the focal length of the objective lens depends on the energy of the electrons; the dependence on electron energy is referred to as chromatic aberration because the wavelength of the electrons depends upon their kinetic energy. Two electrons whose energies, E, differ by a small amount, ΔE, will have a difference in focus given by

$$\Delta Z = \frac{\Delta E}{E} C_c, \tag{5.5}$$

where C_c is the coefficient of chromatic aberration. It is important to note that the same lens is somewhat weaker (or will be underfocused) for the electrons that have higher energies. Since the electrons coming out of the electron gun have a finite range of energies, the images that are formed by electrons with different energies will all have slightly different amounts of defocus. The final result on the image is indistinguishable from what would be seen with truly monochromatic (single-energy) electrons if the focus were changed continuously during the exposure. As equation 5.5 states, however, the smaller the value of C_c, the less is the effect of a given amount of energy spread, ΔE. As was hinted above, both drift and fluctuation of the objective lens focus are, in fact, present to a very small degree due to imperfect stability of the objective lens current. Similarly, if the accelerating voltage is not perfectly stable over the period of time needed to record an image, the result will be the same as if the electrons emitted at any one instant had a range of energies equal to the changes in the voltage itself.

Spherical aberration was mentioned previously in box 3.2, when we referred to the transmittance function of the lens. In light optics, spherical aberration is the aberration that occurs if the surface of the lens is spherical, rather than parabolic in shape. As can be seen by application of Snell's law of refraction, a parabolic shape is the only one that will cause a glass lens to focus all rays (that come from a single point in front of a lens) such that they converge to a single point behind the lens. If a lens with a spherical shape is used, however, rays that hit the outer part of the lens will be focused to a point closer to the lens than those that hit the center part of the lens. By the same token, the effect of (positive) spherical aberration in electron optics is to cause the electrons to be more strongly focused when they pass farther away from the lens axis than when they pass close to the lens axis.

The effect of C_s can also be expressed as a wave aberration function, as we have done for lens defocus in equation 5.3. In the case of spherical aberration, the phase distortion increases as the fourth power of the spatial frequency, rather than as the square of the spatial frequency, as for defocus. More specifically, the phase aberration associated with spherical aberration is expressed in the form

$$\Delta \varphi = 2\pi \frac{C_s \lambda^3}{4} s^4 \qquad (5.6)$$

where C_s designates the spherical aberration constant.

Although spherical aberration can be corrected in light optics by combining a series of azimuthally symmetric lens elements made with different indices of refraction, the same correction cannot be accomplished in electron optics. Instead, it is necessary to resort to time-dependent lens fields, multipole optics, or the placement of field sources/sinks on the optical axis in order to compensate for the positive coefficient of spherical aberration that is physically unavoidable with azimuthally symmetrical lenses (Scherzer, 1947). Practical implementations of aberration correction, based on lenses built with multipole optics, have only recently been realized (Batson et al., 2002; Kabius et al., 2002; Hosokawa et al., 2003; Krivanek et al., 2003).

Spherical and chromatic aberration constants have the dimensions of length, and both constants are approximately equal to the focal length of the lens, even in the best lens design. In the interest of reducing the effect of these aberrations, lens designers have therefore sought to minimize the focal length of the objective lens. This can be done both by increasing the field strength and by decreasing the gap between pole pieces in order to confine the field to a smaller region along the axis of the lens. There are a number of practical constraints that limit the extent to which it is possible to shorten the focal length, however. Because of saturation of the magnetic material that makes up the lens, there is a physical limit to how high one can make the magnetic field strength. Nonuniformities in the saturation behavior of the magnetic material also become especially critical at high field strength, resulting in imperfect lens symmetry. Another constraint in the design of the pole pieces comes from the need to tilt the specimen, as this in turn requires substantial space between the pole pieces, normally a few millimeters. The need to provide a fairly large gap between the pole pieces leads, unfortunately, to a corresponding limit on the extent of field confinement.

The objective lens is normally the only lens for which we need to consider the effects of chromatic and spherical aberration. One way to see why that is the case is to think of the image that is formed by the objective lens as now being the object for the next lens in the system. Because this image is already magnified by a large factor (typically about 100 times) any effect of aberrations from the next lens will be on a very tiny scale compared to the "blown up" (i.e., magnified) effects caused by the aberrations that are present from the objective lens. An alternative way to look at the same question is to be aware that the angles of the electron rays decrease as the magnification increases in successive lenses. With the decrease in angle there is a corresponding decrease in the spatial frequency, s, and the phase aberrations that are written as a function of spatial frequency in equations 5.3 and 5.5 thus become insignificant for the lenses that follow the objective.

The projector lens system

The final electron optical elements, after the objective lens, are commonly referred to collectively as the *projector lens system* (or *intermediate* and *final projector* lenses). The normal function of the projector lenses is to further magnify the image that is formed by the objective lens, and to project the magnified image onto a viewing screen or a detector such as photographic film or a TV camera.

The initial magnification achieved by the objective lens is typically in the range from 50 to 200 times, while the top instrument magnification at the detector, achieved through the use of the projector lenses, may be 5×10^5 or more. The higher magnification ranges, 150,000 to 400,000 times, are especially useful in the initial alignment work and in making a rapid check of the high-resolution performance of the instrument. In biological electron microscopy the images are normally recorded on film at a magnification between 20,000 and 70,000 times, or on a charge coupled device (CCD) camera at a magnification up to 10 times greater than that. The higher magnification range is essential when a CCD camera is used to capture the images because the effective pixel size of such a camera is about 10 times larger than the effective pixel size of photographic film. The increased pixel size encountered with CCD cameras is due mainly to the point spread function of the fluorescent scintillator that is used to convert electrons to photons.

A second function of the projector lens system, extremely important in electron crystallography, is to produce a magnified image of the electron diffraction pattern rather than a magnified image of the specimen. The electron diffraction pattern is itself first produced in the back focal plane of the objective lens. With the appropriate use of the projector lens system, however, one can choose to relay either a magnified version of the image or a magnified version of the electron diffraction pattern down onto the final viewing screen, or onto the detector.

The ability of the projector lens to produce either an image or a diffraction pattern is easily explained by the use of the lens law, written in equation 5.1, and by the two

Instrumentation and Experimental Techniques 121

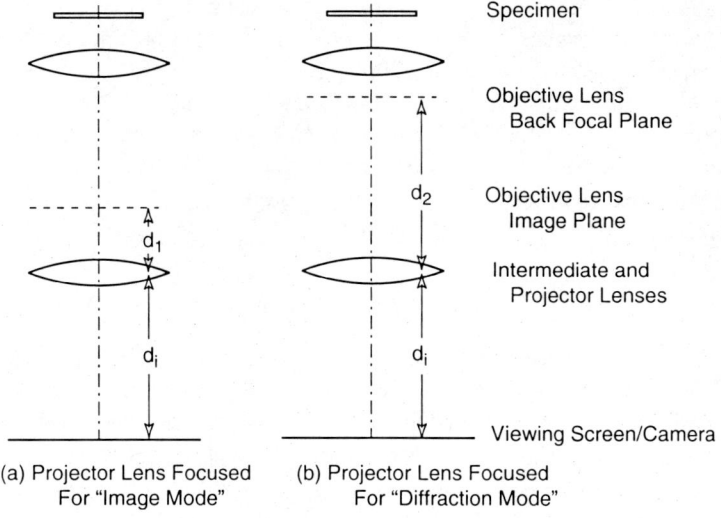

(a) Projector Lens Focused
For "Image Mode"

(b) Projector Lens Focused
For "Diffraction Mode"

Figure 5.4 Illustration showing how the projector lens focus can be adjusted so as to relay a magnified image of either the specimen itself or its electron diffraction pattern onto the viewing screen. The projector lens system is represented here by just a single lens, whereas there are multiple projector lens elements in the real electron microscope. The distance between the projector lens and the detector (designated d_i in both panels) remains the same, but the focal length of the lens is changed, depending upon whether the user wants to obtain an image of the specimen or an image of the diffraction pattern. In the interest of simplicity, however, the focal length of the projector lens is not indicated in the diagram. (a) The objective lens produces a focused image of the specimen, which is shown on a plane that is located at a distance d_1 in front of the projector lens. In order to produce an image of the specimen at the detector, the projector lens focal length must satisfy the relationship $\frac{1}{d_1} + \frac{1}{d_i} = \frac{1}{f_1}$. (b) At the same time, however, the objective lens also produces a diffraction pattern at its back focal plane, which is located at a distance d_2 in front of the projector lens. In order to produce an image of the diffraction pattern at the detector, the projector lens focal length must now satisfy the relationship $\frac{1}{d_2} + \frac{1}{d_i} = \frac{1}{f_2}$.

diagrams shown in figure 5.4. Panel (a) of this figure shows that the focal length of the projector lens can be adjusted to a value that makes the final detector plane conjugate to the objective lens image plane. The projector lens thus "relays" a magnified image of the specimen onto the detector. Panel (b) of the same figure shows, however, the alternative situation in which the focal length of the projector lens has been adjusted to a value that relays the diffraction pattern, rather than a real-space image, onto the detector. The distance between the projector lens and the detector, d_i, is kept the same in both cases. Because the focal length, f, is different in the two cases, however, the plane that is "in focus," conjugate to the detector, must lie at two different positions, d_o, as required by the lens law in equation 5.1. In this way the operator can chose to view either the image or the diffraction pattern simply by changing the focal length of the projector lens system.

While lens aberrations in the projector lens system are not important in determining the resolution of images, the projector lens can be responsible for significant *pincushion* (or barrel) and *spiral distortions* of the image. Known collectively as *field aberrations*, both types of distortion increase as the cube of the distance from the optical axis of the projector lens. Present in any type of magnetic lens, these field aberrations are impractical to correct by electron optics. These distortions may therefore require systematic correction of image data when large fields of view are being averaged. Straightening of such electron optical distortions occurs automatically when "unbending" corrections are applied to images of two-dimensional crystals (section 10.10), an operation that is used primarily to correct for small mechanical distortions of the specimen that occur during sample preparation.

Energy-filter lenses

Electron-optical *energy-filter lenses*, although still not in widespread use, are commercially available for essentially any electron microscope. Electron energy filters allow electrons within a given energy band to pass through to the final image, while blocking the transmission of electrons at all other energies. For applications in electron crystallography, energy filters would normally be set to transmit electrons within an energy window of a few eV to a few tens of eV. Normal usage would also be to center the energy window about zero, that is, to transmit only electrons that have not suffered an inelastic scattering event. In other applications such as elemental microanalysis, of course, one would center the energy window about a particular energy loss of interest. There are two basic design options that are available for use as an energy filter lens.

The first option is to place the energy filter below the column of the electron microscope, effectively adding another projector lens to an existing electron microscope. The advantage of this approach is that such an energy filter can be added to essentially any electron microscope. The limitations associated with the use of a postcolumn energy filter are, however, that (1) one does not have the benefit of the filter when viewing the image on the fluorescent screen or when recording images on film; (2) one must view and record filtered images with a CCD camera, provided as part of the postcolumn filter lens, and thus the number of pixels in the image is somewhat limited (see section 5.5 for further analysis of images obtained with a CCD camera); and (3) the magnification at the CCD camera is at least 10 times greater than that on the viewing screen. This third point, when coupled with the relatively small number of pixels in the CCD output, means that the field of view is correspondingly smaller than one might be accustomed to.

The second option is to build an in-column energy filter into the basic construction of the microscope. This option is currently available on microscopes manufactured by LEO (www.leo-em.co.uk/) and by JEOL (www.jeol.com/), following a design developed by H. Rose (Lanio et al., 1986). The only real disadvantage of an in-column energy filter is that it is not practical to retrofit an existing electron microscope with such a lens. Thus, at the time of purchase, one must make a decision whether to include an in-column energy filter. The benefit of an energy filter for thin specimens is expected to be limited to collecting electron diffraction patterns (sections 9.8 and 15.7), and even in that application an energy filter is not essential in order to collect high-quality data. As a result, it often may not be obvious whether the additional cost of an instrument with an in-column lens is truly justified for work that is limited to thin specimens.

Use of an energy filter lens is certainly important with thick specimens, however. As is discussed in section 15.7, a rather high fraction of electrons is scattered inelastically for specimens thicker than a few tens of nanometers, that is, for specimen thicknesses that are a significant fraction of the mean free pathlength for inelastic scattering. In *electron tomography* (Baumeister, 2004; Lucic, et al., 2005), for example, one even needs to work with specimen thicknesses that are up to two or more mean free pathlenghs for inelastic scattering. Removal of the large number of inelastically scattered electrons that are generated in such thick specimens greatly improves the image quality. At the same time, however, one must realize that there is a limitation to how thick a specimen can be, since removal of the inelastically scattered electrons is of no use if, in the end, there are too few zero-loss electrons left with which to form a statistically well-defined image.

5.4 Specimen stages

The *specimen stage* is an extremely critical part of the electron microscope. Its design must allow for translational movement over a distance of 2 mm or so, with positional control that is as good as 0.1 μm. Both vibrational amplitudes and specimen drift must be much less than 0.1 nm. These specifications are met by surprisingly simple and inexpensive hardware that holds the specimen stationary with respect to the pole piece of the objective lens. The objective lens itself will surely move by distances much larger than are specified above, due to room vibrations and acoustic disturbances. As long as the specimen and the objective lens move together, however, the image will suffer little degradation.

Two fundamentally different design options are offered. The first is the so-called *top-entry* stage, in which the specimen is mounted at the tip of a cone-shaped cartridge like the one shown in figure 5.5. With the help of appropriate vacuum airlocks and transfer rods, the cartridge is inserted down through the bore of the upper pole piece, such that the narrow tip is placed deep within the middle of the objective lens pole piece. The alternative is the so-called *side-entry stage*. In this case the specimen is mounted at the tip of a long rod, also shown in figure 5.5, that can be passed through the gap between the upper and lower pole pieces of the objective lens. In both the top-entry design and in the side-entry design, the final position of the specimen is at a point that is almost midway in the gap between the north and south pole pieces of the electromagnet.

In order to obtain a full three-dimensional data set for specimens prepared as thin crystals, or for any other specimen in which the molecules have a preferred orientation relative to the support film, it is essential that the stage design should permit the user to work with highly *tilted specimens*. The side-entry stage is especially convenient in this regard, while tilting the specimen with the top-entry stage is rather cumbersome.

Further advantages of the side-entry stage are twofold: (1) by rotating the specimen rod about its long axis, the tilt angle of the specimen can be quickly changed to any value that is desired; and (2) by appropriate design of the stage, this rotation axis can be made to be *eucentric*. The term eucentric is used to denote the idea that neither the longitudinal position (i.e., height of the specimen position) nor the transverse position of the specimen should move as the stage is tilted, which requires that the tilt axis must pass through the plane of the specimen at a point that is in the center of the field of view.

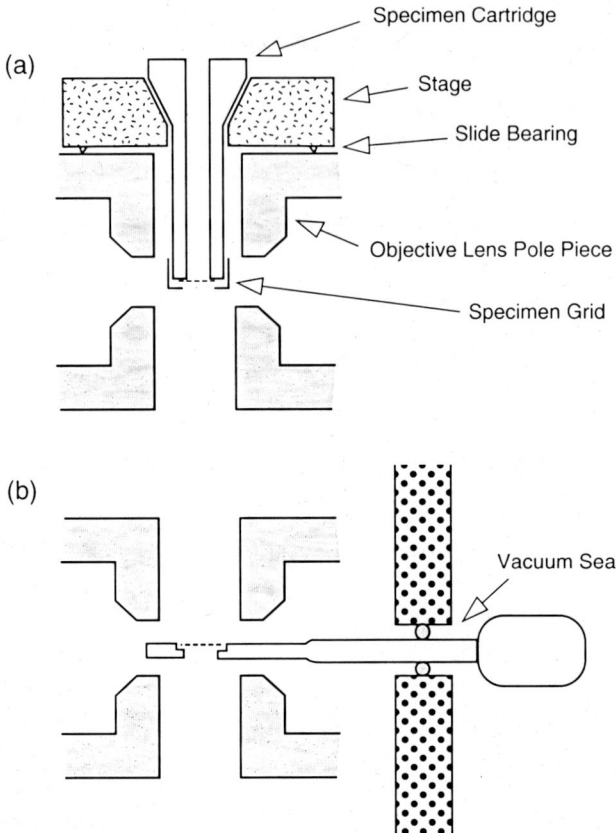

Figure 5.5 Schematic comparison of the design of specimen holders for top-entry and side-entry stages. (a) In the top-entry stage the specimen is mounted at the tip of a cone-shaped cartridge, and the cartridge is dropped down into the main part of the stage, which rests on the top of the objective lens. The specimen is thus lowered through the bore of the upper pole piece, to a position that is within the gap between the upper and lower pole pieces. (b) In the side-entry stage the specimen is mounted at the tip of a long rod, and the rod is inserted through an airlock on the side of the microscope column, so that the specimen is placed midway between the upper and lower pole pieces of the objective lens.

Specimen tilting in the top-entry stage, on the other hand, can be accomplished in two different ways, neither of which is eucentric. The first, "poor man's" solution is to fabricate the tip of the top-entry cartridge so that the specimen grid is mounted at a predetermined, fixed angle of tilt. The second possibility is to build a small, rotating piece at the tip of the specimen cartridge, with appropriate rods and gears to move this piece, thereby providing a way to change the tilt angle while the specimen is in the stage. Top-entry stages of the latter design have the disadvantage that the additional movable parts introduce greater potential for vibration and drift.

Specimen cooling is an essential feature for high-resolution electron crystallography because of the fivefold or better increase in electron dose that can be tolerated by radiation-sensitive biological materials at low temperature (section 4.7). *Low-temperature stages* are also very important in biological electron crystallography because they make it possible to use specimens in which the native, hydrated structure has been "trapped" by freezing a thin, wet sample. Samples can even be frozen so rapidly that the surrounding water solidifies in the vitreous state (see section 6.6). If the stage is then kept cold enough, the ice is prevented from subliming in the vacuum, and the *vitreous* state remains stable for an indefinite length of time, only converting to a crystalline phase if the stage is warmed to a higher temperature. While a temperature of 170 K is sufficient to prevent sublimation, a temperature of 110 K is needed in order to preserve the vitrified state of the ice.

Condensation of ice onto the specimen from residual water vapor in the vacuum system is not generally observed at the higher temperature (170 K). At the lower temperature (110 K), however, the condensation of thick layers of (vitreous) ice occurs very rapidly unless special precautions are taken to keep the partial pressure of water vapor well below 10^{-8} torr. It is necessary, therefore, to use an extremely efficient cold trap, usually referred to as the *anticontaminator*, which subtends a large solid angle around the specimen and acts as a cryopump. The anticontaminator design often consists of a pair of blades above and below the specimen. The temperature of the blades should be below 110 K to prevent sublimation of the water that has been adsorbed onto the cold surfaces. Each blade has a small aperture, typically about 0.5 mm in diameter, which permits passage of the electron beam. The separation between the two blades, one above the specimen rod and one below it, is made small in order to improve the trapping efficiency, but not so small that it limits the extent to which the specimen rod can be tilted. The requirement to allow high tilt angles already puts a constraint on how small the pole-piece gap can be in the objective lens, as we mentioned previously in section 5.3, and the further requirement of having two anticontaminator blades in the gap causes a further limitation. The fact that a larger pole-piece gap must be used to focus high-voltage electrons makes it easier to satisfy the competing design criteria of having high tilt with an efficient anticontaminator while retaining high instrumental resolution.

Vibration and drift are generally more serious problems with a low-temperature stage than they are for room-temperature stages. As a result, the first high-resolution images of 2-D crystals of biological macromolecules were all obtained with top-entry cold stages. Top-entry cold stages developed in Munich (Dietrich et al., 1977), Berkeley (Hayward and Glaeser, 1980), and Kyoto (Fujiyoshi et al., 1991) were used to obtain high-resolution images for the structure analysis of bacteriorhodopsin (Henderson et al., 1990) and the photosynthetic light-harvesting complex (Kuhlbrandt et al., 1994).

Subsequently, however, commercial side-entry cold stages have overcome the factors that previously prevented them from achieving the full resolution of the electron microscope. The principal limitations seem to have been: (1) vibration — caused by the boiling of liquid nitrogen within the dewar — which is transmitted to the specimen rod, and (2) the increased sensitivity of the cold stage to acoustic noise. Liquid nitrogen boiling has been eliminated by careful attention to heat losses in the design of the dewar, so that evaporation at the surface of the liquid suffices to provide the needed cooling (Henderson et al., 1991). Acoustic sensitivity can be reduced by a simple sound-isolation device, which can be mounted as an enclosure around the specimen rod and dewar after the specimen has been inserted into the microscope.

The latest innovation in cold-stage design incorporates two concentric levels at which the stage is cooled. These designs were originally meant to allow cooling the inner part, where the specimen is located, with liquid helium. The purpose of the outer level of cooling is simply to serve as a heat shield for the inner, liquid-helium-cooled part of the stage. Both the inner and outer levels can be cooled with liquid nitrogen, however, if that is desired. This latest generation of stage design also uses a detachable specimen holder, similar in size and shape to the tip of a side-entry stage, that is placed into the cold stage by a side-entry rod. This design provides all of the advantages of a top-entry stage while also retaining the ability for eucentric tilting of the specimen.

A number of sources of mechanical vibration can also limit the image resolution, and these are much more likely to cause problems with a cold stage than with a conventional, room-temperature stage. Normal room vibrations were historically a big concern in electron microscopy, but these are generally not a problem if the site meets the manufacturer's specifications for installations. Vibrations associated with improper flow rate in the cooling water supply, for diffusion pumps as well as for the lenses, is often found to be responsible for degradation of the image resolution. In addition, acoustic noise, from sources as diverse as the ventilation system, mechanical pumps, or the human voice, easily degrades the image resolution.

Small thermal fluctuations in the environment are another factor that cause variable and erratic image drift. Even subtle variations in the temperature of the air circulating within the room will cause tiny thermal effects in the microscope column, which nevertheless are large enough to cause nanometer-sized drifts in the image. The importance of such an effect can be tested by holding one's hand an inch or so from the middle of the projector lens system. It is not uncommon for such a test to show that radiant body heat is sufficient to produce an observable drift when the image is viewed with a TV camera at high magnification.

It is appropriate to add here that *environmental electromagnetic fields* can also seriously affect the microscope's performance. Sources such as power-supply transformers, clock motors, computers, and even solenoids in the microscope itself have all been found to introduce beam deflections, generally at the 50–60 Hz line frequency but with strong harmonics. Ground loops in the microscope and its accessories can add an AC current directly to the deflection coils. The extent to which stray fields are able to deflect the image, and thus limit the image resolution, has been found to vary with the image magnification. This fact suggests that external flux penetration is variable along the electron optical column, and image deflection is greatest for magnifications at which the focal plane of an intermediate projector lens coincides with high values of the flux penetration. Deflection of the incident illumination is especially easy to see with FEGs, because the beam can be focused to a very small spot. Identifying and eliminating ground loops may thus be a difficult but necessary step in taking advantage of the smallest spot that can be obtained with an FEG.

5.5 Detectors that are suitable for observing and recording images and diffraction patterns

An electron microscope that is to be used for electron crystallography should be equipped with three, and possibly even four different *detector systems*. (1) Preliminary alignment and viewing of the specimen is done most conveniently with the standard fluorescent screen. (2) A low-light-level TV camera is then an essential tool for later stages of searching for suitable specimens, and possibly for peeking at their electron diffraction patterns, as will be described in section 5.6. (3) A photographic film camera is normally used to actually record images and diffraction patterns. (4) More advanced technologies for direct electronic readout of data represent a useful alternative to film for final data capture, however, at least for some applications. These detectors currently employ the solid-state CCD camera (Brink and Chiu, 1994; Sherman et al., 1996; Downing and Hendrickson, 1999; Faruqi et al., 1999), but other technologies are also being developed.

Four characteristics are of fundamental importance in determining the usefulness of a particular detector or recording medium. The size (area) of the detector, expressed in terms of the number of pixels that can be usefully resolved one from another, is an important consideration in almost all applications. It is rarely the case that a detector made with as few as 500 pixels on edge will be satisfactory for anything except preliminary observation of specimens and their diffraction patterns. For some applications one might want to have as many as 10,000 pixels on edge, and in this case photographic film is still the only practical medium. The other three important characteristics, which will be discussed in separate paragraphs below, are the detective quantum efficiency (DQE), the dynamic range, and the modulation transfer function (MTF) of the detector.

Detective quantum efficiency

A perfect detector, among other things, should be able to detect every electron that hits it, and it should give zero false counts, as well. A real detector can only approach this ideal to some extent. A simple, quantitative measure of how well a detector performs in this regard is given by the *detective quantum efficiency*, DQE, defined as:

$$DQE = \frac{[\text{signal-to-noise ratio in the readout}]^2}{[\text{signal-to-noise ratio in the input}]^2}. \tag{5.7}$$

In the case where noise from the detector can be ignored, the signal-to-noise ratio in the input will be equal to $\sqrt{N}$, the square root of the number of incident electrons at each pixel. If only a fraction, f, of these electrons produce a detectable signal, then the signal-to-noise ratio of the readout will be equal to $\sqrt{fN}$. In this case the DQE is equal to f, as would be expected.

One cannot always ignore detector noise in the readout, as above, especially when the number of electrons per pixel is small. When detector noise is included in the readout, the overall signal-to-noise ratio decreases to

$$S/N_{readout} = \frac{fN}{\sqrt{fN+n}}, \tag{5.8}$$

where n represents the mean square value of the detector noise, expressed in units of the average readout amplitude produced by an incident electron. As can be seen by substituting equation 5.8 into equation 5.7, the DQE will be only $0.5f$ when the level of additive noise is equal to the number of detected electrons. As is clear from equation 5.8, the DQE decreases even further when the actual signal from the detected electrons is smaller than that from the additive noise.

A number of other factors enter into a more complete determination of the DQE, as discussed by Ishizuka (1993). As an example, direct measurement of the DQE is complicated by the finite point spread function of the detector. Each pixel of the detector receives input from a number of adjacent pixels, while losing some of its signal to the adjacent pixels. If a uniform field of illumination is used as the input, this mixing of signal between pixels decreases the observed variance of the readout below N, the value that would be observed in the absence of mixing. Variations in the readout signal that is produced by each electron, often characterized in terms of the width of the *pulse-height spectrum* for a set of individual events, is another source of noise that adversely affects the DQE (DeRuijter, 1995). This effect may be quite substantial when CCDs are used as detectors for high-voltage electrons.

The DQE does not have a constant value that is dependent only on the properties of a given detector. Instead, the DQE also depends upon the exposure level, as is shown in figure 5.6. At very low exposure (not shown in the figure) the DQE is low because

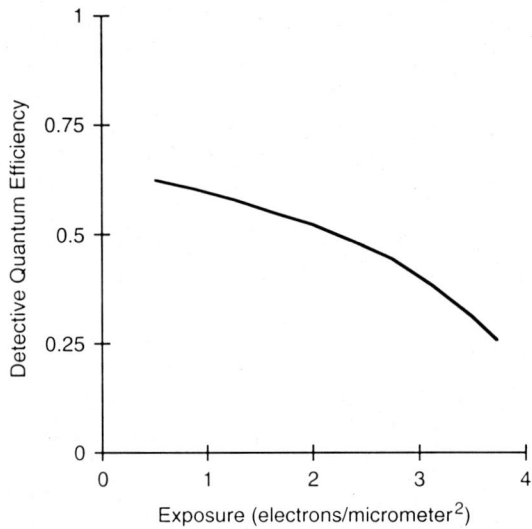

Figure 5.6 The detective quantum efficiency, DQE, of a detector depends upon the mean density, or exposure. At very low exposure, not shown in this schematic graph, the detector noise in equation 5.8 may become large compared to the signal, leading to a low DQE. At higher exposures, the detector noise becomes insignificant relative to the signal, and there is no significant noise other than the shot noise inherent in the image itself, that is, in the input. The DQE may in this case correspond to what one normally means by "the fraction of electrons that produce a detectable signal." At much higher exposures, however, the detector begins to saturate; figure 9.2 gives more details about the saturation of photographic film, for example. Additional electrons cannot contribute any further to the signal at that point. The result of saturation thus is to reduce the fraction, f, of electrons detected, and the DQE again falls to low levels. The curve shown here is redrawn from the data for detection of electrons on photographic film reported by Hamilton and Marchant (1967).

the additive noise generated by the detector is larger than the signal. At high exposures, saturation of the detector causes the fraction of detected electrons to decrease, thereby causing the DQE to again decrease. Between these two extremes there is a range of exposures for which the DQE attains some maximum value, which typically lies in the range from at least 0.2 to as high as 0.8, depending upon the specific detector technology and the electron accelerating voltage.

Few measurements have been presented for the DQE of photographic film exposed to electrons. Figure 5.6, shows a maximum DQE of ~0.6 for Kodak Projector Slide Plates exposed to 100 keV electrons, and the DQE for Kodak SO163 film (currently the standard photographic medium used for electron microscopy) is only slightly better than that, perhaps about 0.7. The maximum DQE for an optimized CCD camera system can be as high as 0.8 to 0.9 for 100 kV electrons, but it is substantially lower than that at higher voltages (DeRuijter, 1995; Daberkow et al., 1996).

A low value of the DQE will have adverse effects on images and on electron diffraction patterns. In the first instance, the effect of a low DQE will be to decrease the amount of signal that one can record, for a given amount of damage done to the sample. The idea is simple: every electron that passes through the sample is equally likely to cause damage to the specimen. Thus, the more electrons that can be registered in the detector, the better will be the signal, for a given amount of damage to the sample. Low-dose images have almost no contrast, so the DQE will be approximately constant everywhere in the image. Electron diffraction patterns, however, consist of diffraction spots with extreme differences in intensity. Especially when using photographic film as the detector, some diffraction spots may reach an exposure level well into the region of saturation, while others may be very close to the fog level of the film. In this case, the DQE will vary in a complex way from one diffraction spot to another.

Dynamic range

The *dynamic range* of a detector is defined as the number of separate exposure levels, or so-called grey-level steps, that can be discriminated before reaching the saturation level of the detector. The amount of additive noise in the readout determines the smallest step-size into which the readout can sensibly be divided, and, together with the saturation level, it determines the dynamic range. As an illustration, if the maximum (useful) blackening of a photographic film corresponds to an optical density (OD) of 3.0, and if the fog level has an r.m.s. OD of 0.03, then the dynamic range would be 100 steps, each of magnitude 0.03. In general, photographic films will have a dynamic range of about 100. A cooled CCD camera (see below), by comparison, can have a dynamic range of 10^4 or more.

Modulation transfer function

Another fundamental characteristic of a detector that must be discussed is its *modulation transfer function* (MTF), which describes the falloff of signal strength in the readout relative to that of the input signal, as a function of spatial frequency. From the discussion of MTFs that was given earlier in section 3.5, it will be clear that the MTF is the modulus of the Fourier transform of the point-spread function (also known as the impulse response) of the detector. Thus, if the point-spread function of the detector is rather broad — and the resolution of the detector is correspondingly rather coarse — then the MTF will approach zero at relatively low frequencies. On the other hand, if the point-spread function is very narrow, and the resolution of the detector is correspondingly high, then the MTF will remain close to unity out to very high frequency.

What is ultimately important about the MTF of a detector is the *number of pixels* that can be used, while still maintaining a high value for the MTF at a spatial frequency equal to the reciprocal of twice the pixel size. If the detector is divided into pixels that are smaller than the point spread function, then the MTF at the *Nyquist limit* — that is, the highest spatial frequency provided by the sampling — will be too small to be of any useful value. This is the case with many CCD camera designs because the point-spread function of the scintillator is much larger than the size of the CCD pixels. On the other hand, if the detector is partitioned into pixels that are much larger than the size of the point-spread function of the scintillator, then the full resolution that could be provided by the detector will not be properly utilized. Roughly speaking, the optimal pixel size of the detector is the same as the width of the point-spread function of the scintillator.

The MTFs of different photographic films were measured for exposure to 100 kV electrons by Downing and Grano et al. (1982). The point at which the MTF falls to a value of 0.5 was found to vary for different types of emulsion. Kodak SO163 has an MTF that falls to a value of 0.5 at a spatial frequency of 0.04 μm^{-1}, while Agfa Scientia 23D-56 has an MTF that extends to 0.2 μm^{-1} before falling to a value of 0.5. A more recent measurement of Kodak SO163 film, based on the spectrum of electron shot noise that was digitized with a precision densitometer, indicates that its MTF currently remains above 0.5 out to a spatial frequency of 0.1 μm^{-1} (Typke et al., 2005). The thickness of the emulsion, the density of silver halide grains within the emulsion, and the grain size itself are all significant factors in determining the MTF of a given emulsion. Since angular straggling of the electrons as they penetrate into the emulsion may also be an important factor in broadening the point-spread function, one can expect the MTF of a given film to be better for 300 kV electrons than it is for 100 kV electrons.

Direct electronic readout

There are many attractive advantages in recording data with some type of direct electronic readout device. As a result, sophisticated "slow-scan" (low-noise) versions of digitizing cameras have replaced photographic film as the detector of choice, at least for some applications. CCD cameras containing 2000 × 2000 pixels (i.e., a 2k × 2k camera) are commercially available at moderate cost, and 4k × 4k cameras are also commercially available, albeit at considerably greater cost. By cooling the CCD chip and by limiting the readout speed (which decreases the output bandwidth), slow-scan CCD cameras can be made to meet or exceed the DQE of photographic film for 100 keV electrons. CCD cameras have the further advantage that they have a much greater dynamic range than film. As a result of the large dynamic range, CCD cameras do not suffer a loss in DQE at high exposures as easily as is the case for photographic film. The practical number of pixels that are available in a CCD camera is very much smaller than the number of pixels on a sheet of film, however, due to the limited size of such electronic "chips" and the large point-spread function of the scintillator (even at 100 keV). The relatively small number of pixels delivered by CCD cameras thus continues to favor film over the CCD for many applications, especially in higher voltage microscopes.

5.6 Low-dose techniques make it possible to record high-resolution images and diffraction patterns even from easily damaged specimens

Organic materials are so sensitive to being damaged by the electron beam that it is essential to limit very strictly the amount of exposure that the specimen receives before its image or diffraction pattern is recorded. With appropriate protocols, it is straightforward to keep the pre-exposure of the specimen below a small fraction of the exposure that is used to record the data. These *low-dose protocols* differ significantly from the procedures that most electron microscopists are accustomed to use, however. Among other factors, it is essential to use an appropriately designed TV system rather than observing the viewing screen by eye when scanning the grid and when peeking at electron diffraction patterns.

Low-dose protocols generally make use of at least three distinct settings of the microscope controls that allow one to (1) search for promising specimen areas, (2) focus on an area adjacent to the chosen area of interest, and (3) record the image. With computer control it is easy to establish the required, preset combinations of illumination size and illumination intensity, magnification, and choice of diffraction or image mode.

The first step in any protocol, when using randomly distributed 2-D crystals, is to search the specimen for areas of interest, using a very low dose rate. Critical exposures for proteins are of the order of 100 to 1000 electrons/nm^2, as we have explained in section 4.7. The exposure rate used during the search for promising areas should therefore be less than about 10 electrons/nm^2-minute. In order to see any specimen features at this low beam intensity, one must work at very low magnification, using an imaging mode that generates as much contrast as possible. These two criteria are usually met by switching the microscope to the diffraction mode, but at the same time overfocusing the projector lens system rather strongly. The central beam of the diffraction pattern is thereby spread out into a low-magnification, bright-field image, which has remarkably good contrast. When a suitable area of the specimen is found in this mode, it is moved to a marked spot on the TV monitor, which in turn corresponds to a point defined previously for use in alignment of the illuminating beam.

With sufficiently large crystalline specimens, covering an area of a few square micrometers, it is possible at this point to peek at the electron diffraction pattern to ensure that the quality of the selected area is adequate for recording high-resolution images. Using the appropriate, preset conditions, the size of the illuminated area is reduced (typically to ~3 μm in diameter), and the illumination is centered on the position of the chosen crystal. These preset conditions should also focus the diffraction pattern, keeping in mind that the desired diffraction-lens (intermediate lens) focus will depend upon the condenser lens settings that are used to form the spot of illumination. With a moderately high exposure rate (approximately 50 electrons/nm^2-second) but a very short exposure (a fraction of a second), the diffraction pattern can be viewed on the monitor of the TV camera, or, even better, it can be captured with a CCD camera. The ability to peek at the electron diffraction pattern, with large enough crystals, can greatly improve the efficiency of getting excellent, high-resolution images, especially when the specimen shows a great deal of variation in either the flatness of individual crystals or how well they are preserved.

Focusing of the objective lens prior to recording an image must be carried out at a magnification that is at least as high as that which will be used to record the image. To do this the incident beam must first be deflected to one side of the area that will be used to record the image. The deflection must be great enough to avoid exposing the area of interest, but not so far from the area of interest that the specimen height (i.e., the defocus) is likely to be very different in the two locations. In those cases when the specimen has been tilted, the beam should be deflected along the tilt axis to avoid a change in focus.

After choosing an area of the specimen and adjusting the objective lens focus, the condenser lens system is changed once again, this time to give a beam with the size and intensity appropriate for recording images. The illumination conditions will be very different, of course, if one wants to expose the entire photographic film at once, or if one intends to record images with so-called spot-scan illumination (section 5.7). Illuminating an area equal in size to the photographic film or larger, which is the usual imaging mode, is referred to as using *flood-beam illumination*. In this case the intensity should be adjusted to a value that will deliver the desired exposure in about one second. If *spot-scan imaging* is to be used, as will be described in section 5.7, it is necessary to use a much higher beam intensity because each area of the film is exposed serially. When increasing the beam intensity for spot-scan imaging, however, it is absolutely necessary to remember that the illumination angle must be kept smaller than about 10^{-4} radians in order to maintain the degree of spatial coherence required for high resolution, as has been explained in section 3.9.

The magnification that is used for recording high-resolution images on Kodak SO163 film should be about 40,000 to 60,000 times, and even lower magnifications can be used for low-resolution work. There are two opposing considerations that need to be satisfied when choosing the best possible magnification for recording the image. A high magnification is preferred in order to ensure that features of the desired resolution are larger than the point-spread function of the film, that is, to ensure a high value of the modulation transfer function (section 5.5) at the desired resolution. A magnification of 60,000 makes it possible to record a spatial periodicity of 0.3 nm with an MTF of ~0.7 on SO163 film, provided that a precision densitometer is used to digitize the film (Typke et al., 2005). A low magnification is preferred, on the other hand, in order to increase the electron intensity at the film. In this case it is the DQE that is at stake. As mentioned in section 5.5, the DQE of film falls steeply as the blackening decreases, such that the fog becomes a significant part of the measured optical density. For a film such as Kodak SO163, developed under conditions to give a speed of 2, and with an exposure of 1000 electrons/nm^2 at the specimen, the density of the film would be 0.67, which is well above the fog.

The total exposure that will produce the best signal-to-noise ratio in a low-dose image can be estimated theoretically, at least within the limits of a number of approximations. One can assume, for example, that the diffraction intensities fade exponentially, in which case the electron exposure that causes the intensity of a given spot to fade to e^{-1} of its initial value is commonly referred to as the *critical dose*, N_{crit}, for that diffraction spot. The signal in the Fourier transform of the image initially increases faster than the noise during the early part of the exposure, when the diffraction intensity has not yet faded substantially. It is perhaps surprising to know, however, that the signal-to-noise ratio continues to improve until the exposure is about 2.3 times the critical exposure (Hayward and Glaeser, 1979). Thus, if N_{crit} for 100 kV electrons is 500 electrons/nm², a value typical of high-resolution reflections at low temperature, it is worthwhile to record images with exposures as high as 1200 electrons/nm², while exposures about twice this value can be used at higher voltages, for example 300 or 400 kV. Practical advice is given in box 5.2, which also describes methods that are suitable to measure the beam intensity.

BOX 5.2 The normal exposure meter can be used for quantitative estimation of the electron beam intensity

Microscopes are always equipped with a system for measuring the beam intensity, to ensure that the images recorded on photographic film are exposed correctly. These systems measure the current of electrons arriving over the whole viewing screen, or, in some cases, over a small sub-area of the screen. In some electron microscopes the image intensity is calibrated in appropriate units of current per unit area, but in other instruments the image intensity is represented only on a relative scale.

In the absence of a built-in calibration, the relative electron intensity scale can be calibrated by the user. This is done on the basis of the known film speed, that is, how dark the film gets for a given exposure reading. Kodak SO163 film, developed in D19 for 12 minutes at a temperature of 20°C has a speed of 2 for 100 kV electrons. This means that the film density is OD = 2 for an exposure of 1 electron/μm², that is, a 1 second exposure at a current density of 16 picoampere/cm².

For conventional (i.e., flood beam) low-dose imaging, one can use the (calibrated) beam intensity system in the same way that it would be used under normal imaging conditions. Low-dose exposures are taken by simply using a proportionately shorter exposure than would be used if radiation damage were not a concern. For spot-scan images, on the other hand, the illumination spot will generally not fill the detector area. In this case one should find that the calibrated intensity measurement does not change as the illumination is spread, as long as the illumination stays within the detector area. Because of this one only needs to scale the measured current per unit area by the ratio of the detector area to the area of the focused illumination spot.

If the current in the illumination spot is too low to be measured accurately with the normal intensity monitor, a larger condenser aperture may be inserted for the measurement, and the result can be scaled by the ratio of the areas of the two aperture sizes. As long as the second condenser is sufficiently overfocused,

(continued)

BOX 5.2 *(continued)*

changing the condenser aperture will change only the beam size, but not the current density.

An additional, supplementary calibration that is valuable to make consists of measuring the exposure-dependent fading of electron diffraction from a beam-sensitive specimen, and expressing the exposure in units of the beam-intensity monitor that is provided as part of the electron microscope. In the end, it is only the amount of electron exposure that the sample can tolerate that is important, and knowing its value in electrons/nm^2 is of secondary importance. Preparing relatively "crude" specimens is all that is required for such a supplemental measurement. Examples that can be recommended are thin crystals of *l*-valine, which can be prepared on formvar films by blotting a half-saturated, aqueous solution of the amino acid (as if preparing a negatively stained specimen) (Glaeser, 1971), or thin crystals of $C_{44}H_{90}$ (paraffin wax), which can be prepared on carbon films by evaporating about two microliters of one-third-saturated solution in hexane (Henderson and Glaeser, 1985). Both of these specimens are about as sensitive to radiation damage as are the high-resolution features of a protein.

To record electron diffraction patterns at low electron exposures, it is important that the beam should be adjusted to a size roughly equal to, or even smaller than, the size of the crystal. The goal here is to limit the amount of unnecessary background scattering that is generated from areas where there is no crystal. A relatively small second condenser aperture and an appropriate combination of the first and second condenser lens excitation must therefore be chosen to provide a beam of the correct size and intensity. The selected area diffraction (SAD) aperture can also be used to limit the area from which the diffraction pattern is taken, as is frequently done in materials science. However, with crystal patches that are only a few micrometers in diameter, the beam size itself can be easily set to limit the area. This approach avoids the need to accurately focus the image of the specimen onto the SAD aperture plane, so as to obtain accurate diffraction intensities. The use of a small condenser aperture (10 to 50 μm in diameter), and a second condenser lens excitation that spreads the beam to no more than a few micrometers in diameter, will allow very sharp focusing of the diffraction spots.

5.7 Spot-scan imaging can minimize beam-induced movement

One of the challenges that still faces electron crystallographers is the fact that the contrast recorded in images at high spatial frequencies (e.g., 2.5 nm^{-1}) rarely matches what one would expect, based upon the measured intensities of spots in the electron diffraction pattern of the same specimen. The signal in the Fourier transform of images falls off with increasing resolution much faster than do the amplitudes of the corresponding electron diffraction spots. The envelope function of the electron microscope (section 3.9) and the MTF of the detector (section 5.5) account for some of this falloff, and these have been well characterized. It is clear, however, that there must still be other defects in high-resolution images of radiation-sensitive specimens that are not completely understood.

Beam-induced specimen motion is believed to be one effect that causes image contrast to fall rapidly with increasing resolution (Henderson and Glaeser, 1985). As the specimen collects sites of radiation damage during the course of recording an image, rearrangement of interatomic bonds and nonbonded contacts must occur at each such site. In addition, volatile molecular fragments such as CO_2, NH_3, H_2, CH_4 and other species can escape from the sample. The resulting mass loss will lead, in turn, to shrinkage-related stress within the sample. One can easily imagine that these different effects produce significant fluctuations in the magnitude and the direction of stresses within the specimen that, in turn, lead to random, small movements (strain) of the specimen.

Spot-scan imaging is a technique that was developed to minimize this type of beam-induced movement (Downing and Glaeser, 1986; Bullough and Henderson, 1987). A spot-scan image is recorded by first focusing the incident beam to a small spot of illumination, of the order of 100 nm in diameter, which is then scanned over the specimen. The goal of spot-scan illumination is to reduce the area of the specimen that is irradiated, and thereby to reduce the overall stress on the specimen. In addition, the idea is that motion of the small area that is irradiated should be better constrained because of the smaller distance to the surrounding area that is not being irradiated. Indeed, the experimental results with spot-scan imaging show significantly higher contrast at high resolution, attributable to reduction of beam-induced motion. In addition, spot-scan imaging has several other advantages, particularly with tilted specimens, which will be discussed below.

There is as yet little experimental data to suggest how small the beam spot should be in order to obtain the greatest benefit from spot-scan imaging. It is fairly clear that any reduction in beam size brings some benefit, and significant image improvements have been shown with a beam diameter around 200 nm (Brink et al., 1992). As the beam size is decreased, however, one must maintain the high degree of spatial coherence that is required for phase contrast imaging, and this means that the largest semi-angle of illumination should be limited to no more than 10^{-4} radian. The diffraction-limited spot size with this illumination angle will be about 40 nm for 100 kV electrons, and about 16 nm for 400 kV electrons. These values represent a practical limit for the minimum size of the focused illumination that can be used in high-resolution spot-scan imaging.

In order to reduce the beam diameter for spot-scan imaging, it is necessary to sacrifice either intensity or coherence. With a conventional thermionic electron gun, maintaining coherence at the expense of intensity may result in such a weak beam that exposure times become unacceptably long. A LaB_6 source therefore gives a significant advantage for spot-scan imaging, and a FEG is even better. For reference, a LaB_6 source operating at 400 kV can provide adequate coherence in a beam of around 30–50 nm diameter while still delivering sufficient intensity to allow an exposure time of about 50 ms per spot.

It is common practice in conventional imaging to overfocus the second condenser lens to spread the beam and enhance coherence. It is also possible to overfocus the second condenser lens in spot-scan imaging, so that the beam is spread to a size that is somewhat larger than the diffraction limited spot, for example 100 nm rather than 40 nm. The *effective* coherence in the processed image will not be improved by this action, however. Referring to figure 5.7, we see that spatial averaging across

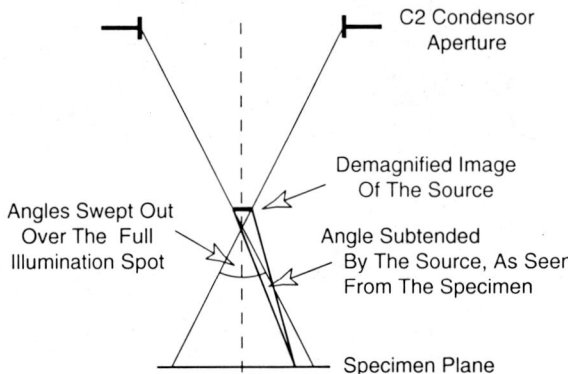

Figure 5.7 The effective spatial coherence of the illumination is not improved by overfocusing the condenser lens when image data are superimposed from the entire spot of illumination. When the condenser lens is overfocused, the crossover is located above the specimen. As a result, the illumination at any one point in the specimen is much more parallel than it would be if the beam had been focused to crossover at that point, as has been explained previously in figure 5.2. However, the local angle of illumination changes continuously within the illumination spot, eventually covering exactly the same range of angles as are present in the beam crossover. When image data are summed together from all areas of the spot, the amount of beam divergence (i.e., imperfect spatial coherence) is therefore not any better than it would be if the beam were focused at crossover on the specimen.

the illuminated spot sums molecular images from areas of the specimen where the illumination has, in the end, covered the same angular range as it would at crossover. Thus there is no benefit gained from the fact that overfocusing the illumination causes the angular range to be locally smaller at each point in the image.

Spot-scan imaging provides a natural solution to a difficulty that is inherent in recording images of highly tilted specimens. At high tilt angles, the defocus value changes continuously over the field of view. This means that the transfer function oscillates (section 3.8) very rapidly at both ends of the defocus ramp, even though the focus may have been set at an optimal value for the middle of the tilted specimen. As a result, the envelope functions may have quite small values over a large part of the image, achieving their optimal values only in a narrow band along the in-focus region of the image. With spot-scan images, however, the illuminating beam can be scanned in a direction parallel to the tilt axis, so that the focus is constant along each scan line, and computer control of the objective lens can be used to reset the focus between scan lines. The resulting protocol provides a *dynamic focus* correction, which results in images that can have optimal values of the envelope function over the whole field (Downing, 1992).

A computational procedure for dealing with the focus ramp in flood-beam images of tilted specimens is described in section 10.11. This same type of focus-ramp correction may also be necessary even when dynamic focus is used, if the transfer function varies significantly within the small illumination spot. For example, if the spot diameter is made as small as 30 nm, the height variation within the spot is still about 52 nm when the sample has been tilted by 60°. The depth of field at a resolution of 0.35 nm, on the other hand, is only 38 nm for the favorable case of 400 kV electron. Thus the highest resolution data must be corrected for the defocus ramp that exists over a single small, illuminated area. With a tilt of 45°, however, no correction would be required within an illuminated spot with a diameter of 30 nm, until the data extended to a resolution better than 0.31 nm.

5.8 Samples prepared as self-supported specimens within (or over) holes require additional precautions in order to minimize specimen charging

Specimen charging occurs whenever a poorly conducting sample is irradiated by the electron beam in an electron microscope. Inelastic scattering of the primary beam causes ionization of the specimen, and some fraction of the secondary electrons that result from this ionization inevitably escape from the sample. Since the primary electrons themselves are too energetic to be stopped within the specimen, the loss of these secondary electrons results in the accumulation of a net positive charge within any poorly conducting area of the specimen. As is known from studies of contrast-generation in the scanning electron microscope, however, the escape of secondary electrons is limited to a depth of only a few nanometers from the surface, i.e. the so-called escape depth for the low-energy secondary electrons. Although specimen charging is greatly reduced when nonconducting samples are prepared on a continuous carbon film, residual evidence of charging — probably present as an electrical dipole layer — can be observed even in this case (Glaeser and Downing, 2004).

The majority of biological samples that are embedded in vitreous ice (section 6.6) are prepared as self-supported specimens within the holes of a carbon film, and samples that are adsorbed to lipid monolayers (section 8.4) are also best transferred to carbon films with holes. While the carbon support film itself is a relatively good electrical conductor, the sample material within (or over) a hole is not. The insulating sample material (and the vitreous ice embedment) will therefore build up a positive electrical charge during the course of irradiation.

Specimen charging is known to degrade the quality of high-resolution images quite severely. Beam-induced specimen movement, image astigmatism, and quite possibly image deflection and other electron optical disturbances are among the obvious consequences of specimen charging that must be minimized in order to obtain the best possible images. Large-scale disturbances of this type are readily observed on relatively large areas of unsupported specimens, such as plastic sections that are picked up on bare grids (i.e., without an underlying carbon film). The most severe manifestation of such charging effects can often be eliminated by inserting an objective aperture. The explanation seems to be that scattered electrons hit the metallic aperture and produce low-energy secondary electrons, which in turn are attracted to and thereby neutralize the positive charge on the sample. The severe manifestation of specimen charging effects can also be eliminated by simply coating the self-supported specimen with a thin layer of carbon.

One of the simplest solutions to control specimen charging is thus to evaporate a thin film of carbon onto the self-supported specimen (Jakubowski et al., 1989;

Brink et al., 1998a). Although coating with evaporated carbon can even be done with frozen-hydrated specimens (using an airlock to insert the frozen sample into a vacuum evaporator), there is a concern that the phase-grain of the carbon film can obscure the low-contrast image of small particles in cryo-EM specimens. The use of an evaporated carbon film has therefore not generally been adopted as a solution for specimen charging. It should be pointed out that it is not possible to use spot-scan illumination (section 5.7) with self-supported specimens unless such a conducting layer has been deposited onto the specimen, because of the extreme degradation of image quality that is caused by specimen charging (Brink et al., 1998a).

An alternative method for reducing specimen charging with self-supported specimens requires only that the illuminating beam should touch the carbon edge of the hole within which the sample is supported (Brink et al., 1998b; Miyazawa et al., 1999). It is preferable that the entire rim of the hole should be illuminated as uniformly as possible, of course, in order to avoid introducing electrostatic asymmetry into the system. Other precautions that are recommended are (1) the use of an absolutely clean, gold aperture in the objective lens, intended to increase the production of secondary electrons by the aperture foil, and (2) pre-irradiation of the holey carbon film before it is used for specimen preparation, the goal being to improve the electrical conductivity of the carbon film. This latter precaution is thought to be especially important when samples are viewed at helium temperature, since the electrical conductivity of the carbon film decreases very strongly at very low temperature.

6

Specimen Preparation

6.1 Introduction

The first step in preparing a sample for electron microscopy (or electron diffraction) is to mount it in an appropriate way, so that it can be put onto the specimen stage in the objective lens. The majority of specimen preparation procedures involve the use of thin metal grids, normally 300 mesh (300 holes per inch) or 400 mesh copper (molybdenum, titanium) grids, which are used to support some type of continuous, thin film (Bozzola and Russell, 1999; Hayat, 2000). Evaporated carbon films, 5 to 30 nm in thickness, are usually the most appropriate support film. Thin polymer films, which in turn should be stabilized by an additional layer of evaporated carbon, are also useful for many routine purposes.

The extremely high scattering cross-section of electrons makes the reduction of background scattering an important factor in the overall plan for specimen preparation. The substrate used to hold the specimen must be thin in order to limit the amount of background structure in the image. Because the biological specimen itself may be only 5 to 10 nm thick, it is easy to see that a support film that is as much as 20 to 30 nm thick will contribute a significant amount of background.

In many cases it is advantageous to use a support film with numerous holes and to trap specimens in a film of suitable embedding material that spans the holes like a thin membrane. Techniques have been developed to make "holey" support films for this purpose, in which the hole size can be varied from 100 nm to 10 μm (Murray and Ward, 1987a; Toyoshima, 1989; Jahn, 1995). The most recent innovation has been the application of microfabrication technology to produce support films with almost any desired size and distribution of holes (Ermantraut et al., 1998), a considerable variety of which are commercially available. The advantages of using holey support films include, and often go beyond, the obvious idea that background scattering should be minimized. The reasons for using holey support films will become apparent in later chapters, when specific types of objects (and associated methods of specimen preparation) are discussed in more detail.

This chapter begins with an overview of the principal methods of specimen preparation that are used in electron crystallography. The purpose here is to give background information about the ways in which these different specimen preparation techniques can influence the image contrast; how they can achieve preservation of the native, hydrated structures; and how they affect the stability of the sample in the electron beam. Negative staining (section 6.2) often proves to be an ideal technique for early stages of a structural investigation, as is true also for metal shadowing (section 6.3). Embedding in glucose (and in several other, easily hydrated solutes) is an alternative to negative staining that can preserve the native, hydrated structure (section 6.4). Contrast matching is an unavoidable — and usually unwanted — fact of life with glucose embedding, however. The conceptual and mathematical basis needed to understand contrast matching is a large enough topic that it is discussed in a section (6.5) of its own. The use of frozen-hydrated specimens (section 6.6), while technically less convenient than glucose embedding, is favored in many applications because specimens do have significant contrast in ice. In addition, freezing in vitreous ice is arguably the best of all methods to preserve the native, hydrated structure of biological specimens. The rapid quenching needed to vitrify the sample can also be readily adapted to allow study of time-resolved, kinetic intermediates.

Three factors need to be considered when preparing the specimen, especially for high resolution studies. The first two, electrostatic charging of the specimen in the electron beam and the mechanical stability of the specimen in the beam, are discussed in section 6.7. The third topic, covered in section 6.8, deals with the need to have specimens that are free of bending or wrinkling when 2-D crystals are used to collect images and diffraction patterns at high tilt angles. As the reader will see, the final word is not yet said in any of these three areas. The issues are defined clearly enough, and progress has been made in designing steps to address each problem, but considerable scope still remains for further improvement in specimen preparation methods.

6.2 Negative staining provides high contrast as well as excellent stability in the electron beam

The technique of *negative staining* involves embedding a specimen within a thin matrix of a salt such as uranyl acetate, neutralized phosphotungstic acid, or ammonium molybdate. Negative staining provides a rapid and simple method of specimen preparation that is of great value for the initial stages of work in electron crystallography. The basic concept of the negative staining procedure is very simple (Hall, 1955; Brenner and Horne, 1959; Huxley, 1963; Harris and Horne, 1994; Massover and Marsh, 1997). The goal is to replace the bulk water in which the specimen is normally "embedded" by a solid matrix of amorphous salt, which itself had previously been dissolved in the water along with the specimen.

At the beginning of a structural study, negative staining provides an easy way to characterize the morphology and overall quality of the specimen. Negatively stained samples can also be used for crystallographic structure analysis at 2 to 3 nm resolution. This sort of low-resolution structure analysis can provide a good estimate of the surface envelope, molecular shape, and even the domain organization of large macromolecules, as well as the packing arrangement of individual subunits within a larger macromolecular assembly.

Samples embedded in stain may, in rare cases, diffract to a resolution as good as 1 nm, but the mass density at this level of resolution is likely to be confusing because of contrast matching or contrast reversal (section 6.5). At a resolution of 1.5 nm the stain is the more strongly scattering (darker) component. At a resolution better than 1.5 nm, however, the protein itself becomes the dominant source of image contrast because the structure of the stain becomes completely disordered at high resolution. The structural interpretation for resolutions between 1 and 2 nm is obviously problematic.

The purpose of using negative stain embedment is threefold. First of all, by using heavy-metal salts, a relatively high amount of scattering contrast can be obtained in the images, making it easy to see the specimen and some of its internal features. Secondly, by replacing bulk water in the available grooves and channels within the structure, the embedment serves to minimize the flattening and collapse of the biological ultrastructure that would otherwise occur when water is removed by air drying. Flattening and collapse cannot be avoided completely (Kistler and Kellenberger, 1977; Kellenberger et al., 1982), but at least some of the worst consequences that would otherwise derive from capillary (surface tension) forces (Anderson, 1952) can be prevented. The third benefit of negative staining is that it provides a kind of "mineral fossil," or inert, three-dimensional replica of the surface of the structure, which is relatively insensitive to the effects of radiation damage.

To be sure, higher resolution features of the structure that may have been preserved in the initial, negatively stained specimen are still sensitive to radiation damage (Glaeser, 1971) even at liquid helium temperature (Glaeser and Hobbs, 1975), and some redistribution of mass can even occur at low resolution (Glaeser, 1971; Unwin, 1974). In the range of 2 to 3 nm resolution, however, the molecular envelope that is seen in negative stain is usually a faithful replica of the original object structure. The excellent quality of structural preservation in negative stain, for most specimens, has been made clear through direct comparisons of frozen-hydrated specimens and negatively stained specimens (Hoenger and Aebi, 1996).

There are many variations in technique that are used to prepare negatively stained samples. We describe only the basic concept here, and refer the reader to a number of references for more detailed information (Kiselev et al., 1990; Harris, 1997; Bozzola and Russell, 1999). The first step involved is to mix the sample with a solution of negative stain, for example a 2% solution of uranyl acetate. One way to do this is to mix 2 or 3 µl of sample with a similar volume of stain solution directly on the electron microscope grid, or it can be done on a small piece of parafilm. After mixing, excess liquid is blotted off of the grid from one edge, leaving behind a wet film that is thin enough to dry by evaporation in a matter of seconds. In most cases it is found that the results obtained are much more satisfactory on grids that have been made hydrophilic, by exposure to a glow discharge ($\sim$150 mtorr air), than are the results obtained with untreated carbon films.

In a more recent variation of the negative staining approach, specimens are prepared with very concentrated solutions of negative stain, such as 16% ammonium molybdate, and these are simply blotted and rapidly frozen just as if one were preparing a frozen-hydrated specimen (section 6.6). Such *cryo-negative stained* samples are normally prepared with holey carbon films, as self-supported specimens (Adrian et al., 1998), but sandwiching the sample between two carbon films has also been used (Golas et al., 2003). The advantage of this hybrid technique is that it provides relatively high contrast at low resolution without the requirement to use a high level of defocus. The preservation of high-resolution structure can be specimen-dependent, and some specimen materials even dissociate under these conditions of sample preparation. The evaluation of image resolution can also be misleading in such specimens, especially at high levels of electron exposure, as the evaluation can be dominated by the preservation of the protein–water interface (as it is in specimens prepared by the conventional methods of negative staining).

As indicated previously, many different heavy-metal salts have been found to be suitable for negative staining. Uranyl acetate is perhaps the most commonly used stain, and it can be recommended as the first one to try on any new specimen. It is quite possible for a specimen to be well preserved in one type of stain but not in another, however. Therefore, when looking at new types of specimens for the first time, it is important to try a few different stains if good results are not obtained with the first stain that is chosen. The reviews cited in the previous paragraph can be consulted for further details.

6.3 Metal shadowing produces stable samples which reveal surface topography

The technique of (unidirectional) *metal shadowing* (Williams and Wyckoff, 1946; Bozzola and Russell, 1999) is used to produce specimens in which the surface topography is revealed as if illuminated from one side by a bright light. Examples that illustrate the contrast effect that is achieved include the image of a bacterial surface-layer protein, shown in figure 6.1(a) and the image of crystalline membrane proteins revealed in the yeast plasma membrane by the freeze-fracture technique, figure 6.1(b).

The three-dimensional, "shadowed" effect is achieved when metal is evaporated from a relatively small source, at some distance from the sample, and when the surface of the sample is inclined at a fairly small angle to the direction of flux of the evaporated metal. As is explained by the schematic in figure 6.2, the evaporated metal piles up on the near side of any bump in the surface, leaving a metal-depleted "shadow" on the opposite side. The reverse occurs at a depression in the surface. The effect is exactly as if a corresponding macroscopic surface, say objects on a table, were illuminated at a low angle by a bright lamp.

The metal-shadowing technique is attractive because of the exceptional contrast and clarity with which it reveals surface topography. The shadowing effect, to begin with, provides unsurpassed contrast of even relatively small undulations in the surface envelope. In addition, only information about the structure of the shadowed *surface* is present in the image, unlike the case of negatively stained specimens in which information about the full, three-dimensional structure is superimposed as a projection. While information about internal structural features is therefore absent in metal-shadowed specimens, the resulting simplification of structural information no doubt contributes

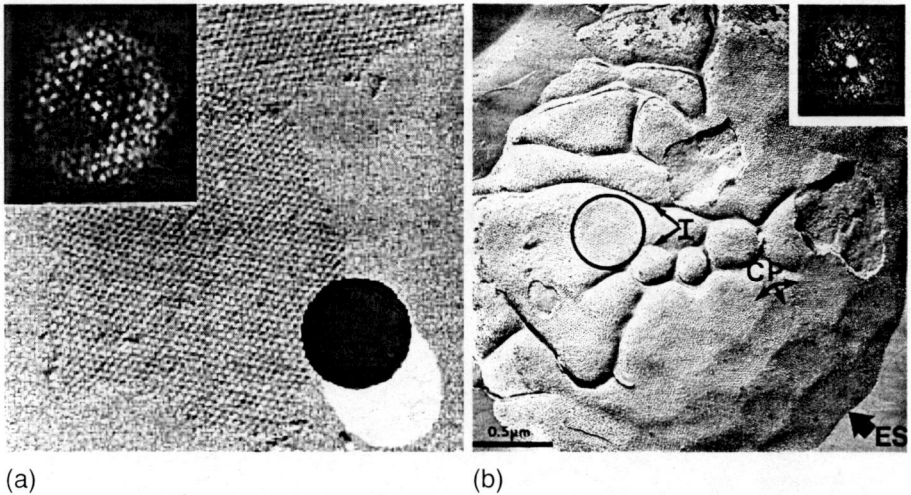

(a) (b)

Figure 6.1 Two examples of shadowed samples. (a) Freeze-dried S-layer of *Methanoplanus limicola* (Cheong et al., 1993). (b) Freeze-fractured "spheroplast" of a yeast cell that had been metabolically starved to induce crystallization of cell-membrane particles (Sosinsky et al., 1986). The metal shadowing technique is very effective in showing the presence of crystalline order in these samples. In freeze-dried samples, the shadowing produces image contrast that is related to the surface topography of the subunits. The freeze-fracture preparation, on the other hand, can reveal internal features that might not naturally occur at the aqueous surface; in the case of freeze-fracture of cell membranes, the lipid bilayer is often peeled apart, revealing integral membrane proteins as freeze-fracture "particles."

to the clarity of the images and the straightforward interpretation which they afford. If desired, the shadowed images can even be converted to quantitative topographs, based upon simple geometric considerations (Smith and Ivanov, 1980; Guckenberger, 1985).

A further advantage of metal shadowing is the fact that it produces specimens that are very stable against radiation damage. The specimen is, in fact, a metallic replica of the macromolecular surface, which had served earlier as a template for the shadowing process. Because of the great stability of the metallic replica, it usually is not necessary to use low-dose imaging conditions to realize the full resolution inherent in the specimen.

It is generally believed that the resolution that can be achieved by metal shadowing is limited by the grain size of metallic microcrystals that form as the evaporated metal condenses onto the surface of the specimen (Woodward and Zasadzinski, 1996). Although considerable effort has been invested in searching for ways to reduce the grain size, the images obtained with the best shadowing materials remain limited at a resolution in the range of 1.5 to 2.5 nm.

Metal shadowing can also result in image contrast whose origin is more complicated than the simple, geometric scheme shown in figure 6.2. The first atoms to land on the molecular surface are often free to diffuse and migrate over the surface. When a preferred binding site is finally found, the first atoms then remain stationary and serve

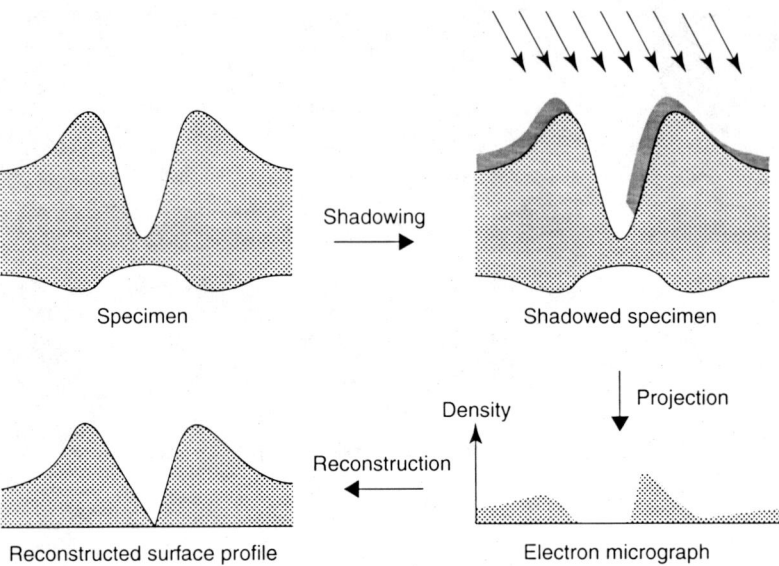

Figure 6.2 Metal shadowing results in the accumulation of evaporated metal on the exposed surface of the specimen. In the simplest case, protrusions with molecular dimensions can act as a physical barrier, when the evaporated metal arrives at an angle to the surface. As a result, the metal accumulates to a greater thickness on one side of the protrusion, and leaves a "shadow" on the other side, where no metal will be deposited. When looked at with the electron microscope, in a direction perpendicular to the surface, the patterns of accumulation and absence of evaporated metal give rise to image contrast that looks similar to the appearance of the same object in a dark room, when illuminated by a lamp at the same angle. This geometrical model of metal shadowing can then be used to reconstruct an estimate of the shape of the surface that produced the observed image contrast. When the amount of evaporated metal is quite low, however, preferred sites for trapping and nucleation can dominate the contrast. Since these sites have more to do with the local chemistry needed to trap the migrating metal atoms, and much less to do with the angle of incidence of the evaporated metal, the metal grains are said in this case to *decorate* the surface in a specific way.

as a site for grain nucleation, as other atoms begin to arrive. The final result is the *decoration of selective sites* in a fashion that is no longer strictly related to surface topography. Although the contrast that is produced by decoration is more difficult to interpret than that which is due to true shadowing, it has been found that decoration can be correlated precisely with known protein structure to atomic resolution. The decoration effect thus has been used to great advantage in determining the symmetry of complex macromolecular assemblies (Weinkauf et al., 1991; Braun et al., 2002).

Numerous other uses have been made of the metal-shadowing technique (and the decoration technique). The preparation of metal-shadowed *replicas* of *freeze-fractured samples* and of *deep-etched samples* plays a significant role in cell biology, and these techniques should not be forgotten in the context of electron crystallography. Additional information can be found in various reviews (Bullivant, 1973; Rash and Hudson, 1979; Bozzola and Russell, 1999). Although restricted to surface information and limited to moderate resolution, metal shadowing plays a very useful role at the earlier stages of structure analysis, much as is true for negatively stained samples.

6.4 Glucose and other "sustains" can preserve macromolecular structures to high resolution

One of the most important developments in high-resolution electron diffraction and electron microscopy of biological macromolecules was the discovery by Unwin and Henderson (1975) that *glucose embedding* could preserve the high resolution, "hydrated" structure of crystalline specimens. The efficacy of glucose embedding is easily tested by electron diffraction, provided that crystals of a suitable test specimen are used and care is taken to avoid radiation damage. An important advantage of glucose embedding is that it is extremely simple to carry out, involving nothing more than to substitute a 2% glucose solution in place of a 2% negative stain solution. A further advantage is that glucose embedding preserves the well-ordered, "hydrated" structure of the specimen in vacuum, at room temperature. This last point meant that it was not necessary for Unwin and Henderson (1975) to use a cold stage or any other specialized equipment for their initial development of high-resolution imaging with catalase and with purple membrane.

The function of glucose embedding is actually twofold. Like negative staining, glucose embedding fills in the available crevices and channels within the structure and thereby prevents the flattening and collapse that would otherwise occur when water is removed by air-drying. It was with this purpose in mind that Eduard Kellenberger first advocated the use of glucose in place of negative stain, and spoke of the glucose as being a *sustain*. Equally important, however, is the ability of glucose embedding to preserve a native, effectively hydrated state for crystalline macromolecules, even after prolonged periods of time in the vacuum of the electron microscope. This latter property is not something that could have been predicted with confidence.

No proper molecular explanation has yet been proposed, and tested experimentally, to understand how specimen hydration can be maintained by glucose-embedment. Direct gravimetric measurement has shown that almost no water is retained in glucose-embedded purple membrane when the sample is exposed to vacuum at room temperature (Perkins et al., 1993), only adding to the mystery of why it is such an effective embedment. Recent work has shown that there are small, systematic differences in the diffraction intensities of glucose-embedded purple membrane, depending upon the water content, which can be controlled by equilibrating the sample at high humidity (Han et al., 1994a). In addition, completely dry, glucose-embedded samples definitely exhibit a greater variability (i.e., noise) in the measured diffraction intensities than do samples that have been equilibrated at high humidity. In spite of these small differences, the important point is that — even when thoroughly desiccated — the glucose- embedded sample has very nearly the same structure as that of the hydrated specimen (Jaffe and Glaeser, 1984; Han et al., 1994a).

Glucose is not the only hydrophilic solute that has been found to be effective in preserving high-resolution structure in "air-dried" specimens. Neutralized tannic acid was used in the high-resolution structure analysis of LHC II, the photosynthetic light-harvesting complex from green plants (Kuhlbrandt et al., 1994). Neutralized tannic acid has also been found to give superior results for 2-D crystals of tubulin (Nogales et al., 1995). Trehalose, another alternative to glucose, is a nonreducing disaccharide which is known to be especially effective in preserving the structure and function of lipid bilayers and of various proteins under otherwise desiccating conditions (Crowe et al., 1992; Sun et al., 1996). Trehalose embedding gives excellent results with purple membrane (Kimura et al., 1997) and aquaporin crystals (Gyobu et al., 2004), and it proved to be more effective than glucose embedding for 2-D crystals of PhoE porin (Walian and Jap, 1990). The range of polar solutes that serve well as "sustains" has, as yet, hardly been explored, and it seems likely that many other compounds could give excellent results. As is known from experience with heavy-metal negative stains, not all specimens will be well preserved in a single type of embedment. Thus, if poor results are obtained with one — say, glucose — a few others should be tried (neutralized tannic acid, trehalose, and even other polar solutes) before it is concluded that the approach of using a sustain is not satisfactory for that specimen.

The effect that the surface properties of the support film can have on the success of glucose embedding is one of the more poorly understood aspects of the technique. In the case of purple membrane, hydrophobic carbon films tend to give the best samples, both in determining the highest resolution that will be obtained (Baldwin et al., 1988; Glaeser et al., 1991) and in determining how flat the specimens will be (Han et al., 1994a). Hydrophilic carbon films are normally unsatisfactory, whether they are spontaneously on the "more hydrophilic" side, as made, or whether they have been made hydrophilic by exposure to an (air) glow discharge. On the other hand, conditioning the carbon film by exposure to a glow discharge in low pressure (i.e., ~150 mtorr) water vapor resulted in excellent results for glucose embedded purple membrane (Glaeser et al., 1991), as well as for trehalose embedded samples of PhoE porin (Walian and Jap, 1990). A very important point about the surface properties of the support film is that the conditions which work best for one specimen (say, purple membrane) will not necessarily be good conditions for the preparation of a different specimen. As a result, each new specimen requires trial-and-error to find the best surface properties and embedment media.

While glucose embedding preserves high-resolution structure in crystalline specimens, it is obvious that it will not provide the high image contrast at low resolution which is given by negative staining. To a first approximation, the specimen will be nearly invisible because the amount of electron scattering by glucose is similar to the amount of electron scattering by protein. Restating this in more technical terms, the mean value of the (shielded) coulomb potential is expected to be quite similar in glucose and in protein. The matching of average coulomb potential will, in turn, result in *contrast matching* (section 6.5) at low resolution, while the high-resolution contrast will not be affected. This low-resolution "contrast matching effect" is a phenomenon that is already well known in x-ray diffraction theory, and even more so in neutron scattering theory. Because contrast matching is a very important phenomenon in electron crystallography as well, we will explain the underlying physics of contrast matching more fully in section 6.5.

A number of hydrophilic compounds have been found that preserve the native hydrated structure of biological macromolecules in the vacuum of the electron microscope, and which, at the same time, provide more low-resolution contrast than that which is present with glucose. The most widely used of these high-contrast sustains is aurothioglucose (Kuhlbrandt, 1982; Kuhlbrandt and Unwin, 1982; Cheong et al., 1996), but metrizamide (Lepault et al., 1983), glucose-6-phosphate (Massover and Marsh, 2000) and other high-density, polar solutes have also been shown to be effective in preserving high-resolution structure, while still giving reasonable contrast for lower resolution features. It should be pointed out that none of these high-contrast sustains are guaranteed to be effective as a specimen embedment in high-resolution studies. Much of the experience that exists in the use of such embedments is not documented in the literature, especially the unsuccessful investigations. It is safe to say, however, that aurothioglucose works well in a more limited set of cases than is true for glucose or (neutralized) tannic acid. Experience with other sustain-type of embedments is still too limited to justify further generalizations.

All samples that are embedded in glucose, aurothioglucose, and similar organic sustains are extremely sensitive to radiation damage (Lepault et al., 1983), because the embedment is as easily destroyed as is the biological macromolecule. In the case of negatively stained samples, on the other hand, the embedment is highly resistant to radiation damage, while the biological macromolecule itself is quickly destroyed by radiation damage (see section 4.7). Since the heavy-metal salts that are left behind represent a kind of "fossil record" of what the structure had once been, it does not matter that only a residue of the organic (biological) material remains within the hard, three-dimensional replica. On the other hand, when the embedment — such as glucose — is itself as easily damaged as the biological macromolecule, then not even a replica of the original biological structure remains after a large electron exposure.

6.5 Contrast matching can be manipulated by using embedding media with different densities

The phenomenon of *contrast matching* is a concept that is easily understood by picturing a macroscopic example, something that one might see with one's own eyes. A clear glass sphere, which has a high refractive index, is easily seen when it is placed into a clear liquid with a lower refractive index, such as water, even though the contrast in this situation may be small. However, if the water is replaced by a clear liquid that has exactly the same refractive index as the glass, then there is no optical difference between the glass sphere and the surrounding liquid, and the sphere becomes invisible. This example illustrates the fact that the image contrast of a phase object goes to zero when the refractive index of the surrounding medium is made equal to the refractive index of the object. The same effect must also occur for the scattered (diffracted) wave, that is, the amount of scattered light must vanish when the refractive index of the medium matches that of the object.

Scattering experiments that exploit the contrast matching effect are especially common in neutron scattering work. As is well known, values of the neutron scattering contrast can be manipulated over a wide range by adjusting the ratio of hydrogen (which has a negative scattering amplitude) and deuterium (which has a positive scattering amplitude). Manipulation of the H/D ratio can be done either in the biological sample or in the aqueous medium, thus giving a great deal of information from the contrast matching effect.

The example given above, of a glass sphere immersed in a clear liquid, is a bit too simple to describe the contrast matching effect as it occurs at a molecular level, however. A clear glass sphere can normally be described as being a homogeneous object whose refractive index is constant throughout. A protein molecule, on the other hand, is made up of discrete atoms, organized in an arrangement of primary, secondary, and tertiary structure. As a result, a protein molecule will have significant fluctuations in internal density, or "refractive index," from one point to another. More specifically, the quantity

$$\eta(\mathbf{R}) \equiv 2\pi \frac{e}{hv} \hat{\rho}(\mathbf{R}), \qquad (6.1)$$

the 2-D projection of which we introduced previously in connection with equation 3.14, represents the function that is equivalent to the refractive index within the theory of electron scattering and image contrast. Recall that in section 3.6 we have explained the reason for using the symbol $\hat{\rho}$ rather than V to represent the shielded Coulomb potential, and the meaning of the various physical constants e, h, and v are given in connection with equation 3.14.

Although the Coulomb potential is not constant throughout the molecule, it is still possible to separate $\hat{\rho}(\mathbf{R})$ into two terms: the average value of $\hat{\rho}(\mathbf{R})$, which is constant, and departures from the average value, $\Delta\hat{\rho}(\mathbf{R})$. In order to understand contrast matching in the context of molecular structure, therefore, it is useful to write the coulomb potential as

$$\hat{\rho}(\mathbf{x}) = \hat{\rho}_0 + \Delta\hat{\rho}(\mathbf{R}). \qquad (6.2)$$

The average value, $\hat{\rho}_0$, is called the *inner potential* in scattering theory and in applications that involve electron interferometry (Spence, 1981). The contribution that $\hat{\rho}_0$ makes to the image contrast and to electron scattering can be "matched out" by varying the density of the surrounding medium so that the inner potential of the medium becomes precisely the same as that of the protein. Deviations from the average value, represented by $\Delta\hat{\rho}(\mathbf{R})$, will not be "matched out" in this way, of course, and therefore they will continue to contribute to image contrast (or to scattering and diffraction), as before.

Contrast matching affects chiefly the low-resolution features of an object, or its diffraction pattern. The deviations of $\hat{\rho}(\mathbf{R})$ from its average value will, by definition, occur on a size-scale that is smaller than the object itself. The largest and most significant deviations, $\Delta\hat{\rho}(\mathbf{R})$, reflect various structural features like the repeating motif of peptide bonds; the rise and fall of density between adjacent peptide strands, as in β-sheet secondary structure; the high densities of α-helixes; the individual side-chain densities; etc. All of these deviations, $\Delta\hat{\rho}(\mathbf{R})$, in the Coulomb potential occur on a size scale that is much smaller than the size of the whole molecule. As we have already pointed out, these "high-resolution" features, $\Delta\hat{\rho}(\mathbf{R})$, are not altered in any way when the inner potential of the embedment is adjusted to match the inner potential of the object, and therefore their contribution to the image (or to diffraction) will not be altered by contrast matching. What is set to zero, however, is the contribution made by a three-dimensional function that has constant value throughout the overall envelope of the macromolecule. This latter term is the same size as the molecule itself, and therefore removal of this term will have an important effect at low resolution.

Contrast matching makes low-resolution features difficult to see, rendering glucose-embedding useless for single particles. The other, higher-contrast embedments mentioned in section 6.4 can be used for single particles, of course, as was demonstrated by Lepault et al. (1983) and more recently by Cheong et al. (1996). However, ice-embedding — which will be discussed in section 6.6 — has been the most widely adopted preparation technique whenever negative staining or glucose embedding have been found to be inadequate.

The fact that the inner potential of glucose almost matches that of protein has a second, important consequence, in addition to the fact that the contrast at low resolution is very weak. It is unfortunately possible that some of the low-resolution and medium-resolution features of the protein may have a density lower than the average density of the glucose, while other protein features may have a density greater than that of glucose. The net outcome in this case is that the lower density parts of the protein show up in the negative contours of an image, while the higher density parts of the protein show up in the positive contours. This situation should be contrasted with the case for negative stain, where all of the protein shows up with lower density than the stain, that is, all of the protein shows up in the negative contours. Because different regions of the protein density can be either positive or negative relative to glucose, however, there can be an unresolvable ambiguity between what is glucose and what is protein in a particular projection map. Precisely this effect was observed by Cohen et al. (1983) with glucose-embedded crystals of a protein, gp32*I. The projection of these glucose-embedded crystals could not be interpreted in terms of the images obtained either in ice or in uranyl acetate, which themselves were essentially the negative of one another (Cohen et al., 1984). Fortunately, ambiguity between glucose and protein does not occur at high resolution, because the Coulomb potential at the "core" of the peptide or at the "core" of a side-chain density is always much greater than the inner potential of the embedding medium. Because of this fact, the contrast matching effect that occurs at low resolution does not affect our ability to fit a peptide chain into density, or even to define the envelope within which high-density peaks (due to the polypeptide chain) are to be found.

In principle the low-resolution contrast-matching effect can be used as a tool to identify different molecular components, each of which have different values of the inner potential. For example, one value of the inner potential of the embedment could be used to match the inner potential of protein while another value could be used to match the inner potential of nucleic acid. The idea here is to compare images of a given macromolecule that are obtained in two different embedments, and thus to match out first one and then the other molecular species. This approach was taken by Kuhlbrandt and Unwin (1982) in order to separately image the RNA and protein components of the ribosome. While this approach is conceptually sound, and would seem to have a broad range of possible applications, it has not been widely adopted.

The resolution that can be achieved in contrast-matching experiments in electron crystallography is comparable to that which is usually achieved in solution scattering with neutrons. Given the fact that phase information is retained in electron crystallography, and the fact that solution scattering data are inherently "spherically averaged" (whereas the electron image data are not), it would seem that intentional contrast-matching experiments could be used much more extensively in electron microscopy than has been the case so far. Neutron scattering experiments have one further advantage, however, which cannot be realized in the case of electrons: separate subunits of a complex assembly can be isotopically enriched for deuterium, giving rise to inter-subunit contrast. Thus, neutron scattering can "label" individual subunits without the risk of structural perturbation that might come from attachment of the heavy-atom clusters or labels needed in electron microscopy.

A formal, mathematical description of the way in which the inner potential changes between a protein and its surrounding embedment can give us a more quantitative understanding of why the contrast matching effect is greatest at lower resolution. Because the term $\hat{\rho}_0$ is confined to the protein molecule, we can represent its effect by the product of a rectangle function with an appropriate three-dimensional shape, $S(\mathbf{R})$, that corresponds to the surface, or envelope of the protein, and a function that is constant everywhere, with a value equal to $\hat{\rho}_0$ (protein). Similarly, the contribution of the inner potential of the solvent (or the embedment) can be represented by $\hat{\rho}_0$ (solvent) $\cdot [1 - S(\mathbf{R})]$. The net contribution made by the respective inner potentials of both components can be represented by

$$\hat{\rho}_0(\mathbf{R}) = \hat{\rho}_0 \text{ (protein)} \cdot S(\mathbf{R}) + \hat{\rho}_0 \text{ (solvent)} \cdot [1 - S(\mathbf{R})]$$
$$= \hat{\rho}_0 \text{ (solvent)} + [\hat{\rho}_0 \text{ (protein)} - \hat{\rho}_0 \text{ (solvent)}] \cdot S(\mathbf{R}). \quad (6.3)$$

By taking the Fourier transform of equation 6.3 we can see that the inner potential contributes two terms to the scattering. The first term, proportional to $\hat{\rho}_0$ (solvent), will be reflected only in the intensity of the unscattered beam. The second term is given by $[\hat{\rho}_0 \text{ (protein)} - \hat{\rho}_0 \text{ (solvent)}]$ times the Fourier transform of the protein envelope, $S(\mathbf{R})$. As would be true for a "rectangle" function of any shape, the Fourier transform of $S(\mathbf{R})$ is largest at the lower spatial frequencies, and the amplitudes of the Fourier coefficients become progressively smaller at higher frequencies. It is worthwhile to emphasize once more that the contribution of the protein shape function, $S(\mathbf{R})$, vanishes (at all spatial frequencies) when the inner potential of the medium matches that of the protein.

6.6 Embedding in vitreous ice is the preferred alternative for the preparation of unstained, hydrated specimens

The primary objective of preparing samples in a thin film of frozen water is to preserve the native hydrated structure of the specimen exactly as it had been in the instant prior to freezing. Once frozen, the sample can be put into a cold stage at such a low temperature that there is no loss of water by sublimation. In practice, a specimen temperature below $-120°C$ is sufficient to prevent sublimation within the vacuum of the electron microscope. Success in the preparation of *frozen-hydrated specimens* therefore requires that the biological macromolecule should not become dehydrated during the freezing step itself, for example by the transfer of water of hydration from the protein to the surrounding ice. Additional hazards arise from mechanical stress due to ice crystal formation, or mechanical stress due to the volume change in the water even when undergoing the liquid-to-glass phase transition. Mechanical stress must be generated during cooling, as well, as a result of differences in the thermal expansion coefficients of vitreous ice and protein (or other macromolecules).

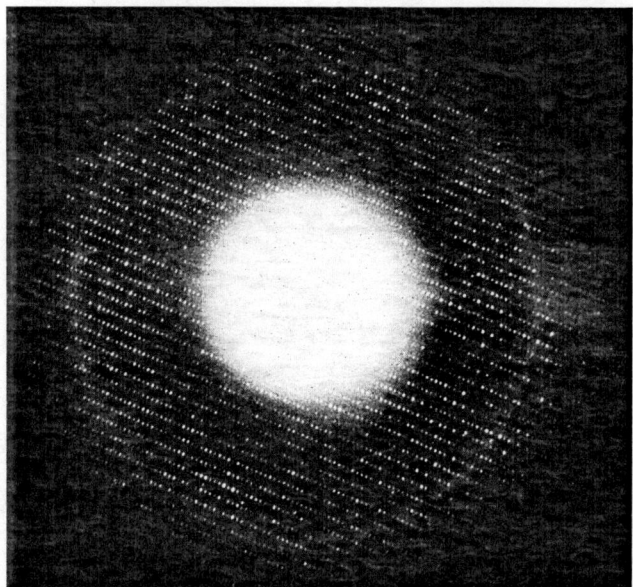

Figure 6.3 Electron diffraction pattern obtained from a frozen-hydrated crystal of catalase by Dr. Kenneth A. Taylor. The high resolution that is retained in such diffractions patterns provides the key evidence that the frozen-hydrated specimen technique can be used for high-resolution structural studies of biological macromolecules.

The fact that high-resolution protein structure could be well preserved in the frozen-hydrated state was demonstrated by Taylor and Glaeser (1974). Thin plates of catalase were examined by electron diffraction, making it possible to show that the protein molecules remained well-ordered to a resolution of 0.28 nm (Taylor and Glaeser, 1976). An example of one of these early electron diffraction patterns is shown in figure 6.3. It is unlikely, of course, that the structure of these protein molecules could have been changed by dehydration or by mechanical stress, while still preserving the degree of highly identical structure, from one unit cell to the other, that is required for such high resolution diffraction. Oriented "rafts" or bundles of filamentous structures have also been used to confirm that high-resolution structure is preserved in the frozen state. Examples include tobacco mosaic virus (Cyrklaff and Kuhlbrandt, 1994; Ruiz et al., 1994), microtubules (Wolf et al., 1996), and bacterial flagella (Ruiz et al., 1994). Prior to these experiments, however, it was not obvious that the high-resolution structure of biological macromolecules could be preserved by freezing. Earlier experience in x-ray crystallography was that protein crystals became disordered, and often were physically shattered as the result of freezing. More recently it has been discovered that very many protein crystals can be frozen without apparent loss of crystalline order, and the preparation of frozen crystals for x-ray crystallography has become the method of choice (Hope, 1990; Rodgers, 1997).

The problem of contrast matching that occurs in glucose-embedded samples is avoided completely if one uses frozen-hydrated specimens. Historically it had been argued that there would be very little contrast between protein and water because the average atomic number in organic materials is so close to that of water. A more careful analysis of the expected contrast must also take the mass density into consideration, however. The mass density of protein is considerably larger than that of water, of course, in the range of 1.36 g/cm^3 for hydrated protein, based upon measurements of partial specific volume of proteins in water (see table 2 in the supplementary material for Durchschlag and Zipper, 1997).

For materials with similar atomic number, the inner potential (see section 6.6) should scale as the density of the material. As a result, protein and ice would be expected to have a 30 to 50% difference in the value of the inner potential, and thus there should be significant contrast between them. This expectation was indeed borne out when catalase crystals and other biological materials were imaged in the frozen-hydrated state (Taylor and Glaeser, 1976). Figure 6.4 shows an image of a catalase crystal in ice in which the crystalline lattice rows are seen with high contrast. Another example of the high contrast afforded by an unstained, frozen-hydrated specimen is shown in figure 6.5, in which the specimen is a periodic surface-layer protein from the outer membrane of *Aquaspirillum serpens*. This micrograph demonstrates that even the bilayer structure of accompanying lipid vesicles can be seen in ice-embedded specimens.

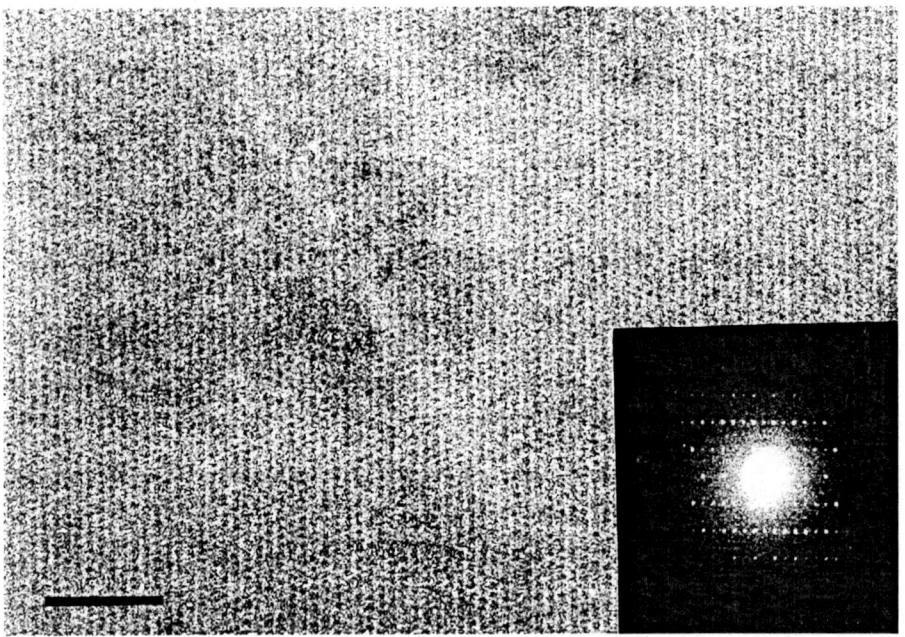

Figure 6.4 High-magnification image of a catalase crystal prepared in the frozen-hydrated state, as in figure 6.3. The scale bar represents 200 nm, and the Fourier transform of the image, shown in the lower right corner, includes several reflections on the third row of diffraction spots. The clear image of the protein lattice (Taylor, 1978) demonstrates that it is not essential to use stain in order to generate images with well-defined contrast. The real limitation in such images is associated with radiation damage, not with the inherent contrast between protein and ice.

Specimen Preparation 153

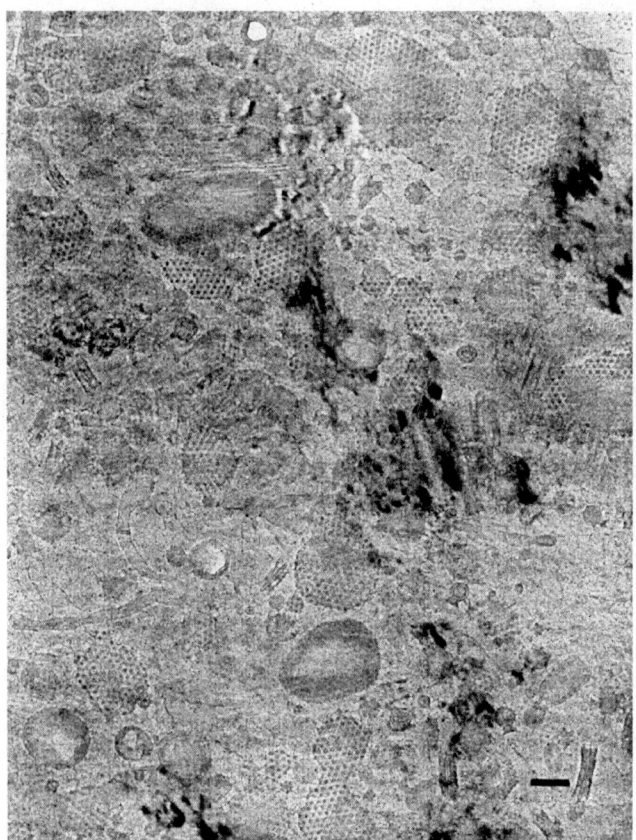

Figure 6.5 High-magnification image of a frozen-hydrated preparation of outer membrane material isolated from *Aquaspirillum serpens* (Taylor, 1978). Various structural features that can be seen in this image include 2-D crystalline patches of a surface-layer protein and profile views of outer membrane lipid bilayers.

A critically significant advance in the preparation of frozen-hydrated specimens was made by Dubochet and colleagues when they introduced techniques to prepare specimens in vitreous, rather than crystalline ice (Dubochet et al., 1982a,b). The preparation of samples in vitreous ice has proven to be extremely valuable for all types of specimen including thin protein crystals; helical filaments such as tobacco mosaic virus, microtubules, or actin filaments decorated with the S1 ATPase of myosin; icosahedral viruses; symmetric protein assemblies such as proteosomes or GroEL; individual macromolecular assemblies such as ribosomes; and proteins with a relatively high molecular weight such as α_2-macroglobulin (Larquet et al., 1994).

The key to *vitrification* of thin, aqueous films (box 6.1) is rapid plunging into liquid ethane at a temperature that is slightly above its melting point and well below its boiling point. Because the liquid ethane does not boil as it cools the sample, one avoids the formation of a gaseous film at the interface between the cryogenic liquid and the sample. Such a gaseous film is very difficult to avoid with liquid nitrogen, even

BOX 6.1 Vitrification of thin, aqueous samples

Vitrification of thin layers of water requires an exceptionally high rate of cooling in order to avoid nucleation and crystallization. The cooling rate required for vitrification has been estimated to be 10^6 degrees/second (Dubochet et al., 1988). A cooling rate sufficiently high to achieve vitrification can, indeed, be realized for thin aqueous films, as is described below. Water — or ice — is itself a poor thermal conductor, however, and the cooling rates achievable in the interior of an aqueous sample become increasingly self-limited as one moves farther in from the surface. As a result, it is difficult to vitrify a sample that is more than 1 μm in thickness (Dubochet and McDowall, 1981). The greatest sample thickness that can be used in electron crystallography is limited to much less than 1 μm by multiple scattering and by inelastic scattering (chapter 15), however. Thus the maximum specimen thickness for which vitrification is possible is not a practical consideration in electron crystallography.

In order to achieve the high rate of cooling required for vitrification, it is necessary to plunge the specimen grid into a liquid cryogen at high speed. The rapid flow of the liquid across the surface of the grid promotes convective heat transfer away from the sample, yet it is gentle enough to not break the support film or the thin aqueous layer that spans the holes in a holey support film. In practice it is found that the plunging rate should be at least 1 m/s, or else vitrification does not occur (Dubochet et al., 1988). Hand-held plunging of the electron microscope grid usually does not achieve the needed speed at entry, and therefore a faster, mechanical device is almost mandatory in order to vitrify the sample.

Several different types of plunging apparatus are able to produce a plunging speed in excess of 1 m/s. Many devices use a simple gravity drop; this needs to have adequate length to accelerate to the required speed, and the plunger needs to be guided by a low-friction rail which does not limit the speed prior to entering the cryogen. Other mechanisms that work well for accelerating the grid into the liquid cryogen include the use of a compressed spring (Murray and Ward, 1987b) or compressed gas to drive the rod that holds the electron microscope grid.

A worthwhile feature to be included in the plunge-freezing apparatus is some type of shutter that is placed immediately above the liquid cryogen, and which opens just before the grid enters the cryogen. The design should accomplish two things. The shutter should prevent water vapor from crystallizing in the cold gas above the liquid cryogen, thereby contaminating the liquid with microcrystals of ice. In addition, the shutter can minimize the distance over which the grid must travel through cold gas, during which the sample could be precooled, prior to freezing. Precooling has been documented to cause structural disassembly of microtubules and to alter the shape of liposomes due to a phase transition when the temperature drops below the melting temperature of the fatty acid tails. There is even the danger that the grid might be cooled below the freezing

(continued)

point of water as it passes through the cold gas above the cryogen. When this happens, the cooling rate is unfortunately too slow for the water to be vitrified.

The use of some type of controlled humidity environment is another feature that is important in many applications (Murray and Ward, 1987b; Battersby et al., 1994). There are several advantages that should be mentioned. A high humidity slows down the final drying process, of course, after most of the excess liquid has been blotted away. The slower drying rate makes the timing between blotting and freezing less critical, and reduces the risk that the sample will be dried out completely. The slower rate of evaporation also reduces the extent of precooling of the sample due to evaporative heat loss. The preparation of microtubules as frozen-hydrated specimens is, again, an example where evaporative precooling had to be avoided. For some specimens it is even necessary to remove essentially all of the excess water by blotting and to completely avoid further thinning by evaporation. This is the case, for example, for samples that are sensitive to increases in the ionic strength, such as myosin filaments.

Plunge-freezing equipment that meets the various requirements described above is now available from a number of manufactures. Photographs of three such instruments are shown in figure B6.1.

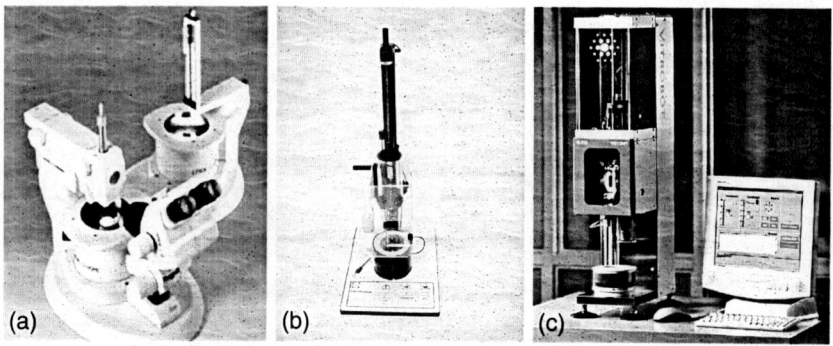

Figure B6.1 Representative examples of the type of plunge-freezing apparatus that is commercially available. (a) This apparatus (Leica) represents a basic unit in which the sample is blotted under ambient environmental conditions and then plunged immediately into liquid ethane. Note the pedestal on which the cup is mounted, into which the ethane is condensed. (b) The instrument shown here (Gatan) includes a housing that surrounds the work area in which the EM grid is blotted before plunging. (c) The most complete system currently available (FEI *Vitrobot*™) includes computer-controlled setting of the temperature and the relative humidity within the work area. Blotting of the grid with filter paper is done mechanically (i.e., by the robot).

when cooled to its melting point. Because of the poor rate of heat transfer across the gas layer, cooling of the sample becomes too slow to achieve vitrification.

While other organic liquids such as propane or freon are easier to handle than ethane, and are similarly effective in facilitating rapid freezing, ethane has one further advantage that makes it the only suitable liquid for the preparation of vitrified thin films. As background, it should be pointed out that the specimen temperature, following initial vitrification, must be kept below ~130 K in order to avoid converting the vitrified ice to polycrystalline (cubic) ice through a *glass-to-crystal phase transition*. On the other hand, once the specimen has been plunged into the liquid cryogen, and then transferred to the cold stage, it will be coated with a very thick film of liquid — or even solid — cryogen. The film of liquid ethane, which initially coats the grid after it is removed from the freezing vessel, will easily evaporate at temperatures below 130 K, but most other cryogenic liquids remain solid and do not even sublime at a significant rate below this temperature. Therefore propane or freon are unsuitable as alternatives to ethane.

Before the sample is vitrified, of course, it is necessary to blot away most of the liquid, leaving just a thin, relatively uniform layer of water in which the specimen is still embedded. A number of different support films have been found to work quite well for this purpose, as long as they are first made hydrophilic by exposure to a glow discharge. The support film that is preferred in most work is some type of holey ("reticulated" or "lacey") carbon film of the type described in section 6.1. The idea then is to look for areas in which the sample is suspended within a thin, aqueous film, spanning a hole. An effective trick is to apply the sample to the top of the holey film and then blot away the excess liquid from the opposite side. The holey film thus acts as a porous filter, or sieve, trapping the sample in the holes (Toyoshima, 1989). Thin, aqueous sample films can also be prepared on continuous carbon films. In this case it is worthwhile to try applying the sample through the grid bars and blotting the excess from the same side, that is, again through the grid bars (Subramaniam et al., 1993). Surprisingly, it is even possible to prepare thin, aqueous films with bare grids, that is, without any other support film (Adrian et al., 1984). Some investigators also advise to mix a surfactant with the sample in order to increase the spreading and stability of the thin aqueous layer. The surfactant can be detergent (Dubochet, private communication; Dubochet et al., 1988) or phospholipid vesicles (Dubochet, private communication), other proteins (filamentous viruses have been found to be quite effective), or even the sample protein itself! The need to add a surfactant may be sample-dependent, and possibly depends upon the type of glow discharge treatment used to pretreat the support film. It has been suggested, for example, that glow discharge treatment in vapor of alkylamine can itself deposit surfactant material on the support film (Lepault and Dubochet, 1986).

A wide range of techniques can be used to blot away excess sample prior to freezing. One of the first techniques used was to pinch the electron microscope grid between two pieces of filter paper. Surprisingly, the anticipated physical abrasion does not seem to damage the delicate support film. Blotting with a cut edge, or point, of filter paper, from one side of the electron microscope grid, is also satisfactory. As mentioned above, blotting on the "back side," that is, through the grid bars, seems to be an especially good approach. Simple trial-and-error experience is as good a way as any to learn how extensively the initial blotting should be carried out, and how long (if at all) to wait after the blotting, before doing the freezing. Whatever technique is used, one wants the result to be reasonably reproducible, and to give a high fraction of grid-windows where the aqueous film has a suitable thickness.

Several laboratories have experimented with the construction of boxes that enclose the grid-blotting work area so that one can control the humidity, and even the temperature during specimen preparation. A sophisticated version is commercially available in which the blotting operation itself is carried out by a robot, and a number of relevant parameters are reproducibly maintained by computer control (www.vitrobot.com/index.htm). In addition to improving the reproducibility of conditions under which the specimen is prepared, precise control of the temperature and humidity makes it possible to hold ("drain") the grid for a prolonged length of time with minimal evaporation.

Inspection of the frozen-hydrated specimen grid in the electron microscope immediately gives the answer as to whether the blotting procedure resulted in the desired thickness of water, prior to freezing. Qualitative criteria such as the transparency of the sample — based on previous experience and the remembered transparency of carbon films or negatively stained samples — or just the clarity of image contrast, are often all that is needed to evaluate the grid. Quantitative measurements of the relative thickness are also easily made by using the exposure meter provided with the electron microscope to compare the beam intensity transmitted through the area of interest with that transmitted through a hole. Converting the relative values of transmitted intensity into an estimate of the absolute ice thickness requires the use of a centered objective lens aperture with a known cutoff angle (Eusemann et al., 1982; Lepault and Dubochet, 1986). Another simple, quantitative method of measuring the ice thickness is provided by drilling a small hole through the ice with a well-focused electron beam. The sample can then be tilted by a known amount so that the hole size appears to become elongated perpendicular to the tilt axis. The thickness of the ice layer can then be determined by simple geometry.

Another critical advance in the use of frozen-hydrated samples involved the development of commercially available cold stages, cryogenic specimen work-stations, and cryotransfer equipment, all of which allow simplified manipulation of the specimen grid. The electron microscope anticontaminator devices that should be used with frozen-hydrated samples must also be designed to cover a larger solid angle with respect to the sample than is required for anticontaminators that are normally provided for room-temperature specimens. The commercial availability of all these accessories has made it possible for almost any laboratory to work with frozen-hydrated specimens without the need to engage in the uncertain task of constructing specialized, one-of-a-kind equipment.

The use of rapidly vitrified samples offers a unique way to trap *kinetic intermediates* with a time-resolution of a few milliseconds. The fundamental idea is to start a given reaction shortly before the sample grid is plunged into liquid ethane. In some cases an intense light flash can be used to start reactions at a time that is synchronized relative to the instant of freezing, giving rise to the "flash and splash" method of doing *time-resolved studies*. A few specimens, like bacteriorhodopsin, carry out biochemical reactions that are based on photochemical stimuli, and synchronization of structural changes by a light flash (Subramaniam and Henderson, 1999) is especially easy to achieve in these cases. Other specimens may utilize substrates, such as ATP, that can be released photolytically from "caged" precursors. A different alternative for initiating reactions is to spray an aerosol of substrate solution onto a previously blotted grid, with the specimen already on it (Berriman and Unwin, 1994), giving rise to what can be called the "puff and splash" method. An easily visible marker, such as ferritin, should be included with the substrate in order to identify areas of the grid where droplets of substrate have mixed with the previously blotted specimen.

Some words of caution should be added about the possibility that specimens might become structurally disrupted due to surface tension effects and denaturation at the air–water interface. The likelihood that proteins in a sample will denature when Brownian motion brings them into contact with the air–water interface should not be underestimated. The formation of such a denatured protein film can even be exploited as a method for forming self-supported thin films that span the holes in a carbon support film (Yoshimura et al., 1994). Many larger, macromolecular complexes, on the other hand, may not appear to denature and spread completely, yet a strong interaction with the interface is evident (Cyrklaff et al., 1994). In some cases interaction with the interface causes macromolecular complexes to take up one or a few preferred orientations relative to the interface (Cyrklaff et al., 1994). Still another example of structural effects that can happen at the air–water interface is provided in the case of helical protein assemblies, such as tobacco mosaic virus. It is not uncommon that the optical diffraction pattern of such specimens shows much stronger intensity from one side of the helix than from the other; structural disorder on the side of the helix that is adsorbed to the air–water interface is the most likely explanation for such differences. Interphase chromatin is a sample that seems to be especially sensitive to disruption at the air–water interface. If chromatin samples are first stabilized by cross-linking with gluteraldehyde, one can easily find the characteristic beads-on-a-string strands of DNA with interspersed nucleosomes, known previously from negatively stained samples. Without cross-linking, however, the nucleosome organization may disappear completely in frozen-hydrated samples, but not in negatively stained samples (Dubochet et al., 1988; Dubochet, personal communication). Of significant importance, however, is the observation that inclusion of phospholipid vesicles in the preparation, as an added surfactant, prevents the denaturation of the unfixed chromatin in frozen-hydrated specimens (Dubochet, personal communication). The likely explanation is that the lipid (i.e., surfactant) monolayer, which is formed on newly created surfaces after blotting, acts as a physical barrier and prevents the denaturation of chromatin at the air–water interface.

Many macromolecules have been found to bind so strongly to the carbon film that little or no material remains left in the vitreous ice within the holes, especially when the concentration of macromolecules is quite low. One (partial) solution to this problem is to prepare samples on continuous carbon film, which is often supported on a holey carbon film. The use of holey carbon as a kind of "microgrid" makes it possible to use much thinner continuous carbon than can be supported on "bare" 400 mesh grids. Adsorption of a macromolecule onto continuous carbon films can actually be used to advantage as it effectively "concentrates" the specimen on the grid, from an otherwise dilute solution. One must then be aware, however, that adsorption may or may not result in preferential orientation of the macromolecule. An alternative solution, if one wishes to avoid adsorption of the macromolecule onto a carbon film, is to use "lacey" or reticulated holey films, so that the fraction of surface area that is made up by the holes is much greater than the fraction that is made up by carbon.

One must also be aware of the possibility that ions and other small molecules in the sample (buffer) will become more concentrated if there is significant evaporation of water after the specimen is blotted, but before the specimen is frozen. One must therefore be cautious to ensure that changes in pH or ionic strength, if they do occur, do not result in structural changes in the macromolecule under investigation. A few specimens, such as microtubules, are also known to be cold-labile to such an extent that they disassemble just because of the sample cooling that occurs due to evaporation. The safest approach, of course, is to blot the specimen under conditions of controlled humidity as was mentioned above, such that little or no evaporation occurs. For some specimens it may be convenient to prepare samples in a cold room, the advantage being that evaporation is slowed considerably, and it is also convenient to achieve a high relative humidity in this case.

The sensitivity of frozen-hydrated specimens to radiation damage is the same as for any other type of biological specimen at low temperature. As has been explained in section 4.7, electron diffraction can be used to show that high-resolution features are destroyed by electron exposures of about 500 electrons/nm^2. Lower resolution reflections fade away more slowly, with the dose for fading of the intensity to e^{-1} of its initial value being approximately 1000 electrons/nm^2. Overall molecular shape, at a resolution of about 5 nm, is retained up to a dose of about 3000 electrons/nm^2. All of these values refer to exposures by 100 keV electrons, and corresponding values at higher voltage are increased by the relatively small factors that have been discussed in section 4.7.

6.7 Charging and mechanical stability vary with details of the specimen preparation method

Both glucose-embedded specimens and frozen-hydrated specimens exhibit beam-induced specimen movement, and/or image deflections. Neither effect, however, has been noticed in negatively stained specimens. In the most extreme cases, the direction of specimen movement can vary from one point to another within a single micrograph, ruling out stage drift as the cause. These *beam-induced movements* even occur at very low electron exposures, corresponding to the low-dose conditions that must be used with radiation-sensitive specimens. Beam-induced movements of this type are especially severe with frozen-hydrated specimens that are prepared as unsupported thin films within holes, and they are even quite noticeable for glucose-embedded specimens that are prepared on carbon films if the carbon is too thin.

In the case of glucose-embedded samples, conspicuous beam-induced movement can be eliminated if thicker carbon films are used. As a result, one of the key recommendations for the preparation of glucose-embedded specimens is to use relatively thick carbon films, in the range 20 to 30 nm in thickness, when preparing samples for high-resolution imaging. For electron diffraction studies, thinner carbon films can be used in order to reduce the background scattering, since diffraction intensities are not sensitive to small translational movements of the specimen.

Specimens that are not prepared directly on carbon films, and which have not been subsequently coated with an evaporated carbon film, do show substantial effects of *specimen charging*, even at electron exposures that are small compared to those which lead to significant radiation damage. The most easily observed charging effects are those that can be seen in the unscattered beam of the electron diffraction pattern. In the least disturbing form, charging can lead to a change in focus of the diffraction pattern from one area of the grid square to another, as well as to small deflections of the diffraction pattern. In more extreme cases the central spot may "bloom" into an irregular figure, which changes its size and shape as the specimen is translated, but which cannot be refocused to a sharp spot. Quite surprising phase-contrast effects are also seen in images at low magnification. A ring of high contrast first develops at the perimeter of the irradiated area for very low dose, the entire irradiated area then acquires nearly uniform contrast relative to the unirradiated areas around it, and at even higher exposure the charge-induced contrast can spread far beyond the area that is directly in the electron beam (Brink et al., 1998b). Fortunately, these more severe forms of specimen charging are eliminated completely by preparing samples on carbon initially, or by evaporating carbon onto the sample after preparation (Jakubowski et al., 1989).

There is a possibility that some of the specimen movement mentioned above is caused by electrostatic forces exerted on the specimen. In addition, charging may cause distortions and deflections of the image itself. The net result, in either case, is to limit the image resolution by causing a steep fall-off of contrast at higher spatial frequencies. Henderson and Glaeser (1985) have shown that this type of movement causes a severe reduction in signal at high resolution for paraffin and for purple membrane, both of which are highly sensitive to radiation damage. On the other hand, the same effect was found to be relatively small for vermiculite, which is known to be quite insensitive to radiation damage. The reduction in signal at high resolution is much less severe once the sample has been irradiated long enough to come to a stable, stationary state. Using extensive pre-irradiation to stabilize a specimen is, of course, something that can only be exploited with samples that are themselves insensitive to radiation damage, such as graphite (Bottcher, 1995) or crystalline ice (Cyrklaff and Kuhlbrandt, 1994). Unfortunately, no method has yet been discovered to fully overcome the apparently chaotic movement that attenuates the image signal during the initial phase of irradiation. The technique of spot-scan imaging, described in section 5.7, is only able to partially reduce the beam-induced movement effect.

Although it is not yet well understood why it should be the case, the image quality is likely to be poor if glycerol is included in the buffer when one prepares frozen-hydrated specimens. It is unlikely that poor image quality is due to contrast matching, because problems are seen even when the concentration of glycerol is much too low to significantly increase the density of the buffer. It is possible that the explanation may be that increased radiation damage occurs in a solid solution of glycerol in water, compared to the same buffer without glycerol. Thin water–glycerol films do show bubbling (discussed in section 4.7) at high electron exposures, and it has even been shown that these bubbles consist of molecular hydrogen gas at high pressure (Leapman and Sun, 1995). One can thus speculate that beam-induced movement, even at much lower electron exposures, may be more of a problem because of increased production of H_2 gas as a result of radiolytic reactions that occur more frequently when glycerol is present.

6.8 Preparing extremely flat specimens continues to be one of the most important challenges when working with 2-D crystals

A certain degree of *wrinkling*, buckling, or *imperfect flatness* is commonly found to limit the highest resolution and the highest tilt angle for which data can be obtained from thin 2-D crystals. Two separate effects are involved, and the extent to which each is sensitive to tilt angle and to resolution is quite similar. The most noticeable effect is a *broadening of the diffraction spots* for Bragg reflections that lie in the direction perpendicular to the tilt axis. This broadening increases both with tilt angle and with resolution. The second effect is an overlapping (i.e., superposition) of values of the 3-D Fourier transform of the specimen that are adjacent to one another on the same reciprocal lattice line, and which need to be measured independently. These two related effects, which affect both images and electron diffraction patterns, are discussed in more detail below.

We will employ a model for calculating the Fourier transform of a wrinkled crystal that is highly simplified, the purpose of the simplification being to more easily illustrate the basic physical principles that need to be understood. This simplified model is only meant to help understand why flat (i.e., planar) specimens are needed, and it will be used to estimate the degree of wrinkling that can be tolerated. However, the model is not meant to provide a basis for dealing mathematically with *imperfect flatness* when it is present in the data.

The model first of all approximates the wrinkled crystal as a piece-wise planar sheet, in which each slightly tilted domain is still large enough for one to ignore the broadening of diffraction spots that comes from its finite size. The Fourier transform of the crystal is therefore approximated as the linear superposition of the Fourier transforms from separate, planar domains, each one tilted in a slightly different way.

Since each of the piece-wise planar domains faces the incident beam with a slightly different orientation, the Fourier transform of each will be sampled on different central sections, each of which cuts through Fourier space at a slightly different angle. This idea is illustrated in two (rather than three) dimensions in the drawing shown in figure 6.6, where the Ewald sphere for two different specimen orientations is represented for clarity by oblique lines, rather than by planes. The two oblique lines correspond to areas of the specimen that are tilted by θ and by $\theta + \Delta\theta$, respectively. In a more complete description of a wrinkled crystal the central sections for different domains would be distributed continuously over an angular range of $\pm\Delta\theta$ around a mean angle, θ.

The vertical lines in figure 6.6 are called *reciprocal lattice lines*. These lines are a unique feature of the Fourier transform of a 2-D crystal, as will be explained in detail in section 7.4. If necessary, the reader should return to this discussion of imperfect specimen flatness after referring to section 7.4 for a more complete description of how the Ewald sphere intersects the Fourier transform of a 2-D crystal. For the moment we only need to say that the reciprocal lattice lines play the same role for 2-D crystals that is played by reciprocal lattice points for 3-D crystals. Every reciprocal lattice line is intersected by the Ewald sphere at some point along its length, regardless of the tilt angle, as shown in figure 6.6.

As is shown in figure 6.6, the range in local tilt angles causes a given reciprocal lattice line to be sampled over a continuous range of values in z^*. Each point along this continuous sample falls at a slightly different scattering angle, of course. As the

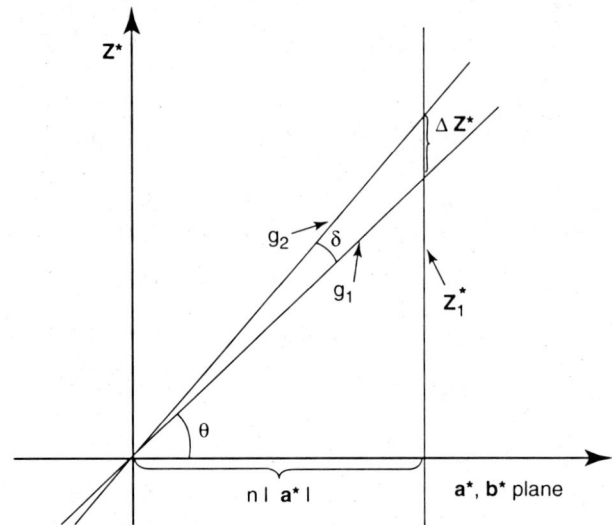

Figure 6.6 Geometrical representation of the way in which imperfect specimen flatness causes (1) broadening of diffraction spots that lie in the direction perpendicular to the tilt axis and (2) smearing of intensities along individual reciprocal lattice lines. Both of these effects are illustrated in this drawing, which is based upon the simplified model of a wrinkled, but piecewise flat (planar) sample. The horizontal axis is a schematic representation of the a^*, b^* plane, while the vertical axis represents the z^* direction in reciprocal space. A single reciprocal lattice line is indicated, running parallel to the z^* axis. The Ewald spheres for two different, piece-wise planar areas of the wrinkled crystal are represented by the two lines passing through the origin at tilt angles of θ and $\theta + \delta$, respectively. The two Ewald spheres intersect the reciprocal lattice line at different spatial frequencies, g_1 and g_2, causing a broadening in the diffraction spot. In addition, the reciprocal lattice line is sampled at positions z^* and $z^* + \Delta z^*$, thus smearing the intensity measured within the broadened diffraction spot (figure and legend after Glaeser et al., 1991)

figure shows, each sampled point along the reciprocal lattice line also falls at a different distance, $g + \Delta g$, from the center of the diffraction pattern. The spread in local tilt angles thus means that a given diffraction spot will no longer be seen as a single sharp spot, at a spatial frequency g, but it will instead be spread out as a continuous smear of intensity, over a range of spatial frequencies centered around g. As stated before, we assume — for the sake of explanation — that the broadening of the diffraction spot that occurs because of the finite size of each domain is only a small effect (see sections 2.6 and 7.6 for additional information on broadening that is due to finite crystal size). In a more complete theoretical treatment, the broadening due to finite domain size should be added to the broadening associated with wrinkling.

The spread in values of g, that is, the amount of broadening, depends upon the resolution, g; the mean tilt angle, θ; and the azimuthal location of the particular reciprocal lattice line relative to the tilt axis. Figure 6.7(a) shows an example of an electron diffraction pattern produced by a crystal with imperfect flatness. Parallel to the direction of the tilt axis, the Ewald sphere always passes close to the point $(h, k, 0)$,

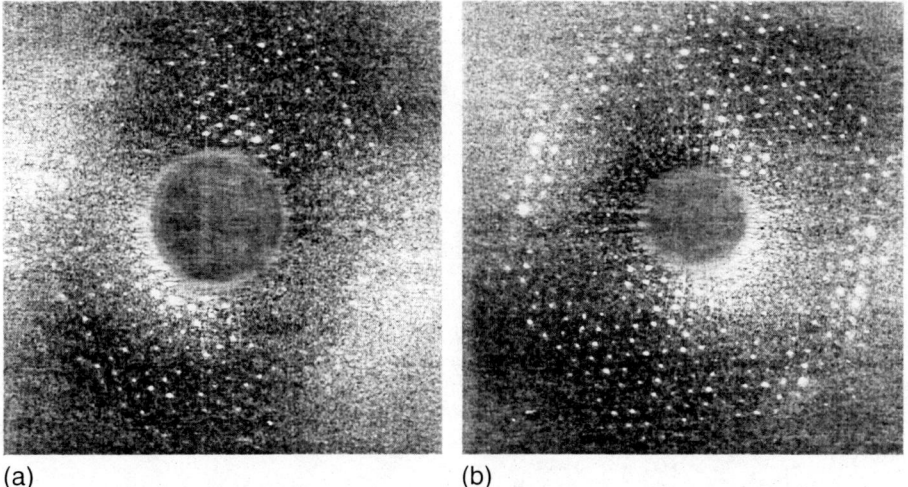

Figure 6.7 Electron diffraction patterns of tilted samples of purple membrane that show (a) imperfect flatness, and (b) nearly perfect flatness (Han et al., 1994b).

regardless of the tilt angle, and the diffraction spots will remain perfectly sharp even though the crystal is wrinkled. In the direction perpendicular to the tilt axis, however, the consequences of wrinkling that are explained in figure 6.6 will have their maximum effect. It is clear that the broadening of the diffraction spots should grow linearly with the value of g, that is, the broadening should get progressively worse at higher resolution. Similarly, the broadening will increase rapidly, and in a nonlinear fashion, as the tilt angle, θ, increases.

It will be intuitively clear that broadening of the diffraction spots makes it more difficult to get accurate values for the background-subtracted, integrated diffraction intensities. Up to a point, however, some degree of wrinkling (broadening) can be tolerated without compromising the data significantly. On the other hand, when the wrinkling is so severe that the diffraction spots begin to touch one another, recovery of accurate diffraction intensities will become difficult, if not impossible.

It is instructive to calculate how much wrinkling can be tolerated before the broadening exceeds a specified value, say 10% of the separation between adjacent spots. Some representative values for bacteriorhodopsin, with a lattice constant of 6.24 nm, are listed in table 6.1. The requirement that the diameter of the diffraction spots should remain smaller than 10% of the spacing between spots at a resolution of 0.35 nm, and a tilt angle of 60°, means that the specimen must be flat (planar) to within 0.2°. This is an exceedingly stringent requirement, of course. It is perhaps surprising, therefore, that the required sharpness of diffraction spots can be achieved, at least some of the time. Panel (b) of figure 6.7 shows a diffraction pattern of bacteriorhodopsin that was recorded at a tilt angle of 45°, in which the diffraction spots are so sharp that one can hardly tell the direction of the tilt axis, unlike the example shown in panel (a).

It is normal practice to integrate all intensities within a diffraction spot, although doing so is not a fundamental requirement of data processing. When the intensities

Table 6.1 The relative broadening of diffraction spots and the length of the interval along the reciprocal lattice lines along which data will be superimposed within the broadened diffraction spot. The values listed here apply to the example of bacteriorhodopsin (Glaeser et al., 1991)

Angular range of bending or wrinkling (degrees)	Relative broadening[a] of diffraction spots	Length of the interval along z^* over which data are averaged[b]
2	0.95	0.90
1	0.46	0.45
0.2	0.09	0.09
0.05	0.02	0.02

[a] Width of the diffraction spot divided by the separation between adjacent spots, which in this example was 1/(5.4 nm).
[b] The distance in z^* over which data are averaged is expressed in units of $1/T$, where T is the thickness (~4.5 nm) of the bacteriorhodopsin monolayer crystal.

are summed together across a spot, however, one will be adding data incoherently (i.e., by intensity rather than by the coherent vector sum of amplitudes) from slightly different parts of the Fourier transform of the specimen. If the range of z^* over which the incoherent summation occurs is small enough, the effect will be insignificant, because the continuous Fourier transform must change slowly with z^*. As will be described in section 7.4, however, the data on a reciprocal lattice line can be described by the convolution of a one-dimensional sinc function with the diffraction pattern of a 3-D crystal. This fact tells us that the continuous function must be sampled in steps that are separated by no more than $1/2T$ in spatial frequency, where T is the thickness of the 2-D crystal.

The drawing in figure 6.6 shows that the spread in z^* over which intensities are recorded, within a single spot, depends once again upon the resolution, g; the tilt angle, θ; and the azimuthal position of the particular reciprocal lattice line relative to the tilt axis. As before, we can calculate how much wrinkling can be tolerated before the interval along z^* becomes larger than a specified value, say $1/T$. Some representative values for bacteriorhodopsin, with a thickness of 4.5 nm, are again listed in table 6.1. The requirement that the z^*-sampling should remain "coherent" at a resolution of 0.35 nm and tilt angle of 60°, that is, that it should not average inequivalent samples of the continuous Fourier transform, means that the specimen must be flat (planar) to about 1°. This is again a very demanding requirement.

It would be possible, in principle, to "index" the sample points *within* each reflection, centered about (h, k, z^*). This indexing would prevent the incoherent addition of nonequivalent measurements from points along the reciprocal lattice line that were separated by more than a desired fraction of $1/2T$. The limitation to doing this, however, is set by the low signal-to-noise ratio in the measurements, as the signal becomes broadened out due to wrinkling of the specimen.

In the case of images, an even easier solution to the specimen flatness problem would seem to be to compute the Fourier transform of progressively smaller areas, until a size is found over which the crystal is effectively piece-wise planar. The limitation to reducing the area used to compute a Fourier transform is once again determined by the signal-to-noise ratio, however. Unfortunately, values of the high-resolution signal (i.e., the computed Fourier amplitudes) from small areas of a 2-D crystal are as much as ten times lower than they could be, due to beam-induced movement (see section 5.7). In order for the use of smaller, piece-wise planar areas of a 2-D crystal to become a practical way to solve the specimen flatness problem, therefore, a way must first be found to reduce beam-induced motion (see section 5.7).

What is the fundamental reason why thin 2-D crystals often fail to be as flat as the example shown in figure 6.7(b)? One answer lies in the fact that a very thin sheet is inherently easy to bend. In fact, the equal partitioning of thermal energy to all modes of bending, at thermal equilibrium, requires that one must expect an angular bending of the sample by about 16° over a distance of 1 μm (Glaeser et al., 1991). In other words, crystals that are thin enough for electron diffraction work do not on their own have sufficient bending rigidity to resist the wrinkling that will be imposed by thermal vibrations, or "Brownian motion." It is certain that samples which are suspended freely in bulk water, for example in thin aqueous films, cannot be planar to the extent required in table 6.1, over lengths of 1 μm. This fact requires that some type of external constraint has to be used in order to achieve the needed specimen flatness.

The traditional carbon film that is used as a specimen carrier evidently can serve quite well as a flat reference plane, as is demonstrated by the electron diffraction pattern of glucose-embedded purple membrane that is shown in figure 6.7(b). Specimens with this degree of flatness are only obtained with great difficulty, however, and the reasons why that is so are not yet well understood. One surprising fact is the finding that the starting stock of carbon that is used to make the evaporated carbon film seems to make a big difference in the success of preparing flat 2-D crystals (Han et al., 1994b). The specimen flatness achieved with continuous carbon films seems to also depend upon the length of time used to "age" the evaporated carbon (Henderson et al., 1990; Ceska and Henderson, 1990; Han et al., 1994b). Some success in improving the flatness has also been obtained by exposing the carbon film to a glow discharge in which water vapor is the dominant gas (Glaeser et al., 1991). Heavy reliance on pure trial-and-error repetitions, however, is still the only way known to eventually get a specimen grid for which there is a reasonable yield of crystals that are flat enough for data collection at high tilt angles. Less success seems to have been had in preparing flat specimens that are embedded in vitreous ice, even when continuous carbon films are used as a specimen support. Little is known about the flatness of frozen-hydrated specimens that are suspended over the holes of holey carbon films.

The problems of specimen bending that have been discussed here for 2-D crystals can be equally troubling for tubes, rods, and filaments, of course. In this case the problem should perhaps be described as one of imperfect linearity, rather than imperfect flatness. In the case of imperfect linearity, separate problems are associated with in-plane bending, which can be compensated to some extent by computational unbending (as is done for 2-D crystals; see section 10.10), and out-of-plane bending, which leads to incoherent superposition of data, as was discussed above for 2-D crystals. These issues are discussed further in section 12.3.

Adsorption and even growth of 2-D crystals at the air–water interface (section 8.4) should also be a promising option for preparing extremely flat specimens. The air–water interface has been shown to have a surface roughness of only 0.3 nm, due to so-called *capillary waves*, and this roughness is reduced even further by the addition of a surfactant at the interface (Daillant et al., 1989). Stabilization of surface roughness by the addition of surfactant would also protect the sample material from denaturation or distortion due to strong interfacial interactions. Whether this idea can be turned into one that is practical for preparing flat specimens remains to be shown, however. How difficult it will be to transfer the sample from an air–water interface, where it may well meet the required specifications for flatness, to the electron microscope grid itself is still unknown.

Even when suitably flat specimens can be prepared on the electron microscope grid at room temperature, care must be taken that they do not subsequently get wrinkled during cooling (or freezing), because of a mismatch in the thermal expansion coefficient of the specimen itself and that of the specimen grid. Copper grids are not a good choice for working with 2-D crystals if the specimen is prepared on a carbon film, because the *thermal contraction* of copper, as the grid is cooled, is greater than that of the carbon film. As a result, the carbon film begins to buckle and wrinkle as the copper grid contracts during cooling, an effect that has been dubbed *cryo-crinkling* (Booy and Pawley, 1993). Not surprisingly, specimens are then observed to lose their flatness. The problem of cryo-crinkling can be avoided, however, by the use of molybdenum grids (Glaeser, 1992) and also by the use of titanium grids, both of which have a much smaller thermal expansion coefficient than copper.

7

Symmetry and Order in Two Dimensions

7.1 Introduction

The repetition of a particular structure, or *unit cell*, in a pattern generated by some form of translational symmetry, is a unique, defining characteristic of crystals. While crystals may also have additional types of symmetry, such as that corresponding to rotation about a certain axis, the periodic repetition of a given structural motif over an extended region of space is something that uniquely defines what is meant by a crystal.

This chapter is concerned with a special class of crystals, namely those that consist of a motif (unit cell) that is repeated in two dimensions but not in the third. We will usually refer to these as *two-dimensional (2-D) crystals*, although it will be understood that the repeating unit cell itself has (nonrepeating) three-dimensional structure. Thus a 2-D crystal is not the same thing as the 2-D projection of a crystal, which also has translational symmetry in only two dimensions, but lacks finite thickness in the third dimension.

A description of the symmetries that are allowed in projections (images) and in 2-D (single-layer) crystals of biological macromolecules is given in sections 7.2 and 7.3, respectively. The information about crystal symmetry discussed in these sections is presented more fully in the International Tables of Crystallography, vol. A (Hahn, 2002), or can be deduced from what is presented there. An excellent description of crystal symmetry, with explicit discussion of objects that have translational symmetry in only one or two dimensions, is also provided by volume 1 of *Modern Crystallography* (Vainshtein, 1981).

The types of symmetry operations that are present in 2-D crystals are the same as those that exist in 3-D crystals. Because of the absence of translational symmetry in the third dimension, however, there are only 17 distinct space groups possible in

2-D crystals of chiral molecules (section 7.2), while there are 65 enantiomorphic space groups that are possible in 3-D crystals (section 7.3).

The Fourier transform of a 2-D crystal has one very important feature that is different from what is found in the Fourier transform of a 3-D crystal. The fact that translational symmetry does not exist in the third dimension means that the Fourier transform of a 2-D crystal is continuous in the third dimension rather than being confined to discrete spots, as is true for a 3-D crystal. Section 7.4 explains how it is that the 3-D Fourier transform can remain continuous in the direction perpendicular to the plane of a 2-D crystal, while still being sampled on a discrete lattice in the first two dimensions of Fourier space.

When discussing diffraction and Fourier transforms for real crystals, it is important to recognize that real crystals are also bound to have a number of imperfections. These imperfections are necessarily reflected in the diffraction data and in the Fourier transform of the image. The nature of the various imperfections that are encountered in 2-D crystals, and how these imperfections are manifested in the diffraction data, is the subject of the final part of this chapter, section 7.5.

7.2 Classes of symmetry in projection

Since we are interested in biological macromolecules, we will ignore all symmetry groups that are inconsistent with chiral structures. We first begin by considering the types of crystallographic symmetry that can exist in a mathematical plane. The description of symmetry operations in a plane is used to enumerate the different types of symmetry that can be present in projections, that is, images of thin crystals. Later, in section 7.3, we will discuss how the symmetries seen in projection are related to the symmetry of a crystal that may be as little as one unit cell in thickness.

There are only five types of symmetrical lattice in two dimensions

A *lattice* is simply a regular, geometrical arrangement of points, and a *crystallographic lattice* is — formally — one that is infinite in extent. By extending the arrangement of points to infinity, the lattice remains indistinguishable from its initial state after suitable, allowed symmetry operations have been carried out on the lattice. A crystallographic lattice is thereby transformed into itself by a *crystallographic symmetry operation*.

An infinite lattice that lies within a single plane is properly called a *net* (Bravais, 1969). According to convention, two real-space vectors are identified as **a** and **b**, and the angle between them is designated by γ. The vectors **a** and **b**, in turn, define a parallelogram, which is the *unit cell* of the net. The lengths of the unit cell vectors are designated by the corresponding symbols without bold face, that is, a and b, respectively. The crystallographic convention is that the vector **b** should point horizontally, to the right, while the vector **a** should point down; in addition, the angle, γ, should not be less than 90°.

The most general net is one in which the unit cell is a *rhomboid*, which is to say that a $\neq$ b and $\gamma \neq 90°$. Unit cells that are *square*, *rectangular*, or *hexagonal* can thus be seen to be special cases of the general, rhomboid net. A net is said to be *oblique* if it does not fall into one of the three special cases just mentioned. The rectangular

Table 7.1 Defining characteristics of the five types of two-dimensional nets

Shape of the unit cell	Values for unit-cell lengths and the angle between them[a]	Common name
Parallelogram (rhomboid)	$a \neq b$ $\gamma \geq 90°$	Oblique or monoclinic
Rectangle	$a \neq b$ $\gamma = 90°$	Rectangular (primitive)
Rectangle	$a \neq b$ $\gamma = 90°$	Rectangular (centered)
Square	$a = b$ $\gamma = 90°$	Square or tetragonal
60-° angle rhombus	$a = b$ $\gamma = 120°$	Hexagonal

[a] By convention, the b-axis (y-axis) is horizontal and the a-axis (x-axis) points downward. γ, the angle between the a-axis and the b-axis, is greater than or equal to $90°$.

unit cell is unique in that it gives rise to the only net that can be *centered*, that is, it is the only net that can have a point at the center of the unit cell as well as those at the corners. As we will see in more detail, the lower-case, italic symbol p is used to designate nets that are *primitive*, in the sense that there are points only at the corners, while the symbol c is used for the centered net. The centered net could as well be described by a primitive net, but doing so would mean that the additional symmetry that is present in a centered, rectangular net would be ignored. The distinctions that can be made between the five different nets are summarized in table 7.1.

There are 10 crystallographic point groups in two dimensions

The concept of *point-group symmetry* is used to describe all operations, except those involving translation, that transform a given object into itself. Rotation of an equilateral triangle by $120°$, about the point at the middle of the triangle, is an example of such an operation. Because translation of the object is not one of the allowed operations, at least one point of the object is not moved during a point-group symmetry operation.

Rotation about a point, as in the example of an equilateral triangle, and *reflection across a line* are the only point-group symmetry operations that are possible in two dimensions. A rotation is said to be *n-fold*, and the operation is identified by the integer, n, if the amount of rotation is $2\pi/n$. The operation of reflection across a line is identified by the symbol in italics, m, also known as a *mirror operation*.

If a 3-D object contains an axis of twofold rotational symmetry, projections of this structure will also have special symmetry. If the projection operation is carried out in a direction perpendicular to the 2-fold axis, then the axis itself becomes a mirror line. On the other hand, if the direction of the projection is parallel to the 2-fold axis, then the projection possesses *inversion symmetry* about the point of the 2-fold axis. A visual definition of each of these symmetry operations, shown in figure 7.1, may help one to understand the terms and concepts, in the event that they are not already familiar to the reader.

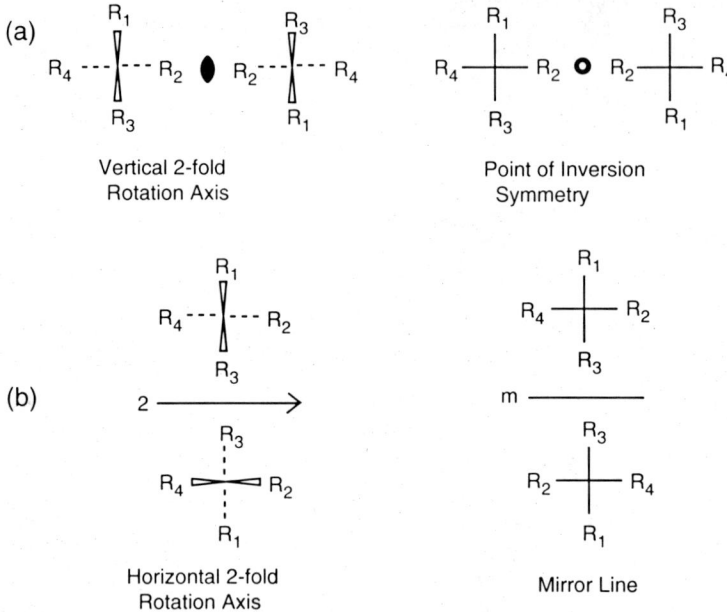

Figure 7.1 A cartoon of a chiral center, with four different residues, R_1, R_2, R_3, and R_4, illustrates the difference between a vertical and a horizontal 2-fold rotation axis, and shows how the two operations lead to inversion symmetry or mirror symmetry in the projection images of the chiral structure. (a) A pair of identical, chiral centers is shown on the left with their positions and orientations related by a vertical, 2-fold symmetry axis, which itself is designated by the filled (black) oval, or almond-shaped symbol. The same pair is shown in projection on the right, where it can be seen that the four residues on one center are now related to the four residues on the other by inversion through the central point, which is marked by the filled circle. (b) A pair of identical, chiral centers is shown on the left with their positions and orientations related by a horizontal, 2-fold symmetry axis, which itself is designated by the symbol "2" and an arrow. The same pair is shown in projection on the right, where it can be seen that the four residues on one center are now related to the four residues on the other by reflection across the mirror line, which is designated by the symbol "m" and a solid line.

The 10 2-D point groups and the system of notation by which they are identified are described in more detail in table 7.2. As already mentioned, only two symbols are used to describe the symmetry operations that distinguish the 10 point groups; an integer describes the rotational symmetry about an axis that is perpendicular to the plane, and the symbol m describes mirror reflection across a line. In the case of the 2mm, 4mm, and 6mm point groups, the combination of the rotation axis and one mirror line already implies the existence of a second mirror line, which is designated as such in the third position of the "full" symbol. Strictly speaking, therefore, the addition of the second mirror in the notation is redundant, and some authors will prefer to use the equivalent (short) notation that is indicated in parentheses. In this case, 2m and the conventionally used mm are equally complete short notations for 2mm.

The five different nets described in the previous subsection can give rise to 10 different objects, depending upon the type of *crystallographic point-group symmetry*

Table 7.2 Classification of the 10 2-D point groups (G_0^2) according to the four possible lattice types in two dimensions

Lattice type	Point-group symmetry[a]
Oblique	1, 2
Rectangular	1m, 2mm
	(m) (mm)
Square	4, 4mm
	(4m)
Hexagonal	3, 3m, 6, 6mm
	(6m)

[a]Integers represent rotation axes with n-fold symmetry, and the symbol "m" represents mirror reflection across a line. The short symbol, if any, for a given point group is given in parentheses, below the full-point group symbol. Symmetry operations that are left out in the short symbol are already implied by the remaining symmetry operations.

with which they are endowed. The restriction to only 10 different classes is the result of requiring that the object is a net, and therefore has translational symmetry, even though translation is not yet one of the operations that we want to discuss. The unique symmetry operations of each of the ten crystallographic point groups are illustrated with conventional symbols in figure 7.2. The requirement that the object is a net immediately limits the type of rotational symmetry that is possible to $n = 1, 2, 3, 4$, and 6. As is shown in figure 7.2, and as is summarized in table 7.2, oblique nets may or may not have 2-fold rotational symmetry, and this is true for rectangular nets as well. Square nets on the other hand must have 4-fold rotational symmetry, while hexagonal nets can have either 3-fold or 6-fold rotational symmetry. Out of these seven options, 10 different point-group symmetries emerge because of the fact that nets with 3-fold, 4-fold, and 6-fold rotational symmetry can exist either with or without mirror symmetry.

There are 17 space groups in two dimensions

The concept of *space-group symmetry* is used to describe all operations, including translation, that transform a crystalline object into itself. When the translational operations are limited to two dimensions, the space group is called a *plane group* (Hahn, 2002).

Interestingly, a new plane-group operation becomes available in two dimensions, in addition to the point group operations and the regular (translational) repeat of the unit cell that is inherent in the 2-D lattice, or net. The new operation is called a *glide*, and it is designated by the symbol g. The glide operation involves translation by one half of the repeat distance in a particular direction, followed by reflection across that line. A glide line occurs in the 2-D projection of a chiral structure only if the parent 3-D structure contains a 2-fold screw axis. Similarly, as mentioned previously, a mirror line occurs in the 2-D projection of a chiral structure only if the parent 3-D structure contains a 2-fold rotation axis. A *screw operation*, as its name implies, combines the operations of rotation and translation.

The glide operation can occur in only a limited number of combinations with the point-group operations of rotation and reflection. As a result there are only 17 combinations (i.e., groups) of crystallographic symmetry operations that are possible in

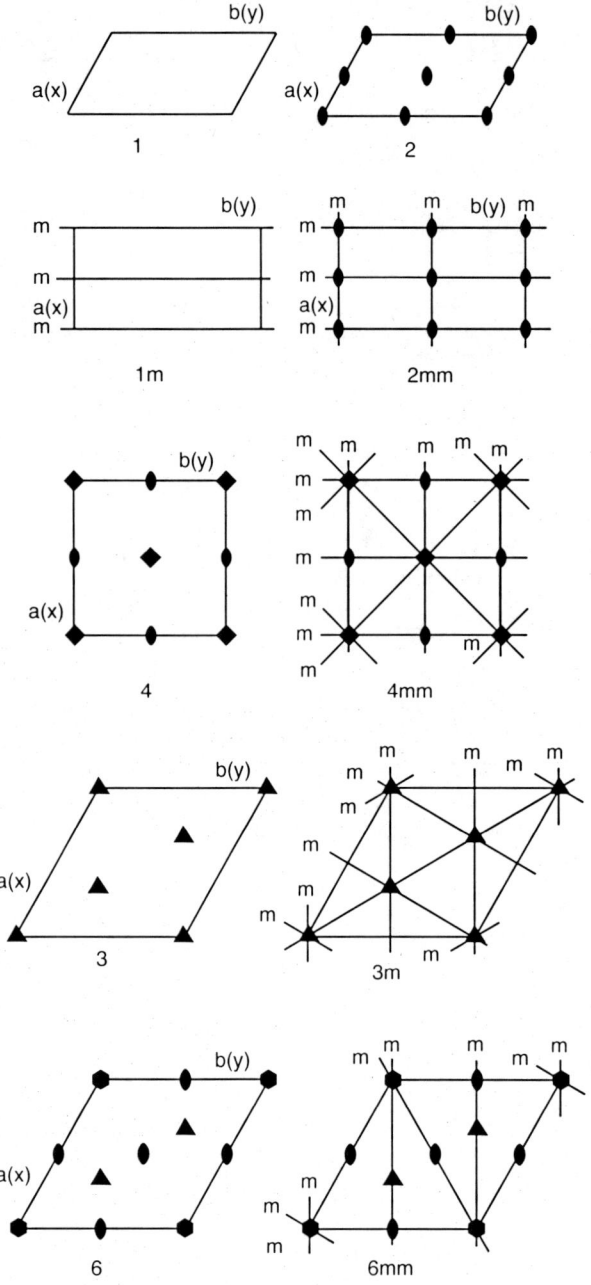

Figure 7.2 Schematic representations of the 10 crystallographic, 2-D point groups (G_0^2). Rotational symmetry axes perpendicular to the plane are shown by the filled symbols that have 2-fold, 3-fold, 4-fold, and 6-fold symmetry to designate the corresponding rotational symmetry of the respective axes. A mirror line is shown by the symbol "m" at one end of a solid line. For clarity, only one pair of mirror lines is shown in the illustration of point group 6mm, the remaining five pairs being generated from the pair shown in a self-evident way by the 6-fold symmetry axis.

Table 7.3 The 17 crystallographic groups (G_2^2)

Lattice type	Point-group symmetry[a]	Full plane-group symbol[b,c]	Short symbol	Conventional space-group number
Oblique	1	$p1$	$p1$	1
	2	$p211$	$p2$	2
Rectangular	m	$p1m1$	pm	3
		$p1g1$	pg	4
		$c1m1$	cm	5
	$2mm$	$p2mm$	pmm	6
		$p2mg$	pmg	7
		$p2gg$	pgg	8
		$c2mm$	cmm	9
Square	4	$p4$	$p4$	10
	$4mm$	$p4mm$	$p4m$	11
		$p4gm$	$p4g$	12
Hexagonal	3	$p3$	$p3$	13
	$3m$	$p3m1$[d]	$p3m1$	14
		$p31m$	$p31m$	15
	6	$p6$	$p6$	16
	$6mm$	$p6mm$	$p6m$	17

[a] The meaning of symbols is defined in the footnote to table 7.2.
[b] The first position uses the symbol p to designate that the lattice is primitive, and the symbol c to designate that the lattice (in this case a rectangular lattice) is centered. Primitive lattices are always defined by the two shortest, independent vectors whose axes satisfy the definitions given in table 7.1.
[c] The second position specifies the rotational symmetry of an axis perpendicular to the plane. The integer 1 is placed here, in the full plane-group symbol, if there is no rational symmetry about an axis perpendicular to the plane. The symbols in the third and fourth positions, if any, designate symmetry elements in a direction parallel to (but not necessarily coincident with) the b-axis (third position) and a second direction (fourth position). The symbol "g" is used to represent a glide operation, defined in the text. The meaning of these symbols is further illustrated in figure 7.3.
[d] Two hexagonal plane groups exist with $3m$ point-group symmetry. Figure 7.3 emphasizes the difference between them that is implied by the notation $p3m1$ and $31m$.

two dimensions. These 17 plane groups, and the notation by which they are designated, are enumerated in table 7.3, and the symmetry operations for each are displayed visually in figure 7.3. As before, some of the symmetry elements identified in the full space-group notation are implied by the presence of other symmetry elements, thereby allowing the alternative, short notation to be used. All 2-D projections of crystalline specimens (electron microscope images, for example) must fall into one of the 17 plane-group symmetries listed in table 7.3.

Reflection across a line and *inversion* through a point (which is equivalent to 2-fold rotation perpendicular to the plane) are allowed operations in images of crystals of biological macromolecules, even though they are not allowed in the parent, chiral objects themselves. The fact that mirror reflection and inversion through a point do not exist in biological macromolecules should not lead to confusion about the presence of mirror and glide operations — or inversion — in the space groups that describe images, since these are projections of the structure. To illustrate this point, consider a protein dimer that is made up of one copy that is facing up and a second that is facing down, that is, is rotated by 180° from the first; a clenched hand would serve as

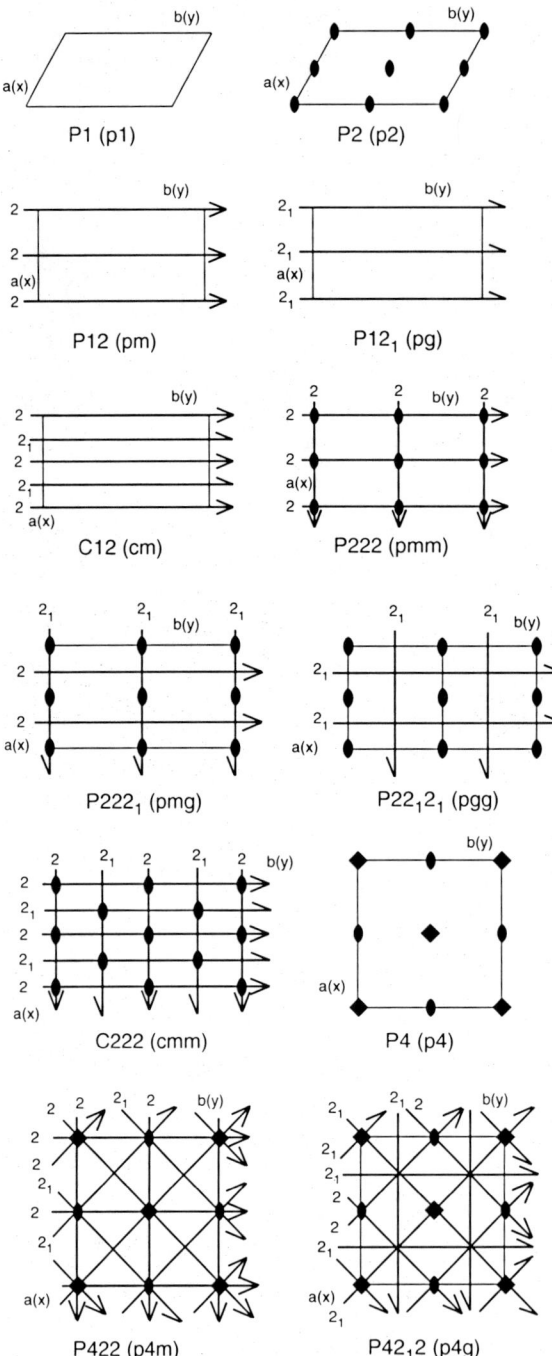

Figure 7.3 Schematic representations of the 17 enantiomorphic layer groups (G_2^3). As in figure 7.2, rotational symmetry axes perpendicular to the layer are shown by the filled symbols that have 2-fold, 3-fold, 4-fold, and 6-fold symmetry, respectively. Two-fold axes parallel to the layer are shown by the symbol "2" at one end of a full arrow, while 2-fold screw axes parallel to the layer are shown by the symbol "2_1" at the end of a half arrow. Note that the 17 crystallographic plane groups (indicated in parentheses), which are generated as projections of the layer groups, can be illustrated by the same diagrams simply by replacing the 2-fold axes with mirror lines, and the 2-fold screw axes with glide lines. For the sake of clarity, the symbols for many of the 2-fold rotation axes have been replaced by mirror lines (represented only by bold lines) in the diagrams for the $P312$ and $P622$ layer groups, and many of the 2-fold screw axes have been replaced by glide lines (represented only by bold dashed lines).

Symmetry and Order in Two Dimensions

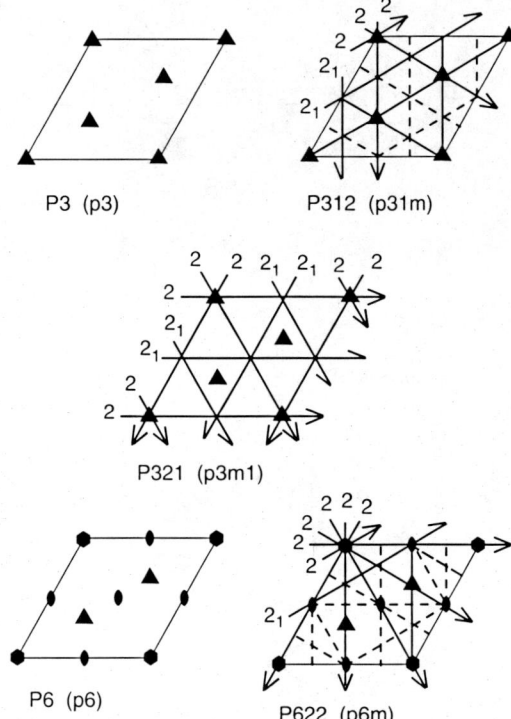

Figure 7.3 *(continued)*

an example. Projection in the direction perpendicular to that 2-fold axis will produce two side-by-side images that are related by reflection across a line, as is illustrated in figure 7.4. In a similar way, projection perpendicular to a 2-fold screw axis will produce a glide line.

7.3 Three-dimensional symmetry classes for monolayer crystals

It is common practice in electron crystallography to refer to a thin specimen as being a "2-D crystal," even though it is understood that the molecules that make up the crystal are, themselves, 3-D structures. In those cases when the sample is really just one unit cell thick, a more accurate description would be to refer to the specimens as "monolayer crystals," as in the heading of this section. Use of the more euphonious "2-D crystal" nevertheless prevails in normal conversation.

Ambiguity regarding terminology can enter, unfortunately, when we wish to speak about the space group of our 2-D (i.e., monolayer) crystal. The problem here is that it may be unclear, when we speak of a 2-D space group, if we are referring to the *plane group* that describes a 2-D projection, or to the 2-D *layer group* of the (3-D) object itself. A more extensive discussion of crystallographic symmetry in the plane and in a 2-D layer is presented in box 7.1. The distinction between plane groups

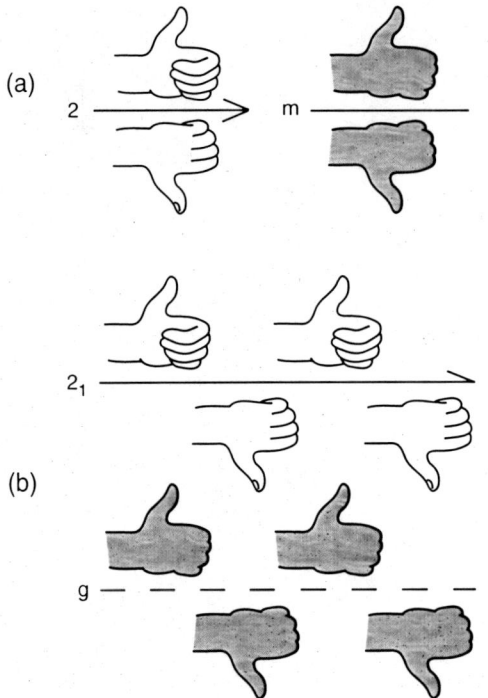

Figure 7.4 A cartoon of a partially clenched hand is used to illustrate how mirror symmetry can arise in the projection (shaded gray) of a chiral structure when there is either a 2-fold rotation axis or a 2-fold screw axis lying within the plane. The drawing in (a) illustrates a packing arrangement in which a left hand is present within a unit cell in two copies, each related to the other by a 2-fold rotation axis. In projection, however, the images of the two left hands are the mirror image of one another. The drawing in (b) shows the two hands packed with a 2-fold screw axis relating one of them to the other. In projection their respective images are related by a glide plane, that is, by the combination of translation and mirror symmetry.

BOX 7.1 Space groups, layer groups and plane groups

Space groups are sets of operations that combine point group symmetry with translational symmetry. Since the translational symmetry operations can be carried out in 1, 2, or 3 dimensions, it is necessary to use terminology that can distinguish between those different possibilities.

We here follow the usage of Vainshtein (1981) in which a clear distinction is made between space groups in which translation is confined to 1, 2, or 3 dimensions, respectively, and (for the first two cases) to objects (unit cells) that may or may not be 3-D. In the notation used by Vainshtein, a specific space group is designated by a symbol, G_n^m, in which the integers m and n can (in principle) have values ranging from 0 to 3, but not in all combinations. The superscript in this notation designates the dimensionality of the unit cell, and the subscript designates the number of dimensions in which translational operations occur. The group G_3^3 thus designates the set of space groups of 3-D crystals, while G_2^3 designates the set of space groups in which 3-D objects are translated in only two dimensions. Continuing with this notation, the symbol G_0^3 represents the point groups for 3-D objects, while G_2^2 represents the set of space groups for 2-D objects lying in a plane.

(continued)

The latter set of space groups, G_2^2, can be called *plane groups*. Images of monolayer crystals, and projections of crystalline objects in general, can be represented by space groups of the type G_2^2, that is, by plane groups. Monolayer (i.e., 2-D) crystals themselves, however, would be represented by groups of the type G_2^3. Vainshtein suggests the clean terminology in which the expression *layer groups* refers specifically to the class of objects whose symmetry is of the type G_2^3. Although not of immediate interest at this point, we mention that the terminology used by Vainshtein can be extended in a natural way to *line groups*, G_1^1, and to *rod groups*, G_1^3, the latter of which would include cylinders and helixes.

Papers in the field of electron crystallography have often used the expression "*two-sided plane group*" (Amos et al., 1982) to describe the space-group assignment for a 2-D crystalline specimen. We prefer, and recommend, Vainshtein's expression "layer group" because — in the context of biological macromolecules — it is intuitively more expressive of the type of object to which it is to be applied. The earlier terminology of "two-sided plane group" for symmetry groups of the type G_2^3 is eminently suitable for objects such as twin boundaries or mathematical planes which are colored differently on the two sides of a plane (Holser, 1958), but its application to monolayer crystals is less appealing. We would therefore discourage use of the terminology "two-sided plane group" in favor of "layer group" when designating the 3-D space group of a 2-D crystal. To be clear on the point, however, space groups of the type G_2^3 are the same thing whether called layer groups or two-sided plane groups. For chiral molecules the operations from which layer groups can be composed include only translation in two dimensions, 2-fold rotation and screw operations within the plane, and 2-, 3-, 4-, or 6-fold rotation perpendicular to the plane.

Following the convention of Holser (1958), who listed the groups in a notation that is consistent with Hermann–Mauguin form of the International Tables for X-ray Crystallography, the first position identifies the lattice type (primitive; centered). The second position identifies an axis of rotational symmetry perpendicular to the plane; the absence of a rotational symmetry axis perpendicular to the plane must be noted explicitly by the symbol "1" in the second position. The third position denotes a rotational or screw axis (if any) lying in the plane of translation, by convention designated as the x-axis, pointing downward. The y-axis then points horizontally and to the right. The fourth position denotes any further rotational or screw axes that also lie in a second direction within the plane of translation.

---

and layer groups is maintained by using lower-case italic letters to designate a 2-D function such as a projection or a net, and upper-case italic letters for 3-D objects, in our case monolayer (or other thin) crystals.

Vainshtein (1981) introduces a helpful notation for designating the different types of symmetry groups that describe different classes of specimen, which we will use in the following sections. The class G_0^2, for example, would refer to 2-D point groups, the subscript zero indicating that there are no translational symmetry operations and the superscript 2 indicating that the objects are 2-D. As a second example, one would designate the 2-D space groups (i.e., layer groups) by the symbol G_2^3 since the objects are 3-D, but translations occur only in two dimensions.

Biological macromolecules can pack in only 65 space groups (G_3^3)

The combination of all possible crystallographic point-group symmetry operations with 3-D translational symmetry operations results in a total of 230 space groups. The majority of these space groups contain mirror symmetry, glide planes, or inversion symmetry, however, and those can all be eliminated as candidates for protein crystals. When the set of allowed space groups is limited to those which contain chiral structures, only 65 of the original 230 space groups remain. The 65 enantiomorphic space groups are enumerated in table 7.4, which is taken from section 4.6 of Blundell and Johnson (1976), where additional details are given concerning the arrangements of the symmetry elements.

Following the standard conventions for *space-group notation*, which are described in the International Tables, the letter in the first position describes the lattice type (*P* indicates a *primitive lattice*, for example). Integers represent rotational symmetry axes, and integers with subscripts indicate screw axes. According to commonly used conventions, the first integer in the space-group symbol refers to the direction of the **b** axis for a monoclinic crystal and to the **c** axis for both square and hexagonal space groups. Additional integers indicate the presence of rotational or screw axes oriented independently of the **b** axis. As is shown in table 7.4, the 65 chiral space groups are classified into 7 systems, which in turn are described by 11 point groups.

Biological macromolecules can pack in only 17 layer groups (G_2^3)

Each of the 17 plane groups has associated with it a single layer group (G_2^3) for which there still are no translational symmetry operations in the direction perpendicular to

Table 7.4 The 65 entantiomorphic, three-dimensional space groups (G_3^3)

Common name (system)	Point-group symmetry	Space-group symbol
Triclinic	1	$P1$
Monoclinic	2	$P2, P2_1, C2$
Orthorhombic	222	$C222, P222, P2_12_12_1, P2_12_12, P222_1, C222_1,$ $F222, I222, I2_12_12_1$
Tetragonal	4	$P4, P4_1, P4_2, P4_3, I4, I4_1$
	422	$P422, P42_12, P4_122, P4_12_12, P4_222, P4_22_12,$ $P4_32_12, P4_322, I422, I4_122$
Trigonal	3	$P3, P3_1, P3_2, R3$
	32	$P312, P321, P3_121, P3_112, P3_212, P3_221, R32$
Hexagonal	6	$P6, P6_5, P6_4, P6_3, P6_2, P6_1$
	622	$P622, P6_122, P6_222, P6_322, P6_422, P6_522$
Cubic	23	$P23, F23, I23, P2_13, I2_13$
	432	$P432, P4_132, P4_232, P4_332, F432, F4_132,$ $I432, I4_132$

the plane of the 2-D crystal. The 17 plane groups map in a trivial way to the 17 layer groups, which are enumerated in table 7.5, by replacing a mirror line in the plane group with a 2-fold axis in the layer group, or by replacing a glide operation in the plane group with a 2-fold screw in the layer group. To illustrate this, the plane group $p2$ maps to the layer group $P2$, while the plane group pm maps to the layer group $P12$. In the latter case the symbol "1" is used in the second position of the layer-group notation, to make it clear that the 2-fold axis (in the third position) lies parallel to the plane of the

Table 7.5 Identifying characteristics of the 17 entantiomorphic layer groups (G_2^3)

Conventional space-group number	Point-group symmetry	Projection symmetry	Layer-group symbol[a,b]	Systematic absences in the ($hk0$) plane	Phase constraints in the ($hk0$) plane
1	1	$p1$	$P1$		none
2	2	$p2$	$P2$		all reflections 0 or π
3	m	pm	$P12$		$k=0$: 0 or π
4	m	pg	$P12_1$	h odd, $k=0$	$k=0$: 0 or π
5	m	cm	$C12$	$(h+k)=$ odd	$k=0$: 0 or π
6	$2mm$	pmm	$P222$		all reflections 0 or π [also extending along ($h0z^*$) and ($0kz^*$)]
7	$2mm$	pmg	$P222_1$	$h=0$, k odd	$h=0$ and $k=$ even: 0 or π $h>0$, any k: 0 or π $h=0$, $k=$ odd: $\pi/2$ or $3\pi/2$
8	$2mm$	pgg	$P22_12_1$	h odd, $k=0$ and $h=0$, k odd	$h>0$ and $k>0$: 0 or π h even and $k=0$: 0 or π $h=0$ and k even: 0 or π h odd and $k=0$: $\pi/2$ or $3\pi/2$ $h=0$ and k odd: $\pi/2$ or $3\pi/2$
9	$2mm$	cmm	$C222$	$(h+k)=$ odd	all 0 or π
10	4	$p4$	$P4$		all 0 or π
11	$4mm$	$p4m$	$P422$		all 0 or π
12	$4mm$	$p4g$	$P42_12$	h odd, $k=0$	$h>1$ and $k>1$: 0 or π h even and $k=0$: 0 or π $h=0$ and k even: 0 or π h odd and $k=0$: $\pi/2$ or $3\pi/2$ $h=0$ and k odd: $\pi/2$ or $3\pi/2$
13	3	$p3$	$P3$		none
14[c]	$3m$	$p31m$	$P312$		$h=k$: 0 or π
15[c]	$3m$	$p3m1$	$P321$		$h=0$: 0 or π $k=0$: 0 or π
16	6	$p6$	$P6$		all 0 or π
17	$6mm$	$p6m$	$P622$		all 0 or π (extending to all z^* as well)

[a] The symbol 2_1 designates a 2-fold screw operation, which can only occur in the plane for 2-D space groups (i.e., layer groups).
[b] The first position of the layer-group symbol designates whether a lattice is primitive or centered. The second position specifies the rotational symmetry of an axis perpendicular to the layer. The third position, if any, designates symmetry elements oriented parallel to (but not necessarily lying coincident with) the b-axis, also known as the y-axis. The fourth position, if any, designates symmetry elements in a second direction, once again parallel to the layer.
[c] The convention for numbering of the layer groups $P312$ and $P321$ is reversed from the corresponding plane group numbering for their projections. Thus the plane group $p31m$ is number 14 in this table, but number 15 in table 7.3.

layer, rather than being perpendicular to it as in the *P*2 layer group. Both layer groups are, of course, derived from a 3-D crystal whose space-group is *P*2, by extracting a layer that is one unit cell thick. The *P*12 layer would contain the 2-fold axis of the *P*2 (3-D) space group within the plane of the specimen, whereas the 2-fold axis of the *P*2 (3-D) space group would be perpendicular to the plane of the *P*2 layer.

As is explained above, the layer group for a 2-D crystal is immediately known as soon as a plane group is assigned to the image of an untilted specimen. Untilted specimens are essential for making this assignment because their projection images exhibit more (i.e., higher) symmetry than is present in the projection of a tilted specimen. To understand this point, one can think about the projection of a square lattice, which is 4-fold symmetric when untilted but becomes 2-fold symmetric when tilted, or of a 3-fold symmetric lattice, which loses its 3-fold symmetry when the specimen is tilted.

The assignment of the 2-D space group of a projection can be made on the basis of the geometry of the lattice (net); the presence of symmetry-forbidden reflections, if any; and the phase residual between the experimental phases and those required for competing space-group assignments. For example, a computer program, called ALLSPACE, has been described which calculates the phase residual to be expected for each of the 17 space group families (Valpuesta et al., 1994). The calculation takes into consideration the experimentally determined signal-to-noise values at each reflection, which themselves generate a finite phase residual even when the correct space group relationships are used to calculate the phase residual. When a wrong space group is tested, it will produce a phase residual that is larger than that which would be generated just from the inherent noise in the data, making it possible to eliminate that particular space group from further consideration.

Use of the full symmetry of a space group maximizes the signal-to-noise ratio

It is important to understand why it is valuable to determine the space group of a crystal. It is true, after all, that one could process data for all 2-D crystals with no more information than a determination of the values of the two unit cell vectors, **a** and **b**, and an estimate of the thickness of the crystal. In other words, all crystals could be treated as *P*1 structures. The only redundancy present in the Fourier transform of a *P*1 structure is that provided by the Friedel relationship (box 3.3) between points at **g** and −**g**. The unique data to be measured would thus be distributed within a hemisphere of Fourier space, centered on the origin.

The presence of higher symmetry increases the redundancy that is present in the measured data. Diffraction spots whose values are redundant, by reason of real-space symmetry in the crystal, are said to be *symmetry-related reflections*. Far from wanting to discard redundant data, our goal can be to average all symmetry-related measurements so as to improve the signal-to-noise ratio in the measurement. Improvement in the signal-to-noise ratio that results from using the space group with the highest possible symmetry explains why face-centered space groups are included in the classification of crystal lattices, when the same packing of molecules could also be described by the primitive subgroup.

A specific example will help illustrate how symmetry in the real-space packing arrangement leads to redundancy in the 3-D Fourier transform. The molecules of bacteriorhodopsin that are found in the purple membrane of *Halobacterium salinarum* (see, for example, section 1.3) form a 2-D crystal with *P*3 symmetry. When such a crystal is rotated by 120° or by 240° about an axis perpendicular to the plane, it cannot be distinguished from the original crystal. It therefore follows that the Fourier transforms in all three orientations must also be identical. The z^* axis in Fourier space must therefore be a 3-fold symmetry axis, that is, the values of the Fourier transform at points related by 3-fold rotation about this axis must have the same value. The unique data are thereby reduced from a hemisphere to a 60° wedge, that is, one-third of a hemisphere. This point is explained again in figure 7.5 and in the accompanying caption. Data measured within each 60° wedge of reciprocal space can be averaged not only with their respective Friedel mates, but also with data from the reflections that are related by 3-fold rotation.

For many layer groups it is possible to ignore some of the real-space symmetry and to treat the crystal as if it had lower symmetry, without having to ignore symmetry completely. A *P*6 crystal, for example, could be treated as a *P*3 crystal, or a *P*4 crystal

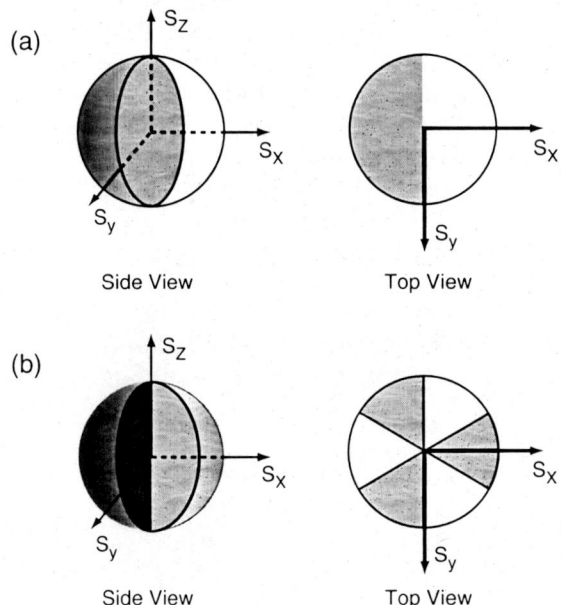

Figure 7.5 Symmetry determines the fraction of data in Fourier space that is unique. In the absence of rotational symmetry, the unique data fill half of reciprocal space, represented by the clear hemisphere in panel (a). Because of Friedel symmetry, the values of the Fourier transform in opposite hemispheres are the complex conjugates of each other. All of the data in the shaded hemisphere is therefore redundant with that in the clear hemisphere. When the object, and thus its Fourier transform, has 3-fold rotational symmetry, however, then the values within a given 60° wedge, shown by one of the clear sectors in panel (b), must be identical to the data in the two other 60° wedges that are also shown as clear sectors. Once again, Friedel symmetry makes the values in the shaded wedges redundant with the values in the clear wedges. Thus, the unique data are confined to a single 60° wedge, such as the one shown facing us in the side view.

could be treated as a *P*2 crystal. In general, any space group that is produced by ignoring one or more of the symmetry operations that are present is called a *subgroup* of the parent. No error can be made by assigning a symmetry subgroup for the purpose of data processing; the only disadvantage is that one is not taking full advantage of the redundancy inherent in the data.

On the other hand, a serious mistake will be made if a space group is assigned that has one or more additional symmetry operations beyond those that are really present in the crystal. A space group that is formed by the addition of an operation to a parent group with lower symmetry is called a *supergroup* of the parent. For example, the space group *P*6 is a supergroup of *P*3, and *P*4 is a supergroup of *P*2. Assignment of a supergroup will result in averaging data that are not, in fact, equivalent or redundant. If the unit cell of a 2-D crystal is an equilateral triangle, for example, and the crystal is correctly assigned *P*3 symmetry, then the result of averaging symmetry-related data will produce an equilateral triangle, as it should. If the crystal is wrongly assigned *P*6 symmetry, however, the result will look, incorrectly, like a six-pointed star.

It is important to know that another type of structural redundancy can be present in real crystals. This additional redundancy occurs when biological macromolecules crystallize as symmetrical oligomers, but the packing arrangement does not "take advantage" of the symmetry of the oligomer. Trimers of porin, for example, can crystallize in an orthorhombic lattice (Jap et al., 1991), which lacks a 3-fold symmetry axis. Crystals in which this happens are then said to have *noncrystallographic symmetry*. It is very advantageous to use this noncrystallographic symmetry to improve the signal-to-noise ratio. Exploitation of noncrystallographic symmetry is more complicated than the simple averaging of symmetry related reflections, however. More can be found on the topic of noncrystallographic symmetry in standard texts of x-ray crystallography, such as the one by Drenth (1994).

7.4 The Fourier transform of a 2-D crystal is sampled at discrete points in two dimensions, but it is continuous in the third dimension

The reader is familiar with the fact that the Fourier transform of a periodic function (i.e., a crystal) is a set of regularly spaced diffraction spots. Furthermore, in section 2.5, we derived the fact that the Fourier transform of a crystal is a set of regularly spaced, discrete samples of what would otherwise be a continuous function, that is, the Fourier transform of a single unit cell. The derivation of this familiar result represents a special case, however: it applies only to diffraction from 3-D crystals. Electron crystallography, on the other hand, requires the use of thin crystals, preferably ones that are only a single molecule thick. We need to understand, therefore, that the mathematical theory is no longer the same for a thin, 2-D crystal.

It should be plausible, by intuition, that the Fourier transform of a 2-D crystal will continue to be sampled — in the same, regularly spaced fashion as before — in the two directions of Fourier space that correspond to the plane of the crystal. At the same time, however, the Fourier transform of a 2-D crystal retains the original, continuous values in the third direction that are found in the Fourier transform of a single unit cell. A formal, mathematical proof of these two points will be discussed shortly.

It may first be helpful, however, to describe what is meant when we say that the Fourier transform of a single unit cell can be sampled in two dimensions while remaining continuous in the third dimension. The answer is given by thinking of a regular array of parallel lines, as in the illustration shown in figure 7.6. Since the lines are parallel to one another and since they are spaced apart at regular intervals in two dimensions, this set of lines can be used to sample the continuous Fourier transform in the same way that a 3-D lattice of points is used to sample the Fourier transform of a 3-D crystal. The values of the continuous Fourier transform that lie on these parallel lines provide discrete samples in two dimensions while retaining the original, continuous values of the Fourier transform in the third dimension.

We will refer to the set of parallel lines shown in figure 7.6 as the *reciprocal lattice lines*, although they are also called "reciprocal lattice rods." The reciprocal lattice lines play the same role in indexing diffraction data from 2-D crystals that is played by reciprocal lattice points in the case of 3-D crystals. In the case of a 2-D crystal, however, the third integer, l, of the traditional Miller indices, (h, k, l), is replaced by a continuous variable, usually denoted by z^*. Thus, each reciprocal lattice line is specified by two, not three Miller indices, and a particular point on a given line is identified by the triplet, (h, k, z^*).

An instructive proof that the Fourier transform of a 2-D crystal remains continuous in the third dimension begins by observing that a 2-D crystal can be represented as the product of a 3-D crystal and a one-dimensional rectangle function, designated $S(z)$ (where the symbol S indicates the support, or the shape provided by the rectangle function), whose width is equal to one unit cell in the z-direction. This equivalence is restated mathematically in the following equation:

$$\hat{\rho}_{2-D\,\text{crystal}}(\mathbf{R}) = \hat{\rho}_{3-D\,\text{crystal}}(\mathbf{R}) \cdot S(z), \tag{7.1}$$

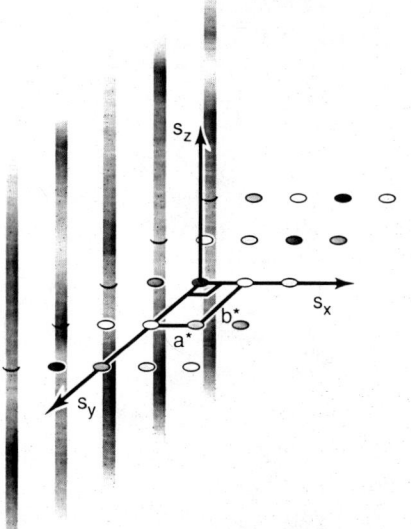

Figure 7.6 The idea of parallel lines arranged on a regular lattice in reciprocal space is shown here by a cartoon in which the lines are parallel to the s_z axis. The independent variation of the value of the Fourier transform on each line is indicated schematically by smooth variations of shading along each line. Only five such parallel lines are shown here, but the positions of others are indicated by variously shaded disks where the lines intersect the s_x, s_y plane. The parallel lines are separated from one another by fixed amounts, which can be different in two independent directions. In this particular example, the lattice is shown as being one with a 90° angle between the **a*** and **b*** basis vectors.

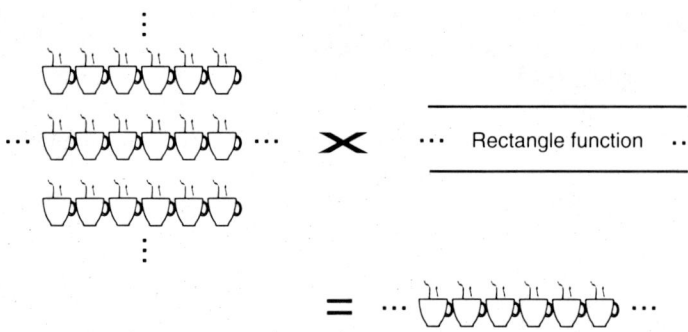

Figure 7.7 A 2-D crystal can be represented as the mathematical product of a 3-D crystal and a rectangle function that extends infinitely far in two dimensions, but which is only one unit cell thick in the third dimension. This idea is illustrated here in two dimensions: the rectangle function has a value of 1.0 within the slab indicated by the box, and a value of zero outside the slab. The third dimension of the 3-D crystal, and the second dimension of the resulting 2-D crystal are implied to be perpendicular to the plane of the figure.

and the concept is illustrated by the cartoon shown in figure 7.7. The convolution theorem (box 2.3) then tells us that the Fourier transform of a 2-D crystal is the convolution of (a) the Fourier transform of the 3-D crystal and (b) the Fourier transform of the rectangle function, which we designate by the function $\sigma(z^*)$, that is,

$$F_{2-D\ \text{crystal}}(S) = F_{3-D\ \text{crystal}}(S) * \sigma(z^*). \tag{7.2}$$

To keep the following explanation simple, we will assume that the **c*** axis of the 3-D reciprocal lattice is perpendicular to the **a*** and **b*** axes. The Fourier transform of the rectangle function, on the other hand, is easily shown to be the *sinc function*, $\frac{\sin 2\pi \Delta z \cdot z^*}{2\pi \Delta z \cdot z^*}$. Because Δz, the width of the rectangle function in real space, is taken to be precisely the unit cell repeat distance along the c-axis, the sinc function will have a width (from the center to the first zero) that is equal to the **c*** repeat of the 3-D reciprocal lattice. Recall that, in other contexts, the Fourier transforms of rectangle functions with various shapes also play key roles in understanding how the objective lens aperture limits microscope resolution (section 3.5), how the resolution of a density map is limited by the maximum spatial frequency to which diffraction data are collected (section 4.6), and how diffraction spots become broadened when the crystal size becomes extremely small (section 2.6).

Convolution of the one-dimensional sinc function with the three-dimensional reciprocal lattice points will leave the sampling unaffected in two dimensions, while generating a continuous function in the third. It is instructive to look more closely at this process. Figure 7.8 shows two of the many situations that occur in the course of the convolution operation. In the first situation, the sinc function is centered at a reciprocal lattice point, identified by the Miller indices (h, k, l). In this instance the zeros of the sinc function fall at the positions of all other reciprocal lattice points in the row, and the value of the convolution integral is equal to the (sampled) value of the reciprocal lattice point. The values of the Fourier transform of a 3-D crystal and the Fourier transform of a 2-D crystal must therefore be identical at the positions $z^* = lc^*$. After the

Symmetry and Order in Two Dimensions   185

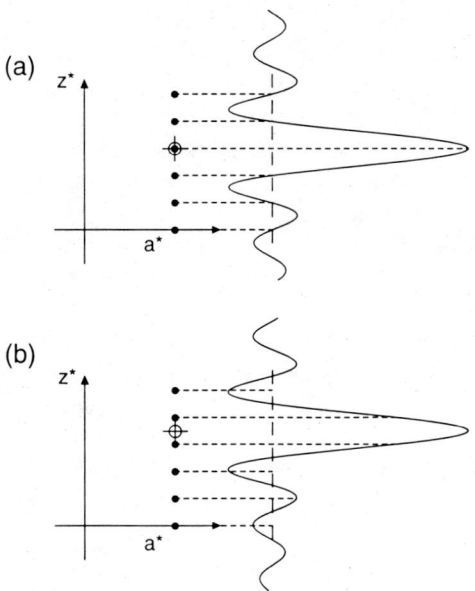

Figure 7.8 Two examples are selected to show the product formed between a 1-D sinc function and the reciprocal lattice of a 3-D crystal. The integral of this product is then computed for each position of the sinc function, when the sinc function is convoluted with the Fourier transform of the 3-D crystal. In (a) the sinc function is shown to be centered at the $(hk3)$ reciprocal lattice point, extending from there in either direction along the (hk) row of lattice points. Because the rectangle function associated with the sinc function was exactly one unit cell thick, the zeros of the sinc function coincide exactly with the reciprocal lattice points on either side of the $(hk3)$ lattice point. In (b) the sinc function is shown to be centered half way between the $(hk3)$ and the $(hk4)$ reciprocal lattice points. The sinc function now has nonzero values at all of the reciprocal lattice points along the (hk) row, and as a result the value of the convolution integral at this point on the reciprocal lattice line will be a linear combination of all of the values at each of the points on the row. The contributions from the $(hk3)$ and the $(hk4)$ reciprocal lattice points are weighted more strongly by the sinc function than are the other points, however.

sinc function has been displaced to a point between two of the reciprocal lattice points of a 3-D crystal, however, the sinc function has nonzero values at the positions of all of the reciprocal lattice points. The value of the convolution integral then becomes a weighted sum of the values found at all reciprocal lattice points in a given row. The weights given to each term in the sum are simply the values of the sinc function at each point. It is clear that the value of the convolution integral will therefore change smoothly and continuously as the sinc function is moved up and down the row.

The values of the continuous Fourier transform that are produced between reciprocal lattice points do not add any new information about the structure within the unit cell. Instead, as we have just described in the previous paragraph, the values at any point, z^*, are merely a linear combination of the values found at all of the points specified by $z^* = lc^*$. The reader should now be able to show that the value of the continuous transform at a point very close to a given reciprocal lattice point is determined mainly by the value of the Fourier transform at that point, with little influence from other reciprocal lattice points. On the other hand, the value midway between two reciprocal lattice points is determined mainly by the Fourier transform at those two points, which are given equal weight.

One more point should be made to close this section. Although the Fourier transform of a 2-D crystal is continuous in z^*, every diffraction pattern that is recorded will always consist of a 2-D array of sampled spots. This remains true no matter how the crystal is tilted, with the sole exception of the case when the crystal is tilted by 90°.

The Ewald sphere construction shown in figure 7.9 explains why the diffraction pattern of a 2-D crystal is always made up of a full set of diffraction spots, unlike that of a 3-D crystal. The fact that the Fourier transform is a continuous function along each of the reciprocal lattice lines guarantees that there will be a finite measurement for every reciprocal lattice line, regardless of the tilt angle, unless the continuous transform of the unit cell is itself zero at a particular point. In the case of a 3-D crystal, however, the Fourier transform is sampled at discrete points rather than along continuous lines, and the Ewald sphere can easily pass by without intersecting most of the reciprocal lattice points.

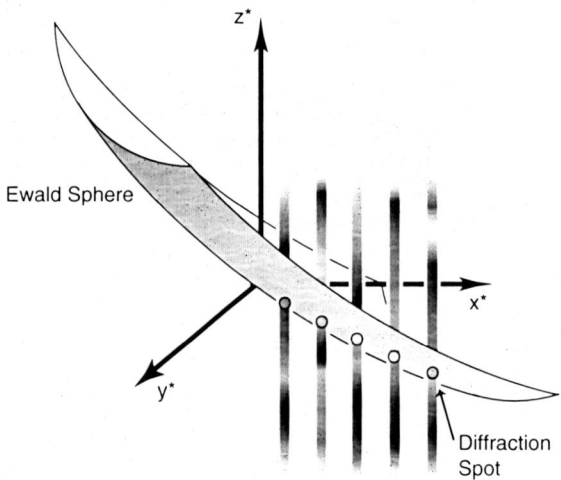

Figure 7.9 The sampling of reciprocal lattice lines that occurs on the surface of the Ewald sphere. For clarity, only a small number of the reciprocal lattice lines are shown (all of which intersect the x^* axis), and the curvature of the Ewald sphere is "foreshortened" (exaggerated) to emphasize the third dimension. Although the 3-D Fourier transform itself is continuous in the third dimension, the intersection of the Ewald sphere with this Fourier transform nevertheless always leads to a diffraction pattern that is made up of a 2-D array of discrete spots. The diffraction spots in this case represent the points where the Ewald sphere intersects the parallel array of reciprocal lattice lines. The independent variation of the intensities along the reciprocal lattice lines is indicated schematically by the independent variation of darkness on each lattice line. The family of reciprocal lattice lines extends in front of and behind the x^*, y^* plane, of course, and thus a 2-D array of spots is produced when the Ewald sphere, of which only a section of the surface is shown here, intersects this lattice of parallel lines.

7.5 Disorder and crystalline defects are an important fact of life

The ideas of mathematical symmetry described in the paragraphs above are an idealization of the degree of symmetry that really occurs in crystals of biological macromolecules. It may be true that a really good crystal, at first glance, seems to fit the description of having perfect symmetry. As we will discuss next, every crystal — when it is examined closely enough — will eventually fail to have the mathematical precision associated with the symmetry operations that are used to describe a perfect crystal. When we reach the point where precise mathematical symmetry is not fully met by our (real) crystal, we need to add new concepts and language to describe the observed deviations from ideality.

We have previously explained in section 2.6 that there are two important ways in which the electron density function of a real crystal is certain to differ from the model of a perfect crystal. The first of these effects occurs when the structure within one unit cell is slightly different from that within the next unit cell; this effect is referred to as *short-range disorder*. The second effect occurs when there are distortions and other types of defects in the crystalline lattice function itself; this effect is referred to as *long-range disorder*. In addition, 2-D crystals, like 3-D crystals, can have various local defects. These local defects will also affect the Fourier transform of the crystal in ways that need to be recognized, should they come up during data processing.

Symptoms of short-range disorder in 2-D crystals

As was described in section 2.6, short-range disorder causes the diffraction spots to be less intense than they would be for a perfect crystal, the effect becoming increasingly severe at higher resolution. Although the sharpness of the diffraction spots is not affected by short-range disorder, the background scattering between the diffraction spots increases, because the intensity lost from the Bragg reflections must still show up somewhere. In other words, the total elastic scattering cross-section is independent of whether the sample is crystalline or amorphous. The increase in background scattering may be difficult to notice, however, unless the disorder itself has sufficient spatial organization to cause the diffuse background scattering to be concentrated in specific regions of the diffraction pattern.

As we said previously in section 2.6, the fall-off of intensities is modeled rather well by multiplying the diffraction intensities for the ideal crystal by a Gaussian function

$$I_{real}(\mathbf{g}_{hkz^*}) = I_{ideal}(\mathbf{g}_{hkz^*}) e^{-\frac{B}{2}g^2}. \tag{7.3}$$

We have introduced the reciprocal lattice vector, $\mathbf{g}_{hkz^*}$, in equation 7.3, with the same meaning as it had in equation 2.2, but we now use the continuous "index," z^*, rather than the integer, l, to indicate that $\mathbf{g}_{hkz^*}$ can point to any continuous position along the (h, k) reciprocal lattice line. By the same token, the diffraction amplitudes are also modified by a Gaussian function:

$$\mathbf{F}_{real}(\mathbf{g}_{hkz^*}) = \mathbf{F}_{ideal}(\mathbf{g}_{hkz^*}) e^{-\frac{B}{4}g^2}. \tag{7.4}$$

A good explanation of the effect of short-range disorder is given by first thinking about what would happen in real space if one were to superimpose and average together all molecules (or all unit cells) within the crystal. Because each molecule in the crystal is a little bit different from its neighbor, the average version will tend to be blurred, or smoothed in comparison to any one individual. A reasonable model of this blurring would be to say that the net effect will be the same as convoluting the ideal structure of one molecule by a Gaussian function. The width of the Gaussian function should be determined by the root mean square displacement in atomic positions that is found when the same point in the structure is compared from one molecule to another.

The corresponding effect on the Fourier transform of the imperfect crystal is immediately apparent: the Fourier transform of the imperfect crystal will be the *product* of a Gaussian function and the Fourier transform of a perfect crystal, as is stated in equation 7.4. From this it follows, in turn, that the diffraction intensity of the real crystal will be the product of the square of the Gaussian function and the intensity of the ideal crystal, as is stated in equation 7.3.

Thermal vibrations are expected to generate a certain degree of short-range disorder, and thereby cause the diffraction intensities to decrease exponentially as the square of the resolution. The Gaussian attenuation factor shown in equation 7.3 is therefore known as the *Debye–Waller temperature factor*, as we have explained previously in section 2.6. The constant, B, properly referred to as the Debye–Waller thermal parameter, is also often called the "temperature factor." The relationship between B and the mean square amplitude of the thermal vibrations pictured in the Debye model was given previously in equation 2.18.

Many other mechanisms can also give rise to short range disorder, in addition to the dynamic disorder associated with thermal vibrations. It is entirely possible for there to be small — or even large — static differences in structure while still maintaining the correct average position of each molecule within the lattice. Static, short-range disorder can be imagined in which the structures of all molecules are identical, but their positions and orientations within the lattice are allowed to vary randomly within some limited range. This type of short-range disorder is likely to dominate when the points of contact between molecules are relatively few or comparatively tenuous in comparison to the number and strength of (noncovalent) contacts within the macromolecule. In this case the intermolecular contacts are easier to deform than those within the macromolecule itself, making the intermolecular contacts the "weak link" in assuring perfect short-range order. In crystals of macromolecules such static forms of short-range disorder are, in fact, normally much more important than the dynamic, thermal disorder.

Symptoms of long-range disorder in 2-D crystals

The hallmark of all forms of long-range disorder is some form of deterioration in the sharpness of the diffraction spots. This is in contrast to short-range disorder, for which the spots still remain perfectly sharp, but the intensities are weaker than they would be for a perfect crystal. Every real crystal will exhibit both forms of disorder simultaneously.

Two models of how long-range disorder can arise in 2-D crystals, just as in 3-D crystals, were mentioned earlier in section 2.6. Although perhaps a bit contrived, one could first imagine a large, 2-D crystal that had been broken up into many small domains. Long-range disorder is then introduced by allowing every domain to separate from its neighbors by small, random amounts. This model pictures the whole 2-D crystal as being, in fact, a mosaic of not-quite-perfectly laid tiles. The second model of long-range disorder pictures a perfect lattice that is first drawn on a thin, rubber sheet, and which is then stretched and deformed in any desired way. This second model differs from the first one primarily in that the long-range disorder occurs in a smooth, continuous fashion, rather than abruptly at the boundary of otherwise perfect domains. In either case, however, the final result is that remote parts of the lattice are no longer found precisely at the positions that the mathematical symmetry operations would predict, hence the name "long-range" disorder.

The "mosaic domain" model allows us to give a simple mathematical explanation of how long-range disorder causes a broadening of the diffraction spots. In order to explain things more simply, we will assume that the individual mosaic domains are not rotated at all from one another, but that they are displaced from each other by random, small amounts. Then, because of these random displacements, we note that the Fourier transform obtained from any one mosaic block experiences random shifts in phase with respect to the Fourier transforms of all other mosaic blocks. The important consequence is that coherent interference is lost between the Fourier transform of the separate pieces. As a result, the intensity of the whole crystal will just be the sum of the intensities of each of the mosaic blocks.

It is now straightforward to explain how this model of long-range disorder will lead to a broadening of the diffraction spots. The Fourier transform of one mosaic block will be the Fourier transform of an infinite crystal, convoluted with the Fourier transform of a rectangle function that corresponds in size and shape to the mosaic block. Because of this convolution, the diffraction spots that result from the incoherent sum of small blocks will be broader than the spots that would be produced from a larger, coherent crystal. A natural generalization of this simplified model of long-range disorder is that the diffraction spots are broadened (convoluted) by a function that measures the distance over which the crystal remains coherent, that is, the "long-range" distance over which the molecules continue to be in their correct, predicted locations. Thus, if the long-range order is quite good, the diffraction spots will be very sharp, but if the long-range order is very limited, then the diffraction spots will be very broad.

Two-dimensional crystals have various types of defects in addition to short-range and long-range disorder

One of the more troublesome defects that is encountered in 2-D crystals is called *twinning*. The idea of the formation of a *twin* defect in a crystal is nicely described by imagining a case where a crystal had been growing for some time in one particular direction, when suddenly it started to grow, from the same face, in a new direction. The cartoon in figure 7.10 shows how a twin could form in the crystalline packing of hard spheres; exactly this type of twin boundary does occur in simple atomic crystals such as gold.

Other, more complicated twins than the one shown in figure 7.10 are also likely to occur in 2-D crystals. The large crystalline sheets of bacteriorhodopsin that are obtained by fusing and annealing smaller patches normally exhibit four types of twins (which perhaps might be called quadruplets!) (Baldwin and Henderson, 1984). The first of the four is arbitrarily assigned to be the normal, $p3$ lattice found in the native, crystalline patches. In the cartoon shown in figure 7.11, this native lattice is represented at the top

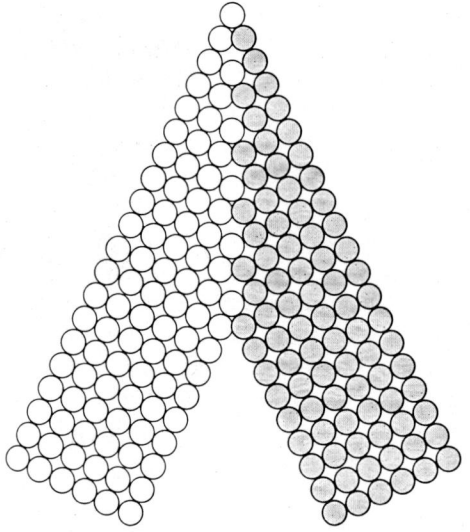

Figure 7.10 Crystalline twins are formed when one crystal grows coherently off the face of another, but in a new direction, as is shown here. Although there is a precise and regular atomic fit between the two crystals, it is clear that the extension of the atomic rows from one of the twins does not extend properly into the rows of the other.

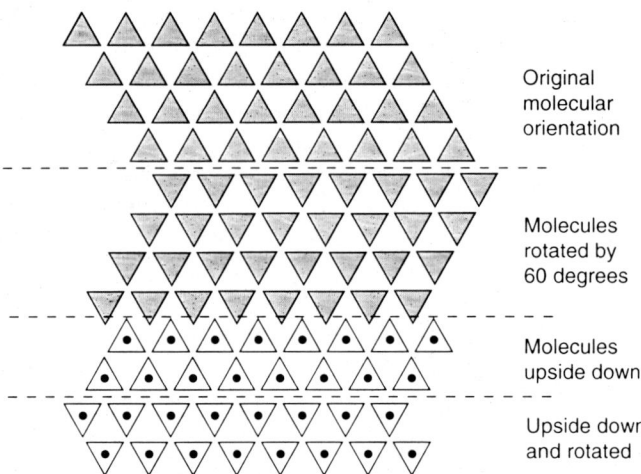

Figure 7.11 The packing of protein trimers into a 2-D crystalline lattice can result in the formation of four distinct twins, shown here as four regions of a single, coherent crystal that are separated by twin boundaries, indicated by the dashed lines. The long-range order is imagined here to be perfect, although that is not required to be the case. Twinning occurs (in this cartoon version) when trimers are placed correctly at lattice points, but with either an incorrect rotation with respect to the preceding neighbors in the lattice, or in an upside-down orientation with respect to the preceding neighbors.

of the figure an the array of shaded triangles that have one vertex pointing upward. A second twin is shown growing from the bottom edge of the native crystal. The shaded triangles in this twin are rotated by 60° from their previous orientation. Once the first such row has been laid down, other trimers are again added in proper orientation to that new row, and therefore they remain rotated by 60° from the native, parent crystal. The result is a crystal twin whose structure is identical to the parent crystal and whose lattice is perfectly aligned with that of the parent, but whose internal structure is no longer superimposable on that of the parent. A third twin is then shown in figure 7.11 as white triangles with a dot at the center, which is meant to indicate that the trimers have been turned upside down. The fourth twin, shown again by white triangles with a black dot at the center, is formed by trimers that have been simultaneously turned upside down and rotated by 60°. All four types of twins are actually found in large, fused 2-D crystals of bacteriorhodopsin.

Even naturally occurring crystals of some macromolecules can exhibit twinning; an example is the surface layer (S-layer) lattice from *Sulfolobus acidocaldarus*. Initial crystallographic analysis of this structure assigned the crystal symmetry as being *P*6 (Taylor et al., 1982), but later investigation, taking advantage of additional information extracted from real-space cross-correlation analysis at the level of individual unit cells (Lembcke et al., 1991), showed this S-layer to be made up of a *P*3 lattice heavily intergrown with twins of the second type described above. In this case, standard Fourier processing of the images — even before imposing 6-fold symmetry — results in an average over areas containing similar fractions of the twins. As a consequence, the Fourier average has nearly 6-fold rather than just 3-fold symmetry, and a wrong interpretation of the structure will result because of the highly interpenetrating twins.

Other, more familiar *point defects* and *line defects* are found in 2-D lattices as well as in 3-D crystals. *Point vacancies* and *line vacancies* can certainly be expected. In crystals of soluble proteins the empty space can simply be filled with water, while in crystals of membrane proteins the empty space could be filled with excess lipid. The ability to discover such defects again shows the virtue of applying cross-correlation analysis at the level of individual unit cells.

Yet another type of familiar defect is called a *dislocation*. A dislocation is the lattice defect that is formed when all molecules (or unit cells) in a row are first removed, starting from a given point in the lattice, as in the formation of a line vacancy. The dislocation is completed, however, when the resulting gap is closed and sealed. A highly strained, "point" defect is then formed at the core of the dislocation, as is illustrated in figure 7.12. In the near vicinity of the core there is a danger that the strain may exceed the elastic limit of deformability and/or that the stresses may be anisotropic. Both of these effects are factors that would, in turn, degrade the short-range order in that area. It is further self-evident, from figure 7.12, that the coherence of the crystalline lattice, and thus the long-range order, is also degraded for some distance on either side of the row that was removed in the formation of the dislocation.

Two-dimensional crystals can have a unique type of defect, called a *disclination*, that cannot be formed in a 3-D crystal (Nabarro, 1987). A disclination is formed when a lattice is cut along two principle directions, as is illustrated in figure 7.13(a). A wedge of the lattice is removed, as is shown in figure 7.13(b), and then the cut is resealed. Since the example used in figure 7.13 is a hexagonal lattice, removal of a 60° sector and resealing the cut edges will result in a single point where five molecules

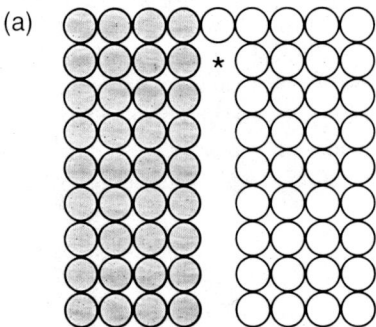

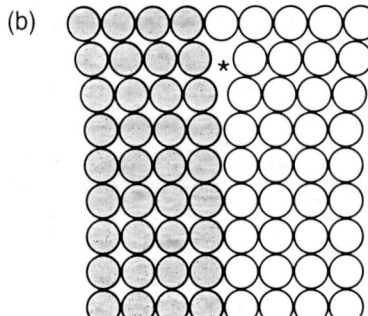

Figure 7.12 Formation of a crystalline lattice dislocation can be thought of as being due to the removal of a whole row of unit cells, beginning at the point indicated by the asterisk in panel (a). The adjacent rows are then brought together, as is shown in panel (b). Near the core of the dislocation it can be seen that the lattice rows are highly curved, and the lattice constant may be somewhat distorted from the bulk value. At some distance from the core, however, the lattice rows again run parallel to those of the bulk and with the same lattice constant.

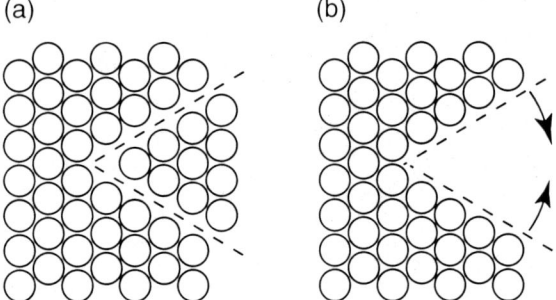

Figure 7.13 A *disclination* is the defect that is created in a 2-D lattice when cuts are made along two of the unit cell vectors. In the example of a hexagonal lattice shown here, a 60° wedge, outlined in panel (a), is removed. Then, when the cut edges are brought together, as indicated in (b), the original hexagonal lattice is restored everywhere except at the core of the disclination. A local 5-fold axis is created at the core, and the previously planar lattice becomes cup-shaped as a result of pulling the previously cut edges together.

are packed together; this is the only pentamer in an otherwise exclusively hexameric lattice. On the other hand, inserting a 60° wedge before resealing results in a single heptamer trapped within a hexagonal lattice. In either case, the result of resealing the cut is that the sheet will buckle, introducing curvature into a previously planar structure.

Disclinations are actually used in nature to build closed shells out of self-assembling proteins that would otherwise go on to make planar, 2-D crystalline sheets. Clathrin coats initially assemble as planar, hexagonal crystals on the cytoplasmic surface of the plasma membrane (Heuser, 1980). Then, in an ATP-driven step, a variety of 5-fold and 7-fold disclinations are introduced, eventually forming a completely closed surface. The icosahedral viruses (see chapter 13) are also described by a hexagonal lattice with 12 symmetrically positioned pentamers, resulting in a symmetrical, closed surface. Disclinations could be incorporated, in principle, in bacterial S-layers, allowing, for example, for curvature around the spherical ends of rod-like cells.

Grain boundaries are yet another type of crystal imperfection, which can be encountered in 2-D sheets as well as in 3-D crystals. As in our earlier discussion of the mosaic model of long-range disorder, a large sheet may be composed of many smaller crystalline blocks. If there is no continuous connection between two such lattices, and if the lattice direction of one mosaic block is pointed in essentially an arbitrary direction relative to that of its neighbors, it is no longer natural to think of the large sheet as being a deformed version of a single, coherent crystal. Instead, it is more natural to describe it as being made up of separate grains, which have distinct boundaries between them. These grain boundaries, in turn, need to be filled with some noncrystalline material that provides mechanical continuity between the separate grains. In the case of crystals of integral membrane proteins this material can easily be excess lipids. In natural 2-D crystals such as bacterial S-layers, these grain boundaries can serve as hinges and flex-points that permit the tiling (closure) of spherical caps. Grain boundaries thus provide an alternative way, other than the introduction of disclinations, to form a closed shell with an otherwise symmetrical net.

8

Two-dimensional Crystallization Techniques

8.1 Introduction

The discovery of efficient methods for growing *2-D crystals* of biological macromolecules is an important research frontier, vital to the further development of electron crystallography. Only a relatively small amount of work in this direction has been done so far when compared to the large effort, over many years, that has gone into understanding the (3-D) crystallization process for x-ray crystallography. Even so, there already are enough publications on this subject to make it impractical to attempt a comprehensive or historically complete review within this chapter.

It is understandable that 2-D crystallization methods have not been developed as intensively as they might have been up to this point. Although there are several examples in which structures of 2-D crystals have been solved at high resolution, collecting the required data still takes considerable effort. As a result, many studies are taken only to the intermediate stage where a map is obtained at a resolution of $\sim$0.7–0.9 nm, which is sufficient to visualize helices within the 3-D reconstruction.

There are some major advantages in working with 2-D crystals, even when the goal is to obtain a density map at only intermediate resolution. The ability to record high-resolution diffraction patterns provides more accurate structure-factor amplitudes than can be recovered from images, and at the same time it gives assurance that the specimen itself does not limit the resolution that can be obtained. In addition, the use of 2-D crystals makes it possible to obtain intermediate- and high-resolution structural information from much smaller proteins than would be possible for cryo-electron microscopy of single particles. Ultimately, of course, we can hope that technical improvements will be found to reduce the challenges of beam-induced movement and specimen charging (sections 5.7 and 5.8) that affect data collection with highly tilted, 2-D crystals. At the same time, of course, these same advances may also greatly reduce the required particle size that can be used for high-resolution work with single particles.

Integral membrane proteins represent a large class of structures where one might suppose that 2-D crystallization could be successful. As is discussed in section 8.2, the two-dimensional character of a lipid bilayer suggests a natural opportunity for 2-D crystallization. Since the lipid bilayer is also the native environment for integral membrane proteins, it can be expected that the native, functional structure of a membrane protein will be at less risk after reconstitution into a bilayer than when it is kept in the detergent-solubilized state.

Numerous soluble proteins also produce both thin, 3-D crystals and true, 2-D crystals. With care in refining the conditions for crystallization, specimens of this type can provide opportunities of considerable importance, as has been illustrated by the structural studies on tubulin (Nogales et al., 1998). Section 8.3 suggests the broader potential for exploiting such crystals, provided that they can be grown with a constant thickness.

Crystallization at an interface is a novel approach by which one can rationally attempt the growth of 2-D crystals of soluble proteins. Section 8.4 surveys a variety of such methods that have been quite successful. Adaptations of the technique have even made it possible to work with detergent-solubilized membrane proteins. It thus is clear that crystallization at interfaces has broad application as a method for growing 2-D crystals.

8.2 Integral membrane proteins represent a natural target for 2-D crystallization

A few integral membrane proteins form 2-D crystals within the plane of the native cell membrane

Bacteriorhodopsin (bR) is the only example of a *membrane protein* that is known to make 2-D crystals in situ that have excellent short-range and long-range order. These crystals can be easily isolated from the parent cell membrane as the so-called "purple membrane fraction." The purified fragments of *purple membrane* can also be fused together into larger 2-D crystals (Baldwin and Henderson, 1984). The excellent quality of these specimens and the ease with which they are obtained has been an important factor that facilitated the development of high-resolution electron crystallography.

As has been mentioned in section 1.3, the atomic-resolution model of bR (Henderson et al., 1990) was the first macromolecular structure to be obtained by electron crystallography. Subsequent work led to a refinement of this model (Grigorieff et al., 1996), and an independent structure determination by electron crystallography has led to an even higher resolution model of the structure (Kimura et al., 1997). Difference Fourier maps were then used to observe structural changes that accompany each step of the proton-pumping photocycle (Subramaniam et al., 1993; Vonck, 1996). After many years of effort, however, 3-D crystals of bR have also been obtained that diffract to high resolution (Landau and Rosenbusch, 1996; Lanyi and Schobert, 2002), ending the monopoly that electron diffraction has had on high-resolution studies of bacteriorhodopsin.

A number of other *integral membrane proteins* are also known to make small crystalline patches within their respective, native cell membranes. None of these proteins have either the excellent short-range order or the good crystal size that is present in purple membrane, however. Nevertheless, spontaneous formation of even small or poorly ordered in situ crystals is an encouraging sign, since better specimens might be obtained when further effort is invested in the biochemistry of such preparations. We note that many bacteria also posses crystalline arrays of surface-layer proteins (S-layers), some of which are quite well ordered (Rachel et al., 1986; Cheong et al., 1996). We will not discuss these natural crystalline arrays further in this section, however, since they are not usually made up of integral membrane proteins.

The gap junction structure is perhaps the best-known example, apart from bR, of a naturally occurring, 2-D crystal of a membrane protein. Considerable work has been done on both native and detergent-modified crystals at low resolution (Gogol and Unwin, 1988). As initially isolated, the gap junction crystals usually suffer from lattice defects and poor long-range order, even to the extent that dislocations and grain boundaries can be recognized in negatively stained specimens. Significant improvement in the crystalline order of gap junctions has been achieved by detergent extraction in the presence of added, short-chain phospholipid (Unger et al., 1999a). This improvement was sufficient to provide a 3-D reconstruction in which the transmembrane helices of the protein are well resolved (Unger et al., 1999b).

Numerous other examples are known of membrane proteins that form ordered arrays within the native cell membrane. Crystalline arrangements of membrane proteins are formed in some photosynthetic membranes, each repeating unit presumably involving a complex of reaction center proteins and associated light-harvesting, antenna proteins. As was shown previously in figure 7.1, the proton-pumping ATPase in the yeast plasma membrane crystallizes spontaneously when exponentially growing cells are shifted to starvation conditions (Sosinsky et al., 1986). The major intrinsic protein of the lens of mammalian eyes (a member of the aquaporin family) forms small, square arrays in the cell membrane, and much larger arrays can be produced after purification and reconstitution of this protein (Hasler et al., 1998). Uroplakin, a protein localized at the apical surface of urothelial cells, is yet another example of a membrane protein that makes 2-D crystals in the native cell membrane. As with the gap junction patches described above, the crystalline order of isolated uroplakin patches can be further improved by detergent extraction (Min et al., 2003).

Other membrane proteins can be crystallized in isolated membrane vesicles

Several examples are known in which integral membrane proteins can be induced to crystallize in preparations of isolated cell membranes, even though the proteins do not show crystalline arrays in the normal cell environment. Such examples of 2-D crystals usually appear in membrane fractions that are already very rich in a single type of membrane protein. One such example is the nicotinic acetyl choline receptor from the electroplax of electric fish. This protein spontaneously forms 2-D arrays upon storage of isolated membrane vesicles (Brisson and Unwin, 1985). Other examples include the plasma membrane Na/K-ATPase, the sarcoplasmic Ca-ATPase, and the gastric H,K-ATPase, all of which are structurally homologous to one another. These P-type ion pumps all form 2-D arrays in the presence of vanadate (Skriver et al., 1981; Taylor et al., 1986; Mohraz et al., 1987; Tahara et al., 1993; Xian and Hebert, 1997), which is a potent inhibitor of their ATPase activity. It is possible that the inhibitor facilitates 2-D crystallization by locking the protein molecules into a unique structural conformation.

In many of these examples the 2-D arrays actually adopt the form of closed cylinders, such as the tubular vesicle of nicotinic acetylcholine receptors shown in figure 1.3. This cylindrical morphology could be a natural consequence of the proteins having a tapered molecular shape (i.e., larger on one end than on the other) and being inserted in the membrane with a single orientation. This is not the only idea that can account for the tubular morphology, however. As is discussed further in chapter 12, any type of thin lattice can be rolled up into a cylinder at rather little cost in terms of strain within the structure. Since proteins that are isolated in vesicles are confined to a closed surface, the tubular morphology may simply represent an efficient way in which a single crystal can propagate and grow, consuming a large fraction of the available area without the need to introduce disclinations or grain boundaries.

If crystallization of a highly enriched membrane protein does not occur spontaneously within the native membrane, another step that can be taken is to reduce the lipid-to-protein ratio in order to crowd the membrane proteins more closely together. One approach is to extract excess lipid by using a detergent that is unable to solubilize the protein itself. Application of the *detergent extraction* approach to gap junctions has already been discussed above.

Extracting membranes with detergents that are not able to solubilize the protein of interest can often remove unwanted proteins of other types, as well as decreasing the lipid content. Detergent extraction might thus be described as being a *negative purification* step, because keeping the target protein in the insoluble fraction is essentially the opposite of purifying a membrane protein by *detergent solubilization*. One of the more extensively studied examples of crystallization following negative purification used either triton detergent or deoxycholate detergent to enrich cytochrome oxidase in isolated membrane vesicles (e.g., Frey and Murray, 1994). Other examples include the photosystem II reaction center (Bassi et al., 1989) and bullfrog rhodopsin (Corless et al., 1982).

Phospholipase treatment represents a second way in which crystallization of membrane proteins can be either initiated in isolated membranes (Mannella, 1984) or improved in reconstituted membranes (Jap, 1988). Hydrolysis of excess lipid by phospholipase A yields free fatty acid and a lysolipid, both of which can be removed by dialysis. It is important to be aware that phospholipase preparations may be contaminated by proteases. Thus it is prudent to check the sample afterward by gel electrophoresis, to verify that the membrane protein remains undamaged.

Purified membrane proteins can be crystallized by reconstitution with lipid

From a biochemical perspective, the most satisfactory way to obtain 2-D crystals of integral membrane proteins is to first *solubilize* the target protein with a *detergent* that leaves its biological function intact. Column chromatography and other (positive) purification methods can then be applied to the solubilized protein. Finally, the purified protein must be *reconstituted* back into a lipid bilayer, using relatively small quantities of lipid, typically in ~1:1 mass ratio with the membrane protein. Although dialysis is the most commonly used technique for removal of detergent, success has also been had in some cases with the use of Biobeads (Rigaud et al., 1997). Good reviews have been written that summarize the major technical variations that have been developed for the technique of crystallization by reconstitution with limited amounts of lipid (Engel et al., 1992; Jap et al., 1992; Kuhlbrandt, 1992; Levy et al., 2001).

There are few rules that determine which detergent to use for purification and reconstitution, and as a result the early stages of any new project must involve a considerable period of characterization of the system. Detergents with a high *critical micelle concentration* (*cmc*), like octyl glucoside or octyl POE, are preferred because they are much more easily removed by dialysis. Solubilized membrane proteins are usually less stable in detergents that have a high cmc, however. Detergents with low critical micelle concentrations must therefore be used when all else fails. In this case, removal of the detergent by dialysis may require frequent changes of the dialysis medium over a period of weeks. Even when *protein stability* is not an issue during solubilization and purification, it may become a problem if the sample must be concentrated prior to reconstitution. The problem here is that the detergent micelles are likely to become concentrated at the same time that the protein is concentrated. For reasons that are not understood in molecular terms, the protein may precipitate at high detergent concentration. It is useful to appreciate that the stability of many detergent-solubilized proteins is greatly improved by the addition of lipid. Stabilization by the addition of lipid can therefore be a better alternative than resorting to a detergent with a lower cmc.

Quite apart from the question of protein stability, the choice of detergent is also constrained by whether the membrane protein being studied will even be solubilized by the detergent. Detergents that have a high cmc are, as a rule, more likely to solubilize membrane components. Nevertheless, detergents with roughly similar critical micelle concentrations can differ in an unpredictable way with respect to their ability to solubilize different types of membrane proteins. When solubilizing a membrane protein, sufficient detergent must be provided to keep the free detergent concentration above the cmc, as well as to allow for the solubilization of other membrane components. As a result, the mass ratio of detergent to total membrane (both lipid and protein) should be more than a factor of two, and possibly more than a factor of 5. The mass of detergent that is referred to in this ratio must be calculated as the amount of detergent that is present in micellar form, that is, the amount of detergent that is in excess of the cmc.

Once a detergent has been found that is effective in solubilizing the target protein, purification can be carried out by the same methods that are used for any soluble protein. Ion exchange chromatography and size exclusion chromatography are routine and generally applicable techniques. When the protein is produced in a gene expression system, it is very effective to add an affinity tag to facilitate purification. Gradient centrifugation is also effective for large, detergent-solubilized membrane complexes.

Many factors go into the decision of which lipid — or mixture of lipids — should be used for *reconstitution*. Some membrane proteins have a biochemical requirement for specific types of lipid. As a default, therefore, it is always worthwhile to try the native membrane lipids, if they can be obtained without extreme effort. Readily obtained natural lipids such as egg lecithin, soy bean lecithin, or whole *E. coli* lipids are also a common choice. Chemically defined lipids have an obvious appeal for crystallization, and *dimyristoylphosphatidylcholine* (*DMPC*) or *dioleoylphosphatidylcholine* (*DOPC*) are among the first that should be tried. If these standard choices of lipids do not result in formation of 2-D crystals of the reconstituted protein, it is possible that the trials should be broadened to include lipids with a variety of types of polar head groups, lipid chain-lengths, and degree of unsaturation of the lipid tail.

Electron microscopy of negatively stained samples (section 6.2) provides the critical, first test of whether a reconstitution has been successful. A variety of morphologies can be adopted by the reconstituted material, some of which are illustrated in figure 8.1. If 2-D arrays have been produced, this is usually apparent when the specimen is viewed at higher magnification. In some cases, however, the computed Fourier transform of the image, or optical diffraction of the photographic negative must be used to reveal crystalline order that cannot be seen in the image itself. Even if 2-D crystals are not formed, much information can be gained about the relative promise of different reconstitution

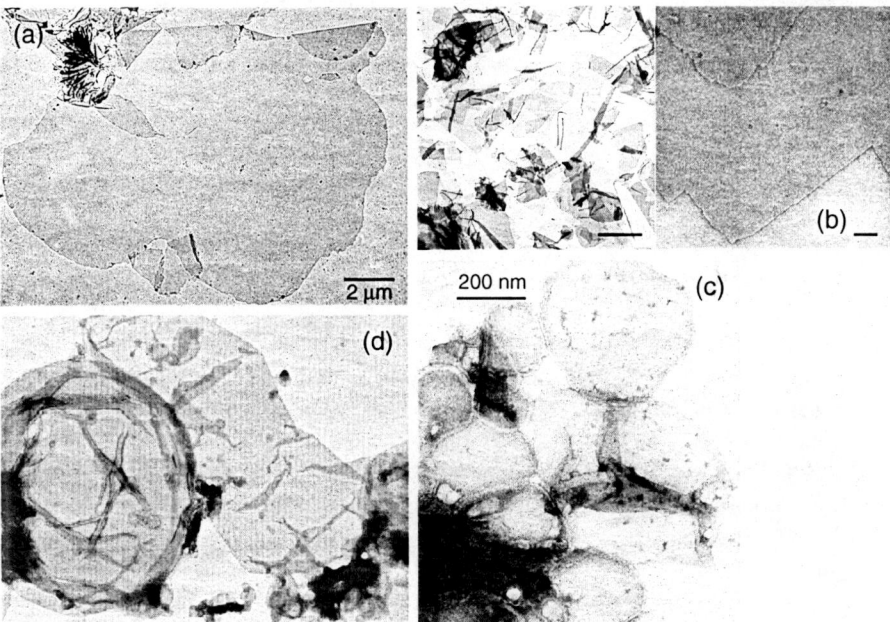

Figure 8.1 Reconstitution of detergent-solubilized membrane proteins with low amounts of lipid can produce 2-D sheets with a variety of morphologies. (a) Large, monolayer sheets of a light-harvesting complex, LHC-II (photograph provided by Dr. Da-Neng Wang), illustrate a nearly ideal morphology, although it is apparent that multiple domains may be fused to one another at their edges. (b) Crystals of aquaporin from *E. coli* (AqpZ); figure adapted from an illustration provided by Dr. Andreas Engel, published previously by Ringler et al. (2000). The scale bar in the left panel represents 2 μm, and that in the right panel represents 100 nm. (c) Vesicles of reconstituted melibiose permease (Rigaud et al., 1997) show no indication of crystalline structure when viewed at low magnification, but nevertheless the crystal lattice becomes apparent in negatively stained samples when the vesicles are imaged at higher magnification. (d) Large proteoliposome vesicles are formed by reconstitution of *Synechococcus* photosystem I with DMPC when Biobeads are used to remove the detergent (LDAO). Figure adapted from Chami et al. (2001). In this particular example the proteins, although densely packed, did not form ordered crystalline arrays. Scale bar represents 1 μm.

conditions from examining the preparation by electron microscopy. A visual inspection of the sample immediately reveals the presence of amorphous precipitates and either large or small proteoliposomes. Similarly, ordered sheets may be monolayers, stacked layers, or irregular clumps. As stated in section 6.2, it is important to use two or three different types of negative stain to characterize samples at this stage, in order to rule out the possibility of stain-induced artifacts. It is always possible for the stain itself to cause aggregation of sample materials, to change the morphology of the sample, or to disrupt ordered arrays that existed before exposure to the negative stain.

When screening the properties of different detergents in any new project, it is worth keeping in mind that the critical concentration for making mixed micelles with lipid may differ from that for making mixed micelles with protein. Slow removal of detergent may thus result in precipitation of protein at a higher detergent concentration than that at which lipid bilayers begin to form, or — conversely — the protein may remain solubilized by the detergent at detergent concentrations below those at which lipid vesicles begin to form. Either way, the possibility to reconstitute the protein into lipid bilayers may be lost if the critical concentration for formation of mixed micelles is not well matched between the lipid and the protein, for the detergent that has been chosen.

Some of the first membrane proteins to be crystallized by reconstitution include cytochrome reductase (Hovmoller et al., 1983), several porins (Dorset et al., 1983; Jap, 1988; Lepault et al., 1988), and photosystem reaction centers (Ford et al., 1990; Dekker et al., 1990; Karrasch et al., 1996; Nakazato et al., 1996). More recent examples in which crystallization has been quite successful include: NhaA, a bacterial secondary transporter (Williams et al., 1999); a ClC-type chloride ion channel (Mindell et al., 2001); bovine rhodopsin (Krebs et al., 2003); the c-ring portion of an ATP synthase (Meier et al., 2003); the mitochondrial ADP/ATP carrier in the attractyloside-inhibited state (Kunji and Harding, 2003); and EmrE, a member of the small multidrug resistance family (Tate et al., 2003). Special mention should be given to the excellent crystals that were obtained with human red blood cell aquaporin (Jap and Li, 1995; Mitra et al., 1995; Walz et al., 1995; Ren et al., 2000), since this is one of the cases in which the structural work was taken to high enough resolution that an atomic model could be built into the 3-D density map (Murata et al., 2000). More recently, crystals of aquaporin-0, which forms a lens-specific water pore and serves, as well, to form cellular junctions between lens cells, have been obtained that diffract to a resolution better than 0.19 nm (Gonen et al., 2005). Other specimens that appear to be of sufficient quality to be used for atomic structure determination include a microsomal glutathione transferase (Schmidt-Krey et al., 2000) and OxlT, an oxylate transporter that is a member of the major facilitator superfamily (Heymann et al., 2003).

Two-dimensional crystals of membrane proteins can also be grown in detergent

Some membrane proteins have been found to form 2-D crystals while still in the detergent-solubilized state, apparently by fusion of detergent micelles. The photosynthetic light-harvesting complex is the most successful example of this type of crystal (Kuhlbrandt et al., 1994). The discovery of these thin, sheet-like crystals was initially an unexpected byproduct of efforts to grow 3-D crystals that would be suitable for x-ray crystallography. Since the crystallization conditions were meant to produce

3-D crystals, it is not surprising that the first crystals were made up of stacked sheets of variable thickness. Such sheet-like crystals can be a promising start, and a systematic variation of growth conditions may later yield true monolayer crystals. The possibility that 2-D crystallization in detergent may occur with some frequency is suggested by the fact that other crystals have been grown under conditions of incomplete detergent removal. Examples include the sarcoplasmic Ca-ATPase (Shi et al., 1995; Lacapere et al., 1998) and the integral membrane domain of the red blood cell Band 3 protein (Wang et al., 1994).

8.3 Many soluble proteins also form very thin crystals

Catalase is without a doubt the soluble protein that is best known to form crystals that are thin enough for high-resolution imaging and electron diffraction experiments (Matricardi et al., 1972; Taylor and Glaeser, 1974; Unwin and Henderson, 1975). Both thin laths (Matricardi et al., 1972) and thin plates (Unwin and Henderson, 1975) are formed by catalase under appropriate growth conditions. The fact that thin catalase crystals grow at very low ionic strength has made these crystals extremely favorable for use as "test specimens" during the development of electron crystallographic methodology. Once used as a standard specimen for the demonstration of negative staining methodology, and even as a standard for calibrating instrumental magnification (Wrigley, 1968), thin catalase crystals played an important role in characterizing the limitations in resolution that are attributable to radiation damage (Glaeser and Hobbs, 1975; Glaeser and Taylor, 1978). Catalase crystals were also of critical importance in the development of methods for the preparation of fully hydrated specimens. Both the frozen-hydrated specimen method (Taylor and Glaeser, 1974) and the use of glucose-embedding (Unwin and Henderson, 1975) were first developed with the help of catalase crystals.

Some soluble proteins may form true 2-D crystals in bulk solution

Crystallization trials with soluble proteins occasionally produce crystals that are large in two dimensions but too thin in the third dimension to be suitable for use in x-ray crystallography. In such cases it is worthwhile to consider optimization of the growth conditions in a way that produces crystals suitable for electron diffraction and high-resolution electron microscope. Some examples of crystals of this type include monolayer crystals of tubulin that are induced by zinc (Larsson et al., 1976) and thin crystals of actin that are induced by gadolinium (Aebi et al., 1981); gp32*I, a proteolytically cleaved form of a single-strand DNA binding-protein (Chiu and Hosoda, 1978); crotoxin, a snake-venom phospholipase (Jeng and Chiu, 1983); monolayer crystals of a stoichiometric complex of influenza neuraminidase and an Fab fragment of a monoclonal antibody (Tulloch et al., 1986); and monolayer crystals of the phage $\phi 29$ connector, a DNA translocating machine (Valpuesta et al., 1994, 1999).

As noted by John Trinick (personal communication), growth of the neuraminidase/Fab co-crystals mentioned above suggests a more general strategy that could yield 2-D crystals of other soluble proteins. The concept would be to exploit monoclonal antibodies (or other bulky ligands) that are directed to binding sites that are exposed on only one face of a 3-D protein crystal. The principle would be to use such a bulky ligand to "poison" the growth of 3-D crystals in the direction perpendicular to that crystal face, thereby limiting crystallization to two dimensions. In some cases this strategy may prove to be successful with a stoichiometric complex between the protein and the bulky ligand, as proved to be true for neuraminidase. In other cases having a low fraction of liganded protein mixed in with the free protein may be all that is necessary in order to spoil growth of a crystal in the third dimension. For completeness we note that monoclonal antibodies have already been used for the opposite purpose, i.e. to promote the growth of 3-D crystals by stabilizing a unique structure and by adding protein mass to the structure of interest in a consistent and reproducible way. This strategy has been especially valuable for the crystallization of certain detergent-solubilized membrane proteins (Ostermeier et al., 1997).

Tubulin is currently the most notable example of a soluble protein that makes excellent, 2-D crystals. The work done with this specimen merits further discussion because of the impact that it has had on cell biology. Tubulin normally assembles into cylindrical microtubules, and as a result this protein has proven refractory to the growth of well-diffracting, 3-D crystals. The addition of zinc prior to polymerization, however, leads to the formation of monolayer crystals (Larsson et al., 1976) in which individual protofilaments alternate in an antiparallel fashion, rather than running parallel to each other as they do in the microtubule.

The original conditions for 2-D crystal growth had to first be optimized before high-resolution electron crystallography could commence (Downing and Jontes, 1992; Wolf et al., 1993). Once suitably large and well-diffracting crystals were obtained, however, the structure was ultimately determined at atomic resolution (Nogales et al., 1998). The atomic model of the two subunits that make up the tubulin protofilament was shown previously in figure 2.5. The model of the protofilament has subsequently been docked into an intermediate-resolution density map of the microtubule (Li et al., 2002). In addition, high-resolution electron diffraction patterns like the one shown in figure 1.4 have been used to visualize the binding of drugs to the tubulin protofilament (Nettles et al., 2004).

Thin crystals of nucleic acids have also been demonstrated to be suitable for use in electron diffraction. High-resolution electron diffraction patterns can be obtained with glucose-embedded crystals of transfer RNA (Fujiyoshi et al., 1984), and Downing has demonstrated that high resolution diffraction patterns (Downing and Glaeser, 1980) and images (Downing, 1984) can be obtained with frozen-hydrated specimens of DNA, crystallized in a chain-folded conformation.

Methodology to solve structures at atomic resolution with thin, 3-D crystals is still in development

Crystals of soluble proteins that are thin enough to be suitable for high-resolution electron diffraction and high-resolution electron microscope are not automatically suitable for use in getting an atomic-resolution structure. In principle, such crystals could be well suited for a high-resolution structure analysis; in practice, new problems in methodology arise because thin, sheet-like crystals of soluble proteins are rarely just a single unit cell thick.

The important issue is not so much that the crystals are more than one unit cell in thickness, but that they tend to be of variable thickness. Variation in thickness makes it difficult to merge data from different crystals to produce a full, 3-D data set. Additional problems are also inherent in any sample that is thicker than 20 to 40 nm; these problems are described in chapter 15, and thus they will not be discussed here.

The 3-D Fourier transforms of crystals with different thicknesses are not identical, of course, even though the structure within the unit cell may be the same. As was described in section 7.4, the Fourier transform varies continuously along reciprocal lattice lines if a crystal is only one unit cell thick. As the crystal thickness increases, however, its Fourier transform begins to look more and more like that of a proper 3-D crystal, which is sampled at regularly spaced reciprocal lattice points. For specimen thicknesses, T, that correspond to just a few unit-cell layers, the Fourier transform is already sharply peaked at regular intervals in z^*, but nevertheless it is still a continuous function. The mathematical form of the Fourier transform of such a "thick" crystal is given by the 1-D convolution of a sinc function (section 7.4), whose width is proportional to $1/T$, with the 3-D array of reciprocal lattice points associated with a true, 3-D crystal. The theory behind this description is identical to that covered in section 7.4 for monolayer crystals. The only difference in this case is that the sinc function is proportionately more narrow when the crystal is thicker than a single unit cell. It is thus easy to see that the Fourier transforms of individual crystals will have quite different values at identical points, z^*, along the lattice lines, if the crystals are not exactly the same thickness. As a result, there is no easy way to merge data from two different crystals.

A reasonable approach in collecting and merging data for samples that are made up of crystals with variable thicknesses is to carefully measure the thickness of each crystal at the time that data are being collected (Leapman et al., 1993). The best strategy is to then collect data for only the thinnest possible crystals, and in any case to merge data only for crystals that are of the same thickness. One advantage of using the thinnest possible crystals is that the continuous Fourier transform changes more slowly in the z^* direction in reciprocal space. Other advantages are related to minimizing the impact of difficulties that are associated with thick samples, which are described in chapter 15. Since thin crystals of soluble proteins tend to be quite large in area, it may be possible to collect many diffraction patterns, at various tilt angles, from a single crystal (Brink and Chiu, 1994). Having multiple diffraction patterns from the same crystal is very helpful when merging the data, provided that the thickness of the crystal does not change across the area that is used to record data. Although many of these techniques were adopted in the structural studies of crotoxin crystals, it was then discovered that this specimen presented yet a further difficulty in that crystals could be noninteger fractions of one unit cell in thickness, and thus crystals of the same thickness could have quite different structural arrangements of the same protein.

8.4 Crystallization at interfaces has potential for wide generality

The adsorption of soluble macromolecules at a 2-D interface provides a unique opportunity in which the crystallization of a macromolecule may be limited to just the two-dimensionally adsorbed layer. The rationale behind this approach is that the favorable binding energy will pay the unfavorable entropy cost that is associated with localizing the macromolecule from a random, 3-D solution to the 2-D plane of the binding interface. If binding occurs in a preferred orientation as well, the favorable binding energy also pays much of the further cost that is required to reduce rotational entropy as part of the crystallization process. If the 2-D concentration of bound macromolecules becomes high enough, crystallization can occur spontaneously under a wide range of buffer conditions. This concept does not in any way preclude the usefulness of screening buffer conditions in order to further assist the formation of 2-D crystals, of course.

Fromherz was one of the first to propose the self-assembly of two-dimensionally ordered protein arrays by binding the protein to amphiphilic monolayers at the air–water interface (Fromherz, 1971). In his original proposal, Fromherz envisioned the binding of proteins through non-specific electrostatic interactions with the polar head groups of lipids. Electrostatic binding has indeed proven to be an effective method for forming monolayer crystals, as well as for producing concentrated fields of single particles. However, another version of interfacial adsorption that has proven to be especially successful is based on the same principles that are used for affinity purification by column chromatography. Reviews covering the development of the lipid monolayer technique (Newman, 1991; Chiu et al., 1997; Asturias and Kornberg, 1999; Brisson et al., 1999; Levy et al., 2001) can be consulted to provide more information than can be presented below.

Other forms of interfacial adsorption and crystallization can, of course, be imagined. Thus it has been demonstrated that ordered arrays of macromolecules can be formed at water–organic solvent interfaces (Aoyama et al., 1995), an approach that may be fundamentally similar to crystallization at the water-air interface, and also at liquid-solid interfaces (Horne and Wildy, 1979; van Bruggen et al., 1986; Cejka et al., 1989; Harris et al., 1992; Zahn et al., 1993). It seems likely, however, that exploitation of these additional approaches will be limited to special cases.

Binding of targeted proteins can be achieved with both natural and engineered lipids

The use of protein-specific ligands to target a protein of interest to the surface of a lipid monolayer was first introduced by Uzgiris and Kornberg (1983). These authors used dinitrophenol, coupled by a linker to phosphatidylethanolamine, as the ligand in order to crystallize an antibody directed against the dinitrophenol antigen. This basic concept was then extended to other proteins and their appropriate ligands (Ribi et al., 1987; Celia et al., 1994). Especially large and well-ordered crystals of streptavidin can be obtained, for example, by using biotinylated lipids to form the surface monolayer (Darst et al., 1991a; Avila-Sakar and Chiu, 1996). Although the use of ligands that are substrate analogs or allosteric modulators for a particular protein of interest clearly has great potential, a difficulty of that approach is that it may require a considerable development effort to optimize the type and length of linker that couples the ligand to the lipid, as well as the type of chemistry used to perform the linkage (Schmitt et al., 1994).

There are a variety of soluble proteins that bind naturally to specific lipid headgroups, and these present good opportunities for 2-D crystallization. Some of these proteins recognize specific polar head groups in order to become attached as peripheral membrane proteins. The cytoplasmic proteins annexin V (Olofsson et al., 1994; Voges et al., 1994; Oling et al., 2001), brush border myosin-I (Jontes and Milligan, 1997), and protein kinase C-delta (Solodukhin et al., 2002) are examples of peripheral membrane proteins that crystallize on monolayers of phosphatidylserine, one of the lipids that form the inner leaflet of eukaryotic cell membranes. A few examples are known of cytoplasmic proteins that even form ordered arrays in vivo, such as clathrin (Heuser, 1989) and dynamin (Zhang and Hinshaw, 2001), but conditions have not been reported to reconstitute this type of assembly in vitro in the form of large 2-D crystals. A variety of protein-toxins also bind to the head groups of specific lipids in order to gain entry into host cells or to create large pores in their membranes. Among the toxins that have been found to form 2-D crystals in a lipid-specific way, one can mention tetanus toxin (Robinson et al., 1988), cholera toxin (Ribi et al., 1988; Mosser et al., 1992), aerolysin (Wilmsen et al., 1992), botulinum neurotoxin (Schmid et al., 1993b), and sticholysin II (Martin-Benito et al., 2000).

Ni-NTA derivatives of lipids provide a generic ligand that can be used with his-tagged recombinant proteins

Affinity purification of recombinant proteins often makes wide use of an added stretch of six or more histidine residues. The his-tag allows the protein to be selectively bound to a solid-phase material that has been derivatized with N-nitrilotriacetic acid (NTA) in order to serve as a chelator for nickel (or other transition element) ions. The first successful application of Ni-NTA derivatized lipids to form 2-D crystals was published by Kubalek et al. (1994). The Ni-NTA affinity ligand has since been shown to work successfully when coupled to a number of lipid hosts, including dioleylphosphatidylethanolamine (DOPE) (Kubalek et al., 1994), 1,3-di-oleylglyceroxyl-acetic acid (DOGA) (Venien-Bryan et al., 1997), derivatized 1,2-di-oleylglycerol (DOGS) (Celia et al., 1999), and an "exotic" fluorinated lipid (Lebeau et al., 2001). His-tagged, soluble proteins that have made well-ordered 2-D crystals on lipid monolayers include HupR, a transcriptional regulator protein, for which a projection map has been published at 0.9 nm resolution (Venien-Bryan et al., 2000), and a soluble form of a major histocompatability complex (MHC) protein, which gave electron diffraction patterns to higher than 0.35 nm resolution (Celia et al., 1999).

When first introduced, it seemed unlikely that the lipid monolayer technique could be used with detergent-solubilized membrane proteins, the concern being that the detergent needed to solubilize the membrane protein would also solubilize the derivatized lipid from the air–water interface. When a mixture of solubilized membrane protein, supplemented with solubilized lipid, was injected beneath a Ni-NTA-DOGS monolayer, however, this concern proved to not always be a limitation. Two-dimensional crystallization of his-tagged membrane proteins can be induced by removal of detergent with Biobeads, which are added to the subphase after micelles containing the his-tagged membrane protein have been bound to the monolayer (Levy et al., 1999). Large crystalline areas were produced with FhuA, a bacterial outer membrane protein, and an ATP synthase from a thermophilic *Bacillus*. Crystallization of other membrane proteins, including bacteriorhodopsin, has also been demonstrated with both positively and negatively charged lipid monolayers (Levy et al., 2001). A further refinement of the technique has used the novel, fluorinated lipids mentioned above. This lipid was introduced because it has superior properties in resisting solubilization by the detergent that is initially present in the subphase. The fluorinated lipids have been used to obtain large crystals of a plasma membrane, P-type H-ATPase, from *Arabidopsis* (Lebeau et al., 2001).

Charged lipid monolayers also provide a generic interface that is suitable for adsorption and crystallization of purified proteins

Just as affinity chromatography provides the biochemical background to understand the binding of proteins to ligand-derivatized lipid monolayers, so too ion-exchange chromatography serves as a model for discussing the binding of proteins to charged lipid monolayers. Positively charged lipids have been used more frequently than negatively charged lipids, consistent with the fact that the majority of soluble proteins have a net negative charge. A number of different positively charged lipids have been used, including stearyl amine and didodecyldimethylammonium. Charged lipids are generally used as guests within a phosphatidylcholine matrix, as is the case for the derivatized lipids discussed above.

The use of charged-lipid monolayers has proved to be particularly effective with large macromolecules. RNA polymerase II, with a molecular weight of over 500 kDa, was the first example in which 2-D crystals were obtained with positively charged lipids (Darst et al., 1988). Other examples of large macromolecules for which good crystals have been obtained include the 50S ribosomal subunit from *Bacillus stearothermophilus* (Avila-Sakar et al., 1994) and the 10S form of myosin (Liu et al., 2003). Charged lipid monolayers seem to be especially effective in binding elongated macromolecules such as alpha-actinin (Taylor and Taylor, 1993; Liu et al., 2004) and smooth muscle heavy meromyosin (Wendt et al., 2001), perhaps because of the increased surface contact that such elongated macromolecules can make with the cationic surface of the lipid layer.

Tubular lipid vesicles are also an efficient substrate for assembling proteins in ordered arrays

Taking note of the fact that various types of lipids form tubular rather than spherical liposomes, the group of A. Brisson demonstrated that helical arrays could be formed by binding proteins to the surface of these tubes (Ringler et al., 1997). As a further development of this approach, galactosylceramide lipids were used as a host in the assembly of derivatized-lipid tubes, and helical arrays of several different proteins have been produced with such tubes (Wilson-Kubalek et al., 1998; Wilson-Kubalek, 2000). Figure 8.2 shows an example of the uniform diameter of the tubular liposomes that can be obtained with galactosylceramide lipids, even when they contain derivatized lipids as a guest species.

The use of lipid tubes offers two important advantages. The first advantage is that electron microscope grids can be prepared with a suspension of tubes, using the same techniques that one employs for a macromolecular particle. As a result one can avoid the difficulties encountered in the transfer of specimens from the air–water interface to the support film of an electron microscope grid (see below). The second advantage is that one does not need to tilt the specimen grid in order to collect 3-D data, an

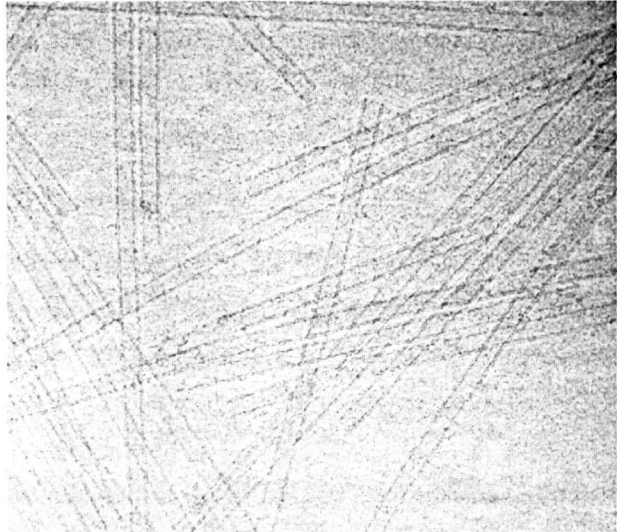

Figure 8.2 Tubular vesicles with uniform diameter are formed by galactosylceramide lipids, even when charged or chemically derivatized lipids are included as guests. In this example, the tubes contain 10% Ni-NTA-DOGS (Wilson-Kubalek et al., 1998). Photograph provided by Dr. Elizabeth Kubalek.

advantage that applies to all helical arrays (see chapter 12). The disadvantage of using lipid tubes rather than 2-D crystals, on the other hand, is that one cannot normally record electron diffraction patterns in order to (1) determine how well-ordered the arrays are and (2) obtain high-resolution structure-factor amplitudes that are more accurate than those which can be obtained from images. In addition, there is a danger that tubular arrays may become partially flattened during specimen preparation, a problem that is expected to be worse for tubes with a relatively large diameter. Nevertheless, recent success in the use of helical specimens to solve structures at the atomic level (Miyazawa et al., 2003; Yonakura et al., 2003) demonstrates that the use of tubular lipids rather than lipid monolayers can be strongly encouraged.

Denatured-protein monolayers are sometimes suitable as substrates for crystallization of a second layer of the same protein

It is well known by biochemists that proteins are readily denatured by adsorption at an air–water interface. It is generally less well known that the same protein may then bind to the denatured surface film in a native conformation, as is determined by the retention of biochemical activity when the amount of bound protein exceeds that of a monolayer. The phenomenon of protein self-absorption onto a denatured (sacrificial) protein monolayer has been intentionally exploited by Yoshimura et al. (1994), who demonstrated the formation of ordered arrays of holoferritin, apoferritin, and (less successfully) 20S proteasomes.

The technique used in this work to deliver the protein to the air–water interface is worth further description. Large areas of adsorbed protein can be prepared by injecting small volumes (~ 1 μl of protein solution) below the surface of a subphase, whose density has been adjusted (by the addition of glucose, for example) to be greater than that of the protein solution. When the rising droplet of protein solution touches the interface, a first, denatured layer spreads extremely rapidly over the entire surface. Rapid formation of the denatured monolayer can be readily observed by eye when using a strongly colored solution of holoferritin. A second layer of protein, which may or may not form as an ordered array, then builds up slowly upon further incubation.

In what may be another example of protein self-adsorption onto a denatured protein monolayer, Cyrklaff et al. (1995) observed 2-D crystals of detergent-solubilized plasma membrane H-ATPase, a P-type ion pump, which formed on the surface of droplets that had been set up in trials for 3-D crystallization of the protein. In a refinement of this technique, it was found that the same 2-D crystals could be grown directly on the surface of a carbon-coated electron microscope grid, which is laid on the surface of a freshly deposited drop of protein solution (Auer et al., 1999). In this case it may be that the interface between the hydrophobic carbon film and the aqueous subphase plays a role similar to that of the air–water interface in the original experiments. As is well known to protein crystallographers, it is not uncommon that a "skin" of denatured and/or aggregated protein forms on the surface of protein droplets during screening of various crystallization buffers. It may thus be worthwhile to test the carbon-film technique of Auer et al. as an auxiliary step in any protein-crystallization trial.

Summary of technical aspects involved in protein crystallization at the air–water interface

Crystallization of proteins on lipid monolayers is normally done with extremely simple apparatus. Small wells are machined into the surface of a teflon block. These wells are typically 4 to 6 mm in diameter and 1 to 2 mm deep. The wells are filled with protein solution, normally to the point that the meniscus is in contact with the rim of the well. The protein concentration can be quite low, less than 0.2 mg/ml, and thus — although the volume is relatively large — the amount of protein required for a single well is similar to that used in screening for the growth of 3-D crystals. Lipids dissolved in organic solvent are applied to the surface of the protein solution, typically using a Hamilton syringe to deliver volumes of 1 μl or less at a lipid concentration of 0.5 mg/ml. Additional modifications of the simple teflon well can be made in order to provide additional "ports" into the subphase chamber, so that additional components can be injected after the monolayer has been spread.

A disadvantage of using just a simple well, machined into a teflon block, is the fact that one cannot control the surface pressure of the lipid monolayer in such a system. There is a danger, of course, that some protein will insert itself into the interface and become denatured if the surface pressure is too low. While surface denaturation of protein is a self-limited process, in that insertion of protein will eventually increase the surface pressure to the point that no more protein can be inserted, the denatured protein that is inserted will nevertheless be an impurity within the lipid monolayer. This type of surface impurity could interfere with the size or long-range order of the 2-D crystals. The general practice used with the microwell technique thus is to apply at least twice as much lipid as is needed to make a monolayer on the surface of the protein solution. The goal here is to have the lipid monolayer be in equilibrium with excess, dry lipid, and thus be at the highest surface pressure that can be achieved (at least under equilibrium conditions).

Protein crystallization experiments have also been conducted with lipid monolayers formed in a Langmuir trough, where the surface pressure can be controlled experimentally. Such experiments are normally practical only with a protein, such as streptavidin, that is available in abundant quantities, because the subphase of a Langmuir trough requires volumes of one milliliter or more. One potential solution that would reduce the sample requirement down to the microliter range lies in the technique used by T. Scheybani (personal communication) to crystallize human C-reactive protein on a positively charged lipid monolayer. The trick used here was to include 2% glucose in the protein-free subphase in order to make its solution density greater than that of the protein sample. After the lipid monolayer had been formed, 1 µl of a 5 mg/ml protein solution was injected under the monolayer. The protein solution then necessarily rises to the surface and spreads as a thin lens covering a relatively large area under the lipid monolayer. In this case the protein solution does not break the lipid monolayer, nor does the protein denature at the interface if the surface pressure is high enough.

Crystallization on a lipid monolayer, if it is going to happen at all, normally is evident within a fairly short incubation time, as little as one to four hours. Since the sample volumes in each well are very small, however, it is prudent to place the teflon blocks in a sealed chamber at high humidity in order to prevent evaporation of water from the protein solution during the period of incubation.

The transfer of monolayer protein crystals from the air–water interface to the surface of an electron microscope grid is perhaps the most challenging aspect of the lipid monolayer technique. The standard technique is to lay an electron microscope grid directly on the surface of the crystallization droplet, wait a few minutes, and then lift off the grid. Successful transfer of the monolayer is indicated if a lens of subphase solution remains attached to the face of the electron microscope grid. This lens of liquid must then be blotted away. Blotting can be done in the same way as one would do it for any sample preparation technique (negative staining, glucose embedding, or the preparation of a frozen-hydrated specimen).

This transfer technique can clearly be a destructive process, however, especially so when one attempts to transfer the sample onto a continuous carbon film. It is not uncommon to find that almost none of the monolayer remains attached to a continuous

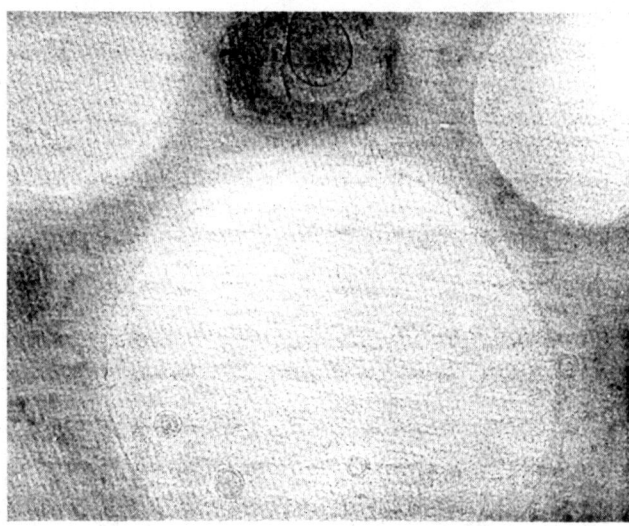

Figure 8.3 The use of holey carbon support films is an effective way to transfer protein crystals, grown on lipid monolayers, from the air–water interface to an electron microscope grid. In this example, a negatively stained crystal of RNA polymerase extends coherently over adjacent holes (Asturias and Kornberg, 1995). Figure adapted from a photograph provided by Dr. Francisco Asturias.

carbon film when the excess, adhering subphase solution is blotted away, indicating that there is little strength in the contact made between the lipid tails and the carbon film. The little pieces of monolayer crystal that do remain attached to the carbon are frequently found in the form of fragments of what evidently had been a much larger, coherent 2-D crystal, as evidenced by the obvious complementarity of the shapes of the edges in adjacent pieces.

Much better success is achieved in the transfer if one uses holey carbon films, such as the one shown in figure 8.3, rather than continuous carbon films (Kubalek et al., 1991). Even a bare lipid monolayer, or a monolayer with randomly distributed single particles can survive the transfer and span the entire area of an individual hole. Not all holes will be covered by the monolayer film, of course, and many holes may show films that contain tears or distortions that were introduced during the transfer.

An alternative way to transfer the lipid monolayer to an electron microscope grid involves a first step in which the monolayer, along with a lens of the subphase solution, is picked up with a small wire loop (Asturias and Kornberg, 1995). The liquid lens is then transferred to the surface of a hydrophilic carbon film. Once again, however, the authors recommend the use of holey carbon films rather than continuous carbon films.

9

Data Processing: Diffraction Patterns of 2-D Crystals

9.1 Introduction

This chapter covers many of the practical details that are associated with digitizing electron diffraction patterns and processing the data. The subsequent use of the diffraction intensities to compute density maps is covered at length in chapter 11. Before getting started with the practical aspects of data processing, however, section 9.2 gives a general overview of how diffraction data are used in the context of electron crystallography.

Electron diffraction patterns are often recorded on photographic film. The subsequent process of converting the diffraction pattern to digital form thus requires precise measurements of the optical density of the film, and this process is described in section 9.3. The measured optical density is then converted to values of the electron intensity by an empirically determined density-exposure curve. Section 9.4 discusses the physical mechanism of photographic blackening in order to provide a theoretical understanding of the shape of the density-exposure curve.

More recently, however, photographic film has been replaced by electronic area detectors as the preferred medium for recording electron diffraction patterns. As an illustration of currently developed technology, section 9.5 describes the properties of charge coupled device (CCD) cameras when used to record electron diffraction data. Direct electronic readout eliminates the time-consuming and labor-intensive step of photographic densitometry, and it also results in a significant improvement in the quality of the data. The improvement in data quality is a result of the low noise level and high dynamic range (i.e., excellent linearity) that can be achieved with electronic devices as compared to photographic film.

Once the diffraction patterns have been digitized, the next step in data reduction is to generate background-subtracted, integrated intensities for each spot. This process requires indexing each diffraction pattern, details of which are described in section 9.6. Individual diffraction patterns are then merged, as is described in section 9.7, to produce a full, 3-D data set of diffraction intensities. Section 9.8 concludes this chapter with a discussion of factors that limit the accuracy of the diffraction data.

9.2 Diffraction intensities are used in a variety of ways in electron crystallography

It is standard practice, when working with well-ordered, 2-D crystals, to obtain the *amplitude* of the *crystal structure factors* from electron diffraction intensities, and to derive the phases from images. In theory it would not be necessary to measure the *electron diffraction intensities*, since the structure factor amplitudes, as well as phases, are already available from the Fourier transforms of the images. However, imperfections in the quality of the image data, especially at high resolution, result in relatively large errors in the amplitudes derived from the Fourier transform of an image. When possible, therefore, it is better to use electron diffraction intensities as the source of structure factor amplitudes.

A diffraction-intensity data-set is also very useful in determining the *contrast transfer function* for high-resolution images. The contrast transfer function modulates the structure factor amplitudes in the Fourier transform of the image, as has been described in section 3.8. As a result, the ratio of the structure factor amplitudes obtained from an image and the corresponding structure factor amplitudes obtained from the electron diffraction intensities can provide a sensitive way of determining the location of the successive "zeros" in the transfer function. This approach has proven to be a very useful part of the full data processing package, as is described more fully in section 10.9.

X-ray diffraction intensities from *isomorphous heavy atom derivatives* of a protein are traditionally used to phase the structure factors, and one might well ask why the same approach has not been used in electron crystallography. The answer is that *atomic scattering factors* for electrons do not increase as rapidly with increasing atomic number as they do for x-rays. As is shown in figure 9.1, the ratio of atomic scattering factors, f, for mercury and carbon is $80/6 = 13.3$ for x-rays, but it is only 5 for electrons. The difference is due to the fact that x-rays are scattered by every electron in an atom, causing the atomic scattering factor to increase in proportion to Z, while electrons are scattered by the shielded coulomb potential, which increases much more slowly with atomic number.

Diffraction intensities are traditionally used in x-ray crystallography to refine a structure, once a molecular model has been built by fitting the known amino acid sequence into the density. The same type of *refinement* can be used in electron crystallography. It is worthwhile to mention that electron diffraction intensities are more sensitive to chemical bonding effects than x-ray diffraction intensities are, as is discussed in section 11.7. This fact may eventually lead to corresponding modifications, unique to electron crystallography, in the way that structures are refined.

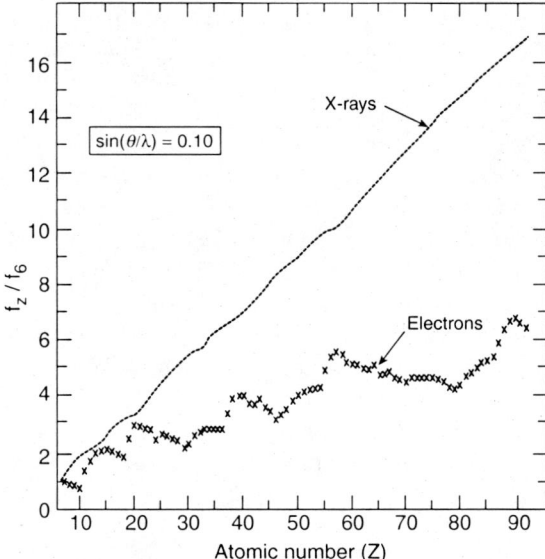

Figure 9.1 Comparison of atomic scattering factors for x-rays and for electrons. The curves for x-ray scattering and for electron scattering are artificially scaled to have the same value for carbon in order to facilitate the comparison between the two curves. The atomic scattering factors for x-rays increase in proportion to the atomic number because x-rays are scattered predominantly by electrons, with virtually no effect of the nucleus. The elastic scattering of electrons is physically due to the shielded coulomb potential of the atomic nucleus, however, and the shielded coulomb potential increases much more slowly than does the atomic number. Electron scattering also exhibits an appreciably greater chemical sensitivity, periodically dropping to a lower value as electron shells become filled.

One final way in which diffraction intensities are used in structural studies is to calculate *difference Fourier maps*, so as to reveal changes in structure between two biochemical states of a protein or other macromolecule. The power of this approach lies in the fact that the phases of the structure factor amplitudes need to be determined only once, for the "native" structure, for example. After binding a ligand or undergoing some other structural change, a new set of diffraction intensities can be measured and used in combination with the old intensities and the old phases to compute a difference map. Examples of this type of work are illustrated in section 11.8.

9.3 Data that have been recorded on photographic film must be converted to digital form with a scanning microdensitometer

Perhaps the most precise instrument that can be used to convert film density into a digital representation is the "flat-bed" *scanning microdensitometer*. The basic idea in this instrument is to use high quality microscope optics to project a demagnified image of a small aperture onto the photographic emulsion. A second lens, on the other side of the photographic film, collects the light that is transmitted through the film and relays it to a photomultiplier. The output of the photomultiplier tube is digitized with an A/D converter as the film is scanned past the stationary spot of light. Taking the logarithm of the digitally recorded, transmitted light intensity then yields the *optical density* at each point in the film.

A representative example of the conditions used to digitize a diffraction pattern is as follows. The focused light spot might be 10 μm in diameter, and digital samples are taken in a 2-D raster at the same spacing. The whole scan typically consists of a 2000 × 2000 array, corresponding to an area of 2 × 2 cm on the photographic film. The PDS 1010 series of microdensitometer, manufactured by Perkin Elmer Corp., is one of the few raster-scanning instruments that can scan with the small aperture and small step size required in electron crystallography. Another type of commercially available microdensitometer that is in common use employs a linear CCD detector rather than the point-illumination system of the raster-scan microdensitometer. In this type of instrument, a low-power "field lens" is used to project an image of the photographic film onto the linear detector, giving instantaneous measurements of the transmitted light intensity in one dimension. The linear detector (or alternatively, the film) is then scanned in the second dimension. The linear detector instruments have the advantage of being very fast, because several thousand pixels are digitized in parallel. Linear detectors have the disadvantage that they will detect light that is scattered in adjacent pixels, resulting in a small level of cross-talk between adjacent pixels. When the highest quality results are required, the PDS 1010 series instrument is preferred over the linear detector instruments for its high precision and flexibility, even though it is very slow and more expensive than other alternatives.

Once the transmittance measurements have been converted to densities by taking the logarithm, the digital record can be compressed to a smaller array size, if desired. In the initial stage of digitizing a diffraction pattern, however, it is important to sample the film with a very fine spacing. The measured transmittance is the sum of the transmittance values at every point within the aperture, of course. If the density of the film changes across the scanning aperture, the logarithm of the sum of transmittance values within the aperture will not be equal to the sum of the logarithms of the respective, point-wise transmittances. As a result, the experimentally accessible logarithm of the measured transmittance might give an incorrect estimate of the desired sum of the point-wise density values within the aperture. The only solution to this problem is to use an aperture that is small enough that the density is nearly constant across the area of the scanned spot.

The error in densitometry that occurs when there is a large range of density values within the scanning aperture has been discovered in many different scientific fields. In x-ray crystallography it is sometimes called the *Wooster effect*, after the author who first described it in the context of digitizing photographs of diffraction patterns (Wooster, 1964). In other areas of microdensitometry and microspectrophotometry it is sometimes referred to as the "distributional error." In spectrophotometry the error is also often referred to as the "absorption flattening" effect, as it results in underestimation of the absorbance (i.e., the optical density) at strongly absorbing wavelengths relative to the absorbance at weakly absorbing wavelengths.

9.4 Density versus exposure characteristics can be used to convert the film density to the corresponding value of electron intensity

The *optical density* of a *photographic film* increases linearly with electron exposure at low exposure. The density range included in electron diffraction patterns generally goes well beyond the range where this linear approximation is valid, however. To understand how the nonlinearity of the photographic response is accounted for, we need to discuss in more detail the response of the emulsion to electrons.

Silver halide grains in a *photographic emulsion* are developed by chemical reduction of the silver halide to the base metal. Any grain is "developable," but the rate at which a grain is reduced to metallic silver is increased by a large factor when nucleation sites are first established before developing. Nucleation sites, called *latent image specks*, are formed when energy is deposited in the silver halide crystal. The energy threshold for creation of a latent image speck is of the order of 20–30 eV, and at least this much energy is deposited in each inelastic scattering event. As a result, a single inelastic scattering event within a grain will make that grain developable. Note that the creation of latent image specks by electrons is very different from the situation with light, where absorption of tens of photons is required to produce a single nucleation site.

The distribution of the number of electrons per pixel on the emulsion is a random process which follows Poisson's law. Thus, if we consider a layer of emulsion thin enough that it should contain no overlapping grains in projection, then the fraction, f, of the grains that will have been hit by one or more electrons, when there are n electrons incident per unit area, will be

$$f = 1 - e^{-nA}, \tag{9.1}$$

where A is the cross-sectional area of a grain.

After the emulsion has been developed, light is absorbed (and scattered) by the grains of silver that have been formed. The increment of light absorbed by a thin layer is proportional to the number of grains in the layer, dn, and the fraction, f, of the grains that had been hit by electrons:

$$dI = (fI)dn, \tag{9.2}$$

where I is the incident light intensity. Integrating this expression through the full thickness of the emulsion, and then taking the logarithm of the transmittance to get the optical density, D, gives the result

$$D = \log\left(\frac{I_{out}}{I_{in}}\right) = D_0(1 - e^{-nA}), \tag{9.3}$$

where the value of D_0 depends on the size and number-density of the grains, and the emulsion thickness. Equation 9.3 tells us, then, that the density on the film should increase linearly with the exposure, n, at low exposure. At higher exposure, however, the density begins to saturate and asymptotically approaches the limiting value, D_0, which represents the maximum optical density achieved when every grain is developed.

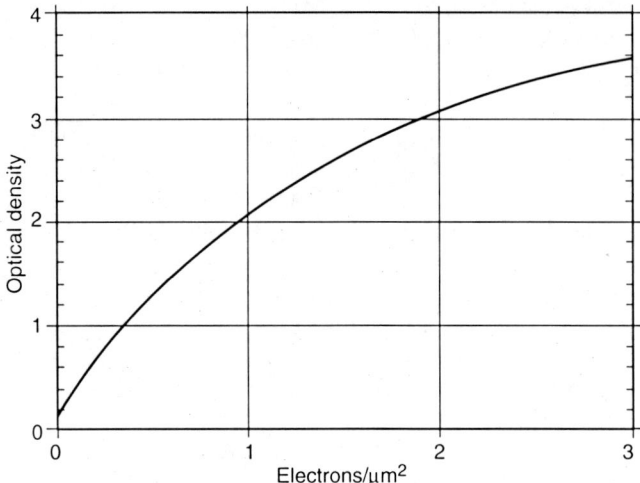

Figure 9.2 Calibration curve for optical density (OD) versus exposure for Kodak SO163 photographic film, commonly used in electron crystallography. This calibration curve applies to films exposed to 100 keV electrons, developed for 12 minutes at 68°F in full-strength D19 (without the addition of antifog reagent). The film density increases linearly with exposure until the emulsion begins to saturate, and then approaches an asymptotic value; this is the behavior expected from equation 9.4. This figure illustrates the further point, discussed in section 9.8, that a certain experimental error in the measured film density (the ordinate) translates into a progressively greater error in the estimated electron exposure (the abscissa) as the film density begins to saturate.

Some grains will be developed even without being hit by electrons. These grains contribute a *background fog* on the film, and add a second term to the measured density, so that the full expression for the density is

$$D = D_{fog} + D_0(1 - e^{-nA}). \tag{9.4}$$

A typical *density versus exposure curve* is graphed in figure 9.2. We see that at a low exposure, the density is linearly related to exposure. At densities up to about $D_0/4$ there is only about a 5% deviation of the curve from an ideal, linear response. Above this exposure the response deviates more and more from linearity as the emulsion saturates. Typical values for D_0 are around 6, corresponding to a transmitted light intensity that is a million times smaller than the incident intensity.

Film densities for low-dose images are generally below one, so the linear response holds quite well for imaging. On the other hand, it is common to have densities well above two in electron diffraction spots and in the central region of the pattern, which becomes strongly exposed by inelastically scattered electrons. Thus the linear approximation is insufficient for accurate determination of diffraction intensities.

Optical densities well outside the linear range can still be used in measurements of the electron intensity, since the response curve can be used to relate the measured densities to the actual electron intensities. In order to generate the required reference curve, exposures that cover a wide range of density values are made in uniform areas on a film. The densities measured in each of the areas, and the corresponding exposure values, can then be fitted to an expression of the form given in equation 9.4. The electron exposure that corresponds to a given value of the optical density in a digitized diffraction pattern can then be calculated from the analytical function that has been fitted to the measured blackening curve.

The slope of the linear region of the response curve is a measure of the *speed* of the emulsion, that is, how rapidly the film density increases with increasing electron exposure. A useful rule of thumb says that the speed is equal to the density (above the fog level) measured when the film has been exposed to one electron per square micrometer.

The characteristics of various photographic emulsions that have been used in electron microscopy vary over quite a wide range (Downing and Grano, 1982). Larger grain size, higher silver content, and greater emulsion thickness are all independent parameters that increase the film speed (signal strength), but it must be remembered that each will increase the fog (noise) as well. If one wants to increase the speed (i.e., sensitivity), it is best to do so by using a thicker emulsion or one with a higher silver content, while keeping the grain size small. Small-grain emulsions have the advantage that their fog level is lower and they have a larger dynamic range in comparison to large-grain emulsions, for the same total silver content per unit area. To some extent the characteristics are also influenced by the development conditions. For example, the size and number of developed grains increases with time, providing greater density and contrast. The fog level, arising from conversion of grains that were not hit by electrons, increases slowly at first during development, but may accelerate with prolonged development. "Antifog" agents may be added to the developer to suppress development of such grains, but these will also inhibit development of weakly exposed grains.

9.5 Data can also be collected by direct electronic readout rather than on photographic film

Although direct electronic readout of images has long been recognized as having significant advantages over photographic recording, the transition to direct readout in electron microscopy has been slow to develop. The reason why *electronic readout systems* have only recently begun to play an important role in data collection has to do with the already superb qualities of film as an electron detector. Indeed, some of these qualities still have not been matched by electronic detectors. Photographic film has quite good performance when it comes to *detective quantum efficiency* (*DQE*; see section 5.5 for further discussion), but its strongest feature is the large number of *pixels per image-frame* that are available on film. While the number of pixels per frame on direct readout systems is rarely a limitation when recording images of diffraction patterns, the standard 1k × 1k and 2k × 2k formats are inconvenient for many applications when one is recording real-space images.

The commercial development of large array-size *CCD cameras* has been a key development, however. Commercially available cameras now have formats as large as

4000 × 4000 (4k × 4k) pixels. This relatively large array size finally makes it attractive to consider such systems as an alternative to photographic film. Formats larger than 4k × 4k are under development and are likely to be available for electron microscopes in the near future.

When adapted for use in electron microscopy, CCD cameras — like other types of video camera — are used as sensors to respond to light that is produced in an appropriate scintillator or fluorescent screen. In many designs the scintillator is deposited onto a fiber optics plate, which is then bonded to the face of the CCD sensor. Fiber-optic coupling is favored as it provides for a more efficient collection of light produced by the scintillator than is possible to achieve with lens coupling (Downing and Hendrickson, 1999).

Advantages of CCD cameras, in addition to immediate conversion of the image to digital form, include the extremely good sensitivity and the low noise of the CCD device. The CCD device also has an exceptionally high storage capacity, which means that the dynamic range can exceed that of film by more than a factor of 100. The high sensitivity and low noise properties of the CCD mean that the DQE will not only be as good as that of photographic film at "high" exposures, where film is already very good, but it can exceed that of film by a large factor at lower exposures, where the fog level is the main limitation of film.

All of these properties — high sensitivity, large dynamic range, and "immediate" conversion to digital form — make the CCD a tremendous improvement over photographic film when recording electron diffraction patterns. At the same time, the inferior MTF that normally comes as a result of achieving maximum DQE (refer to section 5.5) — while a major consideration when recording high-resolution images — poses no limitation when recording diffraction patterns. The difference in this case is that the diffraction spots themselves are relatively large, and thus they place little demand on the MTF of the detector.

The high sensitivity and low noise level of a CCD camera make it possible to record electron diffraction patterns with lower electron exposures. One advantage of reducing the exposure required to record a diffraction pattern is the fact that less radiation damage is inflicted on the sample. The diffraction patterns therefore exhibit lower B-factors (the Debye "thermal parameter," section 2.6) — and as a result the data can extend to higher resolution (Brink and Chiu, 1994). The resulting data also show improved values of R_{sym} and R_{merge} (Downing and Hendrickson, 1999). Improvement in the R-factors can be expected because of the improved linearity of the CCD camera as compared to film, and because any nonuniformity of the detector response — equivalent to local variability of the film speed — can be measured in advance and corrected, unlike the case for photographic film. These issues are discussed further in section 9.8. Finally, a very important advantage of the CCD camera is that the diffraction pattern is immediately digitized as it is read out, in effect making the data instantaneously available for the next steps of data processing. Densitometry of electron diffraction patterns on a Perkin Elmer densitometer, by way of comparison, represents a significant bottleneck in the process by which experimental data are analyzed. The combined effect of faster data output and improved data quality make the CCD camera the unequivocal choice over photographic film for collecting electron diffraction data.

9.6 The digitized diffraction patterns are then indexed and reduced to the final diffraction intensities

In this section we will describe the computational procedures and the software used to extract intensities from digitized electron diffraction patterns. The software most commonly used for this purpose is derived from that developed at the MRC Laboratory of Molecular Biology in Cambridge, England (Crowther et al., 1996). Other software packages for processing diffraction data are likely to employ generally similar operations.

Processing of diffraction patterns begins with subtraction of the smooth background of scattered electrons

Figure 9.3 shows an example of the raw, scanned data for an electron diffraction pattern of a glucose-embedded crystal of bacteriorhodopsin. The central part of the pattern is dominated by the strong, *inelastic scattering* halo, which masks many of the inner diffraction spots because of its high intensity. The high intensity of the small-angle, inelastic background also makes it difficult to print such a digitized image in a way that adequately displays the weaker diffraction spots at higher resolution.

The first step in processing this data is to subtract the strong background intensity. A circularly symmetric, *radial density* function provides a good estimate of the

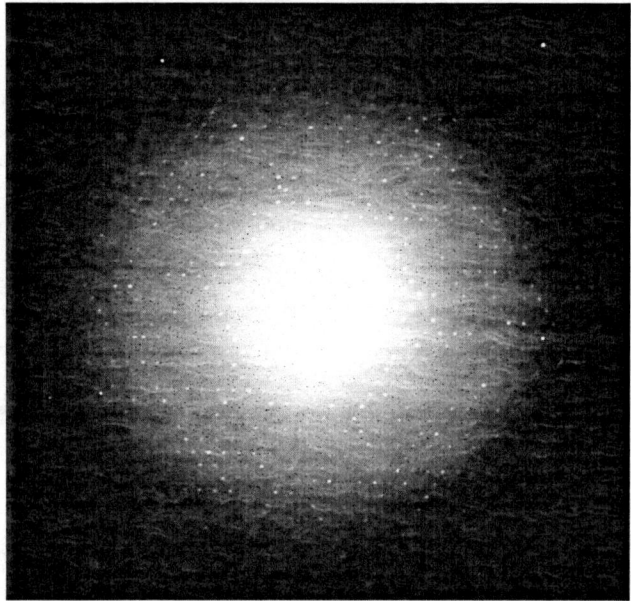

Figure 9.3 Digitized image of an electron diffraction pattern for a 2-D crystal of bacteriorhodopsin, recorded on photographic film. This is an image of the raw data as it first appears after digitization. The wide range of background intensities makes it difficult to simultaneously visualize spots in both the inner and the outer parts of the pattern.

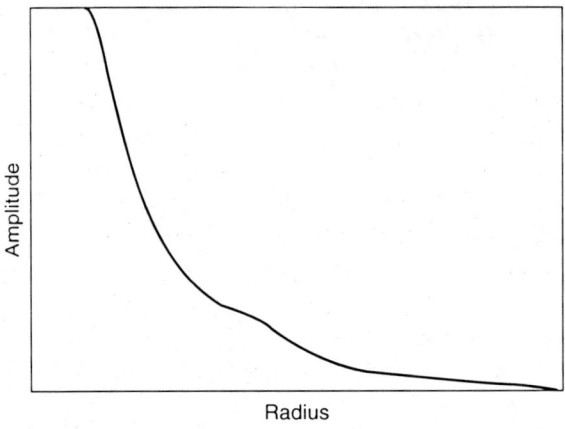

Figure 9.4 Radial background curve produced from the raw image shown in figure 9.3. This is the final background curve obtained after two cycles in which the effects of Bragg diffraction were removed and the background curve was recalculated from the smoothed image.

background that needs to be subtracted in order to better visualize more of the diffraction spots. This background curve is calculated by averaging all densities that are found at each radius from the center of the diffraction pattern. The position of the center of the diffraction pattern can itself be defined from the circular shape of the strong background scattering intensity, by simply calculating its center of mass.

Sharp spikes are initially produced in the radial density function, however, due to the presence of strong Bragg reflections at various specific radii. In order to remove the influence of the diffraction spots on the estimate of the background, a smoothed curve is recalculated through several iterations. In the first pass, all points are used to determine both the curve and its standard deviation at each radius. In the second pass, density values that are more than a set number of standard deviations above the mean for that radius are assumed to be part of a diffraction spot. These values are then excluded in computing the next version of the radial background. Since removal of the strongest spots decreases the standard deviation of what is left, the method becomes sensitive enough to remove weaker spots in the next cycle. Complete removal of all significant diffraction spots requires a few cycles of the algorithm.

Figure 9.4 shows the radial density function that was computed for the pattern shown in figure 9.3. Note that the radial density becomes unmeasurable at low radii due to unreliable performance of the densitometer at an optical density of about 4. Figure 9.5 shows the result of subtracting the smoothed, radial density distribution from the original pattern. Because the remaining background is now relatively flat, the dynamic range of intensities that need to be displayed is greatly reduced. As a result, diffraction spots can now be seen over nearly the full area of the pattern.

Indexing of the diffraction pattern becomes quite straightforward after the radial background has been subtracted

The process of *indexing* the diffraction patterns assigns Miller indices (sections 2.5 and 7.4) to each of the diffraction spots. The process of indexing can be done manually on an interactive graphics terminal by marking the positions of several diffraction spots with the cursor and entering the (h, k) indices of these spots.

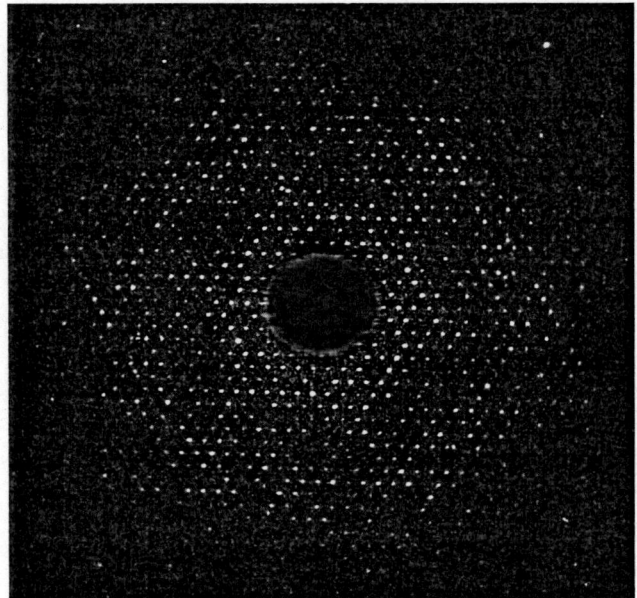

Figure 9.5 Background-subtracted image of the diffraction pattern shown in figure 9.3. The subtraction of the radial background curve (shown in figure 9.4) from the raw image (shown in figure 9.3) makes it much easier to display the Bragg spots that are present throughout the diffraction pattern. Indexing of the diffraction pattern is therefore much easier to accomplish once the radial background has been subtracted.

The reciprocal lattice vectors can then be calculated by finding the values that give the best fit between the observed and predicted positions. Manual indexing is a time-consuming step, however, and the need to process a large number of patterns has led to the development of an automated indexing system, described in the next paragraph.

The strategy used to carry out *automated indexing* is based on the fact that the highest densities in the background-subtracted pattern, figure 9.5, correspond to the diffraction spots. One first computes a density histogram for the background-subtracted data, keeping track of the locations of the highest densities. The indexing algorithm next identifies pixels that are close enough to one another to be part of the same diffraction spot, and then finds the centers of the spots. After a series of candidate spots has been identified, local areas of the diffraction pattern that are centered around the strongest spots (determined by a user-defined parameter) are superimposed in order to increase the signal-to-noise ratio. Figure 9.6 shows an example of this superposition of local areas. The level of definition of the spots that is achieved in the superposition is far better than in the original pattern, so that the centers of the spots can be defined quite accurately at this stage. A search algorithm is then employed along a spiral path from the center of the superposition image, until the two peaks nearest to the center are identified. Vectors between the center point and the two independent peaks give an accurate estimate of the *reciprocal lattice vectors* for the whole pattern. Positions of

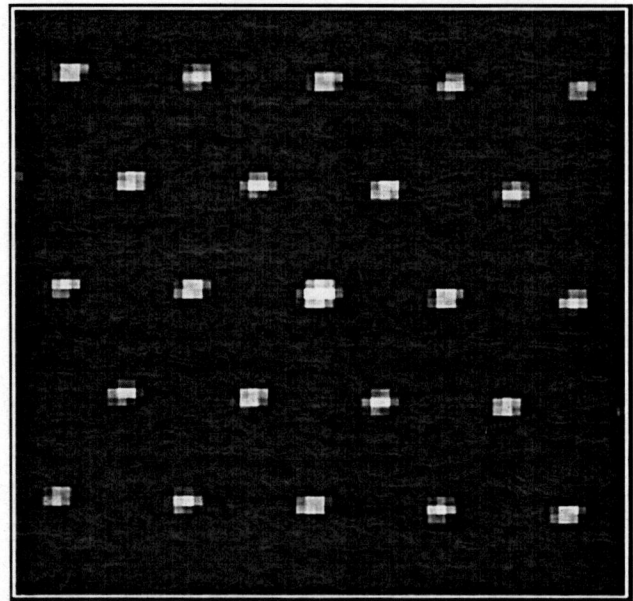

Figure 9.6 Superposition of local areas surrounding strong diffraction spots, a useful step in the automatic indexing of diffraction patterns. Superposition of multiple areas improves the clarity and smoothness of the weaker, neighboring spots. The center of mass of each spot can thus be defined with improved accuracy, thereby enabling accurate identification of the reciprocal lattice vectors.

all spots can then be determined using these basis vectors, starting from the center of the whole diffraction pattern, which was found when computing the radial background density.

There are a few situations in which this automatic search algorithm would fail to determine the true lattice vectors. For example, when the unit cell lengths are different by over a factor of two, or in the case of highly tilted specimens, the nearest two spots could be collinear with the center spot. To deal with such cases, a constraint is included on the angle between the unit cell vectors, to ensure that the correct basis vectors are determined.

The integrated intensity of each diffraction spot must be corrected by subtracting accurate measurements of the local background intensity

Once the reciprocal lattice vectors have been defined, one next proceeds to determine the background-subtracted intensities of all spots. In brief, the intensity of a given spot is determined by summing the intensities within a box surrounding the spot, after which a local estimate of the background intensity within the same area is subtracted. It is important to optimize the size of the small box that is used to isolate each diffraction spot, the goal being to include all the power in the diffraction spot without including too much background. Once the box used to surround each diffraction spot has been

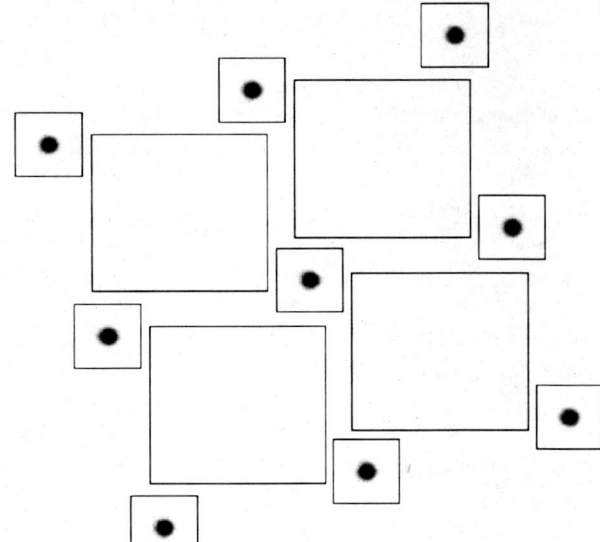

Figure 9.7 Schematic representation of the boxes used for determining integrated diffraction-spot intensities and local background intensities. The box surrounding each spot must be large enough to include all of the diffracted intensity. Background boxes are then chosen to fill as much of the remaining space as possible.

defined, as shown in figure 9.7, one next defines four to six boxes in the background, which are made as large as possible (to increase the accuracy of the measurement), without overlapping the boxes that surround the diffraction spots.

The size of the box enclosing the spots is first determined by examining the radially averaged profile of the sum (i.e., superimposition) of many diffraction spots; an example of the superimposition of many diffraction spots is shown in figure 9.6. The box size is chosen to be large enough to enclose the full diameter of the spot, determined by the radius at which the profile has flattened out to an apparent baseline level. The local background that lies within the area of each diffraction spot is finally estimated by interpolating the background values measured at the four to six points specified by the "background" boxes.

Processing of diffraction patterns is sensitive to minute defects that occur randomly over the field of the pattern. In CCD camera images such defects are thought to be the result of energetic background radiation, while in patterns recorded on film they are known be caused by dust and blemishes in the photographic emulsion. Several criteria have been applied to identify such defects and to exclude them from calculation of either the spot intensity or the background.

Many of the unwanted spots and blemishes can be dealt with on the basis of two criteria. One step is to eliminate any pixel in which the intensity is greater than the intensity at the center of the strongest diffraction spot in the background-subtracted diffraction pattern. Another step is to identify pixels whose intensity differs from that of their neighbors by a user-defined amount. Bad data points of this type should be eliminated if they are confined to a local cluster of pixels that is much smaller in size than the width of a diffraction spot. Such clusters occur at random locations in diffraction patterns recorded with a CCD camera, but similar events are not seen in patterns recorded on film (Downing and Hendrickson, 1999).

A further criterion for rejecting bad data points is based on the comparison of intensities of Friedel-related spots. Since *Friedel pairs* of spots (box 3.3) should have the same intensity, one can properly reject the pair if either the absolute difference or the relative difference between them exceeds some threshold. Unphysical differences between Friedel pairs can be the result of blemishes or image defects that coincide with the diffraction spots themselves, or that have otherwise escaped detection by the two criteria described above.

The accuracy of the background subtraction can be improved, at least in some specimens, by incorporating a model of short-range disorder as a part of the interpolation scheme (Grigorieff and Henderson, 1995). Short-range disorder (section 2.6) adds to the background between diffraction spots, and it usually does so in a way that is correlated with the intensities of neighboring Bragg reflections. Failure to correct for the intensity-dependent diffuse scattering that shows up between spots is likely to overestimate the background for a given spot. The resulting over-subtraction of background apparently explains why negative intensities are frequently produced in regions where the Bragg-sampled values of the continuous transform are low (Grigorieff and Henderson, 1995).

All data must be put onto a common scale, with a common temperature factor

Two important scaling corrections must be made before the data can be properly merged. Variations in the electron beam intensity, exposure time, size of the crystal, and other factors will affect the intensity of the diffraction spots. Thus an overall scale factor must first be determined for each pattern. The second scaling correction takes into account variations in the preservation (short-range order) of the sample, along with other factors that determine how rapidly the spot intensities fade as a function of increasing angle. This second correction normally takes the form of a temperature factor, described in sections 2.6 and 7.5. Both the overall scale factor and the Debye thermal parameter, B, can be obtained from a Wilson plot, as is described in section 2.6. The B-factor is given by the slope of the Wilson plot, while the overall scale factor is given by the intercept at $s = 0$.

In practice it is found that the B-factor is different for each diffraction pattern. Failure to compensate for this difference would make it impossible to merge data obtained from different crystals. In the case of bacteriorhodopsin, for example, diffraction patterns with especially low B-factors were used as a reference, and the B-factors for successive diffraction data were determined relative to the B-factor for the reference data. All diffraction patterns were scaled to the reference patterns by multiplying the measured intensities by a *relative temperature factor*,

$$W(s) = \exp(Bs^2). \tag{9.5}$$

Note that the sign of the argument in the relative temperature factor is positive, rather than negative. Forming the product of $W(s)$ and the measured intensities is identical to dividing the measured intensities by a temperature factor with the conventional, negative sign.

A further modification of the temperature factor correction is useful with tilted specimens, where it is often observed that the diffraction spot intensities fall off more rapidly in the direction perpendicular to the tilt axis than they do along the axis (Ceska and Henderson, 1990). This anisotropy can be described with the use of two thermal parameters, one for the direction parallel to, and one for the direction perpendicular to the tilt axis:

$$W(s) = \exp\left(\{B_{par}\cos^2(\theta) + B_{perp}\sin^2(\theta)\}s^2\right), \tag{9.6}$$

where θ is the azimuthal angle measured from the tilt axis, B_{par} is the value of the Debye thermal parameter parallel to the tilt axis, and B_{perp} is the value of the Debye thermal parameter perpendicular to the tilt axis.

9.7 Intensities from individual diffraction patterns are merged to form a 3-D data set

Each diffraction pattern only gives a 2-D sample of the full, 3-D Fourier transform

Recall that the projection theorem (section 4.3) states that the 2-D Fourier transform of a projection of a 3-D structure is a central section of the 3-D Fourier transform of that structure. This means that each electron diffraction pattern provides data on only one plane passing through the origin of the 3-D Fourier transform of the object. Tilting the specimen at various angles provides separate samples of the 3-D data, each on a different central section (plane). These separate measurements must all be combined in order to form a complete, 3-D data set.

Every diffraction spot corresponds to one point along a reciprocal lattice line

We have explained previously, in section 7.4, that the 3-D Fourier transform of a 2-D crystal consists of a set of lattice lines, along which the phase and amplitude vary in a continuous fashion. Each diffraction spot contributes a measurement of the intensity at a point where the plane of the central section (or more precisely, the Ewald sphere) intersects one of these lattice lines. We have also explained in section 7.4 that the continuous transform on each lattice line can be thought of as being the convolution of a 1-D sinc function with the discrete samples of the Fourier transform that would be found for a 3-D crystal. That description is incorporated in the mathematical equation that is used to merge the data obtained from 2-D crystals. More is said about this later, in connection with equation 9.7.

The whole purpose of merging the separate, discrete samples is to generate an experimental estimate of the continuous Fourier transform on each lattice line. Once the continuous function is available, it can then be sampled at regular intervals. As is described in chapter 11, the regularly sampled values on the 3-D reciprocal lattice are then used as input for a 3-D inverse Fourier transform, which produces the density map in real space.

It is necessary to determine the tilt angle and tilt axis in order to specify the z^* position for each diffraction spot

The position of each diffraction spot along its lattice line is determined by simple geometry, once the tilt angle and tilt axis are known. A preliminary estimate of the tilt angle and azimuth can be made by a straightforward calculation of the way in which the apparent geometry of the reciprocal lattice (i.e., the positions of diffraction spots) becomes distorted as the specimen is tilted (Shaw and Hills, 1981). The reason for the apparent distortion can be explained most clearly with the help of a specific illustration, as is shown in figure 9.8. In the example shown there, the 3-D Fourier transform of an

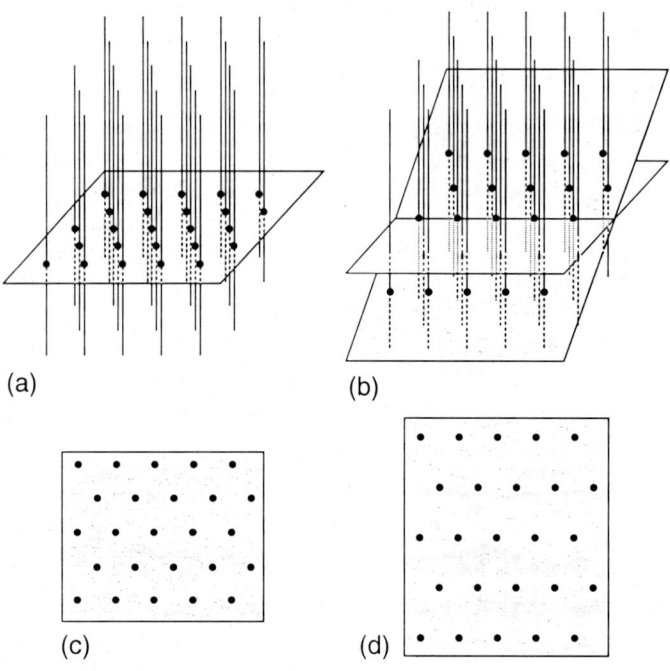

Figure 9.8 Tilting the sample produces an apparent distortion of the reciprocal lattice. The 3-D, parallel array of reciprocal lattice lines shown in panels (a) and (b) describe the 3-D Fourier transform of a hexagonal, monolayer crystal. The Ewald sphere is represented by the plane, highlighted in gray, which intersects the reciprocal lattice lines in both panels. In panel (a) the sample is untilted, and the Ewald sphere coincides with the a^*, b^* plane. Panel (c) shows the hexagonal arrangement of diffraction spots produced by the untilted sample when the Ewald sphere intersects the reciprocal lattice lines in the a^*, b^* plane. In panel (b) the sample is imagined to be tilted by a significant angle; for clarity the tilt axis is chosen to be parallel to the a^* axis. The Ewald sphere now intersects the reciprocal lattice lines at an angle that is equal to the amount of tilt. Panel (d) shows that the corresponding points of intersection between the Ewald sphere and the reciprocal lattice lines generates a lattice composed of distorted hexagons. If the tilt angle is large enough, the geometric distortion of the hexagonal lattice can be used to calculate both the tilt angle and the azimuth of the tilt axis.

hexagonal, 2-D crystal is represented by an hexagonal array of reciprocal lattice lines (section 7.4), as is shown in panel (a) of figure 9.8. If the sample is untilted, the Ewald sphere will correspond to the a^*, b^* plane, which is indicated by the gray plane in the figure. Panel (c) shows, for clarity, the hexagonal array of diffraction spots that is produced in this case by the intersection of the Ewald sphere with the reciprocal lattice lines. If the sample is tilted, however, the plane corresponding to the Ewald sphere will intersect the reciprocal lattice lines at some oblique angle. Panel (b) shows the case in which the sample is tilted about an axis that is parallel to the a^* axis. Because of the tilt of the Ewald sphere, the length of the vectors between the origin of reciprocal space and the points of intersection — the diffraction spots — have to be longer in a direction perpendicular to the tilt axis than they are in a direction parallel to the tilt axis. As a result, the hexagons seen in panel (c) become distorted, as is shown in panel (d). The amount of this distortion obviously depends in detail upon the precise azimuth of the tilt axis, which need not be parallel to the a^* axis, and the precise amount of the tilt angle.

It is sufficient to say at this point that the tilt angle and azimuth, apart from their sign, are related in a deterministic way to the reciprocal lattice vectors that are measured for untilted samples. The solution of appropriate vector equations thus provides initial estimates of the tilt parameters. Once these parameters are known, the points of intersection of the Ewald sphere with each of the lattice rods — i.e., the z^* position of each diffraction spot — can be calculated.

The determination of the specimen tilt parameters from the lattice distortion is not very accurate at low tilt angles, however. As may be easily appreciated, very small tilt angles will give virtually no distortion of the observed reciprocal lattice, yet they may cause the Ewald sphere to deviate from the ($hk0$) plane by more than the inverse of the crystal thickness. As will be discussed below, it nevertheless is helpful to begin a structure analysis by collecting data at low to moderate tilt angles. It is therefore necessary to assign preliminary tilt parameters, realizing that the initial estimates of the tilt angle may have considerable error. As the lattice lines become more densely populated with experimental data, however, the requirement that the measurements should be self consistent makes it possible to refine these relatively inaccurate, initial estimates of the tilt parameters, as will be discussed shortly.

Continuous, smooth curves are fitted to the discrete intensities measured along each lattice line

Merging intensity data obtained from specimens at various tilt angles is accomplished by fitting smooth curves to the discrete measurements that have been made on individual diffraction patterns. A least-squares fitting algorithm is used to obtain the best fit between the observed intensities and a weighted set of evenly spaced sinc functions. The experimental data are thus represented by a sum of the form

$$I_{curve}(z^*) = \sum I_l \, \text{sinc}(c \cdot z^* - l), \tag{9.7}$$

where the weights, I_l, are varied to get the best fit to the observed data; c is the crystal thickness; and the integer, l, has the same meaning as it would have as the third Miller index for a 3-D crystal. The coefficients, I_l, of the set of these sinc functions provide the complete set of merged structure factor intensities, $|F(hkl)|^2$.

Indexing of z^* positions (correct tilt angle and azimuth) can be refined with increasing accuracy

The refinement of the tilt angle and azimuth is achieved by an iterative process in which continuous curves are fitted to the indexed diffraction intensities on each reciprocal lattice line at each stage of the refinement. The first cycle of this iteration, described above, begins by plotting each measured value of the diffraction intensity at the z^* location estimated from the distortion of the reciprocal lattice. Although these initial estimates of z^* must later be improved, they are sufficient to map out the general shape of the continuous transform along each reciprocal lattice line.

The second cycle of the iterative merging process begins by finding new values of tilt angle and azimuth that cause the intensities in each diffraction pattern to have the least discrepancy with the values of the curves obtained in the previous cycle. The new z^* index of each diffraction spot will thus be slightly different from, but consistent with the less accurate value that was first estimated from the distortion of the reciprocal lattice. The second cycle is then completed by fitting new, continuous functions to the re-indexed diffraction intensities. The resulting reciprocal lattice-line curves represent an improved estimate of how to merge the 3-D data.

The iterative merging process can be repeated for three or more cycles. In each cycle there will be some remaining discrepancy between the values of the indexed diffraction intensities and the smooth curve that is fitted to them. This discrepancy is expressed as a *merging R-factor*,

$$R_{merge} = \frac{\sum |I_{obs} - I_{curve}|}{\sum I_{curve}}, \quad (9.8)$$

where I_{obs} are the measured diffraction intensities at each point, z^*, and I_{curve} are the corresponding values for the fitted curve. The summation can be carried out for all spots in a single diffraction pattern, giving a value of R_{merge} for each pattern, or the summation can be carried out for all measurements in the 3-D data set. In usual practice one will monitor the value of R_{merge} for the whole data set to determine when little further improvement is being made.

This refinement process is illustrated in figure 9.9, using diffraction intensities recorded from monolayer crystals of tubulin (Nogales et al., 1998). Panel (a) shows the diffraction intensities as they were initially indexed on the basis of the geometric distortion of the reciprocal lattice. Each measured value of the intensity corresponds to the average value of two Friedel mates, and the error bars for each point correspond to the actual Friedel differences measured between the individual diffraction spots. The continuous, solid-line curve passing through the data represents the fit to the data at the first stage in the process. The dotted-line curve close to the abscissa represents the root mean square difference between the measurements and the curve fitted to them. Figure 9.9(b) shows the same data after refinement. The marked improvement between the fitted curve and the experimental data is reflected in the reduced amplitude of the dotted-line curve after refinement.

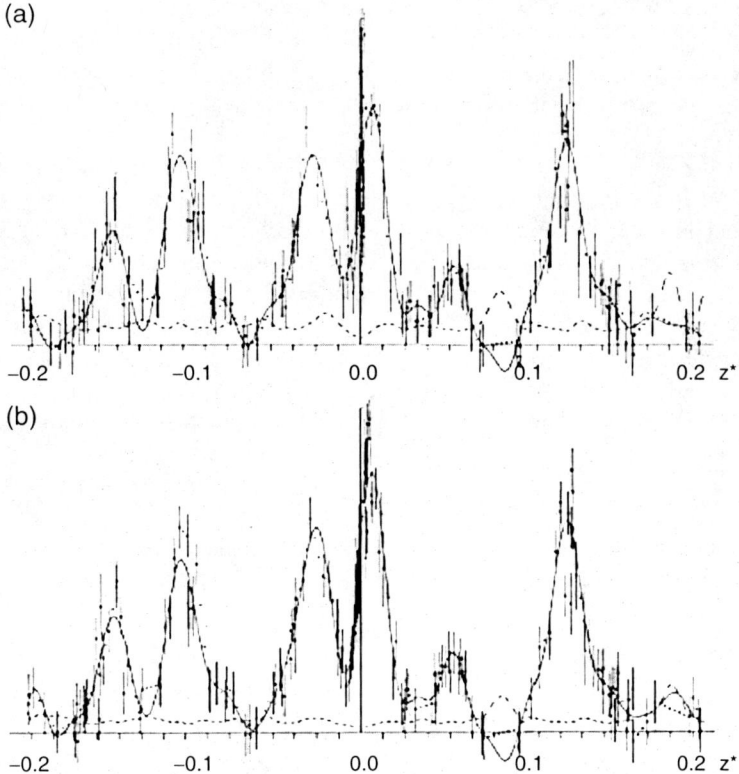

Figure 9.9 Demonstration of two stages in the process of merging data from a large number of diffraction patterns. The process is illustrated here for the (8,13) reciprocal lattice line of 2-D crystals of tubulin (Nogales et al., 1998). (a) Measured diffraction intensities are first plotted at z^*-positions determined from the apparent geometrical distortion of the reciprocal lattice. A smooth function, represented by the solid-line curve, is then fitted to the data. The root mean square difference between the solid line and the experimental measurements is represented by the dashed-line curve, close to the abscissa. Guide points (dots without error bars) are generated by interpolation in regions where experimental measurements are not yet available. (b) In subsequent cycles of refinement, the estimated amount of specimen tilt and the orientation of the tilt axis are treated as adjustable parameters, and an orientation is found that gives the smallest mean square discrepancy between the observed diffraction intensities and the intensities found on the curves at the corresponding z^*-positions. New curves are then fitted to the replotted data. After refinement, the root mean square difference (again represented by the dashed-line curve) between the measurements and the solid-line curve is reduced markedly.

How many diffraction patterns must be merged in order to produce a full, 3-D set of diffraction intensities?

The answer to the question posed in this heading depends upon the sample thickness, the resolution, the space-group symmetry of the crystal, and the highest tilt angle used to produce the 3-D data set. The number of diffraction patterns must be large enough to sample the Fourier transform along each reciprocal lattice line at intervals that are, on average, no larger than the reciprocal of twice the crystal thickness. For crystals that have high symmetry and are no more than 5 nm thick, 20 patterns collected at a fixed tilt angle of 60° can be sufficient to complete the merged data set at a resolution of 0.35 nm.

In practice it is usual to use a number of additional diffraction patterns that are collected at lower tilt angles. For example, a good data set for untilted specimens is essential in order to define the space group (section 7.3) of the 2-D crystals, to determine the lattice constants and their included angle, and to provide a reference data set for the next step in the merging process. In addition, it is common practice (although not strictly necessary) to collect one or two preliminary data sets at relatively low tilt angles, perhaps first one at 30° tilt and then a second at 45° tilt. All of the data should be used, of course, the result being that the merged data will normally be much more heavily sampled at lower values of z^* than it is at points further up the reciprocal lattice lines.

As an example, a total of 94 diffraction patterns were merged into the 3-D data set for tubulin (Nogales et al., 1998). Of the 94 diffraction patterns, 18 were recorded from untilted samples, 57 from samples tilted at 45°, and 19 from samples tilted by at least 55°. A partial representation of the merged diffraction intensities is shown in figure 9.10.

9.8 Factors that affect data quality

A residual discrepancy between the measured diffraction intensities and the final set of reciprocal lattice-line curves cannot be avoided, due to at least three factors: (1) residual errors will remain when defining the tilt angle and azimuth; (2) random measurement errors are inherent in the experimental intensities; and (3) systematic differences are present between individual crystals. The R-factor between Friedel mates (and other symmetry-related diffraction spots, if available) provides a good estimate of how much of the discrepancy is due to random errors inherent in the diffraction measurements. In addition, the values of R_{merge} for multiple diffraction patterns are invariably larger (often by a factor of two) than the corresponding R-factors calculated between symmetry-related reflections within each diffraction pattern. It is possible that systematic differences between individual crystals are the dominant factor that causes R_{merge} to be larger than R_{sym}.

Many types of experimental error can contribute to intensity differences between Friedel pairs

Weak diffraction spots will naturally be inaccurate whenever the measured intensity is not significantly greater than the noise level of the detector. But even when the detector noise is not a limitation, weak spots may have large errors because of the shot-noise, that is, poor counting statistics, that occurs in weak reflections. As an illustration, a diffraction spot that has a film density of 0.2 above background will contain 0.1 electrons/μm^2, assuming that the film speed is 2. If the diffraction spot is 50 μm in diameter, the total number of electrons in the spot will only be 250,

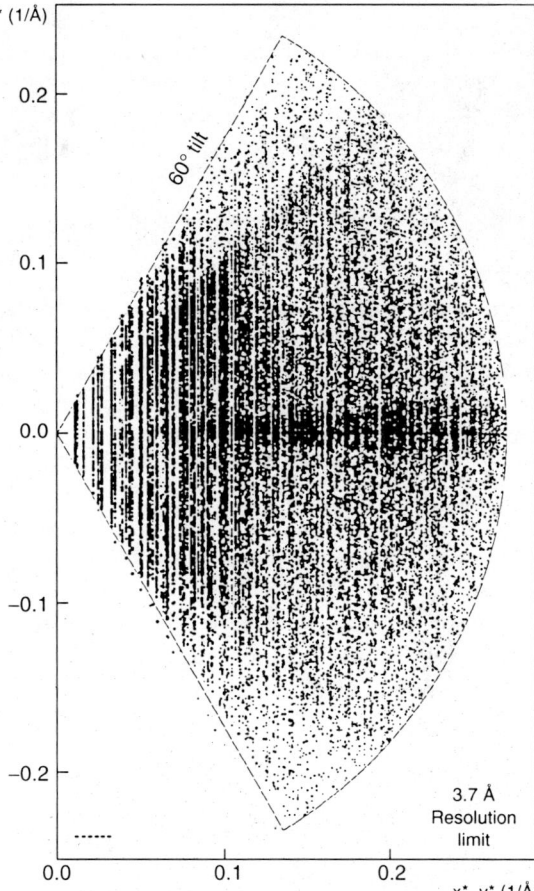

Figure 9.10 Example of the density of data collected throughout reciprocal space in a merged, 3-D data set. The full data set represented here is the one used for the 3-D structure analysis of tubulin at 0.37 nm resolution (Nogales et al., 1998). Individual measurements, originally indexed to specific reciprocal lattice lines, are projected onto one plane at their corresponding z^*-positions and their radial position in the x^*, y^* plane. A total of 94 diffraction patterns were merged to create this 3-D data set.

resulting in a standard deviation of 6% in the expected intensity. The same physical consideration applies when data are collected on a CCD camera, of course.

If recorded on film, strong spots will also have relatively large errors, this time because of saturation of the film. At high exposures the film density begins to increase more slowly as saturation is approached, and the slope of the curve shown in figure 9.2 gets smaller and smaller. This nonlinearity can be calibrated in advance, of course, and the calibration curve can be used to correct the estimated intensity. However, there is always a certain error in the measured density due to variation in the emulsion thickness, statistical fluctuations in the number of silver halide grains within the measured area, etc. Referring again to figure 9.2, it is apparent that a fixed uncertainty in the density value corresponds to a constant uncertainty in the associated exposure over the linear range of the film. However, the uncertainty in exposure becomes progressively larger in the region where the film begins to approach saturation, as it does for very intense diffraction spots. In the case of data collected on a CCD camera, on the other hand, the large dynamic range of the detector can provide a linear response over the whole intensity range of interest. As a result, the quality of data is noticeably better than that which is obtained with film (Downing and Hendrickson, 1999).

High levels of background intensity also give rise to greater error in the measured diffraction intensities. A higher background intensity must automatically result in a larger absolute fluctuation in the shot-noise. A high background will also cause photographic film to saturate earlier than it would due to the intensity of the diffracted spot alone. The resulting "premature nonlinearity" will again result in increased error of the type described in the previous paragraph. The intensity of background scattering is determined in the first instance by the thickness of the support film and any other surrounding material, such as excess ice or glucose. It is therefore important to use a support film and surrounding material that is as thin as possible, in order to keep the background intensity as low as possible.

At small scattering angles the background is dominated by inelastically scattered electrons. Even the crystalline sample itself will contribute to inelastic scattering, so there is a point at which little improvement is made in the background at small angles, no matter how thin the support film is. The inelastically scattered electrons are usually so intense that photographic film becomes saturated for Bragg spacings larger than 2 nm, when the exposure level is adjusted to give densities in the proper range for the rest of the pattern. Such low-resolution reflections are not normally used in high-resolution studies, even in x-ray crystallography, and thus the background from inelastically scattered electrons does not represent an important problem. In any situation where these low-resolution reflections should be measured, however, it is quite easy to do so simply by reducing the exposure to a level such that the inelastic background no longer saturates the film. This new exposure level will typically be a factor of 4 or 5 less than what is needed to record the high-resolution part of the diffraction pattern. Removal of the background of inelastically scattered electrons by use of an energy filter lens (section 5.3) is an even better option, when accurate diffraction intensities are needed for the low-resolution reflections.

The *Friedel difference*, averaged over all spots within resolution zones, or sorted by mean spot intensity, provides a useful measure of the quality of a diffraction pattern. This comparison is expressed as the R-factor (section 2.13) computed between all pairs of Friedel-related diffraction spots. This symmetry-related R-factor is defined to be

$$R_{sym} = \frac{\sum |I(g) - I(-g)|}{\sum \langle I(g) \rangle}, \qquad (9.9)$$

where $\langle I(g) \rangle$ represents the average value of the intensity for each Friedel pair. Typical values of R_{sym} for data that are considered "good" range from 4% for relatively strong spots to as much as 40% for weaker, but still significant, spots.

Curvature of the Ewald sphere, dynamical diffraction, and inelastic scattering are all factors that can produce systematic, rather than random intensity differences between Friedel mates. In each case, the systematic error represents a different facet of how the real data fail to meet the idealized model of the weak phase approximation. The weak phase approximation assumes that the Ewald sphere is a plane, when in fact the radius of curvature is not infinitely large. The weak phase approximation is also a single-scattering approximation, while the real interaction of electrons with any object, even a single atom, will have a small component of multiple scattering. Finally, inelastic scattering arises, in effect, from the imaginary component of the Coulomb potential;

the presence of both a real and an imaginary component of the Coulomb potential will necessarily give rise to violation of Friedel's law. While all of these effects must produce systematic Friedel differences, their contributions to R_{sym} are small for thin samples at high accelerating voltage, for example 100 kV or more. A further discussion of these effects is thus deferred to chapter 15.

Known differences between individual crystals contribute to values of R_{merge} that are systematically larger than values of R_{sym}.

Crystal-to-crystal variation in the amount of short-range disorder can be one of the most important differences that is observed between individual specimens. Variation in short-range order can be caused, among other factors, by the amount of stress experienced by a crystal as it is applied to the support film, or by variations in how well the crystal is hydrated. As was mentioned previously, it is therefore important to estimate the B-factor for each diffraction pattern and to apply a temperature-factor correction before attempting to merge the data. Some degree of error will, of course, remain in making the temperature-factor correction. This error will not be reflected as a difference in intensities of Friedel mates, however, and thus it will not be reflected in the value of R_{sym}.

Varying degrees of twinning (in specimens that are susceptible to twinning, such as fused purple membrane) can be another factor responsible for systematic differences between individual crystals. The relative fraction of the measured intensity that is contributed by each twin can be estimated as one of the parameters that enters into the merging of data (Baldwin and Henderson, 1984). Once the respective twin fractions have been estimated for a given diffraction pattern, the observed measurements can be corrected for the estimated contributions made by the minor twin components, in effect "detwinning" the data. The mathematical process of detwinning, while making an important correction to the measurements, is again susceptible to a certain level of error that will not be reflected in R_{sym}. For this reason it is usual practice to discard data from crystals that are heavily twinned; as an example, one might require that each crystal in the 3-D data set be composed 90% or more of a single crystal orientation.

Variation in the degree of flatness (planarity) of each crystal is another important source of systematic differences between individual specimens. Non-flatness causes the measured diffraction intensities to be an average of the intensity values found along a short interval of the reciprocal lattice line. This issue was discussed previously in section 6.8. Since the extent of averaging varies with the degree of flatness, imperfect specimen flatness is another factor that causes nominally equivalent measurements that have been made on two or more independent crystals to differ from one another and from the true value of the continuous transform.

10

Data Processing: Images of 2-D Crystals

10.1 Introduction

A substantial amount of image processing is required to extract the information contained in high-resolution electron micrographs of biological macromolecules. Descriptions of commonly used software packages for processing electron microscope images have been published in the *Journal of Structural Biology*, volume 116, issue 1 (1996). In this chapter we focus on the methodology for extracting structural information from images of thin, two-dimensional crystals.

We begin with a section on the preliminary evaluation of image quality by optical diffraction (section 10.2) followed by a section devoted to mathematical and practical details of digitizing the images (section 10.3). In section 10.4 we give a brief description of the mathematical principles by which the fast Fourier transform (FFT) algorithm works. The topics covered in these three sections apply in a similar way to images of any type of specimen, whether isolated molecules or two-dimensional crystals. The subsequent discussion in this chapter deals primarily with crystals, however, while noncrystalline specimens are discussed in chapters 12, 13, and 14. In all cases, image processing is fundamentally aimed at combining a large number of separate images and evaluating the quality of the merged images.

The use of crystals provides a powerful approach for merging data from images of a large number of molecules in identical orientations. The information that is contained in the image of a perfect two-dimensional crystal is most easily extracted in reciprocal space because all of the information about the average structure is contained in the relatively small number of diffraction spots in the Fourier transform of the image. The first step in processing such a Fourier transform is to index (i.e., label) each diffraction spot in a systematic way, and this is done (as is described in section 10.5) in the same way that is used for electron diffraction patterns. Extraction of the data from the transform is then described in section 10.6. The next step is to locate a standard phase origin for each micrograph so that data can be merged from different crystals. This part of the data processing is covered in section 10.7.

We next turn to evaluation and processing of the data. The processing steps required for structure analysis would be almost trivial with perfect images of perfect crystals. Real images and real crystals are imperfect, however, and thus require more elaborate treatment. It is normally the case that significant improvement of the data can be achieved by correcting for known, systematic effects that are present in the raw images. At each stage of the improvement, the quality of the data can be evaluated by methods, described in section 10.8, that estimate the peak-to-background ratio for each diffraction spot. Sections 10.9 and 10.10 describe techniques that compensate for image imperfections arising from distortion of both the specimen and the image. The first step at this stage will normally use the technique of quasi-optical filtering, which is discussed in section 10.9. The approaches that are then used to correct lattice distortions in images of crystals, described in section 10.10, are conceptually similar to the procedures that are used to identify and align images of single particles. With crystals, however, a real-space cross-correlation operation is used to estimate the distortions and long-range disorder in each image, and this is followed by an "unbending" operation before completing the Fourier-based processing. Section 10.11 concludes with a discussion of the corrections that are applied for other systematic errors related to the contrast transfer function and imperfect alignment of the illumination, finally producing data that are ready to be used in reconstruction of the specimen structure.

10.2 Optical diffraction is an effective tool for the preliminary evaluation of image quality

Selection of the best image areas is initially based on evaluation of the Fourier transform of the image. Visual evaluation of the image of stained or frozen-hydrated specimens can give a good indication of where crystals or molecules are located in the image, but visual inspection alone is not sufficient to assess the resolution or the information content of the image. Images of glucose-embedded specimens, on the other hand, may have too little contrast to even indicate the boundaries of the crystals. Nevertheless, optical diffraction of a photographic negative can be used to show the location of crystalline areas within the image. The optical diffraction pattern also provides a direct indication of the imaging parameters that affect resolution, such as defocus and astigmatism, and optical diffraction can be used to identify image defects such as specimen drift. Figure 10.1 shows four examples of optical diffraction patterns that illustrate the type of information that can be gained simply by examining a micrograph on an optical bench.

Various designs of optical diffractometers are used in different laboratories, but the principle is the same for all of them. A lens expands the beam from a laser to a diameter that is typically 1–5 cm, and subsequent lenses focus the beam to a crossover. When a photographic film is placed in the beam, the intensity distribution at the crossover represents the power spectrum (the square of the 2-D Fourier transform) of the optical wave function that is transmitted through the photographic film.

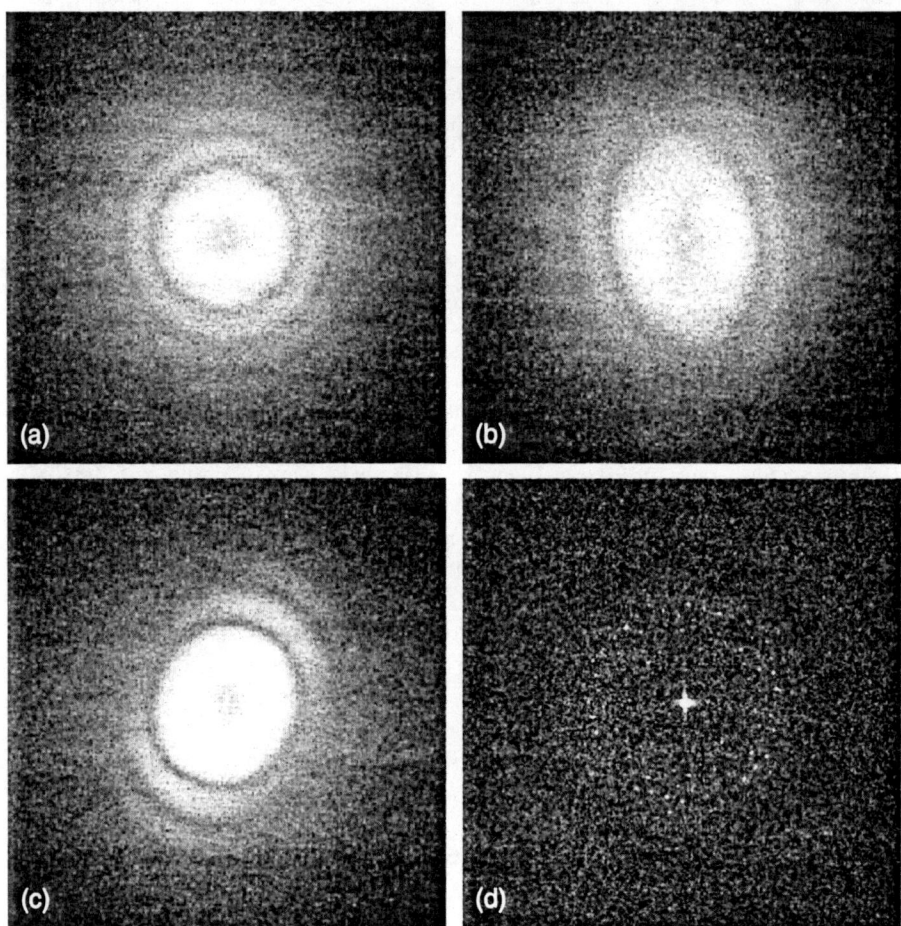

Figure 10.1 Fourier transforms of images display information about the imaging conditions and the specimen. (a) The concentric rings in the transform of an image of an amorphous carbon film arise from oscillations of the *contrast transfer function* (CTF). As a result, the positions of the maxima and minima of these oscillations can be used to determine the defocus and resolution. In this case, the defocus was about 1.2 μm, and the rings extend to a resolution of around 0.5 nm. (b) The elliptical rings in this transform indicate that the image is astigmatic. The defocus values in the directions of the major and minor axes of the ellipses were around 300 and 500 nm, respectively. (c) The disappearance of the rings in one direction of the CTF that is shown in this example indicates a corresponding loss of resolution in that direction, in this case due to specimen drift that has reduced the resolution from around 0.4 nm in the best direction to about 1 nm in the direction of the drift. (d) Transform of an image of a glucose-embedded crystal of bacteriorhodopsin, which is supported on a thin carbon film. Diffraction spots from the crystalline specimen are superimposed on a background arising from the noncrystalline portions of the specimen. The rings from the CTF modulate both the Bragg-sampled and the continuous components of the transform.

One should be aware that the transmitted optical wave is not the same thing as the 2-D optical density of the photographic film. It is not uncommon, for example, to observe that diffraction spots in the optical diffraction pattern of an image of a 2-D crystal do not obey Friedel's law, whereas the Fourier transform of the optical density must always do so. Failure of optical diffraction intensities to satisfy Friedel's law is due to the photographic film acting as a mixed phase-contrast and amplitude-contrast object for the incident optical wave. In practice, however, the intensity in the optical diffraction pattern is qualitatively similar to the power spectrum that is calculated for the digitized optical density of the same area of a film. As a result, the optical Fourier transform of the micrograph provides a useful (preliminary) representation of how well a particular area of the specimen was preserved, as well as how well the area has been imaged.

In the case of a crystalline specimen, the quality of the specimen preservation can be judged by the sharpness of the diffraction spots and by how many diffraction orders are present in the transform. Figure 10.1(d) is an optical transform of an image of bacteriorhodopsin, where diffraction from both amorphous and crystalline components of the specimen are clearly visible. The brightness and sharpness of the spots indicate good specimen preservation; the concentric rings seen in the amorphous background indicate a good choice of defocus without excessive astigmatism. The information content of the image usually extends well beyond what is clearly seen in such an optical transform, though. It is not uncommon to be able to extract accurate phases from crystalline specimens at twice the resolution to which diffraction spots are first seen in the optical transform.

With the addition of direct electronic readout of image intensities, provided by a CCD camera for example, on-line computation of the Fourier transform can be used in much the same way as optical diffraction to control the microscope parameters when recording the images. Computed Fourier transforms of low-dose images can then be used for visual, or even automated (Oostergetel et al., 1998; Potter et al., 1999) screening of thin crystals, which increases the likelihood that the chosen area will produce a high quality image.

10.3 Conversion of the image to a digital form is necessary for computer processing

The first step in image processing is to generate a digital representation of the electron intensity in the image. Most images are still recorded on photographic film, although digital readout systems based on charge-coupled device (CCD) technology are becoming increasingly common. Their continued development should eventually lead to the replacement of film recording, as we have discussed in section 9.5. In the following, however, we will generally assume that the image has been recorded on film.

Digitization involves measuring the film optical density (OD), which is to a good approximation proportional to the image intensity (see section 9.4). The OD is measured at a set of points on a regular grid, thereby converting the analog image (recorded on film) to a set of discrete measurements. There are several kinds of densitometers in use for this purpose. All operate by focusing an image of the object onto either a position-sensitive detector (i.e., CCD) or onto an aperture in front of a detector

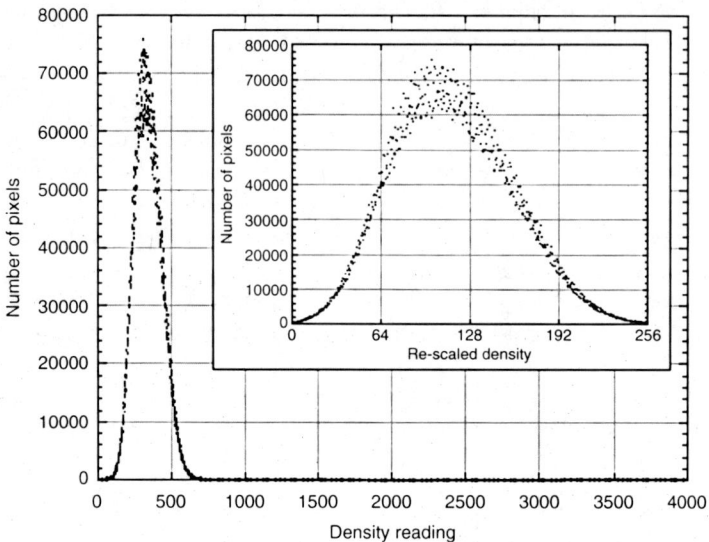

Figure 10.2 Histogram of image densities. A typical low-dose image of a protein crystal, recorded on Kodak SO163 film, has an optical density around 0.5, which corresponds to 500 units in the output from the densitometer. Most of the variation in the measured density values arises from noise rather than from contrast within the protein image.

(e.g., photomultiplier). The measurement at each pixel represents an average of the transmitted light intensity over the aperture (or CCD pixel). The measured intensity is then converted to optical density.

Four parameters are particularly important in describing the performance of a densitometer: grayscale resolution (i.e., the number of bits per OD unit), accuracy of pixel positions, pixel size, and the modulation transfer function of the instrument. Most scanners will produce output with 12–16 bit precision, but the image densities generally fall within a small part of this range. Figure 10.2 shows a histogram of image densities from a typical low-dose image of a bacteriorhodopsin crystal. The original density was measured with 16-bit precision, but the data is more than adequately represented with just 8 bits, as long as the output data are scaled so that the density values fill the 8-bit range.

The precision in the positioning of sampling points is quite important. Random or systematic errors in positioning introduce various types of defects in the transform of the image, not unlike those introduced by short-range and long-range disorder, which have been discussed previously in both sections 2.6 and 7.5.

Digitizing an image also introduces an unavoidable resolution limit related to the pixel size and the modulation transfer function (MTF) of the optical system of the densitometer. Imprecise focusing of the system or the limited aperture of the lens may cause the high-resolution Fourier amplitudes to be transferred at greatly reduced magnitude. Nevertheless, as long as the densitometer itself does not introduce significant noise into the data, the densitometer MTF will not affect the signal-to-noise ratio in the data. Thus the quality of the phases extracted from the image may not be significantly degraded by the densitometer MTF, even though the amplitudes can suffer a significant reduction in scale.

The MTF of the digitization process

The quantitative relationship between the digitized image and the original image becomes clearer when one understands that the sampling process associated with digitization introduces its own MTF. Integration of the original signal over the area of a single pixel of size d, which occurs during digitization, can be described in terms of first convolving the signal with a rectangle function of the same width before sampling the convolution product at discrete intervals. As a result, the Fourier transform of the digitized image is proportional to the Fourier transform of the original image multiplied by the Fourier transform of the pixel. The Fourier transform of the pixel thus constitutes an MTF that is introduced by the process of digitization. This MTF is likely to be well approximated by a sinc function,

$$MTF(s) = \frac{\sin(\pi d s)}{\pi d s} = \text{sinc}(\pi d s), \qquad (10.1)$$

where d is the pixel size (see box 10.1). The sinc function (figure 10.3) crosses through zero for the first time at the spatial frequency $s = 1/d$.

BOX 10.1 MTF of the digitization process

The modulation transfer function (MTF) that is introduced by the act of digitization is clearly explained when the result of digitization is viewed in terms of multiplication and convolution operations. Sampling the image on a regular grid of points corresponds to multiplying the image by a comb function (a set of delta functions at regularly spaced positions separated by a distance d):

$$I_{sampled}(x) = I(x) \cdot \text{comb}\left(\frac{x}{d}\right). \qquad (B10.1)$$

Because of the high noise level in the images, however, it is useful to integrate the signal for each measurement over as large a spot, or "sampling area," as possible, rather than to sample it at discrete points. Integration within the sampling area corresponds to first convolving the image with an aperture function of width d, rect($x; d$), before sampling with the comb function:

$$I_{sampled}(x) = [I(x) * \text{rect}(x; d)] \cdot \text{comb}\left(\frac{x}{d}\right). \qquad (B10.2)$$

Now recall that the Fourier transform of comb(x/d) is another comb function, comb(sd), where d is the spacing between spikes of comb(x/d), while the transform of rect($x;d$) is a sinc function, sinc($2\pi d s$). It then follows that the Fourier transform of the sampled image is given by the product of the Fourier transform of the object and the sinc function, convolved in turn with

(continued)

BOX 10.1 *(continued)*

the comb function:

$$\mathcal{I}_{sampled}(s) = [\mathcal{I}(s) \cdot \text{sinc}(2\pi ds)] * \text{comb}(sd) \qquad (B10.3)$$

Multiplication of $\mathcal{I}(s)$ by the sinc function reduces (modulates) the values in the Fourier transform that are obtained after digitization. Convolution with the comb function causes the modulated Fourier transform, $\mathcal{I}(s) \cdot \text{sinc}(2\pi ds)$, to repeat periodically, centered at the positions of each delta function in comb(sd). Periodic continuation of $\mathcal{I}(s) \cdot \text{sinc}(2\pi ds)$ results in aliasing if $I(x)$ contains Fourier components whose frequencies are greater than the Nyquist limit, a phenomenon that is discussed in box 10.2.

The spatial frequency that corresponds to precisely the distance covered by two successive samples (i.e., where $s = 1/2d$) is known as the *Nyquist limit*, indicated as s_N in figure 10.3. The *Nyquist frequency* has particular significance, as it defines the highest frequency actually calculated in a discrete Fourier transform of the image. As a result, the Nyquist limit is the resolution limit for the digitally sampled data. At the Nyquist limit the MTF has a value of $\sin(\pi/2)/(\pi/2) = 0.64$.

When a single sinusoidal component is sampled at a sufficiently large number of points in each full cycle, or period, the digitized signal will obviously be a faithful representation of the original signal, as is seen in figure 10.4(a). It is only necessary, though, to sample a given sinusoidal component slightly more than twice within each period in order to fully define the amplitude and phase. Stated another way, the Nyquist frequency should always be at least slightly greater than the frequency of any Fourier component of interest within the image.

It is quite instructive to consider a sine wave that is sampled at a frequency of precisely two samples per period. Figures 10.4(b) and (c) show two extremes of the

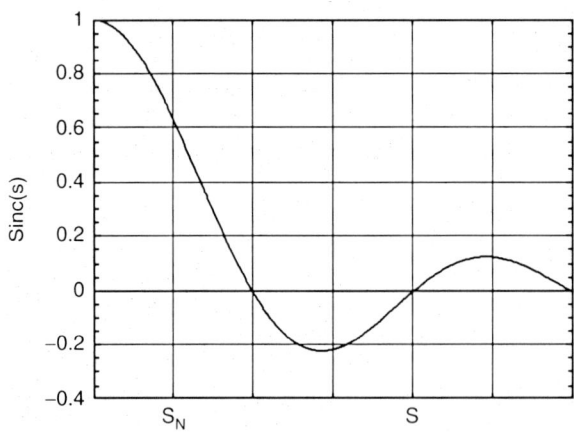

Figure 10.3 Graph of a portion of the sinc function, $\sin(x)/x$. The sinc function shows a maximum value of one at the origin and evenly spaced side lobes with decreasing amplitude. The spatial frequency marked s_N, termed the Nyquist frequency, corresponds to $1/2d$, where d is the sampling interval, and it is the highest frequency at which the Fourier transform is computed. As the graph shows, the value of the sinc function is slightly larger than 0.6 at the Nyquist frequency.

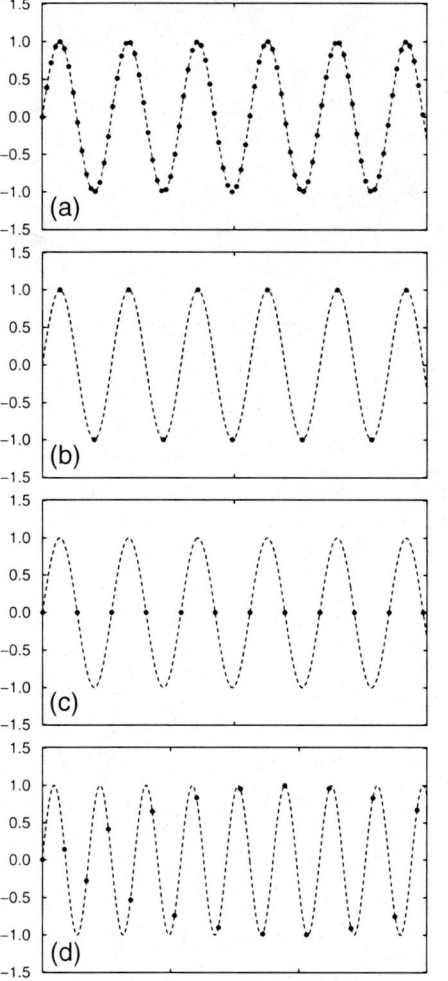

Figure 10.4 The frequency of sampling of a pure sine wave determines how faithfully the wave can be reconstructed from the sampled data. (a) When there are many sampling points in each period of the wave, the function is unambiguously clear. If the sine wave is sampled precisely twice per cycle (i.e., "Shannon" sampling), the amplitude of the sampled data can range between the two extremes of (b) the full amplitude when samples are taken at the peak of the wave or (c) zero, when the samples are at the nodes. (d) When the sampling frequency is just slightly greater than two samples per cycle of the sine wave, however, there is no longer a danger that the samples will get "locked in" at the nodes, and the amplitude and phase of the sine wave can be recovered from a long enough series of samples.

data that could result, depending on the positions of the sampling points within the period. If the sampling positions are at the peaks of the sinusoid, the digitized signal will have the same amplitude as the original signal. On the other hand, if the samples happen to fall at the points where the sine curve crosses the horizontal axis, no signal at all will be recorded. Note that, for any frequency within the object that is even slightly below the Nyquist frequency, as is shown in figure 10.4(d), the sine curve will be sampled at different places in successive cycles. Thus the peak amplitude can, in principle, be determined for any frequency below the Nyquist frequency.

The *Whittaker–Shannon sampling theorem* (e.g., Goodman, 1968) states that a *band-limited image* (one that has no Fourier components above a given spatial frequency) can be completely reconstructed from its sampled values, as long as the sampling interval corresponds to a frequency at or above the band limit of the image. In other words, if the original image contains no spatial frequencies finer than the Nyquist limit, the image can be faithfully reconstructed everywhere, even at points that lie between the sampled values.

Even though the signal from the specimen in an actual micrograph may not extend beyond the Nyquist limit, the spectrum of the noise in the image will generally extend well beyond the Nyquist limit. It is thus useful to see how frequencies higher than the Nyquist limit are represented in the digitized data and how they may influence the interpretation of the image. Note in figure 10.3 that the amplitude of the MTF (i.e., the sinc function) is still nonzero for frequencies higher than the Nyquist frequency, that is, for those Fourier components that are sampled less frequently than twice per period. As a result, these high-frequency components will still be represented in the digitized image. Somewhat paradoxically, these components show up at progressively lower frequencies in the computed transform. In other words, frequencies that are higher than the Nyquist limit appear to get "folded back" from the edges of the Fourier transform This phenomenon, termed *aliasing*, is discussed further in box 10.2. The aliased Fourier components appear with progressively smaller amplitude, however, as determined by the sinc-function MTF that was described in equation 10.1.

BOX 10.2 Aliasing and periodic continuation

It is instructive to consider the effect of digitizing a sine wave whose frequency is slightly different from that of the *Nyquist frequency*, similar to the example shown in figure 10.4(d), in order to understand the effect known as aliasing. Let s_N be the Nyquist frequency. Suppose that the image contains a Fourier component with a frequency $s_N + \Delta$. Referring back to section 2.4 on Fourier transforms, we see when we write

$$a(x) = c \sin(2\pi(s_N + \Delta)x) \qquad (B10.4)$$

that the transform of this component will be

$$A(s) = \frac{c}{2}(\delta(s - s_N - \Delta) + \delta(s + s_N + \Delta)). \qquad (B10.5)$$

But recall that s_N is the highest frequency where the transform is computed; how do we account for an apparent frequency, $s_N + \Delta$ outside this limit? The answer lies in the notion of periodic continuation of the transform, where the part of the frequency spectrum between $-s_N$ and $+s_N$ is seen as only one segment of an infinitely repeated pattern, as illustrated in figure B10.1. The repeated pattern is, itself, the result of the convolution by the comb function discussed in box 10.1.

To illustrate this point further, figure B10.1(a) represents the Fourier transform of some continuous image that is to be sampled with a Nyquist frequency s_N. The transform of the sampled image, which would normally be displayed as only the area within the white lines in figure B10.1(b), contains sections of the continuous transform that appear to be folded back into the region within the Nyquist limit. The computed transform shown in figure B10.1(b) is embedded

(continued)

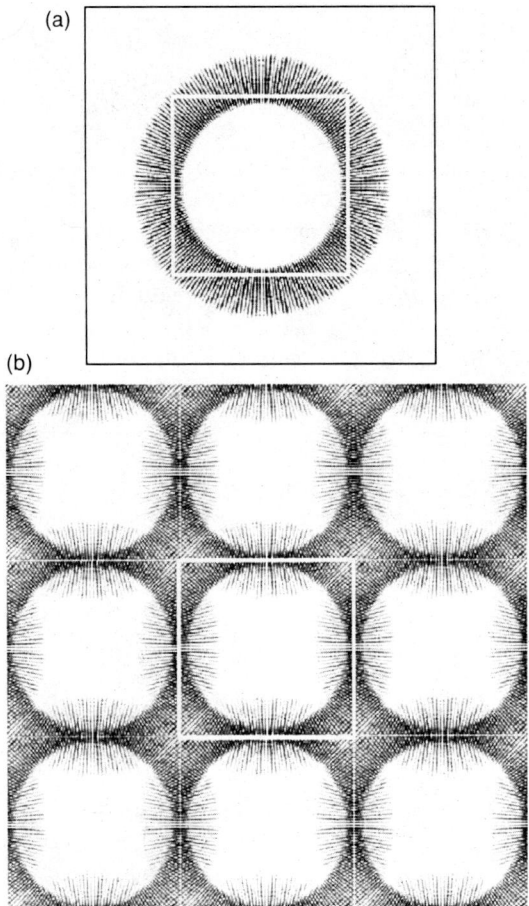

Figure B10.1 When an image contains spatial-frequency components higher than the Nyquist frequency, these higher frequency components in the Fourier transform are effectively "reflected back" at frequencies below the Nyquist limit in the computed Fourier transform of the digitized image. (a) The full spatial-frequency spectrum of a hypothetical image is displayed within the box that is defined by the outer, black lines. The white lines, on the other hand, indicate the Nyquist limit for a given sampling of the original image. Note that this example is constructed such that a significant part of the spatial-frequency spectrum of the image extends beyond the Nyquist limit. (b) The Fourier transform of the sampled image is identical to the Fourier transform of the original object, except for the fact that the latter now is periodically repeated at intervals of twice the Nyquist frequency. Periodic repetition of the full pattern in this way results in an infinite pattern, only a portion of which is shown here. The unique portion of the Fourier transform of the sampled image (together with its Friedel-symmetric complement) is displayed within the white box, which again represents the Nyquist limit. Note that Fourier components of the image that originally lay outside the Nyquist limit (in panel (a)) now intrude into the central, white box from adjacent copies of the periodically repeated Fourier transform. This intrusion of Fourier components that lie outside the Nyquist limit is called *aliasing*.

(continued)

> **BOX 10.2** *(continued)*
>
> in the periodic continuation of (a) (the result of convolution with the comb function), where now it apparent that the continuous transform extends from the central region into neighboring regions of frequency space. Thus all frequencies that were present in the image are represented in the transform, even though they may be higher than the Nyquist frequency s_N.
>
> Two essential points result from understanding the aliasing effect that occurs for frequencies higher than the Nyquist limit. First, the sampling interval must be chosen to be fine enough to sample the highest resolution features of the signal at least two times per cycle. Second, there still may be higher frequency components in the image that are due to noise, that will appear at frequencies below the Nyquist limit due to the effect of aliasing.

It is common practice to sample the image finely enough that there are three or even four samples per cycle of the highest desired Fourier component. This *oversampling* ensures (1) that the MTF is high at the highest frequency of interest, and (2) that the effects of aliasing the noise will be small at the highest signal frequencies of interest.

10.4 The fast Fourier transform is an efficient algorithm for numerical computation

The numerical calculation of the Fourier transform of a digitized image can be described in a straightforward way as a series of multiplications and additions involving complex numbers. The required mathematical operations are summarized in the equation

$$\mathcal{F}\{I(\mathbf{x}_n)\}(\mathbf{s}_j) = \sum_n I(\mathbf{x}_n) e^{-i\,2\pi\,\mathbf{s}_j \cdot \mathbf{x}_n}. \tag{10.2}$$

$I(\mathbf{x}_n)$ represents the digitized real-space image intensity, and the vector $\mathbf{x}_n$ designates the position of the nth pixel in the image. Similarly, $\mathcal{F}\{I(\mathbf{x}_n)\}(\mathbf{s}_j)$ represents the Fourier transform of $I(\mathbf{x}_n)$, and $\mathbf{s}_j$ designates the position of the jth pixel in the 2-D, reciprocal-space image of the Fourier transform. If there are N evenly spaced pixels in the real-space image, then there will also be N evenly spaced pixels in the reciprocal-space (Fourier transform) image.

For a given reciprocal-space position, $\mathbf{s}_j$, the mathematics specified in equation 10.2 requires the following operations: a vector multiplication must be performed, represented by $2\pi\,\mathbf{s}_j \cdot \mathbf{x}_n$; the vector product must then be used to form the complex number $e^{-i\,2\pi\,\mathbf{s}_j \cdot \mathbf{x}_n}$; this complex number must be multiplied by $I(\mathbf{x}_n)$; the whole process must be repeated for each of the N real-space pixels; and all such complex values must be summed for the N real-space pixels. The set of N complex additions and multiplications is then repeated N times, that is, for each of the N reciprocal-space pixels. The net result is that the number of complex arithmetic operations that is required to carry out the computation of a Fourier transform increases as N^2, where N is the number of pixels in the real-space image.

The straightforward evaluation of equation 10.2, as just described, is not the most efficient algorithm for calculating the Fourier transform, however. Much faster algorithms can be devised on the basis of the fact that the Fourier transform is a linear operation, and more specifically that

$$\mathcal{F}\{I_1 + I_2\} = \mathcal{F}\{I_1\} + \mathcal{F}\{I_1\} . \tag{10.3}$$

In equation 10.3 we envision that a large image, $I(\mathbf{x}_n)$, with N pixels, has been divided into two images, I_1 and I_2, each with $N/2$ pixels. The Fourier transform of the image, $I(\mathbf{x}_n)$, can therefore be broken into two smaller Fourier transforms, followed by simple addition of the two smaller transforms. As far as the respective Fourier transforms are concerned, dividing into two smaller pieces results in a large reduction in the number of calculations that must be performed. If the full image is used, there must be N^2 complex multiplications and additions, as has just been explained above. If only a "half image" is used, the number of calculations needed for each is reduced to $N^2/4$. There are two such Fourier transforms that must be calculated, however, and thus the number of calculations is actually reduced to $N^2/2$, which is already a big gain. In addition to the number of arithmetic operations needed to calculate the two, smaller Fourier transforms, one must include the work of adding the two transforms together, as stipulated in equation 10.3. This additional work involves only a small number of arithmetic operations, however, in comparison to the computation of the Fourier transforms themselves.

The fast Fourier transform (FFT) algorithm does not stop with simply dividing the real-space image into two halves, however. Instead, the FFT algorithm takes the process to its logical extreme, dividing I_1 and I_2 themselves into two equal halves, thereby once again reducing the number of complex multiplications and additions by another factor of two. Those halves are themselves divided into subpieces, and the process is continued until no further subdivision is possible. The final algebraic formula (algorithm) for calculating $\mathcal{F}\{I(\mathbf{x}_n)\}(\mathbf{s}_j)$ looks nothing like the right-hand side of equation 10.2, of course, and incorporates the fact that each sub domain of the image is displaced (shifted) in a known, deterministic way relative to all of the others. The Fourier shift theorem (section 10.6) stipulates that the Fourier transform of a function that is shifted by a vector distance, $\mathbf{a}$, is simply the Fourier transform of the unshifted function, multiplied by the additional "phasor," $e^{-i\,2\pi\,\mathbf{a}\cdot\mathbf{s}}$.

Different implementations of the concept underlying the FFT algorithm might be imagined, depending upon how the real-space image is broken down into subimages, and thus how the phasor term is calculated. The usual process is implemented by first creating two interleaved images, in which the first, third, fifth, etc. pixels are assigned to "image 1" and the second, fourth, etc. pixels are assigned to "image 2." Each sub image is then subdivided in the same way, and the process of division is continued to completion. This approach is well suited for images in which the number of pixels, N, is an integer power of 2. The number of arithmetic operations needed to calculate the FFT in this case scales as $N \log_2(N)$, a tremendous savings over the direct calculation, which scales as N^2. A simple comparison of the two numbers, $N \log_2(N)$ versus N^2, makes the point in an impressive way: if $N = 1024 \times 1024$, the gain in speed of the numerical calculation is a factor of $(1024 \times 1024)/\log_2(1024 \times 1024)$ ~ 5000. It is clear that the computational aspects of electron crystallography of large arrays benefits tremendously from the use of the FFT algorithm.

10.5 Images of crystals: indexing the Fourier transform is similar to indexing the electron diffraction pattern

The process of indexing a diffraction pattern requires first that one identify the basis vectors of the reciprocal lattice. As was discussed for 3-D crystals in section 2.5, and for 2-D crystals in section 7.4, each diffraction spot lies at a point in reciprocal space that is defined by these basis vectors. Each spot is identified by its Miller indices, which are integers that specify the number of steps that one must move along each of the axes, starting from the center of the pattern, to reach the spot.

One first identifies the type of reciprocal lattice from its geometry, as was discussed in chapter 7. For example, if the real-space lattice is orthogonal, the reciprocal lattice must also be orthogonal. If the real space lattice has $p3$ or $p6$ symmetry, the angle between **a*** and **b*** must be 60°. By convention, the reciprocal lattice vectors are always the shortest vectors that can be used to define the entire lattice, within the constraints imposed by the unit cell angles.

In some circumstances there exists an ambiguity in the assignment of **a*** and **b***. For example, for crystals with $p3$, $p4$, or $p6$ symmetry, **a*** and **b*** have the same length. In some cases the amplitude of the Fourier transform at specific diffraction spots can be used to distinguish the (h, k) spot from the (k, h) spot, once a convention has been adopted; in other cases, one must rely on phase information to complete the indexing. The final resolution of any ambiguities in indexing can be applied during the merging process, however, and need not interfere with the initial extraction of the data from the transform.

Procedures for indexing the Fourier transform of an image have evolved quite significantly. A description of the earliest procedures, although they are no longer used, is helpful in understanding how the indexing process actually works. In a literally "hands-on" procedure, the transform of the image was, at one time, printed out on paper so that one could use a straight-edge ruler to identify rows of spots that were parallel to each other. The direction of the rows gives the orientation of the reciprocal lattice vectors, and the unit spacing between diffraction spots on the rows gives the lengths of the basis vectors.

A more rapid procedure for manually selecting diffraction spots and assigning Miller indices to the spots used a graphics terminal to display the transform of the image. The task of identifying rows of spots was still required at this point, but that had to be done visually and mentally rather than on paper. A cursor was used to mark the positions of easily recognized diffraction spots, and the Miller indices were typed in. Once a sufficient number of spots were marked, the computer overlaid a lattice that was the least-squares best fit to the marked spot positions. The overlaid lattice served as a visual check that the initial indexing was fundamentally correct. The overlaid lattice also served to identify further spots that were weak or otherwise difficult to index. The accuracy of the result was improved by repeating the process of indexing, using as many spots as possible at the highest distance from the origin. The improved estimate of the reciprocal lattice obtained in this way was once more used as a guide in the next cycle of indexing. While this procedure was much faster than the one based on a print-out, it was still quite tedious to index a large number of images in this way.

Automated implementations of the indexing procedure are now the preferred approach (Schmid et al., 1993a). In these programs, the operator needs to identify as few as two spots, with indices, on the graphics display of the transform. The computer identifies likely candidates for other diffraction spots based on the intensity histogram of the transform, and finds which of these candidate spots lie close to lattice points defined by the first spots. The lattice vectors are then recalculated using the positions of all identified spots. This procedure can be carried through several iterations, providing a substantially more accurate result than is achieved with the manual methods. Of course, the computer could start without any manual input, identifying strong diffraction spots on a regular lattice as is done when indexing electron diffraction patterns (see chapter 9). Because of the lower signal-to-noise ratio of spots in the Fourier transforms of images, however, it is usually worth the effort to start the automated indexing by manually identifying the first two diffraction spots.

10.6 Extraction of amplitudes and phases from the indexed Fourier transform

Extraction of the experimental *structure factors* (see section 2.4) from the indexed Fourier transform is done by making a list of the numerical values of amplitude and phase at the positions of the reciprocal lattice points in the transform. In the initial step of this process, a small array in the vicinity of each reciprocal lattice point is examined in order to provide values for the peak amplitude and phase, as well as values of the local background in the vicinity of the spot. The local noise is determined from values located at positions far enough from the peak to exclude any power from the peak, which may extend several pixels from the center. Often the average amplitude of the noise, $\langle F_{noise} \rangle$, is obtained from the perimeter of a box of 7×7 to 15×15 pixels, the optimal size depending both on how large the scan array is and on the degree of lattice distortion. The noise is subtracted from the peak in quadrature, that is,

$$F_{estimated} = \sqrt{(F_{measured}^2 - F_{noise}^2)} \qquad (10.4)$$

where $F_{measured}$ is the measured amplitude in the Fourier transform of the image, and $F_{estimated}$ is the estimated value of the background-subtracted amplitude of the structure factor.

The value of the peak of the diffraction spot can be extracted either by hand from a display of each subarray, or it can be extracted in a fully automatic way. We shall digress here to say more about the automated procedure, with special emphasis given to analysis of the shape of the expected diffraction spots as well as some other characteristics of the spots. This discussion assumes familiarity with topics such as the convolution theorem that were discussed in chapter 2.

The shape of the diffraction peak is the transform of the crystal shape function

Figure 10.5 shows an example of the type of data one finds in the vicinity of a diffraction peak. The figure contains a 9×9 array of pixels extracted from a Fourier transform, with numerical values of both the amplitudes and phases at each point in this box.

248 ELECTRON CRYSTALLOGRAPHY OF BIOLOGICAL MACROMOLECULES

	Amplitudes								Phases						
37	38	32	50	65	28	25	17	35	132	26	331	100	85	24	81	51	197
29	21	26	19	90	92	9	77	57	146	200	214	293	276	256	245	234	129
68	48	50	84	139	38	53	32	46	340	194	274	100	85	41	104	355	220
26	39	201	236	517	17	91	111	25	218	307	123	274	290	265	268	67	178
66	71	103	419	★726	163	24	30	49	37	308	101	286	★291	114	236	144	225
34	51	42	120	133	21	37	58	71	59	183	101	135	123	78	340	101	298
90	31	27	79	128	46	37	66	25	52	298	148	349	248	358	100	281	93
82	31	20	65	75	87	56	25	57	105	179	249	151	135	108	218	344	65
58	40	75	8	68	51	59	39	11	357	259	135	90	266	261	45	132	298

Figure 10.5 Numerical values in the Fourier transform surrounding a diffraction spot. The amplitudes and phases are displayed separately. The center of the spot, which can be identified from the high amplitudes, is indicated by the star that lies between points where the transform is calculated. The phases at the four adjacent points in the computed Fourier transform, all of which correspond to samples within the central peak of the sinc function, are nearly equal. Phases of the next set of points (out from the center of the peak of the sinc function) are about 180° different from those in the central region, a result of the negative sign of the second lobe of the sinc function.

The actual position of the reciprocal lattice point is marked, and it is seen to fall between points where the Fourier transform was computed.

While we often assume, for the sake of formal mathematics, that our crystals extend to infinity, all real objects are bounded. In the case of a crystal that is represented by the scanned area of an electron micrograph, the object is bounded, at the very least, by the area that has been scanned. It is therefore mathematically convenient to describe the actual image represented in the computer as an infinitely large crystal that has been multiplied by a bounding function, which itself describes the finite spatial extent of the actual crystal, as shown in figure 10.6. It then becomes clear that the Fourier transform of the digitized image is the convolution of the transform of the infinite crystal, which is an array of delta functions at the reciprocal lattice points, with the transform of the bounding function. The bounding function is often referred to in this and other contexts as the *shape function*, and the transform of the shape function has special significance. As we will explain next, the role of the shape function differs in an important way depending on whether the boundary of the crystal is larger or smaller than the digitized array.

When an area of crystalline specimen completely fills the scan area, the bounding function is rectangular in shape, and its Fourier transform is a sinc function. Furthermore, the nodes of the sinc function are located at spatial frequencies that are integral multiples of the reciprocal of the scan width, that is, the zeros of the sinc function have the same spacing as do the points at which the Fourier transform is calculated. If a diffraction spot falls precisely on one of the points where the Fourier transform is calculated, the sinc function will also be centered at the same point, as is shown in figure 10.7. Adjacent points in the calculated Fourier transform will therefore coincided with zeros in the sinc function. As a result, the calculated diffraction spot occupies just a single point in the computed diffraction pattern.

The diffraction spots in the computed transform will all fall exactly on the points of the calculated transform only when there is an integral number of unit cells in

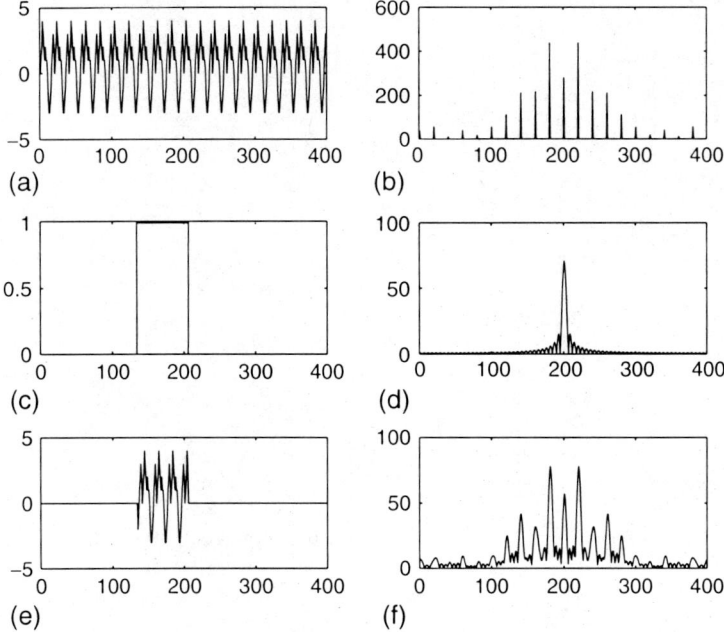

Figure 10.6 Diffraction spot profiles. On the left are functions that represent various real space functions, and on the right are the square of their respective Fourier transforms. (a) A perfect, infinite crystal produces (b) diffraction spots that can be represented as delta functions of different relative heights. (c) A function that defines either the crystal or scan area (the shape function) has (d) a transform that, in this one-dimensional case, is a sinc function (shown here as the function squared). (e) A crystal of finite size (or the finite array that represents the digitized area of a crystal) can be considered as the product of the infinite crystal and the shape function. (f) The diffraction spots are thus the square of the convolution of the Fourier transform of the infinite crystal, shown squared in (b), with the Fourier transform of the shape function, shown squared in (d).

the scanned area. In this case, it is also evident that there are no discontinuities in the periodic continuation of such an image. It may be possible, of course, to trim a digitized image so that the enclosed area contains an integral number of unit cells. If one actually takes the trouble to do this, the structure factors can be read directly from the reciprocal lattice points in the transform.

Vector sums should be used to extract structure factors whenever there is a nonintegral number of unit cells in the digitized image

One does not generally make an effort to have precisely an integer number of unit cells within the scanned area, and thus the reciprocal lattice points (e.g., the true centers of the diffraction spots) will usually fall between the points where the transform is calculated. The central maximum of the sinc function itself will then be sampled at multiple points, as is shown in figure 10.7(b).

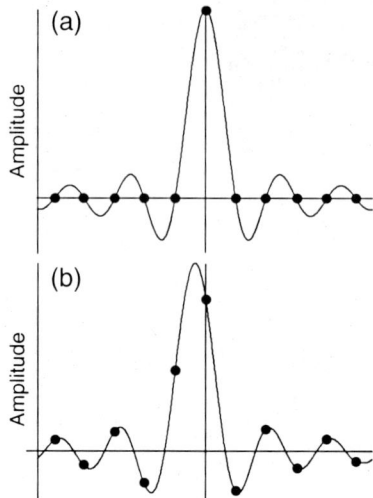

Figure 10.7 Sampling of a diffraction spot. When the crystal completely fills the area of the scan, the shape of each diffraction spot is a sinc function with zeros spaced apart by the reciprocal of the scan width. The Fourier transform itself is sampled at intervals that are also evenly spaced by a distance equal to the reciprocal of the width of the scan. (a) If the unit cell vectors of the crystal are parallel to the scan directions and if there is an integral number of unit cells in both directions, then the peak of each sinc-function-shaped diffraction spot will coincide with one of the sampling points, and adjacent sampling points will fall at the zeros of that sinc function. (b) If the unit cell axes are not parallel to the scan directions or if the digitized area does not contain an integral number of unit cells, the peak of the sinc-function-shaped diffraction peak will generally fall between sampling points. As a result, the central lobe of the sinc function will be sampled twice (in each direction) and the oscillating lobes will also be sampled in the computed Fourier transform.

In this more general case, the actual amplitude and phase should be determined by a vector summation of the transform values over several points, with proper weighting for the sinc function amplitude. Each measurement at a sampling point adjacent to a reciprocal lattice point should have the same phase, but because of noise in the data the phases will not actually be the same (as is illustrated in the example shown in figure 10.5). When these separate measurements are added vectorially, however, the signal increases linearly (coherently) but the noise increases in quadrature. As a result, the signal-to-noise ratio is improved by summing the data vectorially. In practice, only the four pixels adjacent to the actual diffraction spot are added vectorially in order to estimate the amplitude and phase, since the amplitudes at points farther away are low due to the rapidly decreasing value of the sinc function.

A crystal that does not completely fill the scan array will produce broadened diffraction spots

When working with low-contrast images, examination on the optical bench can identify the good areas rather well, but the location of the crystal edges is often not well defined.

Since one would like to use the largest possible area of the crystal, it is common to scan a large enough region to be sure to fully contain the crystal. With stained specimens, the scanned image can be displayed and the good areas can usually be identified visually. With unstained specimens, however, the boundaries of the crystal are best defined by a cross-correlation operation that is described in section 10.10.

When the crystal does not fill the entire scanned area, the digitized image can still be described as the product of an infinite crystal with a bounding function. A specific example, using the scanned image shown in figure 10.8, will help to emphasize this important point. If the new bounding function were rectangular in shape, its transform would still be a sinc function, but its width would be broader than that shown in figure 10.7 (because its real-space size is smaller than the size of the scanned area). As is seen in the real example that is shown in figure 10.8, however, one can usually expect the shape function to be some irregular "mask" rather than having a rectangular shape, and thus its Fourier transform will have a more complicated shape. Strictly speaking, the weighting function used in the estimation of the amplitudes and phases should thus be different from that used with a crystal that completely fills the scan area. It can easily be seen, however, that the errors introduced by assuming that the crystal fills the entire scanned array (when, in fact, it does not) will be small as long as the crystal is large enough to cover most of the scanned area and the weighted vector summation is taken only over, say, a 2 × 2 pixel box closest to the actual lattice point.

Figure 10.8 Example of a crystalline area that is smaller than the full area that was scanned. Because the edges of many crystals are indistinct, it may initially be unclear where the border of a crystal actually lies.

When it is found that the scanned area is larger than the size of the crystal, it is important to ensure that the "shape function" (which describes the size and location of the crystalline area) should be centered within the digitized field. In addition, it may actually be desirable to trim the digitized array to match the area of the crystal as well as possible.

In general, though, as is illustrated in the example shown in figure 10.8, the sides of the crystal will not be parallel to the edges of the array, and frequently the edges of the crystal may not even be straight lines. Furthermore, it often happens that the quality of the image varies within the scanned area. In this case, trimming the array down to include just the best part of the scan could improve the signal-to-noise ratio, but even better results may be obtained with a real-space weighting scheme described in the next section.

Weighting the image to suppress noise

Calculation of the cross-correlation function described in section 10.10 can also be used to assign variable weights to different areas within a crystal in order to account for local variability in the quality of the image. In order to illustrate how such a weighting function is created, figure 10.9(a) displays a map of the locations of strong peaks that were found in the cross-correlation function for the image shown in figure 10.8. Areas where the cross-correlation peaks are strong are given a high weight, while those with weaker peaks are given lower weight, resulting in the gray-scale "mask" that is

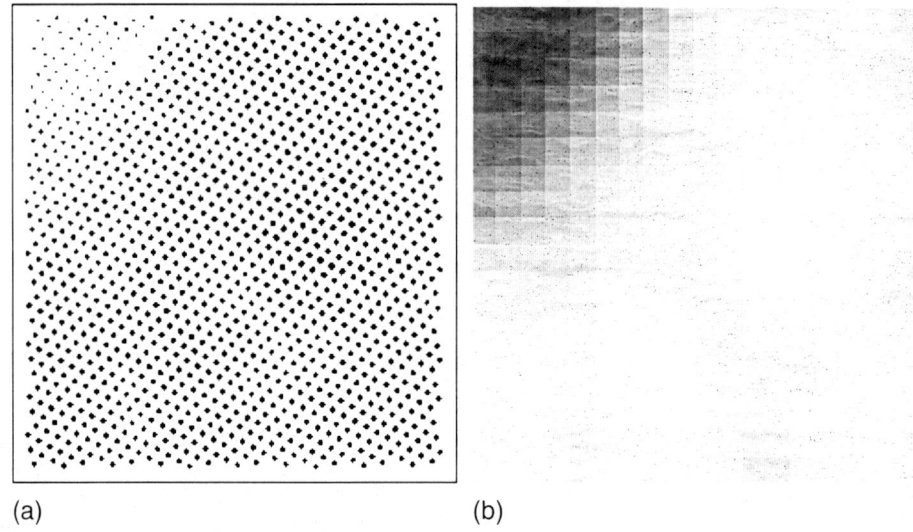

(a)   (b)

Figure 10.9 Weighting function used when the area covered by a crystal is found to be smaller than the full area of the scan. (a) Using the correlation operation described in section 10.10, a map is generated that represents the quality of each area within the scan. (b) From the map in panel (a), a weighting function can then be derived that multiplies the original image to suppress the contribution from areas where the crystal is absent or its quality is low.

shown in figure 10.9(b). This weighting function eliminates contributions from areas where there appear to be no crystal, which will only be noise, and it can also suppress areas where the crystal apparently does exist but its quality has been found to be poor. Use of such a real-space weighting function is especially convenient for crystals of irregular shape.

The real-space weighting function will usually produce a bounded area that fills less than the full scan. If the parameters of the weighting function result in a bounding function that corresponds to a relatively small region within the scan, it must be remembered that the diffraction pattern is then convoluted with a rather broad function, resulting in broad diffraction spots. In an extreme case the transform of the shape function can become so broad that convolution with this function forces the phases at adjacent points within the Fourier transform to be very similar. As a result, the vector sum over adjacent points in the transform will have a reasonably large value and will thus appear to indicate good data. Allowing the weighting function to create a bounded area that is too small will therefore produce strong vector sums even in the extreme case when there is only noise, and no contribution from a crystal, in the transform. Thus one must avoid the application of a real-space weighting function that selects only a small part of the scanned image.

10.7 Establishing a common phase origin allows data from separate crystals to be merged into a 3-D data set

Once the crystallographic structure factors have been extracted from the Fourier transform of an image, these values must be combined with structure factors obtained from many other images. The goal is ultimately to produce a 3-D set of structure factors that can be used to build a 3-D density map in real space. Two problems must be solved in order to merge data from individual images. The first problem is to determine the tilt angle and the orientation of the tilt axis relative to the crystal axes. The second problem is to determine a common origin of the coordinate system in real space, which is referred to in crystallography as the *phase origin*. In the initial stages of a project, one usually builds up a reliable reference using just projection data from nominally untilted specimens. Thus during the first stage one need only be concerned with finding the phase origin.

Why is it necessary to find a common origin in real space before merging the data? Although the amplitudes in the calculated diffraction pattern of the image are independent of the origin chosen for the Fourier transform, the phases are not. Thus, if one wants the calculated phases from different images to have the correct relationship to one another, it is necessary to compute all Fourier transforms relative to a common origin in real space.

The Fourier transform shift theorem

The *shift theorem* of Fourier transforms gives the formal, mathematical explanation of why the calculated phases depend upon the position of the coordinate system. The shift theorem, derived in box 10.3, states that the phases change linearly with the shift of

BOX 10.3 Shifting the origin in real space introduces a linear phase shift in reciprocal space

It might seem at first thought that the Fourier transform of an object should not change when the object is moved from one position to another. The object itself is not changed when it is moved, so why should its Fourier transform change? On further reflection, however, the Fourier transform of an object must contain all the information that there is about the object, including its position. Thus, since moving the object changes its position, it must also change the Fourier transform in some way. As we will show here, moving an object from one position to another changes the phases of the Fourier transform in a systematic, predictable way, while the amplitudes of the Fourier transform are not changed in any way by moving the object.

The Fourier transform of an arbitrary function, $f(x)$, is given by

$$\mathbf{F}(s) = \int f(x) \, e^{-i \, 2\pi \, s \cdot x} \, dx. \tag{B10.6}$$

It therefore follows that the Fourier transform of the same function, after being shifted to a new position, $x = a$, is given by

$$\mathbf{F}'(s) = \int f(x - a) \, e^{-i \, 2\pi \, s \cdot x} \, dx. \tag{B10.7}$$

The relationship between $\mathbf{F}(s)$ and $\mathbf{F}'(s)$ is seen more clearly when equation B10.7 is rewritten after substituting the new variable, $y = x - a$, thus giving the result

$$\mathbf{F}'(s) = \int f(y) \, e^{-i \, 2\pi \, s \cdot y} \, e^{-i \, 2\pi \, s \cdot a} \, dy$$

$$= e^{-i \, 2\pi \, s \cdot a} \, \mathbf{F}(s). \tag{B10.8}$$

Equation B10.8 can be rewritten in symbolic notation:

$$\mathcal{F}\{f(x - a)\} = \mathcal{F}\{f(x)\} e^{-i \, 2\pi \, s \cdot a} \tag{B10.9}$$

The formal result stated in equation B10.9 is known as the *Fourier shift theorem*, and this result is stated again in equation 10.5, using vector notation to generalize the result.

It is apparent from the mathematical derivation given here that the magnitude of the Fourier transform is not affected by moving an object to a new position. The phase of the Fourier transform is changed in a predictable way, however. The rule describing this phase change is a simple one: the new phase of the Fourier transform is equal to the original phase of the Fourier transform plus an additional phase that is proportional to the spatial frequency, s. It is clear, from equation B10.9 that the phases of the Fourier transforms of two identical objects will differ greatly from one another if the origins of the coordinate systems that are used to compute their respective Fourier transforms are not the same.

the origin; in mathematical notation:

$$\mathcal{F}\{I(\mathbf{x}-\mathbf{a})\}(\mathbf{s}) = \mathcal{F}\{I(\mathbf{x})\}(\mathbf{s})e^{-i\,2\pi\,\mathbf{s}\cdot\mathbf{a}}. \tag{10.5}$$

In other words, the phases of an image shifted by the vector **a** differ from those of the unshifted image by $2\pi\,\mathbf{s}\cdot\mathbf{a}$. As one can see from equation 10.5, the calculated phase differences depend not only upon the amount by which the images are shifted, but also upon the spatial frequency at which the Fourier transform is being evaluated.

Untilted (zone-axis) specimens: symmetry constraints identify the space group

The *phase origin* is always chosen to coincide with an axis of high symmetry, if one is present in the crystal space group for the specimen being studied. For example, in bacteriorhodopsin the phase origin that is used lies on the 3-fold rotational axis of the protein trimers. The use of a symmetry axis for the common phase origin is valuable because the position of an n-fold symmetry axis can be found by searching for the origin that most closely satisfies the requirement that the phase values should be n-fold symmetric. The geometrical relationship of the positions of indexed diffraction spots already shows us which of the spots are related to one another by the n-fold rotational symmetry, but the phases will not exhibit the expected n-fold symmetry unless the origin of the Fourier transform coincides with the n-fold rotation axis. Using the phases of symmetry-related diffraction spots to identify the position of the phase origin actually accomplishes two things: (1) a common origin is identified that allows data to be merged from multiple images, and (2) phase relationships (section 7.3) can be used to resolve ambiguities about the assignment of the space group.

Since electron crystallography is able to use only thin ("monolayer") crystals, the only n-fold rotational axis that will be of use to us must be one that is perpendicular to the plane of the crystal. Such an axis is also referred to in crystallography as a *zone axis* because the central section that is perpendicular to the incident beam corresponds to the entire $(hk0)$ set (zone) of diffraction spots. The use of such an n-fold rotation axis to define the common phase origin is easiest for untilted crystals, that is, crystals that are oriented such that the incident electron beam is parallel to the zone axis. This point is self-evident since n-fold rotational symmetry no longer exists in a projection, once the crystal is tilted.

Tilted specimens: the intersection of two central sections produces a common line of data that is sufficient to define a common origin

The Fourier transforms of two different projections of the specimen contain a set of common points that lie along the line of intersection of their respective central sections in reciprocal space, as is shown in figure 10.10. A common origin can thus be defined for the two projections by requiring that the values of the Fourier transform on the common line should all have the same phase.

The use of pair-wise *common lines* of data in reciprocal space to find a common phase origin thus allows all images to be referred to the symmetry origin that is first found in projections of untilted samples. The common line, of course, only constrains the phase origin along the direction parallel to the line. The position of the phase origin

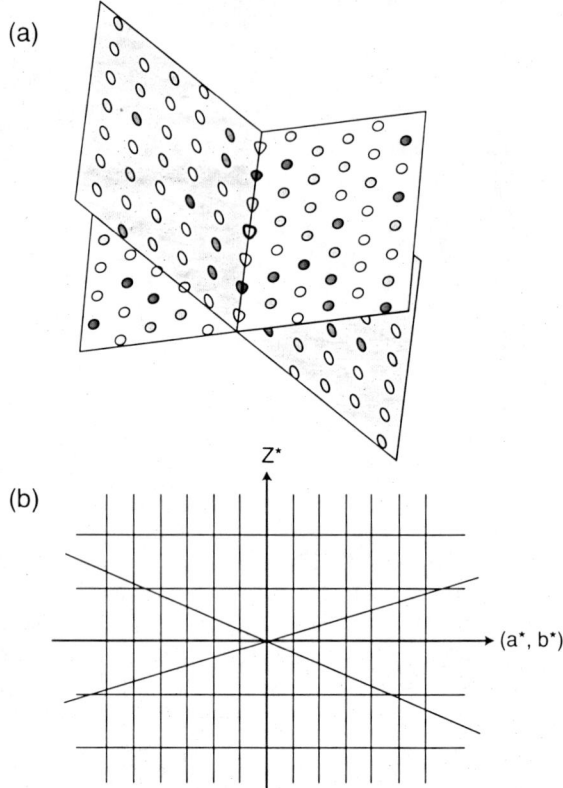

Figure 10.10 The common phase origin for two images can be determined by comparison of data that lie on, or close to, the intersection of their corresponding central sections in reciprocal space. (a) Diffraction spots with different amplitudes are indicated schematically (on each of two central sections) by circles with different relative degrees of darkness. Two planes in reciprocal space, each of which represents a central section corresponding to a different projection of the same crystal, must intersect along one line. Both the amplitudes and the phases measured for the two images must therefore match (within the error associated with noise in the data) for the reflections that lie on this common line. The phases will only match one another, however, when the same phase origin is used for both images. One algorithm for finding the common phase origin is to systematically compute the phase residual for a large set of choices of the origin, in order to find the origin that gives the smallest phase residual. While the use of data on the common line in reciprocal space can determine the relative shift of the two images in the direction parallel to the common line, these data say nothing about the relative shift of the two images in the perpendicular direction. (b) As seen in projection when looking parallel to the tilt axis, the two central sections in (a) are now represented as two lines. If the relative tilt angle between the two planes is small, the values of the amplitudes and phases on these two planes must remain fairly similar for diffraction spots that are close to the line of intersection between two central sections. Comparison of values at points that are only slightly away from the tilt axis can thus be used to determine the phase origin in the direction perpendicular to the tilt axis.

in the direction perpendicular to the common line can be determined in two ways. First, Fourier components that lie close to the line should have similar phases for the two images, even though these components lie at slightly different z^* values in each image (see figure 10.10(b)). Thus points that lie at a small distance from the common line can be used to constrain the phase origin. This constraint is effective only when the relative tilt angle between two images is small. A more general solution is possible when a third image, tilted along a different azimuth, is merged to the first two images. Two new common lines are generated by the intersection of the plane of the transform of the new image with the planes from the previous images. The three common lines in different directions then constrain the phase origin in each of the three images. This concept is discussed more fully in chapter 13 in the context of processing images of icosahedral viruses.

Determination of the tilt parameters and merging of the data follow methods similar to those used with diffraction data

The way in which the tilt axis and tilt angle are determined for electron diffraction patterns has already been described in section 9.7. The same procedure is used for the diffraction patterns that are obtained by computing the 2-D Fourier transform of an image. As was described in section 9.7, tilting the specimen results in a geometrical distortion of the apparent reciprocal lattice relative to that of the untilted specimen. A first estimate of the tilt angle can thus be derived from the distortion that is observed in the computed diffraction pattern.

As with electron diffraction data, smooth curves are fit to the data along the lattice lines, which in this case consist of both amplitudes and phases. The initial estimates of tilt angle and phase origin can then be refined against the curves. Since many comparisons can be made between the measured structure factors and the reference curves, both the tilt parameters and phase origin are determined much more accurately than they were in the initial estimates. As is done with electron diffraction intensities, the refinement of image amplitudes and phases can be repeated iteratively by replotting the experimental data along each lattice line, fitting new curves to the data, and then using the new curves to make the next round of estimates of the tilt angle, tilt axis, and phase origin for each micrograph.

10.8 Evaluation of data quality is based on the signal-to-noise ratio

The ratio of the diffraction peak amplitude to the local noise amplitude near the peak provides a natural and convenient measure of data quality. As was mentioned in section 10.6, the background can be estimated from the values at the perimeter of an array that surrounds each spot (see figure 10.5). The peak-to-background ratio (P/B) provides a useful measure of the reliability of the data, and it can even be used to estimate the phase error in the data.

One frequently used measure of the peak-to-background ratio is called the image quality, or IQ (Henderson et al., 1986). The IQ was derived from an estimate of the phase error associated with the P/B, and as a result it is defined such that

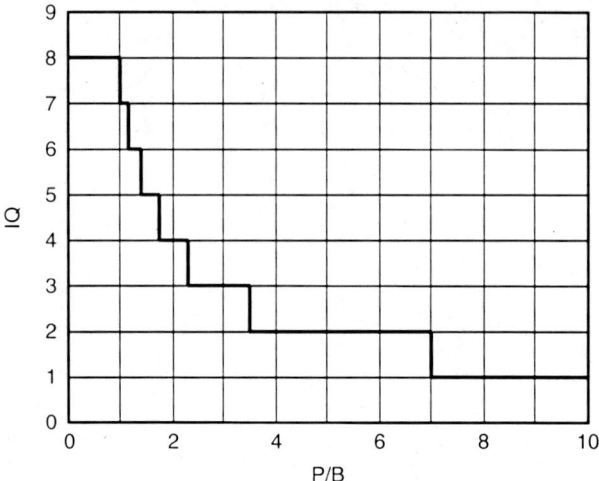

Figure 10.11 The image quality factor, IQ, is a useful measure of the peak to background ratio, P/B, for each diffraction spot in the Fourier transform of an image. This graph represents the relationship between the IQ and P/B.

IQ = 1 + int($7\frac{180}{\pi}\frac{B}{P}$). Note that this definition gives lower (integer) values for the reflections that have high peak values relative to the background. The range of P/B for each IQ value is illustrated in figure 10.11. Diffraction patterns computed from images of 2-D crystals are frequently presented as an IQ plot, such as the one shown in figure 10.12, which graphically shows the IQ (rather than the computed amplitude) for each reflection. These plots give a clear indication of data quality and allow one to visualize the useful resolution range. In some cases these plots can even show the positions of the zeros in the CTF for the image.

The noise amplitude observed in the Fourier transform of EM images follows the Rayleigh distribution (Hayward and Stroud, 1981) (see figure 10.13), indicating that the noise added to the specimen structure factor is a 2-D Gaussian. Figure 10.14 illustrates the relationship, in this case, between the true structure factor and the measured value after the addition of random noise.

One can use the Rayleigh distribution model to calculate expected phase errors for data observed with any given P/B ratio (Brillinger et al., 1990). Expected phase probability distributions are shown in figure 10.14 for two typical values of the P/B, and numerical values of the expected phase error are given in table 10.1 for several values of the P/B.

Considering that a measurement with an expected error in phase of less than 45° will contribute more to the signal than to the noise, we note that diffraction peaks with a P/B as low as 1 (i.e., IQ = 7) could be considered "good" data. However, a number of such peaks will also be present just by chance, even in the absence of signal from the crystal. One can predict the frequency of occurrence of noise peaks with a given peak-to-background ratio by integrating the Rayleigh distribution from any given value to infinity. The probability of occurrence is shown in figure 10.13,

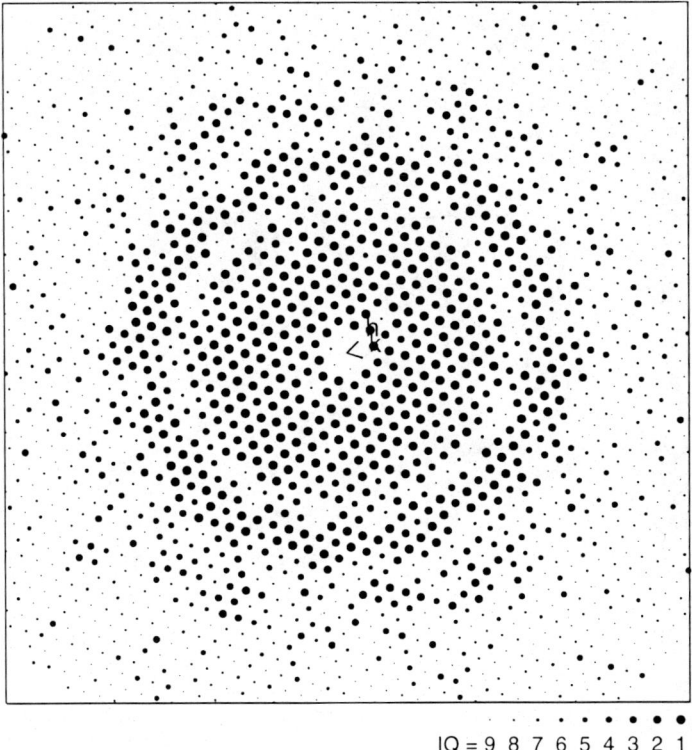

Figure 10.12 An "IQ plot" such as the one shown here gives a good impression of the strength of diffraction spots and the maximum resolution at which data are reliable. Spots of different sizes are placed at each reciprocal lattice point, with the size of the spot indicating the IQ according to the key shown in the lower right.

and some representative numerical values are listed in table 10.1. It is clear that over half of the spots in the Fourier transform of noise (alone) are expected to have a P/B = 1 (IQ = 7). Diffraction spots of IQ = 7 are therefore rather suspect. Diffraction spots with IQ of 4 or less, on the other hand, represent highly reliable data, even when derived from just a single image, since peaks with a P/B this high are very unlikely to occur simply by chance.

10.9 Quasi-optical filtering reduces the noise in the image

The signal-to-noise ratio in the image can be improved, often quite dramatically, by a filtering procedure in which most of the nonperiodic components of the image are removed. After computing the Fourier transform of the image and indexing the diffraction pattern, the transform is masked so that all values are set to zero except those in the vicinity of the diffraction spots. The inverse transform then gives an image in which most of the noise has been removed, leaving a more clear image of the crystal.

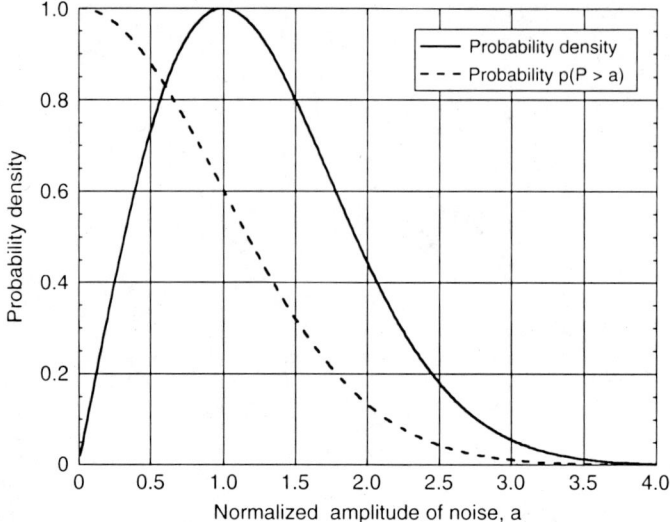

Figure 10.13 The distribution of noise amplitudes within a local region of the Fourier transform of an image is found to follow the Rayleigh distribution, $p(a) = \frac{a}{\sigma^2} e^{\frac{a^2}{2\sigma^2}}$ (Hayward and Stroud, 1981). This distribution has the property that the standard deviation, σ, and the average, μ, are related by $\mu = \sqrt{\frac{\pi}{2}}\sigma$. The curve is plotted for an average amplitude of $\mu = 1.0$ (standard deviation $\sigma = 0.7979$). The probability $p(P > a)$ of finding a peak with amplitude "P" (for "peak") that is larger than some value "a" is given by the integral of the Rayleigh distribution from that value to infinity. This probability is plotted as the dotted line.

This procedure was initially performed with images of proteins in a very simple fashion on a light-optical reconstruction system (Klug and DeRosier, 1966). An opaque mask, with holes at the locations of the diffraction spots, was placed in the diffraction plane of the optical system. The same operation is now generally carried out in a computer and is sometimes referred to as *quasi-optical filtering*. Figure 10.15 illustrates the process.

In order to understand the effects of this masking operation, we return again to the convolution theorem. First we define the *comb function*, comb(sa), that describes the reciprocal lattice of the specimen. This comb function is simply a set of delta functions located at each point where the argument has an integer value. As a result, a delta function occurs whenever s is equal to an integral multiple of $1/a$, where a is the lattice constant of the specimen. The masking function that is used for quasi-optical filtering, $M(s)$, is then formed by convolution of a "hole," $W(s)$, with the comb function:

$$M(s) = \text{comb}(sa) * W(s). \tag{10.6}$$

The masked transform, $F'(s)$, is the product of the object transform, $F(s)$, and the mask,

$$F'(s) = F(s)M(s)$$
$$= F(s)[\text{comb}(sa) * W(s)]. \tag{10.7}$$

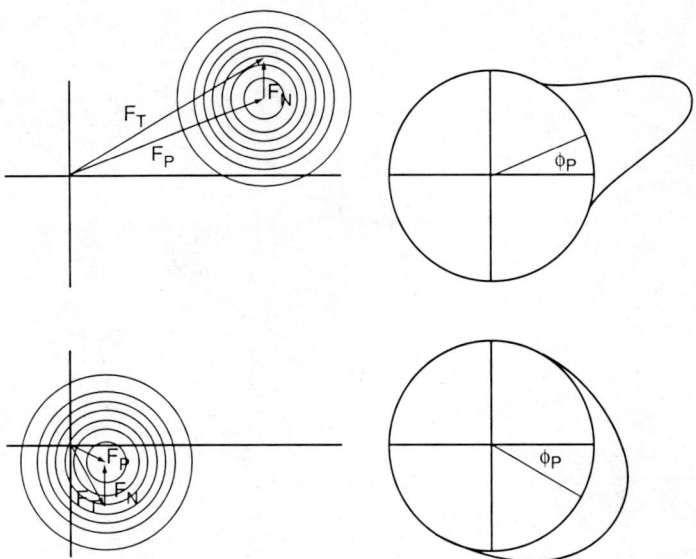

Figure 10.14 Graphical representation of the way in which phase errors can be estimated for different values of the peak-to-background ratio for a diffraction spot. The apparent structure factor measured in the transform of an image represents the vector sum of the true crystal structure factor, $\mathbf{F}_P$, and noise, which has a probability distribution represented by a 2-D Gaussian function. An estimate of the mean amplitude of the background, B, is obtained from the region of the transform surrounding each diffraction spot. The two representative examples show that the phase-probability distribution is quite narrow when the structure factor of the crystal, $\mathbf{F}_P$, is much larger than the noise, but the phase-probability distribution becomes very broad when the structure factor of the crystal is comparable in amplitude to the noise.

Table 10.1 Estimation of the statistical accuracy of phases, based on the IQ of the diffraction spot

IQ	Expected phase error (degrees)	P(IQ)
1	4.3	—
2	8.7	<0.0001
3	14	0.0002
4	22	0.008
5	30	0.066
6	39	0.18
7	45	0.30
8	—	0.41

For each value of IQ, the expected phase error is shown, along with the probability $P(IQ)$ of making a measurement with an apparent IQ in the absence of any true signal. When $IQ = 8$, then $\frac{P}{B} < 1$ and the expected phase error is too large to justify the inclusion of such reflections.

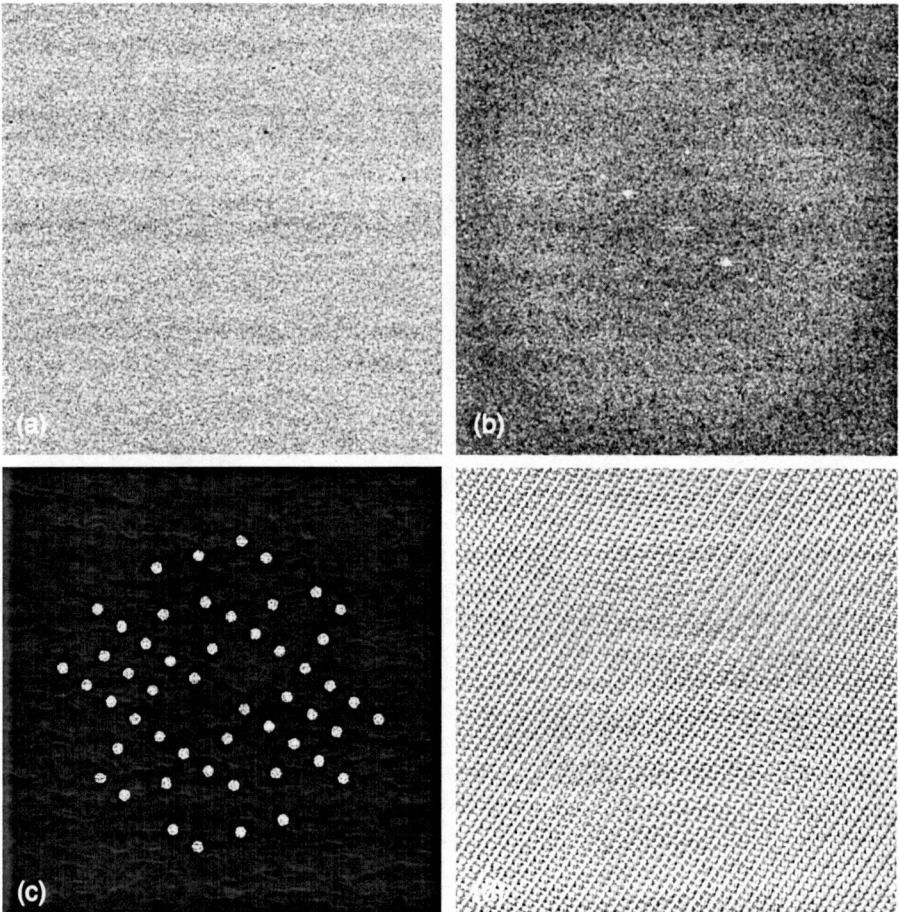

Figure 10.15 Optical filtering can be used to improve the signal-to-noise ratio in an image. (a) An image of a crystal, here an image of negatively stained tubulin, is Fourier transformed to produce (b) the computed diffraction pattern of the crystal. (c) The Fourier transform is masked so that all values are set to zero except in the vicinity of the diffraction spots. (d) The inverse Fourier transform of the masked "diffraction pattern" shows greatly enhanced visibility of the crystal lattice.

The inverse Fourier transform of equation 10.7 gives an expression for the modified real space image:

$$I'(x) = I(x) * [\mathrm{comb}(x/a)w(x)], \tag{10.8}$$

where $I(x)$ is the Fourier transform of $F(s)$, $w(x)$ is the Fourier transform of $W(s)$, and $\mathrm{comb}(x/a)$ is the real-space comb function produced by Fourier transforming $\mathrm{comb}(sa)$. Writing the comb function as an explicit sum of delta functions, we see that each unit cell in the filtered image is a weighted average of adjacent unit cells in the original image:

$$I'(x) = I(x) * \sum(\delta(x - na))w(x)$$
$$= \sum I(x + na)w(na), \quad (10.9)$$

that is, each point in the filtered image is a weighted sum of the values at corresponding points in neighboring unit cells. The operation of masking out the diffraction spots in order to remove noise from the transform is thus mathematically equivalent to averaging the image over a number of adjacent unit cells in real space.

When the masking function that is applied to the computed diffraction pattern is a square-shaped rectangle function, that is, a sharp-edged, square hole, with a width s_0,

$$W(s) = \text{rect}(s; s_0), \quad (10.10)$$

then the weighting in equation 10.9 is given by the values of a sinc function,

$$w(x) = \text{sinc}(\pi s_0 x) = \text{sinc}\left(\frac{\pi x}{x_0}\right), \quad (10.11)$$

where the width, $x_0 = 1/s_0$, of the region that contributes the most to the average (i.e., the central, positive lobe of the sinc function) corresponds to 1/(width of the rectangle function). Consequently, a narrower masking function in reciprocal space produces averaging over a wider area, which includes more unit cells and thus gives a better signal-to-noise ratio. In the limit of strong filtering, when the holes in the mask are limited to a single pixel, the result is an average over the entire image. This is, of course, the result that we obtain by sampling the transform precisely at the reciprocal lattice points.

The signal-to-noise ratio of the filtered image can be further enhanced by retaining only those diffraction spots that have a high enough amplitude to contribute significantly to the signal when computing the inverse Fourier transform. Such a subset of diffraction spots can be generated either manually (by inspection of the computed transform) or automatically, for example based on the IQ value or a similar measure of quality. The amplitudes of all diffraction spots that fall below the user-defined threshold are assigned a value of zero, thereby reducing the noise in the filtered image with minimal loss of the signal.

10.10 Correction for distortions in the image increases the signal quality

It is rare that an image yields diffraction spots that are sufficiently sharp that all of the power is concentrated at the nominal reciprocal lattice points. Two-dimensional crystals are susceptible to distortions when exposed to mechanical forces. As a result, the crystals suffer both elastic and plastic deformations, especially at the time when the specimen is applied to the support film. Off-axis aberrations ("field aberrations") in the imaging systems of the electron microscope can also introduce significant image distortions (section 5.3). The combined effect of these geometric distortions is to

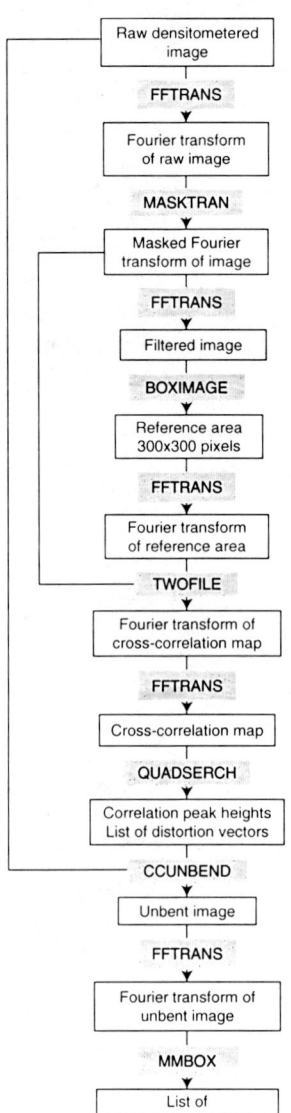

Figure 10.16 Block diagram of the sequence of steps that are carried out in the process of unbending an image. The names of various programs of the MRC suite that are used in various steps are indicated for each step.

blur the diffraction spots and spread them over a number of pixels in the transform. The severity of this problem became particularly clear when larger image areas were processed in order to extend the image resolution (Henderson et al., 1986). Procedures to compensate for real-space distortions are now included as a routine part of the image processing that is used to extract the full information available from 2-D crystals. The steps involved in this computational *unbending procedure* are indicated in the block diagram of figure 10.16, and they are discussed in the following paragraphs.

Cross-correlation is used to determine the distortions that may be present in images of 2-D crystals

The first step in unbending an image is to identify the exact positions of each of the unit cells within the image. These positions would ideally all sit on a perfect lattice, but in fact they are displaced by varying amounts. Unbending of the image involves re-interpolation of the image array such that the unit cell positions are shifted back onto their respective positions in the ideal lattice. The Fourier transform of the "unbent" crystal image will then have diffraction spots that are sharper, and therefore have higher peak amplitudes, than those in the transform of the original image. We note that if one can accurately identify the positions of each unit cell, one could just as well perform the image averaging in real space by superimposing each unit-cell image in the computer. This second approach is known as correlation averaging, and it is discussed in greater detail in chapter 14. In this chapter, we concentrate on the Fourier-based crystallographic approach.

The unit cell locations within the crystal are determined by cross-correlation between the image and a reference motif that is derived from the image, following the procedure schematically outlined in figure 10.16. Figure 10.17 illustrates the results that are obtained at various stages of the process. The reference motif is a somewhat idealized image of a part of the crystal, obtained by quasi-optical filtering as discussed in section 10.9. When this motif is cross-correlated with the entire image, the correlation function contains peaks at the positions in the image that contain a structure similar to the motif. Strong cross-correlation peaks can be produced even when the image of a unit cell appears to be buried in the noise. The positions of the peaks in the cross-correlation function are first found with a peak-identification algorithm in order to locate the unit cell positions. The unit cell positions are then used to determine how far each point of the image needs to be shifted in order to produce an unbent, near-perfect crystal image.

The size of the reference motif, that is, the number of unit cells that it contains, has to be large enough to provide a strong correlation peak. On the other hand, one would ideally like the motif to be the size of a single unit cell in order to precisely determine the positions of individual unit cells. The number of unit cells required in the reference motif depends both on the size of the unit cell and the image quality. It is theoretically possible to obtain a statistically significant CCF peak for a single molecule of $M_r = 40,000$, and in exceptional images of bacteriorhodopsin such peaks have actually been obtained from single trimers with $M_r = 75,000$ (Walz and Grigorieff, 1998). In normal practice, however, especially with unstained specimens, the motif may need to cover an area the size of 10 or 20 unit cells in order to produce a sufficient signal-to-noise ratio in the correlation function. As the motif is made larger, of course, the estimated distortion becomes an average over that which exists for a larger number of unit cells, and the ability to follow sharp or rapid distortions in the lattice decreases.

There is also an interplay between the size of the mask used to generate the filtered image and the size of the motif that is cut out of the filtered image. Towards one extreme, one would like to use small holes in the mask so that the motif will be an average over many unit cells. Creating a motif that is an average of many unit cells will maximize the signal-to-noise ratio of the motif itself, and that, in turn, will increase the signal-to-noise ratio of the cross-correlation function. A higher signal-to-noise ratio in the cross-correlation function then allows one to decrease the area of the crystal image

panel 1

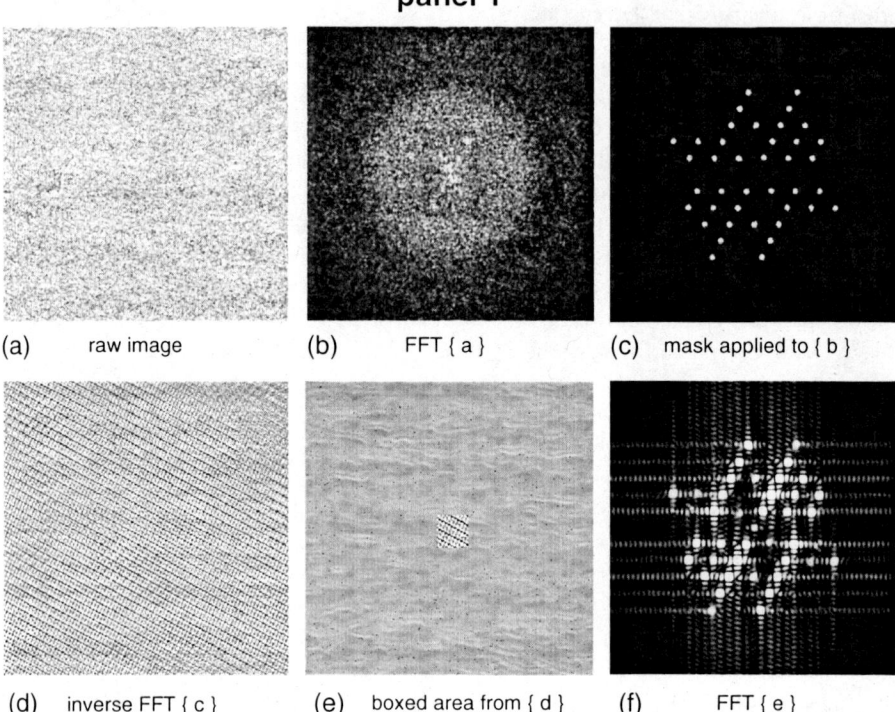

Figure 10.17 Two-dimensional representations of various displays that are produced at successive stages of the unbending procedure. (a) The original, digitized image intensity. (b) The Fourier transform (diffraction pattern) calculated from the image. (c) The Fourier transform of the image after a mask has been applied to isolate the peaks of the diffraction pattern. (d) The "optically filtered" image that is produced by computing the inverse Fourier transform of the masked diffraction pattern. (e) A small "reference motif" that has been boxed out from the center of the optically filtered image. (f) The Fourier transform of the reference motif. The product of this Fourier transform and the one shown in panel (b) is used to compute (by an inverse Fourier transform) the cross-correlation function between the reference motif and the original image. (g) The cross-correlation function, just described, between the reference motif and the original image. (h) A representation of the strengths of the peaks in the cross-correlation function that was shown in (g). (i) A map of the displacement vectors, greatly amplified, which shows the deviation of the positions of the cross-correlation peaks from their expected positions. (j) A map of vectors that shows the smoothed, reverse displacements that are applied in order to "unbend" the original image. (k) The same image shown in panel (a), but now after having been unbent. (l) The Fourier transform of the unbent image.

that is covered by the reference motif, allowing finer distortions of the image to be identified. At the beginning, however, large holes should be used in the mask so that the reference motif will not be an average over areas that are, for any reason, different from one another. The area covered by the motif should also not be larger than the region that is averaged as a result of filtering the image. After distortions in the image have been partially corrected in a first pass, the diffraction spots will be sharper, and

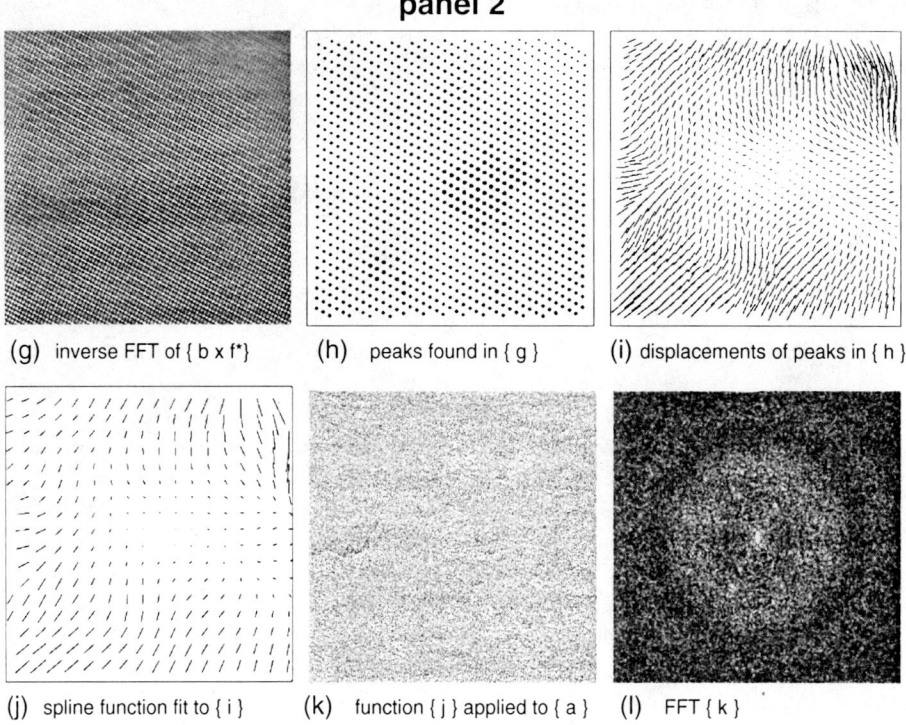

(g) inverse FFT of { b x f* }   (h) peaks found in { g }   (i) displacements of peaks in { h }

(j) spline function fit to { i }   (k) function { j } applied to { a }   (l) FFT { k }

Figure 10.17 *(continued)*

the motif can be safely improved by using a mask with smaller holes, so as to average over a larger area. Successive cycles of unbending can thus be used to improve the unbending process, as is further described below.

The data extracted from the correlation map are the positions of each of the peaks. When the peaks are well above the noise level, there is little difficulty in determining the peak positions, for example by calculating their centers of gravity. Identification of correlation peak positions can also be improved if one knows the expected shape of the cross-correlation peak. The expected peak shape is most simply determined by calculating the autocorrelation of either the reference motif or some small region of the image. The central peak in this autocorrelation function has a signal-to-noise ratio high enough that it can be used as a model for all of the other correlation peaks in the map.

The approximate position of each peak in the cross-correlation function with respect to its neighbors is determined a priori by the real-space lattice vectors. One can therefore start from a peak at the center of the image and step out to the nearest predicted peak position and then search within a small region to find the actual position of the peak. The next peak is then reached by again stepping out from this actual peak position, and the process continues until it has mapped out the entire distorted lattice. Although this algorithm normally works well, it will most likely fail in locations where cracks and other abrupt distortions have shifted the lattice by a distance greater than the search radius. This algorithm may also fail in regions where the short-range order of the specimen is so poor that cross-correlation peaks are very noisy.

Distortions are removed by iterative unbending

It is useful to smooth the function that describes the shift required for each unit cell. This smoothing is done by defining a 2-D function, typically a bicubic spline, that represents the observed displacement data. The complexity of the spline function is determined by the number of knots (which correspond to the nodes in a 1-D polynomial function) that are included. Thus as more knots are included, more complex distortions can be compensated. Accounting for more complex distortions is done at the cost of reducing the amount of averaging within adjacent unit cells, however, which may increase susceptibility to noise.

Once the smoothed correlation peak positions have been determined, bilinear interpolation is used to reverse the distortions that are present in the correlation map. This interpolation amounts to defining the density value at each point in the resampled image by the weighted sum of the values at several points in the original image, as illustrated in figure 10.18.

Figure 10.19 illustrates the improvement in the Fourier-transform data that is achieved by *unbending* the distortions that are present in the original image. Not only are the diffraction spots sharper, as seen in a superposition of the intensities of all spots within a given resolution zone, but data at high resolution, initially buried in the noise, become clearly visible after unbending.

A single pass of unbending is frequently not sufficient to completely correct for distortions in the image, and the results can often be improved with multiple passes of unbending. In the initial pass, the diffraction spots may be so broad that rather large holes must be used in the mask. The filtering will therefore correspond to averaging over a small area, and the signal-to-noise ratio in the motif will necessarily be limited. As a result, the motif must initially be chosen large enough that it will generate significant correlation peaks. This means, in turn, that complex or abrupt lattice distortions may not be detected, so that unbending in the earlier cycles should be performed with only a

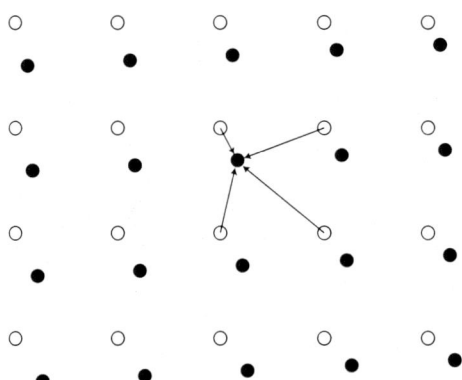

Figure 10.18 In order to resample an image by bilinear interpolation, values at each point in the new image (filled circles) are obtained by computing a weighted sum of values at the adjacent points in the original image (open circles). The weights for each term in the sum depend on the distance from the new point to the adjacent point.

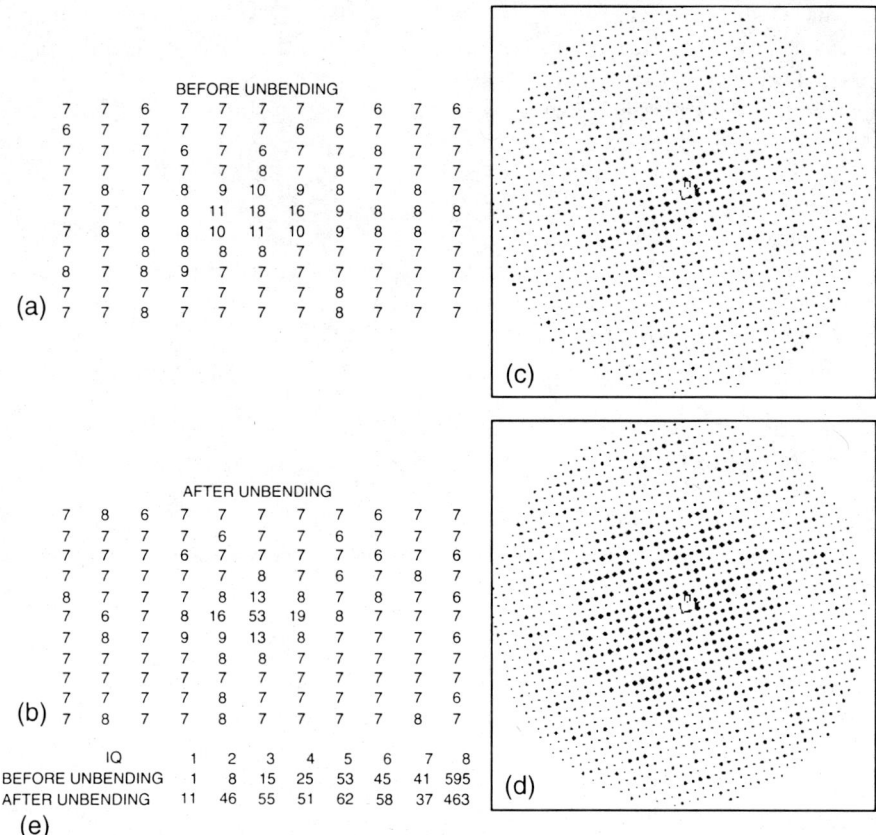

```
                    BEFORE UNBENDING
         7  7  6   7  7  7  7  7  6  7  6
         6  7  7   7  7  7  6  6  7  7  7
         7  7  7   6  7  6  7  7  8  7  7
         7  7  7   7  7  8  7  8  7  7  7
         7  8  7   8  9 10  9  8  7  8  7
         7  7  8   8 11 18 16  9  8  8  8
         7  8  8   8 10 11 10  9  8  8  7
         7  7  8   8  8  8  7  7  7  7  7
         8  7  8   9  7  7  7  7  7  7  7
         7  7  7   7  7  7  7  8  7  7  7
(a)      7  7  8   7  7  7  7  8  7  7  7


                    AFTER UNBENDING
         7  8  6   7  7  7  7  7  6  7  7
         7  7  7   7  6  7  7  6  7  7  7
         7  7  7   6  7  7  7  7  6  7  6
         7  7  7   7  7  8  7  6  7  8  7
         8  7  7   7  8 13  8  7  8  7  6
         7  6  7   8 16 53 19  8  7  7  7
         7  8  7   9  9 13  8  7  7  7  6
         7  7  7   7  8  8  7  7  7  7  7
         7  7  7   7  7  7  7  7  7  7  7
         7  7  7   7  8  7  7  7  7  7  6
(b)      7  8  7   7  8  7  7  7  7  8  7
```

IQ	1	2	3	4	5	6	7	8
BEFORE UNBENDING	1	8	15	25	53	45	41	595
AFTER UNBENDING	11	46	55	51	62	58	37	463

(e)

Figure 10.19 Example of the improvement in data that results from "unbending" an image. The numerical output for the sum of the computed amplitudes of all diffraction spots is shown (a) before and (b) after unbending. The improvement in overall signal-to-noise ratio is evident from the stronger diffraction amplitude relative to the noise, which has been scaled so that the average around the perimeter of each array is 7. The IQ plots for the entire Fourier transforms are shown (c) before and (d) after unbending. The improvement in overall signal-to-noise ratio is immediately apparent from the increased numbers of spots that have low IQ values, which reflect improved peak-to-background ratios. (e) The distribution of IQ values shifts markedly to lower values of the IQ.

rather small number of knots. In subsequent passes, the unbent image is used to obtain an improved reference motif, in which the optical filtering step can average over a larger area and increase the signal-to-noise ratio in the motif. The size of the template can then be reduced, which, in turn, makes it possible to map out and unbend the distortions more accurately.

The statistical significance of a cross-correlation peak improves markedly when the noise level in the reference motif is reduced. The resulting improvement in the correlation map can be used in two ways. First, the motif size can be made smaller in order to more accurately follow any abrupt distortions. Second, the transform of the raw image can be filtered using a mask with larger holes. Use of larger holes reduces the extent to which adjacent unit cells, each following their own distortion vectors, become averaged together. With less averaging over local variations in the distortion function, the filtered image generates, in turn, a correlation map that better reflects the distortions that need to be unbent.

The number of unit cells used for the motif, and the number of knots, as well as the rate at which these can be changed through successive unbending cycles, should be investigated empirically for each new project. In addition, the number of cycles of unbending required to extract all available data depends on the nature of the image distortions as well as the signal strength in the image. It has been found that three to five iterations are almost always sufficient, however.

10.11 Corrections are also required for other systematic image defects

In addition to the image defects associated with distortions, there are a number of other effects that cause systematic errors in the structure factors that are extracted from image data. The contrast transfer function (CTF) can invert image contrast and modulate amplitudes, and the envelope of the CTF causes a progressive decrease in amplitudes at high resolution. Misalignment of the illuminating beam can have a particularly strong effect on phases at high resolution (section 3.12). Finally, the value of the CTF changes across the image of a tilted sample, and these changes affect both amplitudes and phases. Fortunately, all of these defects can be described analytically, each with only a few parameters. These parameters, in turn, can be determined from the much larger number of structure factors produced from each image. As a result, it is possible to correct each of these image defects as a standard part of image processing.

Correction for the contrast transfer function

The mathematical description of the CTF has been developed earlier in chapter 3. When the illumination is aligned parallel to the axis of the objective lens, the CTF is affected principally by defocus and spherical aberration, whose effects depend on s^2 and s^4, respectively. The amount of defocus may nevertheless vary with azimuth, due to the image defect known as axial astigmatism. In addition, image defects that require correction at high resolution are introduced by even very small misalignment of the illumination.

As was discussed in chapter 3, in the absence of defocus or lens aberrations there is no contrast in the image of a phase contrast object. The combined effect of lens defocus and spherical aberration introduces a wave aberration that provides some phase contrast, but it does so with a nonuniform CTF. When the value of the defocus is large enough, the CTF may oscillate multiple times within the range of interest, as is illustrated by the Fourier transform of an image of a thin carbon film that was shown earlier in figure 10.1(a). The rings that are observed in this type of optical diffraction pattern were first discussed in the context of electron microscope images by Thon (1966), and the rings are frequently referred to as *Thon rings*. The dark rings in the transform correspond to spatial frequencies where the CTF crosses through zero, while Fourier

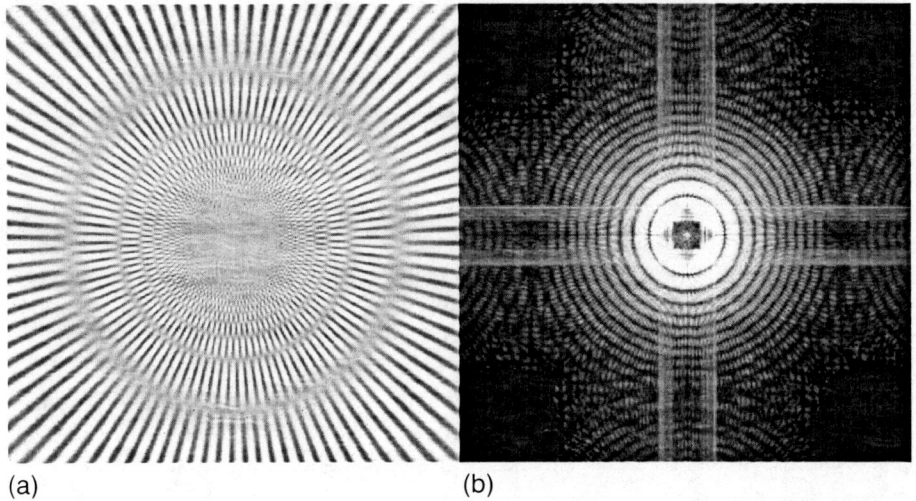

Figure 10.20 Demonstration of the spatial-frequency-dependent contrast in an out-of-focus image. (a) An image of a test pattern consisting of radial spokes illustrates the effect of the CTF in real space. At each radius, the pattern contains a characteristic spatial frequency. Thus a zero in the CTF at a given frequency will produce a lack of contrast at the corresponding radius in the pattern. Contrast reversals can be seen as the alternation of light and dark contrast as one traverses a line out from the center of the pattern. (b) The Fourier transform of the test image clearly shows oscillations in the CTF.

components whose spatial frequencies lie within the bright rings are imaged with either positive or negative contrast, the contrast normally being reversed in adjacent bands.

The effect of contrast reversal is clearly seen in the real-space image of a test pattern of radial spokes, figure 10.20. The spatial frequency increases continuously toward the center of the spoke pattern, and zones of high and low contrast are seen in the defocused image. Dark lines in one zone line up with bright lines in adjacent zones, corresponding to reversal of the CTF in adjacent zones. Rings with no contrast occur at spatial frequencies where the CTF has a value of zero.

The Fourier transform of an experimental image of a crystalline specimen, such as the one shown in figure 10.1(d), sometimes contains a strong enough component from the amorphous part of the specimen that the Thon rings are quite evident. The Thon rings may also be apparent in the pattern of the diffraction spots from a protein crystal, provided that the density of spots is sufficiently high. Thus the positions of the zeros in the CTF can be measured directly on the Fourier transform, and these positions can then be used to calculate the value of defocus.

To achieve the best accuracy in refining the CTF parameters, one can compare structure factor amplitudes from the image to a set of reference diffraction amplitudes that have been obtained by electron diffraction. To test a given set of parameters for the defocus and astigmatism, each of these reference amplitudes is multiplied by the value calculated for the CTF at the corresponding spatial frequency, and the resulting values are compared to the experimental image amplitudes. The trial defocus can then be varied to optimize the fit between these two sets of amplitudes. It has been found that this procedure works quite well even if the actual electron diffraction amplitudes are unknown, and a constant value is used instead for all of the reference amplitudes. Even with such a crude approximation, the fit between the CTF-weighted reference amplitudes and the experimental image amplitudes will improve when the correct CTF is used for the weighting.

Once the parameters of the CTF have been well determined, both the amplitudes and phases of the data measured from the Fourier transform of the image can be corrected for CTF effects. For several reasons, correction of the phase is of particular importance, while correction of the image amplitudes is less critical. First of all, correct phases are more crucial than correct amplitudes in determining the general appearance of a reconstructed image. In addition, electron diffraction data may be available to provide amplitudes for the reconstruction, making it unnecessary to attempt a correction of the image amplitudes. Diffraction amplitudes are found to be substantially more accurate than those determined from the images, since they are not affected by the CTF, various envelope functions, specimen motion, and other effects that may be difficult to determine and correct.

Correction of the phases is simply a matter of adding 180° for all reflections within the spatial frequency zones where the CTF has a negative value. This procedure is equivalent to changing the signs of the amplitudes in the zones where the signs were improperly reversed by the CTF.

In the absence of electron diffraction amplitudes, correction for amplitude effects of the CTF will yield a better reconstruction than that which is obtained using the raw amplitudes determined directly from the image. In a noise-free system, simply dividing the image amplitudes by the value of the CTF would correct both the phase and amplitude, except where the CTF has a value of zero. In practice, this division may greatly amplify noise when the CTF value is small.

The best correction to use, at least on theoretical grounds, is known as the *Wiener filter*, for which the weighting factor depends on the signal-to-noise ratio. The measured Fourier amplitudes of a given image, $F_{measured}$, are modified to produce corrected amplitudes, $F_{corrected}$:

$$F_{corrected} = F_{measured} \frac{CTF}{CTF^2 + (N/S)^2}. \quad (10.12)$$

In the ideal case, the noise-to-signal level, N/S, should be known as a function of spatial frequency (as is the case for the CTF), but in many applications one must settle for a single value of the estimated value of N/S. When the noise is small, that is, $N \ll S$ (or, more precisely, $N/S \ll CTF$), this formula reduces to simply dividing by the CTF. In the limit of large values of the noise, ($N/S \gg CTF$), the role of the CTF in the Wiener filter changes, however, such that

$$F_{corrected} \sim F_{measured} \, CTF, \quad (10.13)$$

that is, multiplying the transform amplitude by the CTF provides the best amplitudes for the reconstruction. It is clear that multiplication by the CTF corrects the sign of the amplitudes, just as dividing by the CTF does. In addition, however, multiplying rather than dividing gives a low weight to reflections where division by the CTF would amplify the noise, while retaining the full weight where the CTF is near its maximum value.

CTF correction with tilted specimens

Tilted specimens pose a particular problem for data recovery in that the CTF is not constant over the whole image area, as has been assumed implicitly in the previous discussion. Since the specimen height changes from one end of the image to the other, the defocus changes as well.

As illustrated with the coordinate system shown in figure 10.21, the defocus at any point in an image of a tilted specimen varies linearly with the distance of the point from the tilt axis. Defining the y axis to be perpendicular to the tilt axis, the defocus is given by

$$\Delta z'(y) = \Delta z + cy, \qquad (10.14)$$

where $c = \tan(\theta)$ and θ is the tilt angle. Strictly speaking, applying the formalism of transfer functions to images requires that they be *isoplanatic*, which, in our case, requires that the CTF is the same at all points in the image. Although this condition no longer holds when the specimen is tilted, we can nevertheless define functions that still allow compensation for the defocus.

The transfer function for a given spatial frequency, g, (section 3.8) is $\sin \gamma(g)$, where $\gamma(g)$ now depends linearly on $\Delta z'$. Thus when the specimen is tilted, the transfer function has a sinusoidal dependence on the specimen coordinate. This effect is illustrated in figure 10.22(a), where the image of a periodic pattern is broken into parallel stripes with alternating contrast. The alternation of contrast is made more apparent by laying a straight edge parallel to the lattice, or alternatively by sighting at a low angle to the page, parallel to the lattice.

To distinguish the function that is used to correct for contrast transfer in tilted specimens from the CTF, which is independent of specimen position, the term *tilt transfer function* (TTF) has been introduced (Henderson et al., 1986):

$$TTF(\mathbf{x}|s) = \sin\left[2\pi\left(\frac{C_s \lambda^3 s^4}{4} + \frac{\Delta z \lambda s^2}{2} + \frac{cy\lambda s^2}{2}\right)\right]. \qquad (10.15)$$

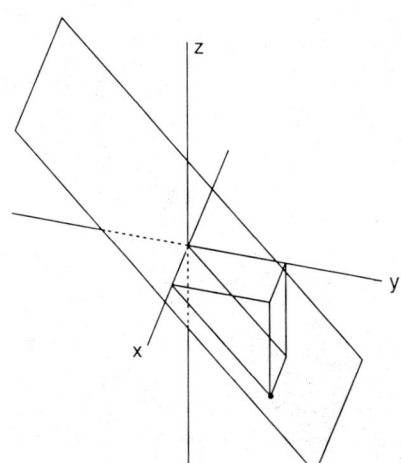

Figure 10.21 The coordinate system used to describe the gradient of defocus values that occurs with a tilted specimen. The tilt axis is taken to be the x-axis, and therefore the amount of defocus is proportional to y.

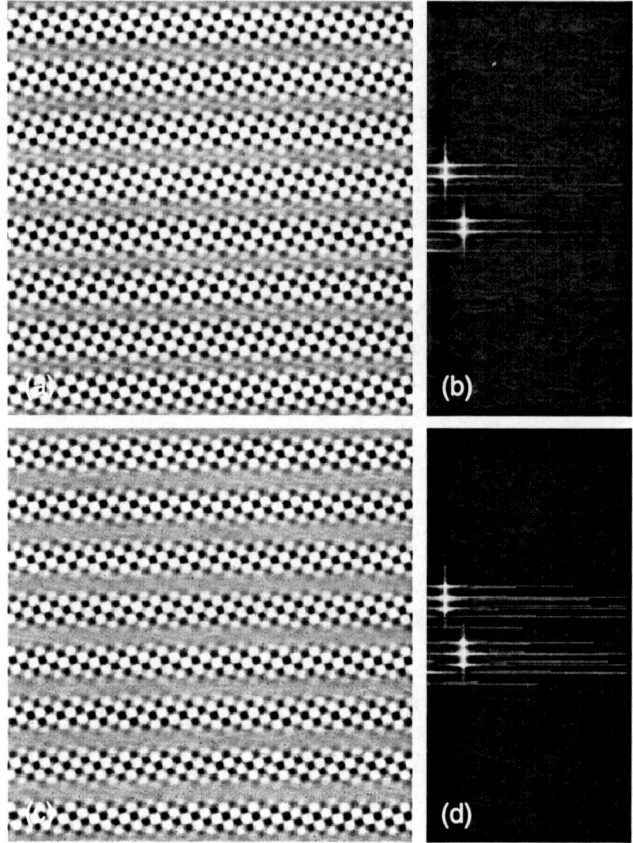

Figure 10.22 Schematic representation of the systematic splitting of diffraction spots that occurs due to the presence of a defocus ramp in images of tilted specimens. (a) A simulated image of a 2-D grating that has been tilted about the horizontal axis. The contrast transfer in the image varies sinusoidally between +1 and −1 due to the defocus ramp in the vertical direction. (b) The Fourier transform of this image shows diffraction spots that are split by the effect of the defocus-dependent contrast modulation. There is little diffraction intensity at the "correct" position, which is half way between the pair of "split" diffraction spots. (c) When the original image is multiplied again by the position-dependent CTF, the previous contrast reversals are removed, although the original bands of high and low contrast necessarily remain. (d) The Fourier transform of the corrected image now has diffraction spots at the correct positions, although these are still accompanied by satellite spots that arise from the remaining modulation in image contrast.

Note that $\mathbf{x}$ can be any vector in the x–y plane. For a given spatial frequency, $\mathbf{g}$, equation 10.15 can be written

$$TTF(\mathbf{x}|g) = \sin(C + c'y), \qquad (10.16)$$

where

$$C = 2\pi \left(\frac{C_s \lambda^3 g^4}{4} + \frac{\Delta z \lambda g^2}{2} \right)$$

and $c' = \pi c \lambda g^2$. The image intensity for this frequency will then be

$$I(\mathbf{x}) = 2F(\mathbf{g})\sin(2\pi \mathbf{g} \cdot \mathbf{x} + \alpha(\mathbf{g}))\sin(C + c'y). \tag{10.17}$$

The choice of the amount of defocus, Δz, depends simply on the choice of the origin of the coordinate system in figure 10.21, and is thus quite arbitrary. For each spatial frequency, the contrast varies sinusoidally across the image at a rate that depends on both the spatial frequency and the tilt angle.

Because the image intensity is expressed as the product of two functions in equation 10.17, the Fourier transform of the image is the convolution of two functions. Ignoring, for the moment, the constant phase terms α and C in equation 10.17, the transform of the image intensity is

$$\mathcal{I}(\mathbf{s}) \propto [\mathbf{F}(\mathbf{g})\delta(\mathbf{s} - \mathbf{g}) + \mathbf{F}^*(\mathbf{g})\delta(\mathbf{s} + \mathbf{g})] * [\delta(c' - s_y) + \delta(c' + s_y)], \tag{10.18}$$

where s_y is the component of $\mathbf{s}$ perpendicular to the tilt axis. The pair of delta functions within the first square brackets represents the Fourier transform of the gth Fourier component of the specimen, imaged as if the CTF had a value of one. The pair of delta functions within the second square brackets represents the Fourier transform of the sinusoidal modulation arising from the defocus ramp, $\sin(C + c'y)$. The transform of this latter term is a pair of delta-function peaks, spaced apart by a number of pixels equal to the number of cycles of the corresponding sine function within the image.

The convolution indicated in equation 10.18 thus results in each diffraction spot in the transform of the crystal being split into two peaks, as illustrated in figure 10.22(b). Note that the separation of the split peaks increases with the number of oscillations of the TTF across the image. The effect of this diffraction-spot splitting is especially serious when it comes to extracting data from the transform, as it results in there being almost no diffraction power precisely at the reciprocal lattice point.

One solution (Henderson et al., 1990) restores the power to the appropriate reciprocal lattice point by using what is, in effect, a real-space operation to compensate for the oscillations of the TTF across the image. For each spatial frequency in turn, one should imagine that every point in the image is simply multiplied by the value of the TTF at that point, thus squaring the TTF. Squaring the TTF ensures that the contrast modulation is positive at each image point, as is illustrated in figure 10.20(c). Note, for example, that the contrast in figure 10.20(c) no longer alternates between the parallel stripes, as can be seen by sighting parallel to the rows of the lattice.

The operation just described can also be thought of as convoluting each set of split diffraction spots by the Fourier transform of the TTF that is appropriate for the spatial frequency of each respective diffraction spot. In practice, this convolution is carried out by boxing out each diffraction spot as an array of, say, 30×30 pixels. Each such array is Fourier transformed to produce a small real-space representation of the data. This real-space representation is then multiplied by the appropriate TTF. The inverse Fourier transform of this product is then computed. This series of operations takes advantage of the fact that the FFT operation is computationally the most efficient way to perform a full convolution operation.

The convolution of the original diffraction pattern of the image of a tilted crystal with the Fourier transform of $\sin 2(C + c'y)$ produces a new diffraction pattern that has a peak at each spatial frequency of the crystal, as it should. The diffraction peak for each spatial frequency is now accompanied by a pair of satellite peaks, however, which are separated from the true diffraction peak by the same distance that previously separated the split peaks in the Fourier transform of the image. The advantage of performing the convolution operation is that one can then extract the structure factors from the central peaks, which are found at the appropriate reciprocal lattice points.

Correction for beam tilt

Misalignment of the beam produces a systematic defect known as axial coma, which is usually important to correct at high resolution. Beam tilt effectively results in a phase shift that increases with the cube of the spatial frequency (see equation 3.38). To a first approximation, a small amount of beam tilt affects the appearance of the image, and its power spectrum, in the same way as astigmatism (Smith et al., 1983). There is thus a possibility that the microscope operator may adjust the stigmation to minimize the apparent defect rather than properly correcting the beam alignment. Unfortunately, even when a compensating amount of astigmatism has been applied, beam tilt still produces an unwanted phase distortion.

The effect of beam tilt can actually be described as producing a resolution-dependent shift of the phase origin of the image (see section 10.6 for discussion of the phase origin). This description of the effect leads naturally to a means of determining and compensating for the beam tilt. The computer program that is used to determine the phase origin can determine the best origin for each spatial frequency by using just the data within separate resolution zones. The position found for the phase origin can then be fit to a cubic dependence on the spatial frequency, so as to match the theoretically expected dependence on beam tilt. Once the direction and amount of tilt are determined, the phase shift for each reflection, due solely to beam tilt, can be subtracted from the experimental phase.

The importance of the effect of beam tilt can be seen by an example. Using manual beam tilt correction, for example with the high-voltage wobbler on the microscope, it is often difficult to correct the tilt to better than a few tenths of a milliradian. If we suppose that $\theta = 0.0002$ rad, we find (by the use of equation 3.38) that the systematic error in the phase will be around $\pi/2$ when $s = 3.0$ nm^{-1}, corresponding to a resolution of 0.33 nm.

Methods have been developed for more accurate correction of the beam tilt, which take advantage of the effective astigmatism introduced by the tilt (Zemlin et al., 1978). Briefly, if the astigmatism is properly corrected to begin with, then the distortions introduced in the power spectrum by tilting the beam in opposite directions from the optical axis will be the same. On the other hand, if the beam is misaligned and at the same time the stigmation is adjusted to obtain a symmetric power spectrum, the apparent astigmatism will become quite different when the beam is tilted by equal but opposite amounts. Automatic microscope alignment schemes utilize this difference in astigmatism to calculate and correct the beam tilt.

11

High-Resolution Density Maps and Their Structural Interpretation

11.1 Introduction

Experimental structures are determined in what is effectively a two-stage process: first a density map is computed from measured structure factors, and then the density map must be interpreted by constructing a model of the structure. The discussion in this chapter focuses on computing and interpreting three-dimensional (3-D) density maps obtained from two-dimensional (2-D) crystals. Additional chapters then follow that are devoted to other specific cases, namely helices (chapter 12), icosahedra (chapter 13), and isolated, single particles (chapter 14).

When sufficiently large 2-D crystalline samples are available, structure factor amplitudes are best obtained from electron diffraction patterns (chapter 9), and these are combined with structure factor phases obtained from high-resolution images (chapter 10). In all other cases, both the amplitudes and the phases of the structure factors are obtained from the Fourier transform of the experimental images.

The steps involved in generating a density map from the 3-D set of structure factors are described in section 11.2. Different options can be used to display the computed density maps, and some of the common methods are discussed in section 11.3. One of the important points to emphasize is the fact that the 3-D data obtained from 2-D crystals will almost always be missing a cone-shaped part of the 3-D set of complex structure factors. Section 11.4 addresses the important issue of the extent to which this cone of missing data can affect the appearance of the high-resolution real-space density map.

If the experimental structure factors are obtained at a resolution of ~ 0.4 nm or better, it may be possible to build an atomic-resolution model of the structure into the 3-D density map. A general overview of the process by which this is done is described in section 11.5. Previously determined atomic structures of individual macromolecular components can also be fitted into the density map of a larger complex, an approach that is commonly referred to as the *hybrid method* of structure determination. The fitting (docking) of known structures into electron microscope density maps can be done with considerable accuracy, even at ∼2 nm resolution. A brief overview of the "hybrid" approach to structure interpretation is given in section 11.6, and further examples are shown in chapters 12, 13, and 14.

In the future we can expect other tools of crystallographic structure analysis to become more widely used within the field of electron crystallography. Refinement of the initial chain-trace model of a structure, for example, is an important step that normally follows the initial construction of an atomic-resolution model. As we discuss in section 11.6, however, it is likely that refinement in electron crystallography will be improved if atomic scattering factors are developed that account for variations in the Coulomb potential that reflect the state of chemical bonding. The use of diffraction data to calculate difference Fourier maps is another tool that has considerable potential in electron crystallography. This rapid and powerful method, described in section 11.7, can be used to visualize subtle changes in conformation, or the binding of small ligands to a known structure.

11.2 Three-dimensional density maps are computed from discrete samples of the complex structure factors

Chapters 9 and 10 have emphasized the fact that the Fourier transform of a 2-D crystal is confined to parallel reciprocal-lattice lines, which sample the continuous Fourier transform of the unit cell. Each such reciprocal lattice line is assigned its appropriate Miller indices (h, k), according to its position in reciprocal space. The amplitude (modulus) of the structure factor is measured either as the square root of the diffraction intensity (chapter 9) or as the amplitude of the computed Fourier transform of the image (chapter 10), and continuous curves are fitted to the experimentally sampled measurements along each reciprocal lattice line. The phases of the structure factors, obtained from the Fourier transforms of images, are similarly represented as continuous curves that are fitted to the experimentally measured values.

A discrete set of complex structure factors, $\mathbf{F}(hkl)$, equivalent to those obtained in x-ray crystallography of 3-D crystals, can be obtained by sampling the continuous curves of amplitude and phase at regular intervals in the z^* direction. In other words, the continuous curves fitted to the data along the (h, k) reciprocal lattice line can themselves be sampled computationally at regular intervals. Sampling thus produces a discrete set of structure factors at points $l \cdot c^*$, where l is an integer and $c^* = 1/c$, for some artificially chosen value of the c-axis unit-cell length. The value of the c-axis unit-cell size can be chosen to be anything that is larger than the thickness of the 2-D crystal. In most cases one will use a value of c that is 1.5 to 2 times greater than the best available estimate of the sample thickness.

The discrete set of complex structure factors, $\mathbf{F}(hkl)$, then serves as input data for any of the standard programs that are used in x-ray crystallographic studies of biological macromolecules. As was explained in section 2.11, these programs compute the inverse Fourier transform of the diffraction data, thereby producing a 3-D density map for the real-space unit cell.

11.3 Options for the display of 3-D density maps

Once a 3-D density map has been obtained, a range of options can be used to examine, interpret, and communicate its important features. Each type of display has its own characteristics that may make it better suited for some applications than for others. For example, interactive computer graphics, with the option for presentation of images in stereo, will be used when fitting an atomic model into the density. Figures for publication, on the other hand, need to rely on static 2-D images to convey the 3-D information as effectively as possible. In addition, characteristics such as the resolution and complexity of a structure can play a role in selecting the display mode. As a result, a combination of display types is often needed to convey the full information. A few examples will illustrate some of the possibilities and their applications.

One of the simplest methods for displaying density maps is to present 2-D contour maps for a series of adjacent sections. These sections can be displayed side by side, as is shown later in figure 11.6, or they can be stacked as transparent sheets, one above the other, with appropriate spacing between them, as in the example that is shown in figure 11.1. The use of stacked contours is quite useful for simple structures, but as both the depth of the structure and the number of distinct features increases, the growing complexity of such a display can obscure its interpretation. Contour plots with multiple contour levels have the advantage that they display the gradient of density within the section in terms of how close one contour level is to the next. The use of multiple contours in one plane quickly leads to confusion, however, when several such planes are stacked above one another. When a series of contoured sections is used to represent the density within a volume, therefore, the contours are commonly drawn at a single density level, such as the density that would enclose a given molecular weight of protein, or a density that is one standard deviation above the mean value.

There are numerous other ways in which one can display the 3-D shapes that are defined by specific contour levels. One of the most commonly used is a type of *wire frame* representation, like that shown in figure 11.2. Contours that are displayed as continuous surfaces that curve through space are referred to as *isosurfaces*. The wire frame representation of an isosurface makes it possible to visualize both the near and far sides of features within the map. Particularly when viewed in stereo, this type of display gives a very useful 3-D impression. Alternatively, the ability to rock the angle of view back and forth in real time, for example on a computer monitor, also gives a much better understanding of the 3-D structure than does a simple static view. Unfortunately, even the wire frame display can easily become so cluttered that interpretation is difficult. As a result, it is often necessary to define a slab, limited by near and far planes, that restricts the volume shown at any one time.

Displaying a solid surface that corresponds to a particular isodensity value is another way to show the important features of a 3-D map. At low resolution such a surface gives a good impression of the overall molecular shape or envelope, which may be more important than are the density variations within the surface. At a resolution of the order of 0.6–1 nm, for example, α-helices are well resolved as sausage-shaped features while regions of β-sheet form more irregularly shaped masses. These elements of

Figure 11.1 A stack of contour plots (prepared with tools in the CCP4 crystallographic suite) that represent the 3-D density of the nicotinic acetylcholine receptor at a resolution of about 2 nm. Figure kindly provided by Dr. Nigel Unwin.

secondary structure are well separated from one another at this resolution, and thus they are clearly seen in surface views. The solid surface obscures any detail behind or within the outermost isosurface level, of course. Internal densities and voids are often important parts of the structure, and in order to show these one can slice the display partway through the structure. One example of such a presentation is shown in figure 11.3.

At a resolution where helices are resolved, internal density variations are less of an issue, but the restricted depth over which a map can be interpreted is still a limitation. Figure 11.4 is an example of a surface view that is effective in showing the secondary structure of the protein, but one needs to view such a map from several directions in order to appreciate the structural organization within the full volume.

It is also possible to make the surface partially transparent so that internal features can be seen. Partially transparent surfaces are particularly useful when one wants to dock other structures, for example other density maps or atomic models, into the density. An example is shown in figure 11.5 where the resolution of the

Figure 11.2 A wire frame view (prepared with the program "O", Prof. T.A. Jones, Institute of Cell and Molecular Biology, Uppsala University) of a density map representing a section of the microtubule wall (Li et al., 2002). The resolution allows identification of helical segments, and the crystal structure of tubulin has been docked into the density. The view is presented in stereo to provide a sense of the depth obtained with the open structure of the wire-frame representation.

density map is high enough that sausage-shaped features can be interpreted as helices, and ribbon models of such helices have been fit into the density to emphasize this interpretation.

Solid, transparent and/or wire frame surfaces are often combined in a single figure to show multiple features, such as more than one isosurface level or boundaries between identified subunits. The use of color, particularly in differentiating among separate components within a map or between different density levels, provides an effective increase in the interpretability of the display as well as a significant aesthetic appeal.

Once a structure has been interpreted at "atomic" resolution, better than about 0.4 nm, the complexity of the model structure becomes problematic. In this case it is traditional to present the overall structure as a ribbon diagram, such as the example shown in figure 2.5. Such a simplified, global representation of the structure may then be accompanied by detailed, chemical representations of specific residues that

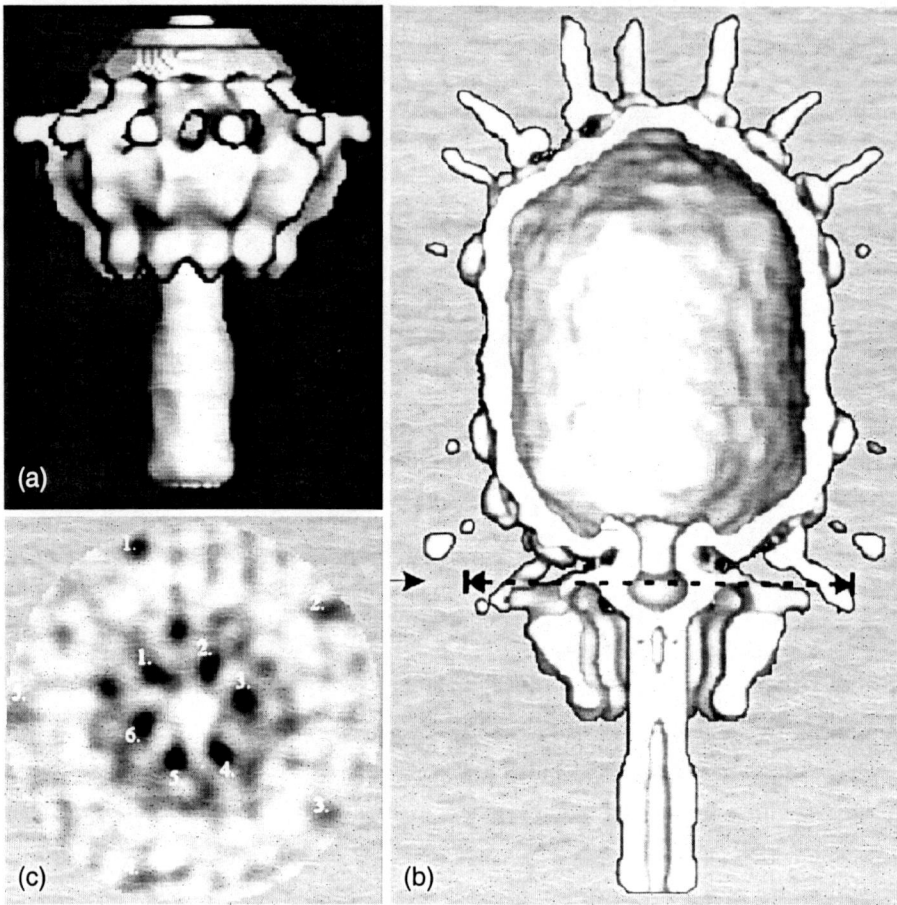

Figure 11.3 A solid-surface representation of the φ29 phage and its tail structure. The tail with its DNA packaging machinery is shown in (a). The intact phage is shown (b), cut through the center to show the shell of the empty head as well as the interior of the hollow tail. In this case a symmetry mismatch between the head and tail was of particular interest. A grayscale view of the density in a plane at the interface between the head and the tail (c) is numbered to show subtle features that were difficult to present as part of the surface view. Reproduced with permission from Morais et al. (2001).

are of interest, which may also be enclosed by a wire frame representation of the density map for those residues.

11.4 The missing cone of data results in poorer resolution in the direction perpendicular to the plane of the 2-D crystal

The theoretical reason why resolution is degraded in the direction perpendicular to the plane of a 2-D crystal has already been discussed in section 4.6. As is explained there,

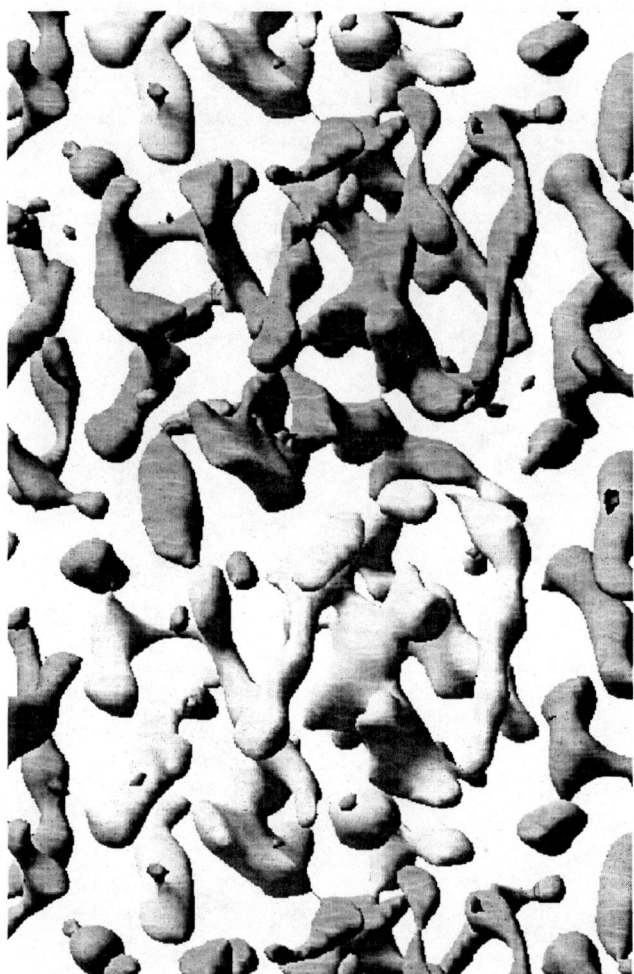

Figure 11.4 An isosurface view (prepared with the AVS software package (Advanced Visual Systems, Inc., Waltham, MA, USA)) of the density map of tubulin obtained at a resolution of about 0.65 nm. Elongated densities are alpha helices, as was confirmed once the resolution had been extended to around 0.35 nm. At the lower resolution shown here, the alpha and beta monomers can already be discriminated, and these are shown in different tones of gray.

the inability to collect data at the highest tilt angles leads to a cone-shaped region in reciprocal space, within which it is not possible to collect experimental data. The diffraction data are, in effect, truncated by a cone-shaped *aperture function*, a term that we have here taken from its usual 2-D context and extended to the context of three dimensions. Truncation of the data in reciprocal space causes, in turn, a loss of resolution in real space, in the same direction as the cone-shaped truncation in Fourier space.

It is natural to be concerned that the anisotropic resolution in real space may create unusual structural artifacts, and that such artifacts may compromise interpretation of the 3-D density map. This very real possibility is commonly referred to as the *missing*

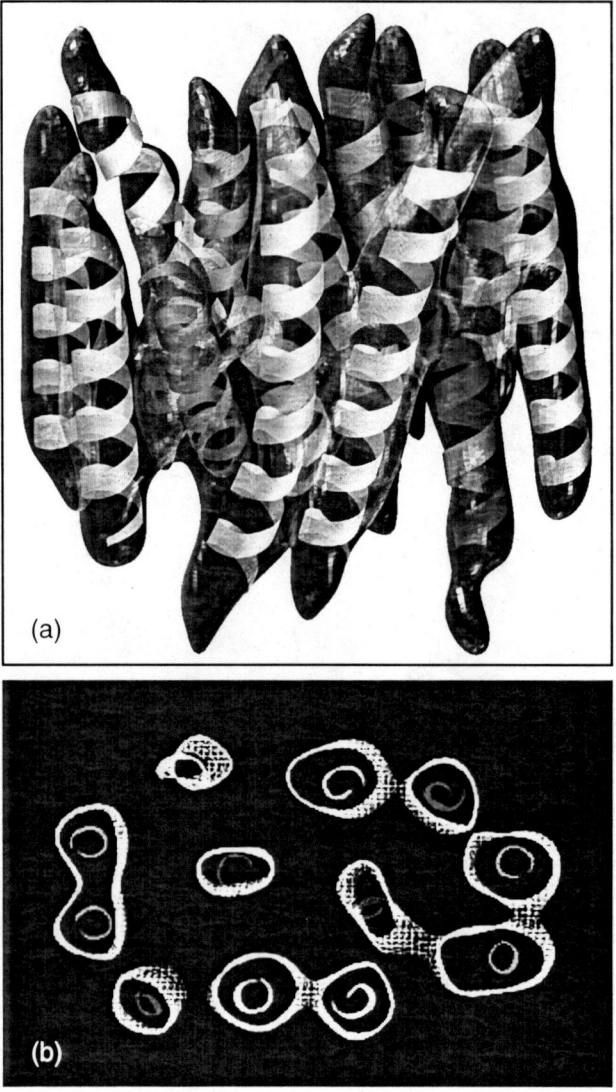

Figure 11.5 A transparent surface view (prepared with "3dsMax", Autodesk, Inc.) of the oxalate transporter, shown in (a), reveals a ribbon model of the helical structure that has been docked into the elongated, transmembrane densities. In order to show the fit of the model within the density even more clearly, a thin cross-section of the structure is shown in (b). Here the density is represented by a wire-frame surface. Reproduced with permission from Hirai et al. (2002).

cone problem. Before proceeding with further details on the interpretation of the density map, therefore, we review what the practical limitations are on collecting data at very high tilt angles.

Fortunately, as is now known from theoretical simulations and from practical experience, high-resolution density maps can still be interpreted accurately in terms of an atomic model, in spite of noticeable anisotropy in the resolution. Nevertheless, it is extremely important to realize that the effects of the missing cone can be severe unless data are collected over a range of at least ±45°, and preferably over a range of ±60° or more.

Practical factors limit the maximum tilt angle that can be used for data collection

In many cases a limitation occurs on how far a sample can be tilted because some part of the specimen grid, or even the specimen holder, blocks the electron beam at high tilt angles. Mechanical considerations of that type can be overcome, if desired, by the use of a properly oriented, rectangular specimen grid (or a single-hole, "slot" grid), and by careful removal of excess metal from the sides of the specimen holder. Other alternatives include the use of intentionally bent, or crimped grids, which allow one to start tilting the specimen from an angle well above 0°. Specially designed specimen rods are also available that hold the specimen grid at only one end, like a tweezers, and allow free rotation of the grid around a full 360° (Chalcroft and Davey, 1984). This type of specimen rod has been used to collect 3-D data from negatively stained samples at tilt angles up to ±80°, for example (Baumeister et al., 1986).

Use of extremely high tilt angles is rarely practical for a number of other reasons, however. The most important limitation often proves to be due to the fact that the specimen is never perfectly planar (i.e., flat). As we explained in section 6.8, broadening of the diffraction spots therefore becomes progressively worse as one tilts the specimen to increasingly higher angles. When the broadening of the diffraction spots becomes so severe that one spot begins to overlap its neighbors, there is no point to tilt the specimen to even higher angles.

The maximum amount of specimen tilt must ultimately be limited by the fact that the thickness of the specimen increases rapidly at very high tilt angles. Tilting to higher angles therefore becomes unproductive for one or more of the following reasons: (1) the specimen thickness exceeds the depth of field of the objective lens, (2) too much of the beam is lost due to inelastic scattering, or (3) multiple elastic scattering renders the data useless for crystallographic structure analysis. The various limitations associated with the use of thick specimens are discussed further in chapter 15.

Moderate anisotropy in the resolution has no significant effect on interpretation of the density map

The effect that the *missing cone* has on our ability to interpret a high-resolution density map can be understood through computer simulations that are carried out for real protein structures. Such simulations have shown that the interpretation of density maps is not compromised if the missing cone is as small as ±30°. As an example, figure 11.6 compares three contour maps, calculated for a resolution of 0.36 nm in the best direction, for equivalent sections of a 3-D density map. These maps have been calculated with (a) isotropic data, (b) a missing cone of ±30°, and (c) a missing cone of ±45° (Glaeser et al., 1989). Differences between the map obtained with isotropic resolution and that obtained with the 30° missing cone are hardly noticeable, but the differences produced with a missing cone of 45° are more apparent. Further details of

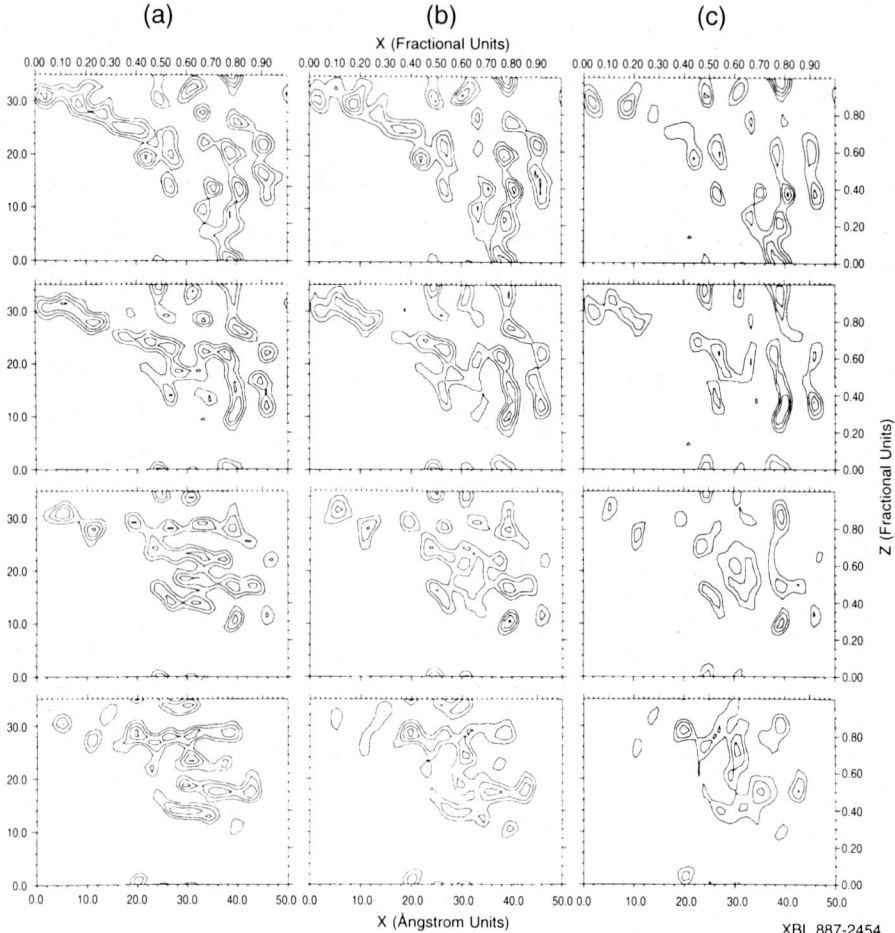

Figure 11.6 Comparison of four identical, successive sections (represented as successive rows in the figure) from the 3-D density maps obtained with isotropic data (column a), data with a 30° missing cone (column b), and data with a 45° missing cone (column c). The fold of monelin, the small protein used for this comparison, consists of one α-helix and five strands of β-sheet. The resolution of the density maps was limited to 0.36 nm in order to simulate a case that is realistic for "high-resolution" electron crystallographic density maps. Nearly all important features that are represented in the density map obtained with isotropic data are still accurately reflected in the density map produced with a missing cone of 30°, but this is no longer completely true when the missing cone is as large as 45°. The conclusion is that the interpretability of a high-resolution density is not seriously affected if the missing cone is only 30°, but interpretation may become more problematic if the missing cone is as large as 45°. Reproduced with permission from Glaeser et al. (1989).

the simulations, shown in that paper, established that the direction of the missing cone relative to the direction of helices and β-sheet does not affect the ability to correctly interpret the density map, as long as the resolution is as good as 0.36 nm.

The number of reflections lying within the missing cone is actually a surprisingly small fraction of the total, less than 13% of the full set of data in the case of a $\pm 30°$ missing cone. This fact helps explain why the density map obtained with a missing cone is not much different from that produced with isotropic resolution. It is true, of course, that artificial functions can be constructed in which Fourier components that lie within the missing cone are the strongest ones present in the test object. For such functions it is obvious that the effect of the missing cone could be severe.

The simulations with real protein structures nevertheless demonstrate that a density map that is computed at a resolution of 0.36 nm, and a missing cone of 30°, can be readily used to fit the peptide chain, regardless of the orientation of the protein structure relative to the missing cone. From this experience one can conclude that the Fourier components of a high-resolution protein structure are widely distributed in reciprocal space, rather than being concentrated in a localized region, and as a result the 3-D reconstruction is relatively insensitive to the direction of the missing cone.

The effect of a missing cone can be further understood by calculations of the 3-D *point spread function* in real space. The 3-D point spread function is simply the inverse Fourier transform of the cone-shaped, 3-D "aperture function" that has, in effect, been applied to the complete, 3-D set of structure factors. Section 4.6 can be consulted for further background about this way of understanding the effect that the missing cone has on the resolution of the density map.

Figure 4.3 shows two representations of the 3-D point spread function for a missing cone of $\pm 30°$ (Glaeser et al., 1989). In panel (a) a contour plot shows the point spread function in the x, z plane, the full 3-D function being given as the figure of rotation about the z-axis. Panel (b) shows line-traces through the same point spread function in the x-direction and in the z-direction. In both representations it is apparent that the resolution in the z-direction is only about 1.25 times worse than it is in the x, y plane.

It should be noted that the region of missing data will be cone-shaped only if the tilt axis in different measurements lies at many different azimuths relative to the unit-cell axes of the 2-D crystal. The region of missing data will be a wedge rather than a cone, of course, if the tilt axis lies at just a single angle relative to the unit-cell axes. The effect of a missing wedge is considerably more severe (and markedly more anisotropic) than that of a missing cone, for the same value of the maximum tilt angle. As a result, the point spread function shown in figure 4.3 should not be invoked in the context of single-axis tomography, or for any situation in which the range of azimuths for the tilt axis is not uniformly sampled. A second caution that is worth mentioning is the fact that simulations of the type shown in figures 11.6 and figure 4.3 show more severe effects of the restricted tilt angle when they are done in two dimensions instead of three, since the 2-D simulations can only sample a single azimuth for the tilt axis.

The relative insensitivity of the high-resolution density map to the effects of a 30° missing cone, for real protein structures, can also be understood on the basis of the modest anisotropy shown in the computed point spread function above. Individual atoms are either covalently bonded to one another, at distances of 0.15 nm or less, or they are in van der Waals contact, with typical distances between the centers of side chain residues that are ~0.45 nm. At a resolution of 0.36 nm the separation between covalently bonded atoms is not resolved in any case, and thus the density values within covalently bonded features are hardly affected by the anisotropic resolution. For side chain residues that are in van der Waals contact, on the other hand, the densities for each residue are well separated with respect to the resolution, even in the direction of poorest resolution. As a result, even though the density of each residue is somewhat elongated in the direction of the missing cone, the densities still remain just as interpretable as they would be in the case of isotropic resolution.

The conclusion that data collected at angles of ±60° are sufficient to provide an interpretable density map is borne out by a number of experimental examples. These include the structure analysis of the photosynthetic light-harvesting complex from chloroplasts (Kuhlbrandt et al., 1994), tubulin (Nogales et al., 1998), and aquaporin (Murata et al., 2000; Gonen et al., 2004), as well as bacteriorhodopsin (Grigorieff et al., 1996; Mitsuoka et al., 1999).

11.5 Interpretation of the high-resolution map involves building the known chemical structure into the 3-D density

Different degrees of interpretation of the density map become possible at successively higher stages of resolution. At a resolution of 0.9 to 0.7 nm it is usually possible to visually recognize segments of secondary structure, such as long α-helices or large segments of β-sheet. For example, all of the transmembrane helices in bacteriorhodopsin (Henderson and Unwin, 1975), the light-harvesting complex (Kuhlbrandt and Wang, 1991), aquaporin (Li et al., 1997; Walz et al., 1997; Cheng et al., 1997), and the P-type ATPases (Zhang et al., 1998; Auer et al., 1998) could be recognized at ~0.7 nm resolution. Significant progress has also been made in using computational approaches to reliably discriminate between helices and other features of secondary structure when the resolution is at least 0.8 to 0.9 nm (Jiang et al., 2001a; Zhou et al., 2001a). Some success has also been demonstrated for the identification (Kong and Ma, 2003) and analysis (Kong et al., 2004) of regions of β-sheet.

At a resolution slightly better than 0.5 nm, however, individual strands of β-sheet begin to show up as separated strings of density. At this resolution there is no longer any ambiguity whether a density feature is α-helix or β-sheet. Even at a resolution of 0.45 nm it is still too early to fit the polypeptide chain to the density, however. Loops in the peptide chain, if they are sufficiently well ordered, may nevertheless start to be seen at this resolution.

At a resolution of 0.4 nm and better, the densities corresponding to large, bulky side chain residues finally begin to emerge. These densities become increasingly well defined as the resolution improves. As an example, figure 11.7 shows a portion of the density map for the α, β-tubulin dimer that was computed at a resolution of 0.35 nm. With a map of this quality, even though it is far from perfect, it becomes possible to build a model at atomic resolution, in exactly the same way that was described in section 2.12 for x-ray crystallography.

Once the resolution of a 3-D density map is high enough to see at least partial densities for a large fraction of the side chains, the task of fitting the known amino acid sequence to the map becomes a rather unforgiving one. A large side-chain density

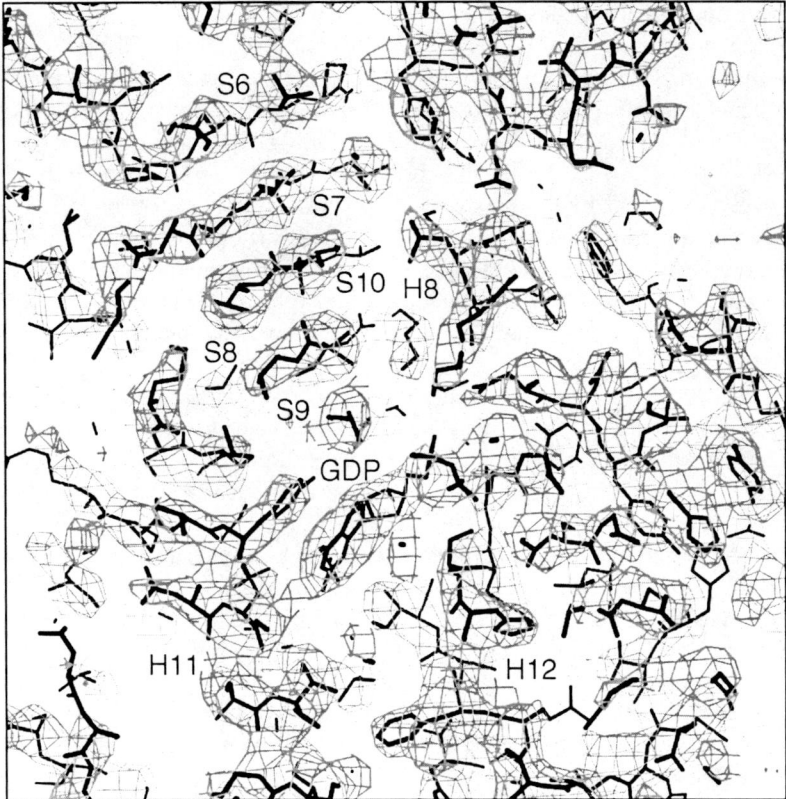

Figure 11.7 A thin slab through the 0.35 nm resolution map of tubulin is shown near the interface between alpha (above) and beta (below) monomers. The axis of helix H8 is perpendicular to the section, while H11 and H12 are roughly parallel to the plane of the section. Beta strands S6–S10 pass through the section with a range of angles. The nucleotide (GDP) is roughly in the center of the image. This display was prepared with the program "O" (Prof. T.A. Jones, Institute of Cell and Molecular Biology, Uppsala University).

means that a bulky residue such as tryptophan, tyrosine, phenylalanine, methionine, lysine, or arginine is located at that position. To get started, then, any arbitrary part of the amino acid sequence that has such a bulky residue can be built into the density. The rest of the sequence is then fitted into the density with the correct phi-psi angles, according to whether the backbone density in the map corresponds to a region of α-helix or β-sheet. When the next, large side-chain density is encountered in the density map, there must either be a corresponding bulky residue in the sequence, or the initially chosen interpretation is wrong. Obviously, many mistakes can be made in guessing the possible structure during the initial stage of interpretation, but as more and more wrong guesses are eliminated, the chance of finding the correct interpretation will increase. The job of getting started is often made simpler if there is additional spectroscopic, biochemical or genetic information about the structure that can narrow the range of starting guesses.

Interpretation of a high-resolution density map is frequently made more difficult by the fact that some features may be structurally disordered, and thus cannot be seen in the map. Thus, if there is a bulky residue at a certain point in the amino acid sequence and there is no corresponding side chain density in the map, either the current interpretation of the structure is wrong, or it may only be that this particular side chain residue is disordered. As a result, the absence of a bulky density feature in the map where one is expected from the known amino acid sequence cannot be used to reject a possible interpretation of the structure. Loss of the main chain density can be even more frustrating, because one does not know for certain where one should go next in order to extend the main chain. In effect, the whole process of interpretation must start again, leaving in place only that section of the sequence that may have already been built into one part of the map.

In the end, the two constraints that (1) the main chain must progress through the density map with phi-psi angles that are in the allowed region of a Ramachandran map, and (2) there must be a bulky residue in the amino acid sequence everywhere that there is prominent side-chain density in the map usually suffice to lead one to the correct interpretation. Even so, one cannot be too careful at this point. Several examples of wrong structural interpretations have been encountered in the field of x-ray crystallography, as has been reviewed by Colovos and Yeates (1993). A very detailed discussion of the types of error that can be made during chain-fitting has been published by Jones and Kieldgaard (1997). Surprisingly, errors as severe as building the chain into density in the reverse direction, or interpreting density as the wrong type of secondary structure, have occurred in unfavorable situations.

Fortunately there are a few additional "reality checks" that can be used to tell whether the interpretation of the density map is likely to be correct. The first step is to compare the calculated Fourier transform of the model structure to the experimentally measured diffraction intensities. The difference between the measured and calculated diffraction amplitudes, usually called the *crystallographic R-factor*, is a traditional measure of how good the model is. The R-factor, defined previously in equation 2.33, is written here again for convenience:

$$R = \frac{\sum |F_{obs} - F_{calc}|}{\sum F_{obs}}, \qquad (11.1)$$

where F_{obs} is the measured diffraction amplitude, and F_{calc} is the amplitude calculated for the model structure. When the R-factor is first calculated for the atomic-resolution model that has just been built into the density map, it is normal to get a value of 0.4 to 0.5, not much different from the value that one would get if the atoms were put into the unit cell in random positions. As is explained in further detail in section 11.7, very small movements of the atomic positions can then be made, by distances in the range 0.1 to 0.2 nm, subject to the restraint that proper bond angles and bond lengths must be maintained. These adjustments in atomic position, although small, can lead to dramatic improvements in the R-factor, provided that the initial atomic model is quite close to the true structure.

Obtaining a low value for the R-factor is not, by itself, a strong proof that a refined structure is fundamentally correct. As mentioned by Jones and Kjeldgaard (1997), the peptide chain can intentionally be fitted into density in the reverse direction, over the entire length of a protein, and still the structure can be refined to an R-factor as good as 21.4%. A far stronger reality check is given by the *free R-factor* (Brunger, 1997), discussed previously in section 2.13. In the case of the intentionally wrong structure just mentioned, R_{free} rose to 61.7% at the end of the refinement process. In x-ray crystallography, R_{free} should ideally be below 30%, and in addition it should not be too much higher than the R-factor. If that is not the case, there is good reason to be concerned that significant errors remain in the interpretation of the density map. In electron crystallography there is still not sufficient experience to say how low the value of R_{free} should be in order for it to be a reliable guide as to the quality of the model structure.

Another important reality check, developed by Eisenberg et al. (1997), makes a quantitative comparison between the type of environment — defined by its amino acid neighbors — that is found for each residue in the 3-D model structure, and the type of environment that is statistically likely to occur for each type of residue. This statistical comparison is based upon what has previously been found within the database of all solved structures. Empirical tests show that Eisenberg's quantitative metric is quite good at identifying chemically incorrect models, even when these models correspond to good fits between sequence and density, and have good R-factors. Similar computational tools have been developed by other authors. One of the more recent of these is the MOLPROBITY server, which accepts PDB files and performs a structure validation on the basis of C_α and C_β geometry as well as phi and psi angles (Lovell et al., 2003). MOLPROBITY adds hydrogen atoms to the model structure according to the stereochemistry of the atomic model, thereby improving the evaluation of all-atom contacts, the analysis of packing density, and the analysis of hydrogen-bonded networks. MOLPROBITY thus facilitates a more accurate rebuilding in areas where selection of an alternative rotamer conformation improves the atomic model.

11.6 Accurate atomic-resolution models can also be obtained by docking atomic models of individual components into the 3-D density map of a macromolecular complex

One of the most important applications of electron microscope to structural biology involves the fitting of atomic models of one or more subunits into specific locations within the lower resolution density map of a large macromolecular assembly. The paradigm involved in this *hybrid approach* takes advantage of the fact that it may be much easier to use electron microscop to obtain a 3-D density map of a large macromolecular complex (at a resolution of 2 nm or better) than it is to obtain well-ordered crystals of the same complex, at the same stage of purification of the target material. At the same time, it is very likely that crystallography or NMR spectroscopy can obtain atomic-resolution models of at least some of the individual subunits much more easily than one can obtain an atomic-resolution density map of the intact complex by electron microscop. Indeed, atomic-resolution models of individual subunits may in some cases be already available in the PDB. The hybrid approach thus combines the best features that are offered by electron microscopy and either x-ray crystallography or multi-dimensional NMR spectroscopy, thereby making it possible to obtain quite accurate atomic models for structures that could not be obtained by any one technique on its own.

The basic idea of *docking* the atomic model of a protein subunit into a comparatively low-resolution electron microscope density map is plausible as long as the protein subunit has a distinctive size and shape. As an example, the fitting of atomic structures of actin filaments and the S1 ATPase fragment of myosin into S1-decorated thin filaments has been shown previously in figure 1.5. In the first stage of such a docking process, it is reasonable to treat the individual subunits as rigid bodies with six degrees of freedom. In this case one can use molecular graphics software to manually translate and rotate the atomic model and visually judge when a best fit has been achieved. The goal is to find a position for the atomic model in which as little of the model as possible lies outside of the envelope of the density map, while at the same time as little of the volume of the density map as possible is left empty.

In many cases, however, individual protein subunits have an indistinct, globular shape. When that is the case, the rather subjective, manual docking procedure described above can produce results that are ambiguous and unconvincing. As a remedy, software tools have been developed that provide a quantitative evaluation of how well an atomic model fits into a density map as its position and orientation are changed. As reviewed by Volkmann and Hanein (2003), a global matching of the entire volume (i.e., not just the surface contours) of a density map to that of the subunit that is being docked is computationally tractable, and doing so provides a more accurate solution to the docking problem. In addition, statistical tools have been derived that provide confidence intervals for the alignment (docking) parameters. These quantitative tools represent a major advance in the interpretation of electron microscope density maps in cases when manual docking is not able to provide a convincing solution.

The docking of known, atomic-model structures that correspond to one or more of the components of a larger complex is normally done by treating each of the known structures as rigid bodies. Docking of individual subunits as rigid bodies of course assumes that the structures adopted under conditions of protein crystallization (or NMR structure determination) are the same as those in the intact complex. A better approach would be to allow the conformation of a known atomic-model structure to adapt itself during docking, so as to give a best fit to the electron microscope density for the corresponding subunit in the complex. One approach to implementing the desired structural adaptation allows the existing, atomic model to be deformed according to the normal (i.e., linearly independent) modes of vibration of the atomic model (Hinsen et al., 2005). As implemented by Hinsen et al., the force that drives (or "pulls") a change in conformation is determined by the discrepancy between the experimental density and that calculated for the initial atomic model of the subunit. The deformation is then allowed to proceed in small, iterative steps, for which the normal modes are constantly updated as the atomic model is pulled into conformity with the experimental electron microscope density map.

Accurate docking fortunately becomes quite unambiguous, even for globular subunits, as soon as the resolution of the electron microscope density map becomes high enough to visualize individual elements of secondary structure, such as helices and regions of β-sheet. Achieving a resolution somewhat better than 1 nm is especially helpful in visualizing α-helices as rod-like densities whose positions and orientation are quite well defined, even though the pitch of the helices cannot be visualized at that resolution. Individual polypeptide chains, including strands within β-sheets, are not well defined until the resolution approaches ~0.4 nm (Nogales et al., 1997), however.

As an example, the docking of an atomic model of tubulin into an 0.8 nm resolution density map of microtubules (Li et al., 2002) made it possible to improve the description of interprotofilament interactions that had been proposed on the basis of a previous docking that was done with a 2 nm resolution map.

The visualization of α-helices as rod-like "sausages" also makes it possible to accurately dock individual globular domains of a multidomain subunit as separate rigid bodies, thereby allowing one to model conformational changes that involve the rotation of domains about hinge-like connections. In a recent application of this capability, Ludtke et al. (2004) were able to demonstrate that the conformation of the GroEL chaperone in solution involves domain positions that are clearly different from those that appear under the conditions required to crystallize the same protein. The ability to accurately model the "repacking" of domains makes it especially important to strive for a resolution better than 1 nm when docking individual protein subunits into larger macromolecular complexes. It can be expected that the formation of a macromolecular complex will often involve biochemically important shifts in tertiary structure relative to what is seen in the crystal structures of individual components. Such shifts in tertiary structure can be expected both during the formation of the multisubunit complex and during the progression of the complex through intermediate stages of its biochemical cycle.

11.7 Refinement of an atomic-resolution model may proceed in a different way for electron crystallography than is traditionally done in x-ray crystallography

Refinement improves the values of the adjustable parameters that describe the structural model

The basic ideas and mathematical principles of structural *refinement* have been described previously in section 2.13. In brief, the positions of atoms in the atomic-resolution model are adjusted so as to minimize the R-factor (equation 11.1) between the measured structure factors and those calculated for the model. If the ratio of independent observations (i.e., diffraction spots) to parameters is high enough, each atom is also assigned its own value of the Debye thermal parameter (B-factor; see section 2.6). At the same time that the atom positions and other adjustable parameters are being varied, an empirical energy function may also be used to estimate the effects of covalent bond distortions and nonideal van der Waals contacts (Brunger, 1997). A global "residual" can then used to balance the desire that the model structure should account as closely as possible for the measured diffraction intensities with the desire that the resulting model should be a stereochemically sensible one.

At a resolution close to 0.25 nm, substantially more independent measurements are available. Refinement can therefore accommodate the increased number of parameters that are required in order to model well-localized water molecules, salt ions, ligands, and other nonprotein features of the structure (Drenth, 1994). By including elements in addition to the polypeptide chain that really are present in the crystal structure, and thus should be accounted for when comparing F_{calc} to F_{obs}, the refined atomic-resolution model produces improvements in both the R-factor and R_{free}

(Kleywegt and Jones, 2002). As was mentioned in section 2.13, the changes in model parameters that result from refinement actually represent an improvement in the accuracy of the structural model, as long as the new values of the parameters are accompanied by a reduction in the value of R_{free}. The value of R_{free} thus provides an important measure of how well the model really agrees with the data.

Refinement can over-interpret the available data, however

As more and more parameters are added to the model, there is always the danger that the R-factor will improve merely because there are more parameters, and not because the model has become closer to the truth. The R-factor will inevitably continue to decrease in each round of refinement as the parameters are adjusted so as to produce values of F_{calc} that give a best fit to the measured data, experimental error and all! In order to be sure that the model is actually continuing to improve, it is important to monitor the progress of refinement by calculating a *free R-factor*. As was described above and in section 2.13, the value of the free R-factor will continue to decrease as long as the changes made in the parameters actually bring the model closer to the real structure. On the other hand, if the adjustable parameters migrate to values that make the model look less like the real structure, or if inappropriate parameters are introduced (that might describe nonexisting water molecules, for example), the free R-factor will immediately increase even though the refined R-factor will inexorably continue to decrease.

It is appropriate to use all of the diffraction data that are available

Electron diffraction intensities can usually be collected at higher tilt angles than images, and in addition, electron diffraction data are often collected to higher resolution than is available in the images. The use of these additional diffraction data during refinement increases the ratio of observations to parameters, thereby improving the legitimacy of the refinement.

Phase extension, a routine aspect of x-ray crystallographic refinement (Drenth, 1994), can also be used to further improve the resolution of the map itself. *Phase extension* is an iterative process in which the model structure that is available at any one stage is used to calculate the amplitudes and phases at all diffraction spots, including those for which there are no experimental measurements of the phases. During each cycle of phase extension a new density map is calculated from all previous amplitudes and phases, along with a small number of new diffraction spots for which there are experimental amplitudes but no phases. The phases calculated for the current model structure are used to compute the phases for the newly added Fourier components, and then a new density map is computed. A new model structure is then rebuilt in conformity with the new density map, and the rebuilt model is used to repeat the process in another cycle of phase extension. In the case of bacteriorhodopsin, Grigorieff et al. (1996) employed phase extension for diffraction intensities that had been measured at tilt angles up to 60°, for which no images phases were available. As is shown in figure 11.8, the anisotropy of resolution that was present in the original map was greatly reduced in this way, allowing the correction of some minor errors in the initial atomic-resolution model (Grigorieff et al., 1996).

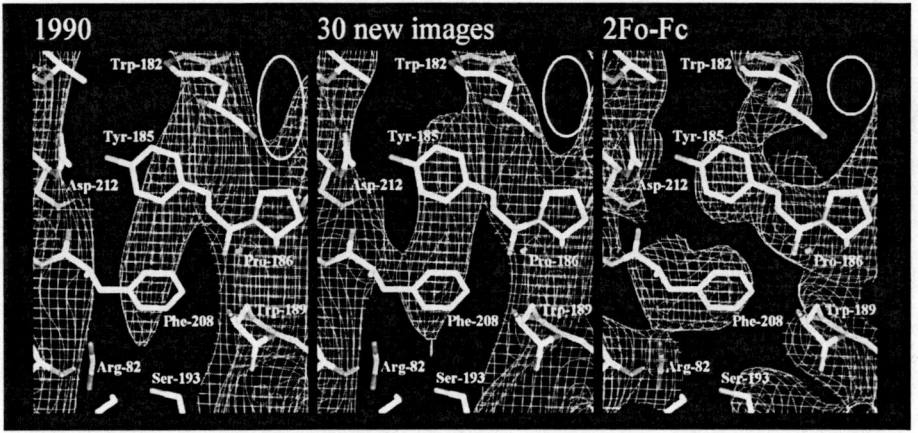

Figure 11.8 The quality and interpretability of the 3-D density map of bacteriorhodopsin progressively improved both as the accuracy of the data improved and as the angle of the missing cone was reduced. The estimate of the anisotropic point-spread function is shown in the upper right-hand corner of each panel. Left panel: A portion of the map obtained with 2750 measured phases and a missing cone of ±45°. Center panel: The same portion of the map obtained with a more complete data set consisting of 3766 measured phases within the same missing cone of ±45°. Right panel: The same portion of the structure, showing the $2F_{obs} - F_{calc}$ map produced with 4743 phases obtained after refinement of the atomic model against the complete set of electron diffraction intensities. Refinement extended the phases beyond the tilt angle at which image phases had been obtained, and as a result the missing cone was reduced to ±30°. Figure kindly provided by Dr. N. Grigorieff, adapted from Grigorieff et al. (1996).

Refinement can also be driven, in principle, by requiring the phases calculated from the model structure to agree as closely as possible with the experimental phases. In the refined structure published by Grigorieff et al. (1996), the experimental phases were used, instead, to calculate a *phase residual* that was used in place of R_{free}, thus releasing all of the measured diffraction amplitudes for use in calculating the R-factor. Mitsuoka et al. (1999), on the other hand, used both the diffraction amplitudes and the experimental phases as observed data in comparing the calculated and the observed structure factors, and about 5% of the amplitudes were held out in the usual way, to be used in calculating R_{free}.

The R-factor and R_{free} are larger for electron crystallography than for x-ray crystallography

Experience in x-ray crystallography has established that a model structure is very likely to be accurate if R_{free} is as low as 0.20 to 0.25. As R_{free} increases above this value, it can still be that the atomic-resolution model of the structure is quite good, but the confidence that one can have in the model gets progressively lower as R_{free} gets larger. In x-ray crystallography of proteins, a model structure that gives a free R-factor as large as 0.3 is now regarded as one that requires significant improvement before details such as the contacts between amino acid side chains or the number and location of water molecules can be taken seriously. Further background on these points is provided in a detailed analysis of the values of R-factors and free R-factors for structures published in the period 1991 to 2000 (Kleywegt and Jones, 2002).

It is currently the case that the free R-factor may not fall below 30% for an electron diffraction structure, however. The R-factor for the bacteriorhodopsin structure refined by Grigorieff et al. (1996) was about 29%, and, as was mentioned above, a free phase residual rather than a free R-factor was used to validate the structure. For the independently built and refined model of Mitsuoka et al. (1999), the R-factor after refinement was about 24% while R_{free} was still 33%. Similarly, the R-factor for the atomic model of LHC II, the photosynthetic light-harvesting complex, was 33% and R_{free} was 38% (Kuhlbrandt et al., 1994). Even higher values of the R-factor and R_{free} were obtained for rather poorly diffracting crystals of aquaporin-1 (Murata et al., 2000), but values of 30% and 34% were obtained for higher diffracting crystals of aquaporin-0 by Gonen et al. (2004). Perhaps the best electron crystallographic refinement to date is that of tubulin (Lowe et al., 2001), in which the R-factor was 23.2% and R_{free} was just under 30%.

One reason why electron crystallographic R-factors are higher than x-ray R-factors quite probably has to do with the significantly higher error in the measurement of electron diffraction intensities. The larger experimental error can be seen, for example, in the experimental values of R_{merge}, which are in the range of 12 to 16% for the best electron diffraction data sets (Faruqi et al., 1995, 1999; Downing and Hendrickson, 1999), as opposed to half or even a third of that value for typical x-ray diffraction data sets. The reasons for the higher values of R_{merge} in electron diffraction data are not yet fully understood; the problem of imperfect specimen flatness, and the tendency to use electron exposures that produce substantial radiation damage may both be significant contributors. Uncorrected experimental errors in F_{obs} are therefore the first factor that may contribute to the relatively high R-factors. Other issues that may contribute to higher R-factors for electrons include failure to account for chemical bonding effects in modeling the atomic scattering factors, described more fully below, and dynamical diffraction, described in sections 15.2–15.4.

Chemical bonding effects may lead to technical differences in electron diffraction refinement relative to x-ray diffraction refinement

In x-ray scattering it is quite legitimate to approximate the scattering from each atom in a molecule by the scattering that would occur for isolated, free atoms. Due to *chemical bonding effects*, however, different atoms will no longer have a completely neutral charge. Oxygen atoms will be slightly negative, for example, and nitrogen atoms will be slightly positive. The development of partial charges on the atoms in the polypeptide backbone is quite pronounced, as is known from the high dipole moment of the peptide bond. Even carbon atoms will acquire a small fractional charge, especially those in the indole ring of tryptophan and the imidazole ring of histidine. These chemical bonding effects will tend to make the R-factor for electrons be somewhat larger than it would be for x-rays, everything else being equal. It therefore is desirable for correction factors to be developed for electron scattering that take bonding effects into consideration.

Electron scattering arises from the interaction of the incident electrons with the shielded Coulomb potentials of the atoms in the specimen (sections 1.1, 3.6, and 4.2), rather than with just the electron charge density. The shielded Coulomb potential is much more sensitive than is the charge density itself to the small shifts in electron charge distribution that occur when two atoms form a covalent bond. For example, removal of 0.1 of an electron charge from a carbon atom would change the total electron charge density on the atom by less than 2%, but it would have a much greater effect on the shielded Coulomb potential. As a result, accurate refinement of model structures against electron diffraction intensities, yielding free R-factors lower than 0.25, will probably require a better description of F_{calc} than can be obtained when using independent, neutral-atom scattering factors.

The influence of chemical bonding effects has even been noticed in experimental density maps. In a number of cases the side-chain density of specific aspartic acid groups is either weaker than expected, or absent altogether (Kuhlbrandt et al., 1994; Kimura et al., 1997). Weak side-chain density might simply be due to structural disorder, of course, but it could also be due to deprotonation and acquisition of a net negative charge. To distinguish between these two possibilities, Kimura et al. have used a clever approach in which the experimental density map obtained with just the data between 0.7 and 0.3 nm is compared to the density map obtained with all data. The rationale behind this approach is based on the fact that the atomic scattering factor of the oxygen anion is reversed in sign (i.e., is negative) at low resolution, and thus the density of a negatively charged aspartate group should be weaker in the map that uses lower resolution data. This effect was indeed found to occur for several of the aspartate residues in bacteriorhodopsin that are known to be negatively charged, but not for Asp 96, which is known to be protonated in the light-adapted, resting state. A further modification of this same approach has also made it possible to identify positively charged side chain residues and density features that are candidates for being either inorganic ions or highly polarized water molecules (Mitsuoka et al., 1999).

An important effort was made by Grigorieff et al. (1996) to correct the atomic scattering factors for chemical bonding effects. In their refinement of the bacteriorhodopsin structure, they added parameters to describe the apparent occupancy of each chemically distinct atomic species. Thus CH_2 carbon atoms, for example, were allowed to adopt a different apparent occupancy than carbonyl group carbon atoms. The rationale for this approach was that deshielding of the nuclear charge (i.e., development of a small, fractional positive charge) would increase the atomic scattering factor and result in an apparently higher apparent occupancy for that type of atom, while a fractional negative charge (e.g., on oxygen atoms) would more completely shield the nuclear charge and lead to an apparently lower atomic occupancy. This approach to accounting for chemical bonding effects clearly worked, in that apparent occupancies (or relative scattering strength after refinement) were found to range from 0.67 for backbone carbon atoms to 1.60 for $-NH_3^+$ groups, and the direction of the effect is exactly what would be expected from chemistry. The approach nevertheless failed to improve the R-factor by even one percentage point.

The simple model of partial occupancy may not, in fact, capture some of the most important effects of chemical bonding. The Coulomb potentials of charged atoms that are represented with occupancies either greater than one or less than one still lack the long-range character of the Coulomb potential associated with real, partial charges.

Similarly, the long-range, orientation-dependent dipole potential of the backbone residues (and many of the polar side chain residues) is not captured by a model based on neutral atoms.

An alternative approach for modeling chemical bonding effects was taken by Mitsuoka et al. (1999), in which the atomic scattering factors for the carbon atom and the oxygen atom of the backbone carbonyl groups of bacteriorhodopsin were both modeled as a linear combination of the scattering factors for neutral atoms and singly ionized atoms. R_{free} was indeed sensitive to the partial charge assigned to the two atoms, but even so the improvement that occurred at the optimal charge of about 0.4 electrons was less than 1%. Unfortunately, there were no available charged-atom structure factors that the authors could use to model the partial charge of the amide NH group of the backbone, since this would have resulted in an even better representation of the partial charges on each atom within the peptide bond and a more accurate orientation for the electrostatic dipole moment of the bond.

A sophisticated level of investigation of chemical bonding effects has been initiated by Chang et al. (1999), in which molecular orbital calculations were used to get accurate Coulomb potentials for some representative molecular fragments. Preliminary calculations first confirmed that chemical bonding effects can make substantial contributions to the R-factor. Table 11.1 shows the R-factors calculated in different resolution shells for two different versions of the molecular structure factors for a few representative functional groups. One version of the structure factor is calculated in the usual way, that is, with neutral, independent-atom structure factors, while the other version is calculated with structure factors derived from shielded Coulomb potentials that are obtained from a full molecular orbital calculation of the electron charge density.

An important point that table 11.1 illustrates about the chemical bonding effects (and charged atom effects) is the fact that discrepancies with neutral-atom scattering factors are limited primarily to scattering at smaller angles. The fact that scattering at higher resolution is quite insensitive to chemical bonding effects is well illustrated by the comparison of the scattering factors for atomic oxygen and for the negative

Table 11.1 Values of the R-factors (expressed as a percentage) calculated for data within resolution shells for four highly polar molecular species with structures similar to structural fragments found in proteins.

1/s (nm)	Formamide	H-bonded pair	Aspartic acid	Ion pair
∞	14.8	18.6	14.9	42.9
1.28	15.8	19.2	13.6	50.9
0.64	11.7	14.4	12.2	20.8
0.42	10.9	11.2	10.1	13.5
0.32	8.2	9.2	9.4	14.6
0.26	7.8	9.3	10.9	11.8
0.21	7.0	9.3	9.6	10.1
0.18	5.2	6.6	6.3	6.9

The R-factors have been computed between molecular structure factors obtained from quantum-chemical calculations, to capture the effects of chemical bonding, and those obtained with the spherical, neutral atom model.

Adapted from Chang et al. (1999).

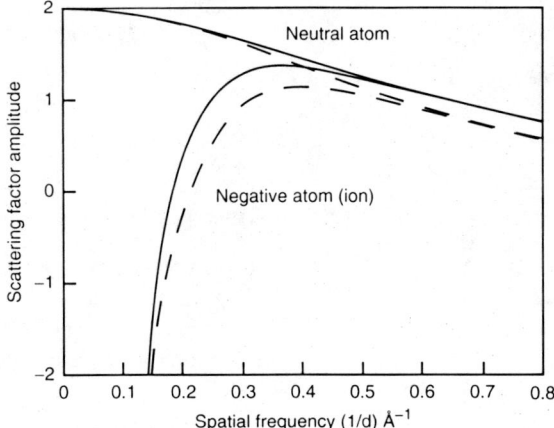

Figure 11.9 Comparison of the atomic scattering factors for neutral oxygen and its negative ion. Solid lines are calculated from the published values in the *International Tables for Crystallography* (1974), while the dashed curves were recalculated with the same computational methods that were used by Chang et al. (1999) to estimate chemical bonding effects for several representative molecules. This comparison illustrates the fact that chemical bonding effects can be large at low resolution (and especially so for isolated charged-ion species), but they become quite small at high resolution. Reproduced with permission from Chang et al. (1999).

oxygen atom, respectively, shown in figure 11.9. Although the structure factors of neutral atoms and chemically bonded atoms can be very different at lower resolution, and may even take on the opposite sign in the case of negatively charged ions, the effect becomes quite small beyond about 0.25 nm resolution. Since the majority of the data measured in electron crystallographic structure determinations may be at lower resolution than 0.25 nm, however, it seems likely that accounting accurately for the effects of chemical bonding on the shielded Coulomb potential will significantly reduce the R-factor.

The effort to characterize structure factors derived from quantum-chemical calculations was then extended by Zhong et al. (2002) to address the question of whether the phi, psi angles of the peptide backbone would have a noticeable influence on the atomic scattering factors. The answer to that question is that the dihedral angles do, in fact, have a relatively large impact that affects the value of the R-factor. For Bragg spacings larger than 0.25 nm, the R-factor between the quantum-chemical calculation and that using neutral, spherical atoms is greater than 8%, while that between the quantum-chemical calculation with any specific geometry versus that for an arbitrary ("standard") geometry is still about 4%. Since these values of the R-factor refer to noise-free simulations, they must be understood as being a measure of the "unnecessary" increase in the R-factor that must be due to the use of atomic scattering factors that are inaccurate for electron scattering. It seems reasonable to assume that, in the most favorable case, this contribution to the R-factor might add in quadrature to contributions from model error and the experimental error in F_{obs}. Nevertheless, the conclusion is that accurate electron scattering factors for backbone atoms should be parameterized in advance, to correspond to the secondary structure of each residue in the atomic model. Since this parameterization would be done on the basis of the quantum-chemical calculations, it would not put any burden on the ratio of independent measurements versus adjustable parameters.

The importance of accurate refinement will depend upon how the model structure is to be used in subsequent work

The exceptional accuracy of the phases that are derived from electron microscope images results in density maps that are unusually clean and easy to interpret, even at a resolution of 0.35 nm. Thus it is unlikely that the atomic models derived at this level of resolution have a significant number of errors when viewed as a ribbon diagram of the structure, even though their R-factors are high by x-ray standards.

Nevertheless, it now seems likely that the coordinate error in a poorly refined electron diffraction structure may be rather high. A feeling for the likely magnitude of coordinate error has been presented by Luecke et al. (1999), who were able to obtain an x-ray crystal structure of bacteriorhodopsin at 0.155 nm resolution. The root mean square (rms) difference in atomic positions between the refined structure of Grigorieff et al. and that of Luecke et al. was 0.257 nm for all atoms in the structure, and 0.171 nm for just the main-chain atoms. The refined structure of Mitsuoka et al. was closer to that of Luecke et al., however, with an RMS of 0.136 nm for all atoms and 0.108 nm for the main-chain atoms. In a few parts of the structure, mechanistically important features were not yet correctly modeled in either of the structures obtained by electron crystallography. For example, the higher resolution x-ray structure revealed one turn of the uncommon π-helix conformation in the region of the backbone of helix G, which included the critical lysine216 residue. This turn of π-helix plays an important role in the way that the protein responds to isomerization of the retinal group, which forms a protonated Schiff base with the ϵ-amino group of lysine216. This example illustrates the point that understanding many of the biochemical details of macromolecular function is likely to require atomic models with smaller coordinate error than can be expected from structures that are good enough to give a first idea of the chain-trace.

11.8 Difference Fourier maps

Difference Fourier maps are density maps that are calculated with structure factor amplitudes that represent the differences between amplitudes obtained under two different conditions. Only a single set of phases is used to compute difference Fourier maps. These phases can be the experimental phases obtained for one of the conditions, or they may even be the phases calculated for a refined model of the structure. Mathematically, the difference Fourier map is expressed as

$$\Delta\rho(\mathbf{R}) = \sum_{hkl} \{F_2(\mathbf{g}_{hkl}) - F_1(\mathbf{g}_{hkl})\} \exp i\alpha_1(\mathbf{g}_{hkl}) \exp -i(2\pi \mathbf{g}_{hkl} \cdot \mathbf{R}), \quad (11.2)$$

where the subscripts 1 and 2 denote two different states of the protein. Note, however, that the difference amplitude, $\Delta F(\mathbf{g}_{hkl}) = \{F_2(\mathbf{g}_{hkl}) - F_1(\mathbf{g}_{hkl})\}$, is the algebraic difference and not the modulus of the difference.

In a typical case, the structure of a protein would have been determined for one initial set of conditions, which might correspond to the "native" state, or the "resting" state of the protein. The next stage of an investigation would then be to determine what conformational changes occur when a ligand is bound, or when some other biochemically meaningful change occurs in the structure. If these changes in structure can be made to occur in the crystal, without causing changes in the unit cell parameters and without significant changes in short-range order, then a new set of diffraction intensities can be collected. The experimental effort required to collect a new set of intensities is far less than that which is required to collect the phases, of course. Thus a difference Fourier map can be calculated with minimal effort.

The theoretical reason why it is sufficient to use just one set of phases to calculate a difference Fourier map is easily understood. The phases for two closely related structures must be very similar if the amount of structural change is small enough. The difference Fourier map is thus most valid for comparing two structures in which the differences are small. In most cases involving the binding of a ligand or an allosteric change in conformation, for example, the changes in phases may even be less than the experimental error with which the phases are first determined for the "native" structure.

Almost paradoxically, it has been shown (Henderson and Moffat, 1971; Blundell and Johnson, 1976) that difference Fourier maps provide good representations of structural changes even when the difference in structure factor amplitudes is smaller than the statistical error with which the amplitudes can be measured, provided that the structural difference is concentrated in one or only a few places in the density map. The explanation is straightforward: the weak signals that are present in each of the diffraction spots all add up coherently in the map, while the noise adds in quadrature. The signal thus grows in proportion to N, the number of diffraction spots used, while the noise grows as only the square root of N. Clear benchmarks for the accuracy of the difference Fourier map and the extent to which the accuracy depends on the magnitude and multiplicity of structural changes have been explored in computer simulations by Lindahl and Henderson (1997).

As was shown by Henderson and Moffat (1971), the peaks in a difference Fourier map have only half the height that is present in the real structural change. The reduction in apparent peak height is a consequence of using one set of phases in equation 11.2, that is, it is a consequence of calculating a difference Fourier map rather than a difference between two Fourier maps. While it may thus seem sensible to multiply the difference Fourier map by a factor of 2 in order to get the correct peak heights, doing so will also multiply the noise by a factor of two. Since the signal-to-noise ratio is the same either way, tradition has held that the difference Fourier map should be defined as it is in equation 11.2.

Difference Fourier maps lend themselves best to convincing interpretations when the structural changes are localized in one or a few places. In general the interpretations can be of two types. In the first case, the difference map may show a single, isolated peak. This peak may represent, for example, the binding of a ligand. Single isolated peaks can also represent the transformation of a local feature of the structure from a disordered state to a well-ordered state. A second case that lends itself well to interpretation is one in which there is a matching pair of positive and negative peaks, located close to one another. A density "dipole" of this type is interpreted as being due to the shift of some element of the structure from one position to another. The positive peak thus represents the position to which the structural element has moved, and the negative peak represents the position where it had previously been.

The difference Fourier map of the N-state intermediate in the photocycle of bacteriorhodopsin shows both types of features (Vonck, 1996, 2000). As is shown in figure 11.10, a positive–negative pair of peaks in the region of helix F suggests that the cytoplasmic end of this helix tilts out into the lipid layer between individual protein trimers, thereby opening an area for increased access of water into the cytoplasmic half of the protein during this point in the proton-pumping photocycle.

Early attempts to use difference Fourier maps in electron crystallography were not very successful. The reasons why this was the case have not been characterized in detail. It may be that the choices of problems to be investigated, and a relatively high noise level in the earlier data, may both have made it difficult to see meaningful structural

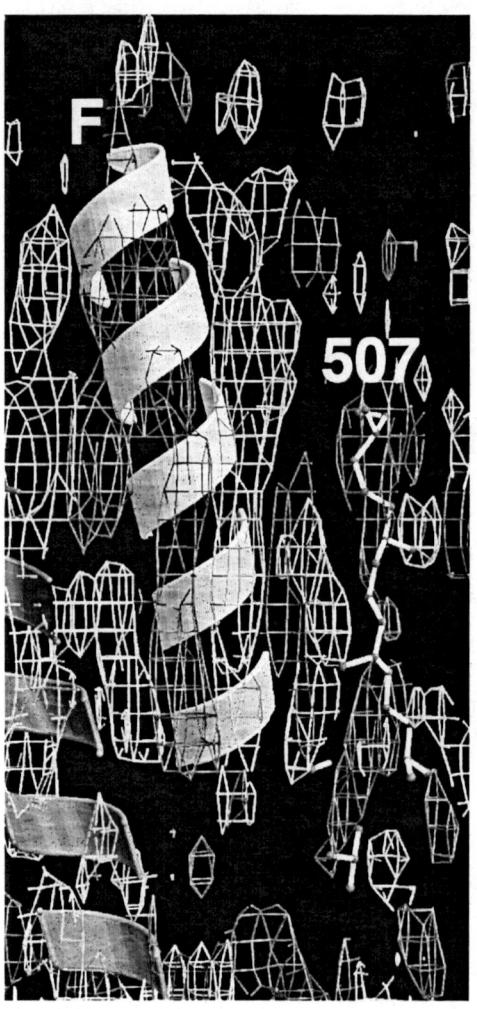

Figure 11.10 A key structural change that occurs in bacteriorhodopsin between its resting state and the N-intermediate in its photocycle is shown in this 3-D difference Fourier map. The position of the cytoplasmic half of helix F in the "resting state" is represented by a ribbon diagram, and a portion of lipid molecule 507 is represented by a stick diagram. The N-bR difference Fourier map shows negative density in the region of helix F that is accompanied by an adjacent region of positive density, suggesting that the cytoplasmic end of helix F moves towards the position of lipid 507 when the protein advances into the N-state intermediate. In addition, a region of negative density appears over the position of lipid 507, suggestion that this lipid molecule is moved (becomes disordered) as a result of the tilt of helix F. The functional significance of the tilting of helix F is believed to be that it opens a "channel" that provides access for a few water molecules to penetrate into the interior of the protein, thereby facilitating light-driven proton transport across the membrane. Figure kindly provided by Dr. Janet Vonck, adapted from Vonck (2000).

changes in difference maps. The use of fully hydrated samples (Subramaniam et al., 1993; Vonck et al., 1994) has itself apparently reduced the noise level in difference maps, and the introduction of CCD cameras to record diffraction patterns has further reduced the errors (Downing and Hendrickson, 1999; Faruqi et al., 1999). The most recent work clearly indicates that difference Fourier maps calculated with electron diffraction data have similar sensitivity to those calculated with x-ray diffraction data.

Careful attention must be given to proper scaling of the two data sets, prior to calculating the difference amplitudes. If the scale factor of one data set is not the same as that of the second data set, then the difference map will look like a weak "ghost image" of the structure itself. Fortunately, this effect, encountered originally in x-ray crystallography, does not occur for electron crystallography when the reciprocal lattice-line curves obtained for the first structure are used as the reference for merging and scaling the data for the second structure, thus placing both data sets on the same scale.

The ready availability of images in electron crystallography naturally opens the possibility to calculate difference images rather than the difference Fourier maps that have been discussed above. Difference images are, in principle, superior to difference Fourier maps because they are free of both the increased noise and the halving of the signal that can be traced to the approximation of using the same values of the phase for the two structural states. In other words, difference images do not rely on the approximation that the phase differences are small for two structures that are being compared. As will be seen with numerous examples that are discussed in chapters 12, 13, and 14, difference images are used very commonly for the comparison of structures at lower resolution, for example to locate antibody binding sites, binding sites for large ligands or the binding sites of macromolecular binding partners. Since applications of this type require only modest image resolution, it is relatively little work to obtain the images (or 3-D reconstructions) that are needed to subtract one experimental structure from another. At high-resolution, however, it is much more difficult to collect phases from images of 2-D crystals than it is to collect diffraction intensities when the sample is tilted at a high angle. As a result, the use of difference Fourier maps rather than difference images remains the preferred approach for high-resolution visualization of relatively small structural changes.

12

Electron Crystallography of Helical Structures

12.1 Introduction

Helical structures are essential for life. They connect and integrate function between distantly separated structures within a cell. Actin and microtubules are examples of helical structures that are part of the cell's cytoskeleton and that define cell shape. They also act as the railroad tracks along which molecular motors tow subcellular material between locations within the cell. Helical structures also occur in pathological states such as the paired helical filaments associated with Alzheimer's disease or fibers of hemoglobin in sickle cell anemia. Helical structures can serve mechanical roles; for example, the bacterial flagellar filament converts torque into thrust to propel the bacterium. In addition to naturally occurring helical structures, some molecules can be induced to form helical structures after biochemical isolation, as in the case of the nicotinic acetylcholine receptor.

Prior to the 1960s, helical structures were studied primarily by x-ray fiber diffraction. Such studies were limited by the need to have well-ordered fibers, and they provided only general information about subunit packing and substructure. The small-angle part of the pattern provided information about distances between neighboring subunits and thus estimates of the size of the subunit. The high-angle part of the pattern revealed whether the subunit had aligned alpha helices or beta sheets. In one case, however, that of tobacco mosaic virus, fibers of superb order allowed Holmes and others to embark on the analysis of the structure to atomic resolution using the general method of isomorphous replacement (Barrett et al., 1972; Holmes et al., 1972; Namba and Stubbs, 1986). This project was a much more demanding undertaking than the production of such a map from crystalline samples, and it stands today as the only example of a helical structure solved by x-ray fiber diffraction using the general method of isomorphous phasing.

In the 1960s, electron microscopy overtook x-ray diffraction as the method most frequently used to analyze helical structures. Electron microscopy is now used routinely to determine the helical symmetry of particles as well as to obtain three-dimensional (3-D) maps showing the subunit arrangement and location of subdomains within the subunit. Most maps are at about 2 nm resolution; a few are at 0.7–1.0 nm resolution; and two are at ∼0.4 nm resolution. Structures examined by electron microscopy include: actin, myosin, microtubules, bacterial flagellar filaments and hooks, bacterial pili, 10 nm filaments, helical bacteriophage tails, helical viruses, the acetylcholine receptor, sickle-cell hemoglobin fibers, paired helical filaments, the calcium-ATPase, filaments from the sperm of Dictyostelium, filaments of LexA or RecA with DNA, chromatin fibers, and fibers of ribosomes.

Helical structures are well suited to analysis by electron microscopy and electron diffraction. High-quality electron diffraction requires well-ordered fibers, but these need only be rafts of particles a few micrometers in width whereas for x-ray diffraction one needs fibers having a diameter at least 100 times bigger. To analyze images, ordered fibers are not needed. One effectively constructs ordered arrays by aligning images of individual filaments in the computer. In addition, having an image allows one to measure the particle diameter directly, and this is important in determining the helical symmetry, which can be a necessary step for further analysis. From the images one can also determine the hand of a particle and the order of each layer line in its transform. Access to such information, which is readily obtained from images, makes determination of the symmetry of helical particles much simpler than when one has only x-ray diffraction patterns.

The most amazing consequence of helical symmetry, however, is that an image of a single helical particle can provide an entire 3-D data set (DeRosier and Klug, 1968). This is in contrast to the case with two-dimensional (2-D) crystals in which one needs to tilt the specimen in the microscope to get a 3-D data set (Henderson and Unwin, 1975). The reason is that the subunits making up the helical structure are arranged around the helix axis in different orientations. For example, in tobacco mosaic virus, there are 49 subunits in an axial repeat. Thus, in a single image, one has 49 differently tilted views of the subunit structure. To make use of these different views, one must determine the helical symmetry of the structure. Then one can use the power of helical diffraction theory to extract a 3-D image of the structure.

We will present helical diffraction theory, as it applies to specimens in the electron microscope, in three stages. The first stage, section 12.2, describes the diffraction pattern of an ideal helix. At this stage we are concerned with establishing the consequences of the fact that any helix can be generated by rolling an appropriate 2-D lattice into a cylinder. Section 12.3 describes the effects that are caused by deviations from ideal helical geometry such as bending of the helical axis, short-range and long-range disorder in subunit arrangement, and partial flattening. Finally, in section 12.4, we describe the conventions and the well-known tools that are used to index the diffraction pattern of a helix.

There are several methods by which one can generate 3-D maps of helical particles by electron microscopy. Two involve the use of images and another the use of electron diffraction, but so far the latter has been applied only to double-walled, carbon nanotubes. In the diffraction-based method (Zuo et al., 2003), over-sampled electron diffraction patterns are obtained. Initial random phases assigned to the amplitudes are refined by an iterative process in which the finite width of the particle and a real positive scattering potential are used as constraints. How applicable this method might be to biological structures remains to be determined. The first of the two image-based methods involves cutting the image of each helical particle into segments and treating each segment as a single particle (Egelman, 2000; Li et al., 2002). While chapter 14 describes the general methodology of the single-particle approach, section 12.8 provides a brief overview of the details peculiar to helical structures. The second method, the so-called helical method, requires that the amplitudes and phases of the Fourier transform of each particle be extracted and merged (averaged) from many particles, after which an inverse Fourier–Bessel transform is used to generate a 3-D map (DeRosier and Moore, 1970). Practical guidance for performing these latter operations is provided in section 12.5.

Both single-particle and helical methods have advantages. The near-atomic-resolution structural studies of helical particles have used the helical method because the Fourier transform of the corrected, averaged particles provides a simple way to determine resolution and, thereby, how well the various corrections and alignments are working. The advantage of the single-particle method lies in its avoiding the assumption that all segments in an intact helical filament are identical in structure and are in identical physical environments. For example, in the case of F-actin, different segments have different helical symmetries and the subunits have different conformations (Galkin et al., 2002). If one enforces helical symmetry, one averages over the structural variations. The single-particle method has also been successfully applied in situations were the helical symmetry is unknown and, even better, where the Fourier transforms of the particles gave such weak layer lines that it was hard to identify and index them.

Some structures, such as microtubules, are helical structures with a seam. Some of these seamed structures can be described by a superlattice, but the application of helical methods in such cases has not been productive. The single-particle method is also possible but has not yet been explored. Recently, Kikkawa (2004) has devised a scheme for analysis of helical structures with a seam as long as the seam is nearly parallel to the helical axis. The method, which is similar to the helical method, allows one to reconstruct the side of the particle not containing the seam. The method is presented in brief in section 12.8.

Helical structures by their nature are weaker scatterers than 2-D crystals (Morgan and DeRosier, 1992). Although in principle one should obtain a complete 3-D data set more easily from a helically symmetric structure than from a 2-D crystal, in practice the data obtained with helices so far have only provided resolutions of 0.6–1.0 nm, except for the helical crystals of the acetylcholine receptors (AchR) (Miyazawa et al., 2003) and the filament of the bacterial flagellum (Yonekura et al., 2003), which are at 0.4 nm resolution. The AChR map contained data corresponding to images of about one million subunits whereas the filament map resulted from images of only about 50,000 subunits. The difference in the number of subunits required for these structures most probably arises from the differences in the order of the two specimens; that is, the flagellar filaments are inherently better ordered helical crystals than those of the acetylcholine receptor. There is undoubtedly still room for improvement in the crystallinity of specimens and the electron microscope images that can be obtained from them, and section 12.7 concludes this chapter with a brief summary of how structure determination with helical specimens can itself be expected to improve, as well.

12.2 Ideal helices and their diffraction patterns

There are excellent review articles on helices and their Fourier transforms by Moody (1991) and Stewart (1988) and, in addition, the original articles by Cochran et al. (1952) and Klug et al. (1958) are valuable resources.

Helical lattices are related to 2-D lattices

All *helical lattices* can be derived by rolling a corresponding 2-D plane lattice into a cylinder. The only condition to ensure that the cylinder is a helical lattice is that at least one pair of points on the 2-D lattice overlap when the lattice is rolled up to form a cylinder. The line between the pair of points that overlap is called the *circumferential vector*. The 2-D lattice and the circumferential vector uniquely describe the helical lattice (Baker and Caspar, 1984). Each unique circumferential vector describes a different helical lattice. It follows that there are an infinite number of unique helical lattices.

As a simple example, let us take a square lattice. Let us choose the line from $h, k = 0, 0$ to $h, k = -11, 1$ (one lattice point up and eleven to the left) as the circumferential vector (figure 12.1(a)). To form the helical lattice we draw two lines that are perpendicular to the circumferential vector and that pass through the points $0, 0$ and $-11, 1$, respectively. We cut out an infinitely long strip along the two lines (figure 12.1(b)) and join the edges to form a cylinder, which is the helical lattice (figure 12.1(e)). Having produced the helical lattice, we can describe it by specifying the unit cell vectors **a** and **b** and the circumferential vector.

Helical symmetry can be described in a different way, which is more often seen. We can generate positions in a helical lattice by a simple stepwise operation in cylindrical coordinates: namely, as a rotation by an angle ϕ_0 about the helix axis and an axial translation of P_0 (figure 12.2). (It is necessary to add an n-fold rotation about the helix axis if the lattice has n-fold rotational symmetry — see below.) If repeated, these operations (i.e., the *screw operations* and the *n-fold rotation operations*) will generate every point in the helical lattice. The radius of the helical lattice is not specified because one does not change the helical symmetry by changing the radius. The set of symmetry operators describe equivalent positions at any radius in a helically symmetric structure.

Helical lattices with rotational symmetry require an additional operation to specify every point

What happens if we double the circumferential vector so that it extends to the points $-22, 2$? Clearly we generate a tube twice the diameter, and ϕ_0 is cut in half to $16.2295°$, but P_0, the axial rise per subunit, remains the same. The helical lattice line along **a**, however, passes through only half of all points. To generate the other half, we include an axial twofold rotation, which relates the half generated by the helical or screw operation to the remaining half of the points. Thus, to generate this helical lattice we first generate a set of points having a rotation of $16.2295°$ and an axial rise of $0.0905a$

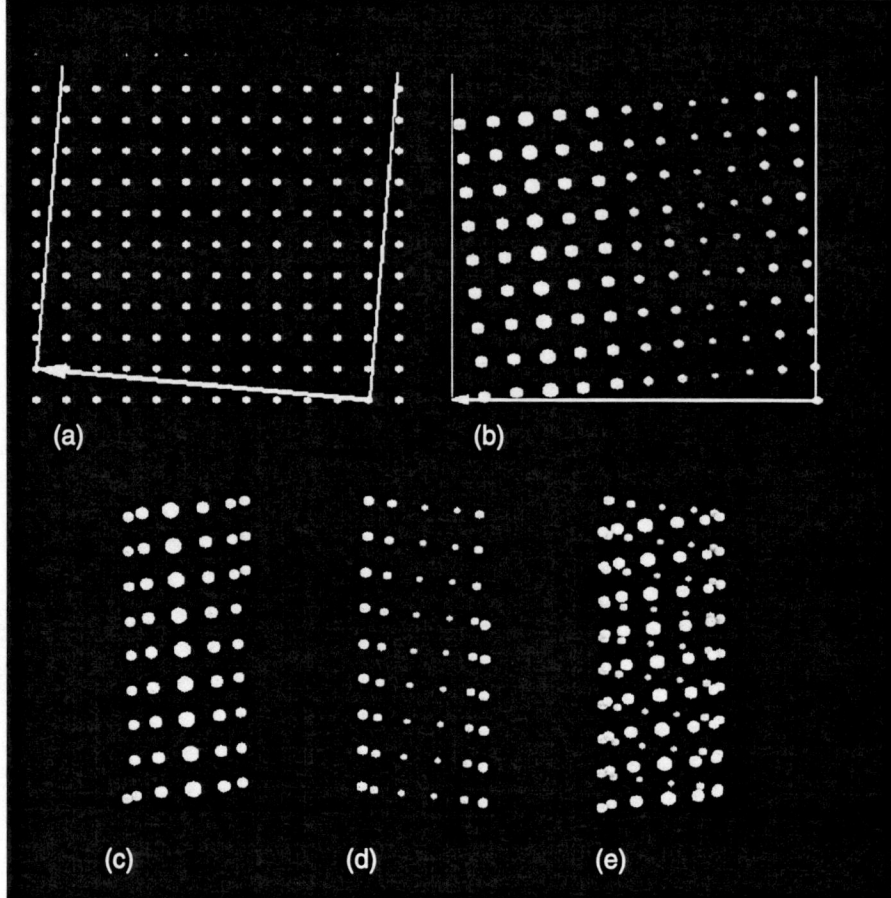

Figure 12.1 Generation of a helical lattice. (a) A square lattice. We can generate families of helical lattices from this 2-D lattice. Each unique member of the family corresponds to a unique circumferential vector. Such vectors must begin and end on a lattice point. The arrow represents one such circumferential vector. In order to generate the unique helical lattice corresponding to that vector, we draw two lines perpendicular to the vector and passing through the lattice points at each end. (b) A strip of the 2-D lattice is cut out along the two lines perpendicular to the circumferential vector. The figure shows a portion of the cutout strip. We then roll the strip into a cylinder and join the two cut edges to generate the helical lattice. The lattice points are shown in different sizes indicating those points that will be closest to the viewer when the strip is rolled up into the helical lattice. (c) The front half of the helical lattice, that is, the portion closest to the viewer. The different sizes of the lattice points provide a depth cue. (d) The back half of the helical lattice. Again the sizes of the lattice points are used to provide a depth cue. (e) The complete helical lattice. Again the size of the lattice points is used to indicate depth.

and then generate the other half simply by adding 180° to every angle in the first set. If we had used the circumferential vector −33, 3, we would have generated a threefold axis of symmetry. The number of lattice points having the same axial position but different angular positions is equal to the rotational symmetry of the helix about its

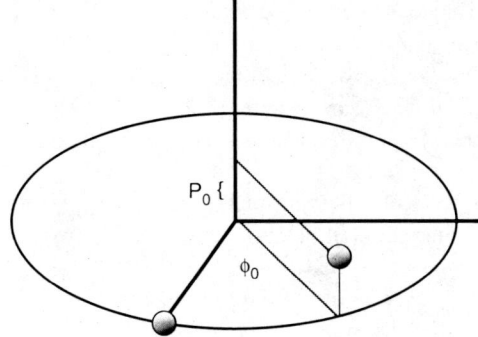

Figure 12.2 A screw operation that generates the same lattice as that shown in figure 12.1. Each point in the lattice is generated by a rotation of ϕ_0 and a translation by P_0.

axis. The tail of the T4 phage is an example of a structure with a sixfold rotational symmetry. It consists of rings of six subunits that are stacked one upon the other but with a rotation of about 100° between adjacent rings (Moody, 1967b).

There is another symmetry operation that all helical lattices (but not all helical structures) possess. It is a twofold axis perpendicular to the helix axis. DNA is an example of a structure that has such a perpendicular twofold axis. The two ends of such a structure are equivalent. For structures lacking a *perpendicular twofold axis*, the two ends are necessarily different in structure. Actin lacks a perpendicular twofold axis. It is therefore a *polar structure*, and the two ends have different properties (Huxley, 1967). For example, one end, the plus end, grows more rapidly than the other upon the addition of actin monomers to a solution of filaments (Pollard, 1986).

In summary, there are three helical symmetry operations: the first is the screw operation characterized by a combined rotation about and a translation along the helix axis; the second is an n-fold rotation about the helix axis; and the third is a twofold rotation about an axis perpendicular to the helix axis.

The Fourier transform of a helical lattice is related to the Fourier transform of the 2-D lattice from which it was derived

The relationship between helical and 2-D lattices leads to a special relationship between their Fourier transforms. The easiest way to see why there is such a relationship is to consider the projection image of the front or near side of a helix (i.e., the side facing toward the observer). The central part of the near side (figure 12.3(d)) looks almost exactly like a vertical strip of the 2-D lattice from which it was derived (figure 12.3(a)). The Fourier transform (figure 12.3(j)) of this part of the helix should and does look similar to the Fourier transform (figure 12.3(g)) of the 2-D lattice. As one moves left or right of the center of the near side, however, there is a progressive foreshortening of the lattice spacing horizontally caused by the curvature; the vertical spacings are unchanged (compare figure 12.3(a) with figure 12.3(d)). The Fourier transform of the foreshortened parts will be stretched horizontally but will be unchanged vertically. Thus, where the Fourier transform of the 2-D lattice has sharp reflections (figure 12.3(g)), the Fourier transform of the near side of the helix has horizontal streaks (figure 12.3(j)), the streaks being caused by the horizontal foreshortening. We now add in the far side (figure 12.3(e)).

310   ELECTRON CRYSTALLOGRAPHY OF BIOLOGICAL MACROMOLECULES

The near side looks essentially like the far side but rotated by 180° about the helix axis (i.e., flipped over). The Fourier transform of the far side, then, should look like the Fourier transform of the near side but flipped about the vertical axis, which is known as the *meridian* (figure 12.3(k)). To obtain the complete helical lattice and its Fourier transform, we add the near and far sides (figure 12.3(c)), and therefore we add their Fourier transforms (figure 12.3(l)). The 2-D transform of the projection of a helix is effectively the sum of the transform of one side and that transform flipped by 180° (figure 12.3(i)). Thus every helix has two independent data sets: one for the near side and one for the far side.

This way of building up the transform of a helical lattice shows how the curvature of the helical lattice gives rise to the horizontally streaked reflections called *layer lines*. It also makes it evident why the distribution of amplitudes of the Fourier transform of a helix has a vertical mirror line of symmetry, with the near side of the particle contributing predominantly to one half of each layer line and the far side to the mirror-symmetric half. Note that on some layer lines, the near side gives rise to the left half of the layer line whereas on the remainder it generates the right half, but more will be said about this in section 12.4.

In descriptions of the Fourier transforms of helices, one often encounters the terms *meridian* and *equator*. The meridian is a line through the origin of the Fourier transform and parallel to the helix axis. Thus, if the helix axis is vertical, then the meridian is also vertical. The equator is the plane through the origin of the 3-D Fourier transform and perpendicular to the meridian. In 2-D central sections of the 3-D transform (such as one finds in transforms of electron micrographs or projections of helical objects), the equator appears as a line rather than a plane. In the Fourier transforms shown in figure 12.3, only the top half of each transform is shown since the bottom half is the same as the top half. In these figures, the equator lies along the bottom edge of each

---

Figure 12.3 The relationship between the Fourier transform of the 2-D lattice and a helical lattice derived from it. (a–c) Strips of the 2-D lattice corresponding to the front, the back, and the complete helical lattice (the superposition of (a) and (b)), respectively. The difference between these figures and those in (d), (e), and (f) is that there is no foreshortening of the lattice points corresponding to the curvature of the helical lattice. (d–f) The front half of the helical lattice, the back half, and the two halves together. Note that here the lattice lines are closer together as one approaches the edge of the lattice. This is due to foreshortening that occurs because the lattice is curved. (g–i) The top halves of the Fourier transforms of the lattices shown in (a), (b), and (c), respectively. Note that in i, the transform consists of two transforms related by a vertical mirror line. Each pair of mirror-symmetry related reflections lie on a line perpendicular to the mirror line. In the transform of a helix, these pairs of reflections will become a *layer line*. The reflection on one side of the line arises from the lattice in (a) (the near side) and the other point from the lattice shown in (b) (the far side). Note that the reflections corresponding to the near side are sometimes on the left and sometimes on the right sides of the vertical mirror line. (j–l) Fourier transforms of the lattices shown in and (d), (e), and (f). Note the similarities of the transforms shown in figures (g), (h), and (i). Note that the point-like reflections in the transforms in (g), (h) and (i) turn into streaks in (j), (k), and (l). Note in (l) that there are pairs of strong reflections related by a vertical mirror line. These pairs define the layer lines. Again, one half of the layer line arises from the near side of the helical lattice and the other half from the far side.

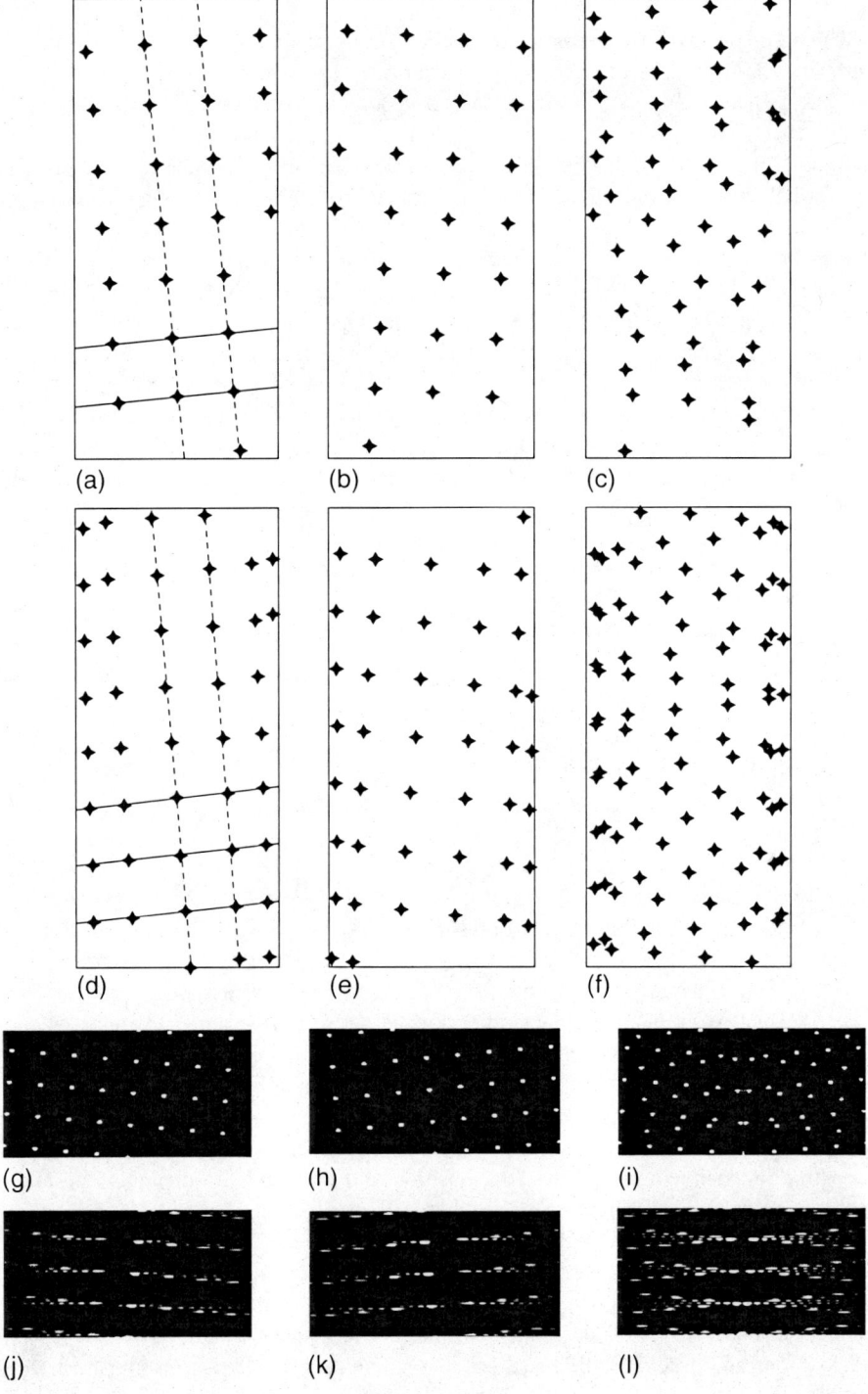

transform while the meridian corresponds to a vertical line through the middle of each figure.

The spots on the Fourier transform of the 2-D lattice can be thought of as arising from the alignment of subunits along the various sets of lattice lines. For example, the spot, indicated by the arrow in figure 12.3(g), and its missing centrosymmetrically related mate, which lies in the missing bottom half of the transform, arise from the alignment of subunits along the set of solid lattice lines in figure 12.3(a). The same is true for helical lattice lines; that is, each pair of centrosymmetrically related layer lines can be associated with a set of helical lattice lines (figure 12.3(d)). On the front half of the particle, this set of lattice lines appears to slope up and to the right. The slope of these lattice lines by itself will generate the upper left and lower right halves of the centrosymmetrically related half-layer lines; the upper one is marked with an arrow. The back side generates the upper right-hand and lower left-hand sides. This is true for all right-handed helical lattice lines. The reverse is true for left-handed helical lattice lines. For example, the set of lattice lines (see dotted lines in figures 12.3(a) and (d)) that are nearly vertical, generate a near equatorial layer line in which the near side generates the upper right-hand and lower left-hand halves of the centrosymmetrically related layer lines.

Each layer line has an order, n, and an axial position, Z, which are related to the number and pitch of a set of lattice lines

The height of the layer line is inversely proportional to the axial distance between the lattice lines. The single (or *one-start*) lattice line (set of solid lines in figure 12.3(d)) has a *pitch* of $1.0041a$. The corresponding layer line will be found at a spacing of $Z = 1/(1.0041a)$. The set of eleven (or eleven-start) nearly vertical lattice lines (set of dotted lines in figure 12.3(d)) have a pitch of $v_c/(11 \cdot \tan \theta) = 11.0454a$; the notation v_c and θ used here is explained in box 12.1. The layer line will appear at a spacing of $Z = 1/(11.0454a)$, about one-eleventh the spacing of the layer line corresponding to the one-start lattice line. We would then refer to these two layer lines as the first and the eleventh layer lines based upon their relative axial positions in the Fourier transform.

In addition to axial height, each *layer line* has an *order*, n, associated with it. The order is equal to the number of members in the set of lattice lines. The eleven-start lattice lines give rise to an order $n = -11$ layer line, and the one-start set of lattice lines gives rise to an order $n = +1$. The sign of n is determined by the *hand of the helical family*. The eleven-start lines are left handed, and by convention its order is negative ($n = -11$). The one-start line is right handed, and its order is positive ($n = +1$). Given these two independent layer lines, the position and order of all other possible layer lines can be calculated. The pairs of numbers (n, Z), if plotted, generate a 2-D lattice (figure 12.4(b)). This lattice is the reciprocal lattice for the 2-D (real space) lattice from which the helical lattice is derived, again showing the intimate relationship between helical and 2-D lattices (figure 12.4(a)). The point, then, is that having indexed two of the layer lines ($n = -11$, $Z = 1/(11.05a)$ and $n = 1$, $Z = 1/(1.004a)$), we can predict the positions and orders of all layer lines that can appear in the Fourier transform of a particle having that helical symmetry. We will use this fact in the processing of images of helical structures.

BOX 12.1 The determination of the axial translation and rotation parameters for a helical lattice

We can determine the rotation and translation that connect neighboring lattice points in our example. It is easiest to work on the unrolled helical lattice as shown in figure B12.1(a). The circumferential vector for the helical lattice is the hypotenuse, $|AC|$, of the right triangle, $\triangle ABC$, having one side, $|BC|$, of length $11|\mathbf{a}|$ and the other side, $|AB|$ of length $a = |\mathbf{a}|$. The length, $v_c = |AC|$, of the circumferential vector is:

$$v_c = \sqrt{11^2 a^2 + 1^2 a^2} = a\sqrt{122} = 11.0453a. \tag{B12.1}$$

The angle that the circumferential vector makes with the unit cell vector **a** is simply the arctangent of the ratio of the two sides of the right triangle, which has the circumferential vector as its hypotenuse:

$$\theta = \tan^{-1}\left(\frac{1a}{11a}\right) = 5.1944°. \tag{B12.2}$$

We have described the circumferential vector in terms of the square lattice. We now wish to recast the vector **a** as an angle and an axial rise. To do so we need to determine the length, v_a, of one turn of this helical path. The length of v_a is equal to the line segment $|AC|$ (figure B12.1(b)). Consider the right

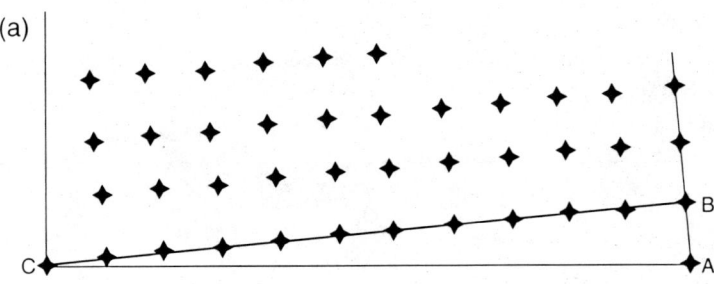

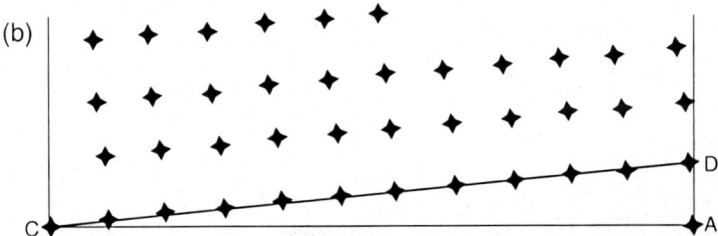

Figure B12.1 The determination of the rotation and translation of the screw operator. (See the text in box 12.1).

(continued)

BOX 12.1 *(continued)*

triangle, $\triangle ACD$. One leg, $|AC|$, is the circumferential vector, and hence $|CD|$, the hypotenuse, is equal to the length of the circumferential vector divided by $\cos\theta$:

$$v_a = \frac{v_c}{\cos(\theta)} = \frac{\sqrt{122}a}{\cos(5.1944°)} = 11.0909a. \quad (B12.3)$$

Since there is a distance $|a|$ between lattice points along the helical path, there are 11.0909 lattice points in one turn of the helical line. The helical line $|CD|$ corresponds to one turn of the helix and therefore to a 360° rotation. So the angle between lattice points is:

$$\phi_0 = \frac{360°}{11.0909} = 32.4590°. \quad (B12.4)$$

The pitch, p_a (which is the axial rise per turn), of the helical lattice line is the leg, $|AD|$, of the right triangle $\triangle ACD$ (figure B12.1(b)):

$$p_a = v_c \tan(\theta) = a\sqrt{122}\tan(5.1944°) = 1.0041a. \quad (B12.5)$$

The axial rise, P_0, per lattice point is the total rise, p_a, divided by the number of lattice points per turn (11.0909):

$$P_0 = \frac{1.0041a}{11.0909} = 0.0905a. \quad (B12.6)$$

Thus if we put down a set of points rotating about the axis 32.4590° in a counterclockwise sense (looking down the helix axis) and translating by 0.0905a (up the helix axis), we will generate the exact same helical tube that is produced by rolling up the square net about the circumferential vector $-11, 1$.

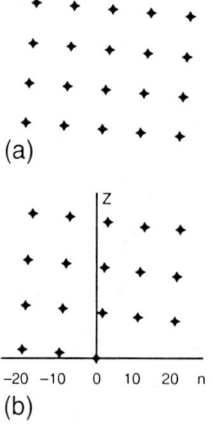

Figure 12.4 The relationship between the reciprocal lattice and the n, Z plot for a helical lattice. The n, Z plot is also referred to as an n, l plot, where l is an integer that identifies the layer line at the position Z. (a) The reciprocal lattice of the 2-D lattice. (b) The n, Z plot of the helical lattice. The n, Z plot is completely specified by two vectors in the same way as is the reciprocal lattice in (a).

Determining the *indexing* of each *layer line* is a prerequisite for implementing the helical method of image analysis. It is, in that respect, the equivalent of indexing the Fourier transform of a 2-D lattice. It is a trivial matter to determine the value of Z for each layer line but it can be difficult and frustrating to determine the values of n. We will take up this point in section 12.4.

The Fourier transform of a 2-D helical structure

The intimate relationship between the Fourier transform of a helical object and the 2-D crystal extends even deeper (DeRosier et al., 1999a; DeRosier et al., 1999b; Klug et al., 1958). Imagine that we had a truly 2-D crystal made up of point "atoms." When this structure is rolled up into a helix, all the atoms have the same radius. (Of course in a real 2-D crystal such as the purple membrane, atoms do not all lie in the same plane.) There is a direct relationship between the Fourier transform of the 2-D sheet of point "atoms" and the same sheet rolled up into a helix. If $F_{n,Z}$ corresponds to the amplitude and phase for a reflection at the coordinates n, Z of the Fourier transform of the 2-D sheet, then for the helical structure, the corresponding amplitudes and phases along the layer line are given by:

$$F(R, Z) = F_{n,Z} J_n (2\pi R r) \qquad (12.1)$$

The variable R is the distance out along the layer line from the meridian; r is the radius of the cylindrical surface that contains all the atoms; and J_n is a *Bessel function* of order n. Note that n is the order of the layer line. Equation 12.1 shows that the amplitudes and phases of reflections from the Fourier transform of a 2-D sheet are preserved in the Fourier transform of the related helix but the reflection is stretched out along the layer line by the curvature of the helical lattice. To help the reader visualize Bessel functions, figure 12.5(a) shows some plots of Bessel functions of different orders as a function of argument.

For 3-D structures, the situation is only slightly more complicated. Imagine a structure consisting of two planes of atoms rather than one plane. When the structure is rolled into a helix, one set of atoms lies at an inner radius, r_1 while the other set of atoms, of course, lies at an outer radius, r_2. The distance between the two radii is the distance between the two planes in the starting structure. The Fourier transforms of this pair of structures is given by:

$$F_{n,Z} = F_{1,n,Z} + F_{2,n,Z} \qquad (12.2)$$

$$F(R, Z) = F_{1,n,Z} J_n (2\pi R r_1) + F_{2,n,Z} (2\pi R r_2) \qquad (12.3)$$

Equation 12.2 shows the amplitude expected for the 2-D crystal and equation 12.3 that for the helix. In the case of the 2-D crystal seen in projection (as is the case in the electron microscope), the diffracted amplitudes from the two layers add but remain as a single spot, and all the information about the distance between the two planes is lost. The lost information can only be recovered by carrying out tilts in the microscope. In the case of the helix, however, the diffracted beams from the two planes, although they overlap, are modulated by Bessel functions of the same order, n, but of different arguments ($2\pi R r_1$ and $2\pi R r_2$). Bessel functions of the same order but of different arguments are orthogonal. The mathematical property of *orthogonality* means that we

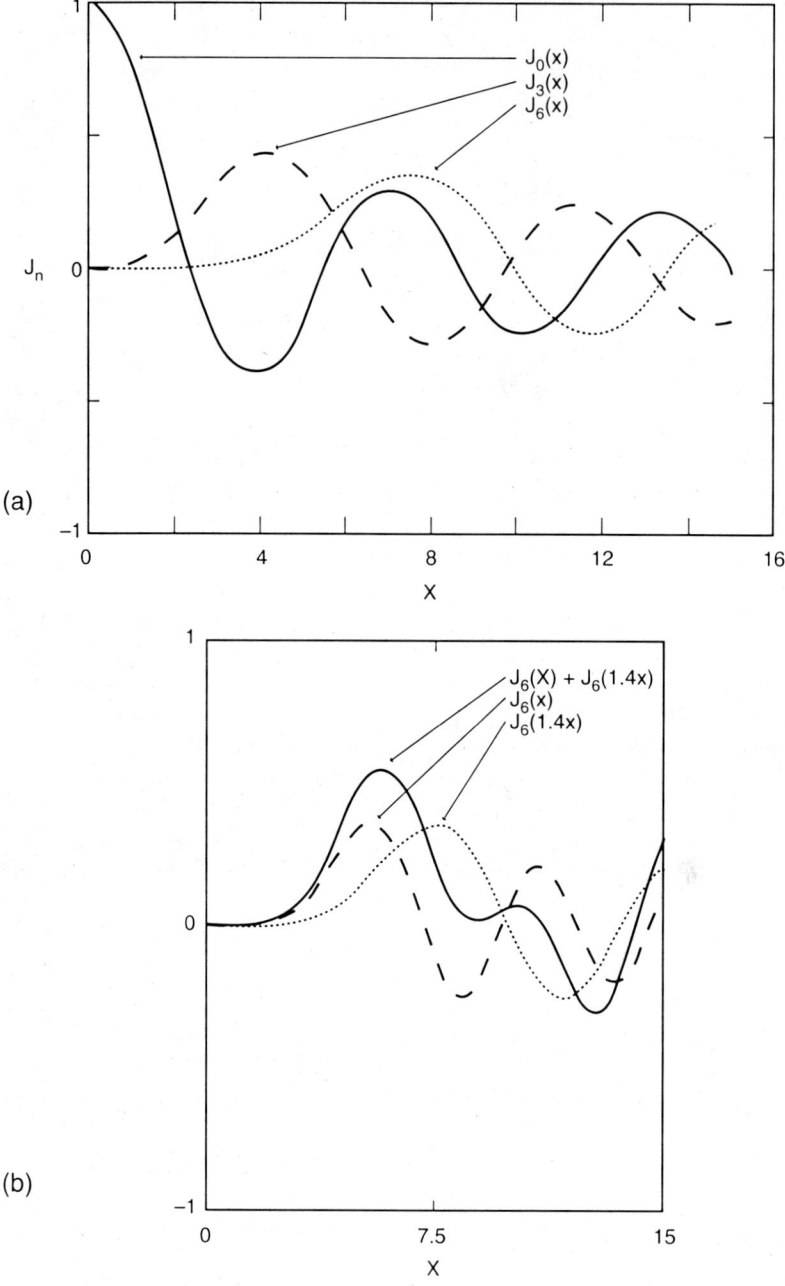

Figure 12.5 Bessel functions. (a) Plots of the Bessel functions of order 0, 3, and 6. The Bessel function of order 0 has its first peak at $x = 0$. The secondary peaks that ensue become smaller in amplitude as x increases. The Bessel function of order 3 has its first peak at $x = 4.2$ while that of order 6 has its first peak at $x = 7.5$. (b) Plots of two Bessel functions of the same order (6) but with different arguments and a plot of the sum of these two Bessel functions. These curves apply to a helical structure made from two atoms at two different radii. Each atom contributes a Bessel function of the same order but different arguments to each layer line. Here we look at a layer line of order 6. Note that positions and heights of the peaks can be altered in the sum.

can separate out F_1 from F_2 and thereby recover the 3-D information from a single view. Figure 12.5(b) depicts the curves corresponding to the situation in equation 12.3.

The Fourier–Bessel transform

We can recast the Fourier transform in cylindrical coordinates. Here we present the general result for the Fourier transform of any object, helical or otherwise:

$$F(R, \Psi, Z) = \sum_{n=-\infty}^{\infty} e^{in(\Psi+\pi/2)} \int \left[\iint \rho(r, \phi, z) e^{-in\phi} e^{2\pi i zZ} d\phi dz \right] J_n(2\pi Rr) \, rdr \tag{12.4}$$

where $\rho(r, \phi, z)$ is the density distribution and $F(R, \Psi, Z)$ is the Fourier transform both expressed in cylindrical coordinates. The presence of helical symmetry imposes a *selection rule* that specifies which terms of equation 12.4 are nonzero. The selection rule is just the n, Z plot (see sections above and figure 12.4), which shows the allowed values of n and Z. Thus, for a unit cell described by the density distribution $\rho(r, \phi, z)$, one would only calculate terms in $F(R, \Psi, Z)$ for values of n and Z given by the n, Z plot. This is analogous to the transform of a 2-D crystal, which contains the transform of a single unit cell sampled by the reciprocal lattice. In this case, the sampling by the reciprocal lattice is determined by the selection rule. The same is true in the case of a helix.

The form of equation 12.4 is just an extension of equation 12.3. The term inside the square brackets is the transform of the sheet of atoms at a radius of r. The transforms at each radius are multiplied by the appropriate Bessel function (as in equation 12.1) and then integrated over all radii (as the sum in equation 12.3). Equation 12.4 describes the case in which one knows the structure and wishes to calculate its Fourier transform. What one wishes to do in practice is the inverse; that is, one wants to use the images and Fourier transforms to determine $F(R, \Psi, Z)$ and then calculate a 3-D map of the structure.

We divide the inverse operation into steps. The aim is to generate the 3-D density from functions that have the helical symmetry of the object. These functions are denoted $g_{n,Z}(r)$ (Klug et al., 1958). The density distribution within the object is then:

$$\rho(r, \phi, z) = \sum_{n=-\infty}^{\infty} g_{n,Z}(r) e^{in\phi} e^{-2\pi i zZ} \tag{12.5}$$

where the only allowed values of n and Z are those points in the n, Z plot.

To determine the values for $g_{n,Z}(r)$, we define a quantity $G_{n,Z}(R)$, which serves as a convenient bridge between $F(R, \Psi, Z)$ and the desired terms of $g_{n,Z}(r)$:

$$G_{n,Z}(R) = \int \left[\iint \rho(r, \phi, z) e^{-in\phi} e^{2\pi i zZ} d\phi dz \right] J_n(2\pi Rr) \, rdr \tag{12.6}$$

The relationship to F, using equation 12.4, is simply:

$$F(R, \Psi, Z) = \sum_{n} e^{in(\Psi+\pi/2)} G_{n,Z}(R) \tag{12.7}$$

where again the only allowed values of n and Z are those points in the n, Z plot. If the helical symmetry is such that there is only one value of n for each value of Z, then $G_{n,Z}(R)$ is directly obtained from F, the measured Fourier transform. In the case that there is more than one value of n for each value of Z (a situation termed *layer line overlap*), then one needs at least one independent view for every overlap. If there are two overlapping orders and one is odd and the other even, then the near and far sides correspond to the two independent views; if the orders, however, are both even or both odd, then an additional image corresponding to a different view of the structure is needed. If every layer line has three overlapping orders, then one needs three independent views to extract the data for each order, and so on. Equation 12.7 represents a set of linear equations, which can be easily solved for the values of $G_{n,Z}(R)$ (Crowther et al., 1985). Generally, helical objects orient about their long axis randomly so that obtaining the required views does not require tilting but merely selecting the appropriate views from a field of particles.

Having obtained the required set of $G_{n,Z}(R)$, we can directly generate the set of $g_{n,Z}(r)$:

$$g_{n,Z}(r) = \int G_{n,Z}(R) J_n(2\pi R r) \, 2\pi R dR \qquad (12.8)$$

With the set of $g_{n,Z}(r)$, we use equation 12.5 to produce a 3-D map of the helical object.

12.3 Real helices and their diffraction patterns

If the good news is that a single image of a helical particle can contain a complete 3-D data set (assuming no layer line overlap), the bad news is that helical objects produce much weaker values of $F(R, \Psi, Z)$ than do 2-D crystals. As a result, the signal-to-noise problem is more severe. The weakness of the Fourier transform of the helical structure relative to the 2-D crystal can be determined easily from equation 12.1. For every reflection, the ratio of the amplitude from the 2-D reflection to that from the helical layer line is simply J_n. If one can see reflections to, say, 0.3 nm in purple membrane, what would happen if we examined the same structure rolled into a tube? The most favorable ratio occurs at the maximum of each Bessel function. Figure 12.6 shows a plot of the peak height, for the first maximum, of J_n as a function of n. If we assume that the level of noise is comparable for helices and 2-D crystals, then we would expect that the amplitudes calculated from helices would be much weaker depending on the order n of the Bessel function. For a structure that requires layer lines having orders as high as $n = 80$, the amplitudes for the highest order layer lines would be over six times weaker. Thus, if one needs as many as 1000 unit cells to obtain 1 nm resolution for a 2-D crystal, one would need more than 6000 unit cells with the same subunit in a helical lattice.

Real helical structures deviate from ideality. The most obvious deviation of a helical structure from its ideal geometry is curvature of its helical axis. Curvature need not only be the consequence of mishandling preparations, it is also driven by Brownian forces in solution: a solution of filamentous actin contains rods that tumble and drift but also flex. When one puts a helical specimen down onto a microscope grid, the curvature of the filaments is evident in the images and is related to that in solution.

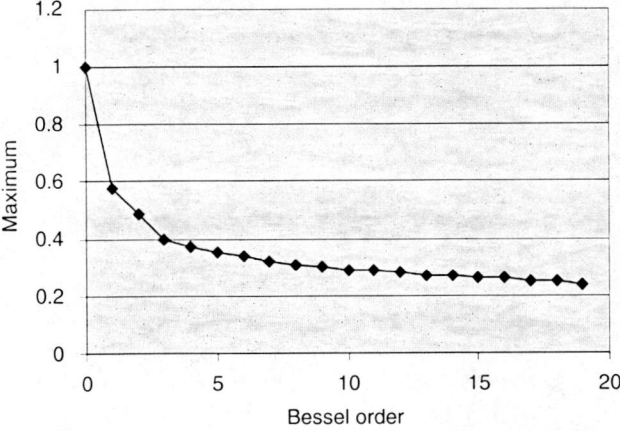

Figure 12.6 A plot of the maximum amplitude of the Bessel function as a function of its order. A Bessel function of order 0 has a maximum value of 1.0 (when its argument is zero). The higher the order of the Bessel function, the lower the amplitude of it first (largest) peak.

The stiffer the filament, of course, the less the curvature that will be present. In analyzing the structure of filaments, one must either select only those filaments that are perfectly straight, as was done initially for the acetylcholine receptor, or unbend them.

For small amounts of curvature, Egelman (1986) proposed to approximate the structural changes by the rules governing the bending of thin beams. In this approximation, the structural features at the outside of the bend are stretched while those on the inside of the bend are compressed. Egelman used a spline to parameterize the curvature of the axis and an interpolation routine to straighten images. He showed that straightening increases the layer line amplitudes, and hence the signal-to-noise ratio and resolution, relative to those from unstraightened images.

It might be, however, that data from straightened images are not as good as data from images of particles that were straight to begin with. Morgan and DeRosier (1997) investigated this issue for the bacterial flagellar filament. In this study, they straightened the images of all the filaments and generated an averaged data set. They then compared the average to each individual, straightened image to see whether agreement between the average and the individual filaments was correlated with the initial amount of curvature. They showed that there is only a weak correlation between the initial amount of curvature and the phase residual (a measure of agreement as described in equation 12.18). Thus, from this limited testing, straightened particles are almost indistinguishable from straight particles.

Helical structures can also be bent out of the plane of view such that even though they appear straight in the image (i.e., in projection), they are still curved. Beroukhim and Unwin (1997) have shown that it is important to correct for such curvature.

Disordering of subunits represents another deviation from ideality. The intimate relationship between 2-D crystals and helical structures carries over into disorder. One can imagine characterizing helical disorder by simply unrolling the helix into a flat sheet and comparing the result to types of disorder seen in 2-D crystals. *Disorder of the first kind*, also called *short-range disorder* (figures 12.7(h), (i), and (j)), is a term

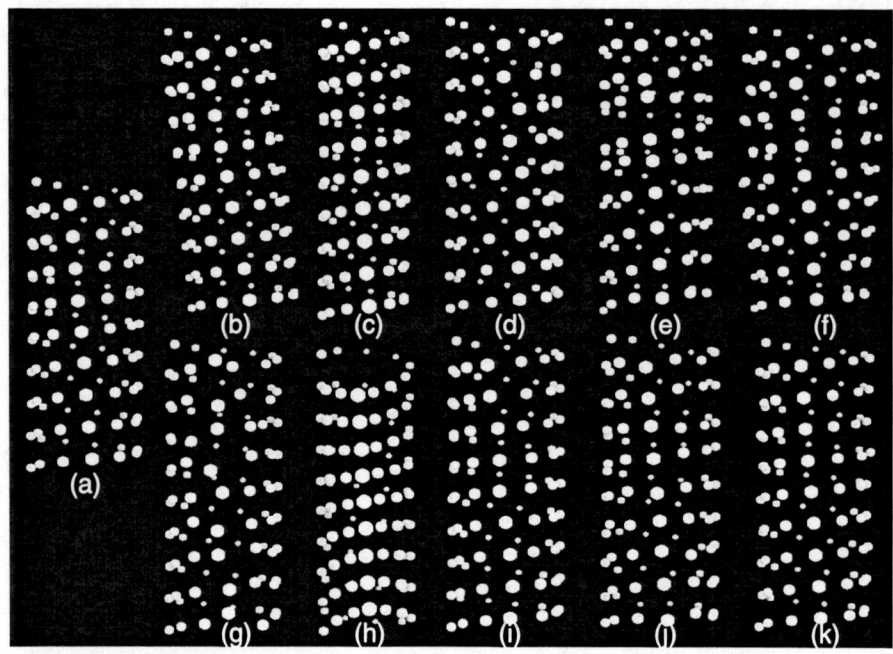

Figure 12.7 Distorted and disordered helices and their Fourier transforms. *Images:* (a) An ideal helix. (b) A helix with shear distortion. The distortion shifts the subunits progressively in the lateral position. (c) A helix with lateral shrinkage. The distortion makes the structure appear too thin. (d) A helix with cumulative angular disorder. The angular position of each subunit varies from the ideal position relative to its immediate neighbors. As one moves along the helix, the absolute deviation from an ideal helix becomes progressively larger. (e) A helix with cumulative axial disorder. The axial position (or height) of a subunit relative to its neighbors varies. Its absolute axial position of each subunit deviates more from the ideal as one moves along the helix. (f) A helix with cumulative radial disorder. The idea is the same as in the preceding two cases but the deviation is radial rather than angular or axial. (g) A helix with a disordered near side. Such a case might arise if the near side interacts with a surface such as a carbon film or the air-water interface. (h) A helix with a shrunken near side. This is sometimes seen in negatively stained samples in which the surface of stain away from the carbon support film shrinks relative to the layer next to the carbon film. The idea is that the shrinkage is proportional to the distance from the carbon support film. (i) A helix with noncumulative angular disorder. In this form of disorder, the angular position of the subunit varies relative to its ideal position in the helix. Since the deviation is always relative to the ideal position, the deviation does *not* increase as one moves along the helix. (j) A helix with a noncumulative axial disorder. The idea is the same as that in the preceding example but the deviation is in axial position. (k) A helix with a noncumulative radial disorder. The idea is the same as that in the preceding example but the deviation is in radial position. *The Fourier transforms:* (l) The Fourier transform of the ideal helix. The arrows point to a set of layer lines that are tightly clustered. There are similar sets above and below. (m) The Fourier transform of the helix with shear distortion. Note that although the meridian is vertical, the layer lines slope from lower left to upper right and thus appear rotated counterclockwise in the plane of the transform. This case is distinguishable from a helix that is rotated counterclockwise. If the helix were rotated instead of sheared, then both the meridian and the layer lines would be rotated (and by the same amount). (n) The transform of a helix with lateral shrinkage. The effect is to stretch the transform in the lateral direction. Note that the strong maxima near the meridian are more widely separated here than are those in the transform of the ideal helix. (o) The Fourier transform of a helix with

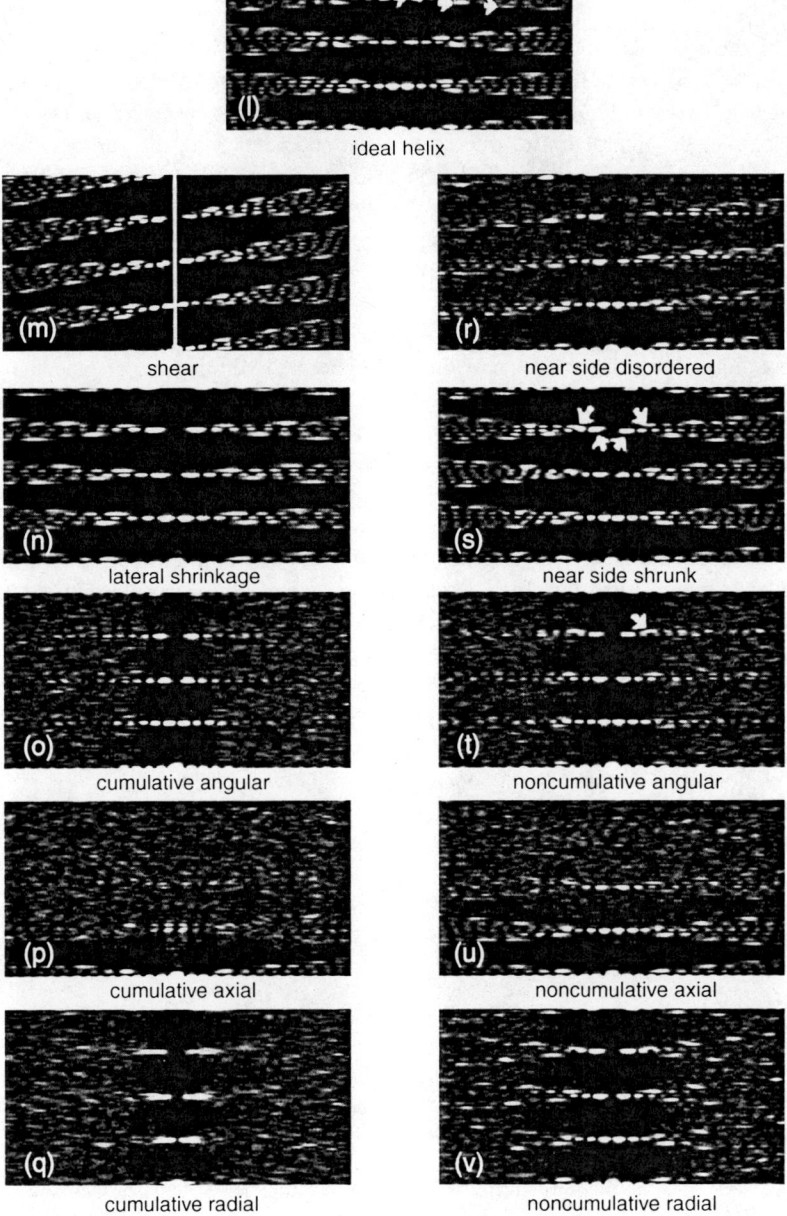

Figure 12.7 *(continued)* cumulative angular disorder. In this form of disorder, layer lines with higher orders, n, become progressively broader and weaker. Hence the layer lines with peaks near the meridian, which are low order layer lines, are stronger and sharper than those with peaks further from the meridian. Note that this applies to the subsidiary peaks of the layer lines as well as the main peak. The order of the layer line applies to these subsidiary peaks as well as the main peak. (p) The Fourier transform of a helix with cumulative axial disorder. In this form of disorder, the layer lines at larger axial spacing become progressively weaker. Thus while the layer lines near the equator look nearly perfect (including the subsidiary maxima away from the meridian), the layer lines far from the equator are unrecognizable.

that applies to noncumulative displacements of matter from its lattice positions. In 2-D crystals, these displacements result in a loss of amplitude without any broadening of the sharp reflections. In helical structures, the corresponding displacements result in a loss of amplitude without a widening of the layer lines (figures 12.7(r), (s), and (t)). For example, the interaction of one side of the specimen with the carbon support or with an air–water interface can result in a local loss of order (figure 12.7(f)). This disordering will weaken those halves of the layer lines arising from the disordered side of the particle (figure 12.7(p)). Choosing specimens over holes (figure 12.8) eliminates interaction with the carbon film, but the particle might instead interact with the meniscus of the fluid layer.

Disorder of the second kind (*long-range disorder*) is cumulative. Displacements of the subunits perturb the positions of their neighbors. Given that the displacements are randomly negative and positive, the expected position the nth subunit is the same as that for a structure without disorder because the accumulated deviations can be negative or positive with equal probability. The expected root-mean-squared (r.m.s.) deviation, however, is not zero. Instead, the r.m.s. deviation increases as the square root of the subunit separation. Cumulative disorder introduces a lack of long-range coherence into a regular structure. In helical structures, it results in a loss of amplitude and a widening of the layer lines. Correction for disorder of the second kind can be an important part of extracting data from images of 2-D crystals and helices.

Cumulative angular disorder, a disorder of the second kind, is depicted in figure 12.7(c) and is seen, for example, in actin filaments (Egelman et al., 1982). This type of disorder means that the subunits of the helix are rotated axially off their ideal position and that the deviation increases with distance. Thus if the average angular deviation from the ideal screw operation is $1°$ between subunits 1 and 2, it will be $1\sqrt{100} = 10°$ between subunits 1 and 101.

Cumulative angular disorder causes a broadening of layer lines, which increases with increasing order, n (figure 12.7(m)). Another disorder of the second kind is axial disorder. In a structure with axial disorder (figure 12.7(d)), subunits within the structure are shifted axially from their ideal positions. In Fourier transforms of structures with

---

Figure 12.7 (*continued*) (q) The Fourier transform of a helix with cumulative radial disorder. Here only the peaks near the meridian appear (relatively) unaffected. Those further from the meridian appear indistinct. This change in the subsidiary maxima distinguishes this case from that of angular disorder. (r) The Fourier transform of a helix in which the near side is disordered. In this case the transform appears one-sided because it is like the transform of just the far side of the helix. (s) The Fourier transform of a helix in which the near side is shrunk relative to the far side. The layer lines appear rotated, but unlike those arising from a helix that is rotated or a helix with shear, the tilt of the layer lines is different for different layer lines. The pairs of arrows show neighboring layer lines that appear tilted in opposite directions. The apparent tilt is due to stretching of the near side transform relative to that of the far side. (t) The Fourier transform of a helix with noncumulative angular disorder. The effect is that the pattern gets weaker and noisier but the layer lines do not become diffuse. The arrows reveal higher order reflections that are lost in the transform with cumulative disorder. (u) The Fourier transform of a helix with noncumulative axial disorder. Again, the layer lines remain sharp but become weaker. (v) The Fourier transform of a helix with noncumulative radial disorder. Again the layer lines maintain a sharpness but the subsidiary peaks fall off in intensity radially.

Figure 12.8 Cartoon of a filament over a hole: a side view. The hole is represented as the space between the shaded boxes, and the filament is shown schematically by the repeated word "filament." Note how the filament is bowed in the middle. Such a bow, viewed in projection from the top, would not be apparent. Hence micrographs of filaments that appear perfectly straight may still be bowed out of the plane of view.

this kind of disorder, the larger the value of Z, the broader and hence weaker the layer line (figure 12.7(n)). Another form of disorder is radial disorder, in which there is a radial modulation of subunit position (figure 12.7(e)). The effect of this form of disorder on Fourier transforms (figure 12.7(o)) has not been well explored.

An early example of correction for angular disorder was used in the analysis of sickle cell fibers. Carragher and co-workers (Bluemke et al., 1988; Carragher et al., 1988) determined the rotational orientation of segments within a single fiber and then treated the segments as a set of different views of the unknown structure. A 3-D reconstruction was then carried out using a real-space reconstruction algorithm, which combined the different segments (views) of the structure. This is one way to treat such disorder. Beroukhim and Unwin (1997) on the other hand, similarly determined the rotation, axial shift, etc., for segments within a single helix, but they then corrected each segment for tilt, rotation, and shift, and reassembled the segments back into a distortion-corrected image. This method is analogous to that used by Henderson et al. (1986) for 2-D crystals.

In dried, embedded preparations, changes in the specimen upon drying can be a problem. The detailed changes vary from one specimen to the next but include flattening, isotropic shrinkage, shrinkage of the side of the particle away from the support film relative to the side next to the film, and shearing of the filament. The effect of these structural distortions on the Fourier transform is known in a general way, although it is not, in every case, directly apparent from the transform whether the distortion is present.

The effect of total flattening is to turn the helical structure into two 2-D crystals so that the diffraction pattern goes from that shown in figure 12.7(l) to that in figure 12.7(i). Moody (1967a) has given a thorough analysis of the intermediate stages. Unless one has images and transforms of unflattened particles, it is hard to tell if a particular particle is partially flattened. As a result, images of tilted specimens have been used to assess the degree of flattening. By measuring the change in diameter with tilt, one can measure the degree of flattening (Crepeau, 1980):

$$\text{Degree of flattening} = 1 - \frac{a}{b} \tag{12.9}$$

where a is the minor axis of the elliptical cross-section and b is the major axis. Seymour and DeRosier (1987) have analyzed the problem quantitatively to determine the accuracy with which flattening can be measured using tilts. The conclusion is that for a given degree of tilt, the measured degree of flattening is a lower bound to the true amount of flattening (i.e., the specimen could be flatter than it appears).

Shrinkage in the plane of the image affects the transform in a reciprocal manner. If the structure shrinks by 20%, the transform is stretched by 20% (figures 12.7(b) and (l)). This assumes of course that the particles deform elastically. This approximation holds for the relatively low-resolution analysis carried out on negatively stained preparations.

Shrinkage can be anisotropic. Shrinkage of the side of the specimen away from the grid is seen in some negatively stained preparations (Moody, 1971). In the case of a helix, anisotropic shrinkage of this kind will cause the side of the specimen away from the grid to become smaller relative to the side next to the grid (figure 12.7(g)). This kind of shrinkage has a distinctive effect on the transform, which was first recognized by Linda Amos (personal communication). The parts of the transform arising from that shrunken side appear stretched so that the half-layer line corresponding to the shrunken side will have a slightly larger axial spacing compared to the axial spacing of the other half. The layer lines will thus appear tilted, and the apparent tilt will be opposite for layer lines whose orders (hands) are of opposite sign. The radial positions of the peaks will also be increased (figure 12.7(q)).

Shearing is another kind of distortion. Shearing in effect causes an increasing lateral displacement with no change in axial position (figure 12.7(a)). The effect of shearing on the Fourier transform is that the meridian and equator are not at right angles (figure 12.7(k)).

There is a simple criterion by which we can judge the success of corrections to the images. The shorter the length of particle, the broader will be the layer lines in its Fourier transform. For a perfect helix, the widths of the layer lines are inversely proportional, and the amplitudes are directly proportional, to the length of the segment transformed. Both of these assertions assume that the particle is an ideal helix. The presence of disorder of the second kind (long-range disorder), for example, invalidates these rules. Therefore, one criterion for the successful correction of images for disorder is whether the correction produces a sharpening of the layer lines and/or an increase in their amplitudes.

DeRosier and Moore (1970) analyzed the effect of a tilt of the helix out of the plane of the image. Tilt of this type results in a tilting of the true meridian of the structure out of the plane of the computed Fourier transform. There is a slight shift of the axial positions of all the layer lines to larger values of Z, but a more important effect is seen in the loss of symmetry or antisymmetry of phases across the layer lines. The phases of mirror-symmetric peaks across the meridian should differ by 0°, if the order of the layer line is even, or 180°, if the order is odd. Tilt of the axis out of the plane perpendicular to the direction of the electron beam by an angle, ω, causes a deviation in phase by $\Delta\alpha_{tilt}$ from the expected value (0° or 180°) where:

$$\Delta\alpha_{tilt} = -2n \times \tan^{-1}\left(\frac{Z \sin(\omega)}{R}\right). \tag{12.10}$$

The sign of ω is taken as positive if the top of the particle is tilted away from the direction of view (i.e., away from the source of the electron beam).

Like tilt, displacement of the phase origin from the helix axis affects the phases but not the amplitudes. This displacement arises because one's initial estimate of the position of the helical axis is made by eye; the origin is taken as the center of a box placed about the particle. Displacement of the axis affects the phases differently from tilt, however, because a shift of Δx results in a phase change of $\Delta \alpha_{shift}$:

$$\Delta \alpha_{shift} = 4\pi R \Delta x. \quad (12.11)$$

In order to differentiate tilt from shift of the axis, one needs at least two pairs of reflections symmetrically placed about the apparent meridian of the transform. Given a set of symmetric pairs of reflections, one searches for a tilt and shift that minimize the differences in phases as described in section 12.5. Generally, tilts are within a few degrees and shifts within a few tenths of a nanometer of the initial estimate of the position of the helix axis.

Curvature of the Ewald sphere (or, equivalently, insufficient depth of field) also results in a loss of the symmetry or antisymmetry of phases because the meridian of the transform does not follow the curvature of the Ewald sphere. Because the wavelength of electrons is very small, this effect is generally negligible until one gets to high resolution. A rule of thumb to estimate the resolution at which this effect becomes important is approximately:

$$\text{Resolution} = \sqrt{\lambda D/1.4} \quad (12.12)$$

where λ is the wavelength of the electron beam and D is the thickness of the specimen.

For example, in the case of a structure with a diameter of 80 nm, the resolution limit is about 0.46 nm at 120 kV. However, even at lesser resolution, it may be advantageous to correct for the curvature of the Ewald sphere. The phases can be corrected relatively easily for helical structures (DeRosier, 2000).

12.4 The hardest step: indexing the diffraction pattern

To produce a 3-D reconstruction, we must determine the height, Z_l, and the order, n_l, for each layer line. The former is relatively easy and the latter more difficult.

Layer line heights, Z, obey a simple rule

All *layer line heights* can be related to each other using only two independent parameters. Take for example the diffraction pattern of actin. The pattern has strong layer lines at $Z_a = 1/(36 \text{ nm}) = 0.0278 \text{ nm}^{-1}$ and $Z_b = 1/(5.9 \text{ nm}) = 0.169 \text{ nm}^{-1}$. These two are known as the first and sixth layer lines, respectively, because the latter is about six times the former. The positions of all other layer lines are derived from the sums and differences of integral multiples of these two layer lines:

$$Z_m = h Z_a + k Z_b \quad (12.13)$$

where h and k are integers. For example, the seventh layer line ($h = 1, k = 1$) is found at $1 \cdot 0.0169 + 1 \cdot 0.00278 = 0.197 \text{ nm}^{-1}$. This is the 1/(5.1 nm) layer line. Sometimes, different linear combinations (e.g., $h = 5, k = 0$ and $h = -1, k = 1$) generate identical or almost identical layer line positions: $5 \cdot 0.00278 = 0.0139$ and $1 \cdot 0.0169 - 1 \cdot 0.00278 = 0.0141$. This is the condition of layer line overlap described by equation 12.7 and surrounding text. These two overlapping layer lines will have different orders and, in the event only one of the two is visible, the reciprocal lattice helps tell us which one we are seeing. The layer line positions should obey the rule in equation 12.13, which acts as a guide in deciding which streaks in a diffraction pattern are noise and which are true layer lines.

The diffraction pattern contains two data sets: one for the near side of the particle and one for the far

After locating the layer lines, the next step is to divide the diffraction pattern into two mirror symmetric parts (Klug and DeRosier, 1966). The transform of a helical lattice (seen in projection) is very similar to the transform of a 2-D lattice which has been mirror-symmetrized about a vertical line through the origin (see figures 12.3 and 12.4 and the accompanying text). Just as one can draw a reciprocal lattice through the transform of a 2-D lattice, one can draw a pair of mirror-related lattices through the diffraction pattern of a helical lattice. Layer lines have multiple maxima rather than a single strong reflection like a 2-D lattice. Usually, the reciprocal lattice points will lie very close to the first maxima on either side of the meridian. One lattice will pass through the maximum to the left of the meridian and one through the maximum to the right. For an ideal helical structure the two lattices will not pass exactly through all the maxima. The failure is due to the curvature of the helix, which tends to push the maxima out to higher radius in the diffraction pattern. The two lattices group the collection of half layer lines into two sets of data: the data corresponding to the near side and that to the far side of the particle. Which is near and which far depends on the hand of the helical structure.

The order, n, of each layer line obeys the same rule that governs its height, Z

If we know the heights of two layer lines and their orders, then the same integers h and k that determine the height, Z_m (equation 12.13), also determine order, n_m:

$$n_m = hn_a + kn_b \tag{12.14}$$

where n_a and n_b are the orders of the layer lines corresponding to Z_a and Z_b.

Estimates for $|n|$ can be determined for each layer line using the distance, R, from the meridian to the first peak of intensity on the layer line and the radius, r, of the outer edge of the particle. The radius of the particle can be a surprisingly hard quantity to measure accurately. It is easy to underestimate the diameter of a particle, especially if the particle is in negative stain where the temptation is to measure from the darkest part of one edge to the darkest part of the other. Generally, the true edge of the particle is farther from the axis and, if one looks at the structural features along the edge, one sees that the periodic part extends beyond the darkest point. It is better to use an estimate that is a little to the high side than to err on the low side. Some particles, like actin, show a variation in apparent width along their length. The correct radius is measured at the widest point.

Flattening of the particle also introduces errors in estimating the helix diameter. If the particle is a thin tube and is totally flattened, then the width is half the circumference rather than the diameter. As discussed in the section on distortions, one way to estimate the amount of flattening is by tilting and observing the change in particle width.

The next step is to measure the horizontal distance, R, from the meridian to the first peak on either side of each layer line. For real particles these two measurements are often not the same. One-sided flattening or differential staining, for example, can cause the two peaks to lie at slightly different radii. One can keep two sets of R, a near-side set and a far-side set, for each layer line using the two reciprocal lattices to group measurements.

An approximate value for $|n|$ is obtained for each layer line by calculating the quantity $2\pi Rr$ corresponding to $J_n(2\pi Rr) = $ max. Table 12.1 gives values of the argument where J_n has its first maximum, for various $|n|$.

The values of $|n|$ that are determined from the radius of the particle and the position of the peaks in the transform are approximate because the features that give rise to a particular layer line may not be at a radius corresponding to the outside of the particle. If the radius r used in the calculation is truly at the outside edge of the particle, then the expression $2\pi Rr$ is a maximum and hence the value given in table 12.1 represents an upper bound on $|n|$. The smaller the value of $2\pi Rr$, the more likely one is to get the correct value of n. Suppose that there is a 10% level of error. If one's calculation for n yields a value 2.9, the uncertainty due to the 10% error is about 0.3. One would be reasonably sure that $n = 3$. If, on the other hand, the calculation yields a value of 14.2, the error in this case is 1.4. The correct value for n could be from 13 to 15.

We can narrow down the uncertainty by determining if $|n|$ is odd or even. If $|n|$ is even, then the phases of the pair of mirror-symmetric peaks across the meridian should be the same. If odd, they should differ by 180° (DeRosier and Moore, 1970). In the case in the previous paragraph, if n is determined to be even, we would then be reasonably certain that $n = 14$.

The order of each layer line can be either positive or negative. Those corresponding to a left-handed helical family have a negative order, $n < 0$, and those corresponding

Table 12.1 $2\pi Rr$ when J_n has it first maximum, as a function of $|n|$

$|n|$	$2\pi Rr$
0	00.0
1	01.8
2	03.1
3	04.2
4	05.4
5	06.4
6	07.5
7	08.6
8	09.6
9	10.7
10	11.8
11	12.8
12	13.9
13	14.9
14	16.0
15	17.0
$|n|$	$\sim|n| + 2$ or $\sim 1.03|n| + 1$ for n large

to a right-handed family have $n > 0$. The hand of the corresponding helical family can be determined from electron micrographs of shadowed particles so one is looking at only one side of the structure. This method was used in determining the handedness of the flagellar filaments of *Rhizobium* (Trachtenberg et al., 1987).

Another method of hand determination is to take an image of the structure with its helical axis tilted out of the plane perpendicular to the direction of the electron beam. The effect of tilting a helix is to produce a cycloid pattern for each set of helical grooves in the particle (figure 12.9(a)). One side of the image will have a gentle scalloping (figure 12.9(b)) and the other side will be serrated (figure 12.9(c)). Assume the top of the helix is tilted away from the observer. A right-handed helix will be serrated on the right side whereas a left-handed helix will be serrated on the left. In principle, one needs only to look at the image and assign the hand. In practice, it can be difficult to see the serration, but there is a trick that uses diffraction patterns to detect differences in the right and left halves of a tilted helix, and seems to work very well.

To understand how the trick works, we must recall that a helical lattice contains sets of helical lattice lines; in the example in section 12.2, we considered a structure having a left-handed, eleven-start set of helical lattice lines and a right-handed, one-start helical lattice line, and we established that each set of helical lattice lines gives rise to a particular layer line in the diffraction pattern. If one diffracts from the right-hand half of a tilted helical structure (the top is tilted away from the viewer), one will obtain the transform of the serrated half of all the right-handed helical lattice lines. The serrated features are strong features and will give rise to strong intensity on the corresponding layer lines (figure 12.9(f)). If one diffracts from the left-hand side of the image, that same layer lines will be weaker (figure 12.9(e)). Thus, what one does in practice is to obtain a diffraction pattern of the right-hand side of the particle and one of the left-hand side. Those layer lines that are stronger on the diffraction pattern of the right-hand side compared to that of the left-hand side are right-handed and vice versa. The hand of only one set of lattice lines is needed; the hands of the rest are fixed by the helical selection rules. Examples of the use of this method to determine the hand of a helix are found in papers by Finch (1972) and by Wagenknecht et al. (1981).

Using equations 12.13 and 12.14, one can now decide which values of n are correct by trial and error. One should check various combinations of n within the range of possible values. Remember that the value of n determined from table 12.1 is an upper bound. Smaller values are allowed but not on every layer line. Suppose $n = 16$ for one layer line (as derived from measurements of R and r) but according to the values of n for other layer lines (equation 12.14), it should be 12. There may not be a discrepancy because $n = 16$ is an upper bound. It would be a problem, however, if the value obtained from the other layer lines was 20, since the diffraction maximum would come from a feature that was at a radius larger than the outermost radius of the structure. If one is lucky, there will be only one solution that fits all layer lines.

When the solution is found, one can summarize the results by plotting the points, n, Z. These will form a 2-D lattice that describes the helical lattice. Z is plotted vertically and n horizontally. One can also produce a layer line selection rule of the form $l = tn + um$ where l is the layer line number and n is order of the layer line (Cochran et al., 1952). If the structure has rotational symmetry along its helix axis, then the values of n that are allowed are multiples of the rotational symmetry

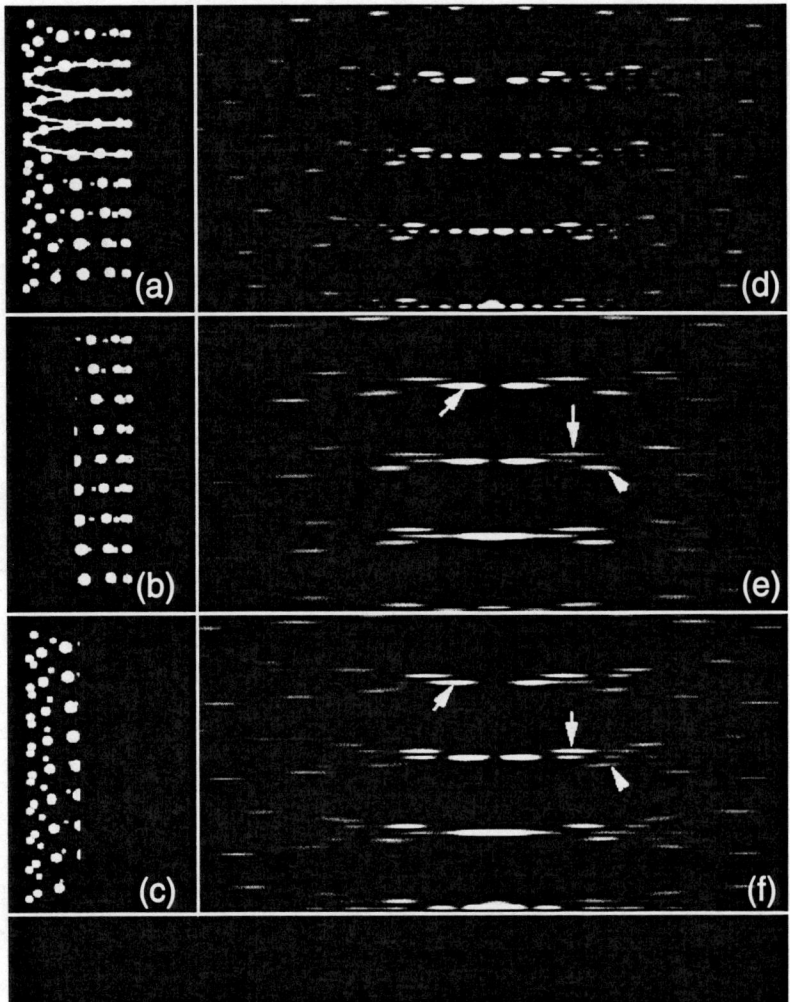

Figure 12.9 A helix tilted out of the plane of view. The upper end of the helix is tilted away from the viewer. (a) A tilted helix. The serration is accented by joining with lines, a consecutive set of some of the points making up the helix. (b) The right-hand side of a tilted helix. Note how the edge is deeply serrated. (c) The left-hand side of a tilted helix. Note that this half lacks the serration. (d) The transform of the tilted helix. The transform looks very similar to that of an untilted helix. One notable difference, compared to the transform of the untilted helix (figure 12.7(k)) is that the strong low-order layer appears to have its maximum on the meridian rather than a pair of maxima with a minimum at the meridian. This is a subtle effect. (e, f) The Fourier transforms of the right and left halves as shown in (b) and (c). Note that near meridional layer line indicated by the left-side arrow is stronger in (e) than in (f). This means that this layer line derives from a right handed set of helical lattice lines. (The assumption is that the top end of the helix is tilted away from, not toward, the viewer. The result would be reversed if either the hand or the sense of the tilt is reversed.) Note that the layer line denoted by the arrow in the middle right is stronger in (f) than in (e). This layer line corresponds to a left-handed set of helical lattice lines. The layer line marked by the arrowhead is slightly stronger in (e) than (f), indicating a corresponding right-handed family of lattice lines. As one moves away from the meridian, the difference in layerline intensity becomes smaller. This method of tilting is an excellent way to determine the absolute hand of a helical structure.

(Klug et al., 1958). The layer line selection rule is a less useful form of the information than that which is found in the n, Z plot.

12.5 Gathering amplitudes and phases is the next step in the reconstruction process

Collecting data from single particles

Before amplitudes and phases are obtained from the Fourier transform, the particle is normally masked off from its surroundings in order to maximize the signal-to-noise ratio. It is important not to mask too tightly, of course, thereby trimming off the edges of the structure. In masking off the particle from the background, however, one introduces sharp changes in density at the edge of the mask. These abrupt changes in density cause aliasing in the calculated Fourier transform and must be minimized. *Floating* (subtracting a constant density which sets the average value of density at the masked image perimeter to zero) and *apodization* (providing an extra rim around the particle in which the density values at the particle perimeter are gradually reduced to zero) both reduce edge effects (DeRosier and Moore, 1970; Stewart et al., 1981).

Prior to calculating the transform, there is an additional step that must be taken to ensure that the layer line data can be extracted at full amplitude. If the image is transformed as is, the layer lines will not necessarily lie precisely at the positions at which the transform is calculated. One solution is to re-interpolate the image on a slightly different grid so that all layer lines coincide with lines on which the transform is calculated. Smith et al. combined this strategy with their DKZ transformation technique (Smith and Aebi, 1974; Smith et al., 1976). An alternative strategy is to pad the image with a surrounding sea of zero density so that the image array is about three times the dimension of the particle. The effect of this is to sample the transform three times more finely. One can then recover the desired amplitudes by interpolation.

Usually several layer lines are obvious in the transform of an image, and one can pick out their positions by eye. This is best done using a graphics terminal and cursor. Recalling that the positions of layer lines are related, one can obtain a best fit for the positions, Z, for all layer lines. One measures the heights of all the obvious layer lines. Then one chooses two independent layer lines heights, Z_a and Z_b, to define the helical unit cell. At this point, Z_a and Z_b are the two independent unknowns in equation 12.13 and Z_m is the observed set of layer line heights. One then has a set of linear equations from which one can determine a least-squares fit for Z_a and Z_b (Owen and DeRosier, 1993).

Once the best estimate for the positions of the layer lines is known, the values for the amplitudes and phases can be extracted from the Fourier transform. In general, the layer lines will not fall on integral pixels of the transform. Assuming one oversamples the Fourier transform as described above, one can extract the amplitudes and phases by bilinear interpolation. The bilinear interpolation should be carried out on both the real and imaginary parts of the Fourier coefficients, not on the amplitudes and phases. The reason is that the phase can change abruptly as the real or the imaginary part goes through zero. The other condition that is important is that the origin used to calculate the phases should be precisely at the center of the box of the image and not, for example, some convenient point on the particle axis. The reason is that one wants to minimize the spatial variation in phase across a peak in the Fourier transform in order to get an accurate interpolation.

If one follows the procedures set out by DeRosier and Moore (1970), one next selects, on each extracted layer line, pairs of reflections that are symmetrically placed about the meridian. These pairs of reflections are used to refine the position of the helix axis and determine the tilt of the particle out of the plane. The position and tilt of the axis are those values that minimize the phase differences between the pairs of symmetrically related reflections. A range of values for tilt and shift are tried and at each trial the phases of the transform are adjusted according to equations 12.10 and 12.11. If the particle is an ideal helix and the tilt and shift are correctly determined, then the phase difference of symmetrically related reflections across the meridian is 0° if the order of the layer line is even, or 180° if the order is odd. Since many points are included in the fit and since the images have noise, one uses an *amplitude-weighted phase residual* $\langle|\Delta\alpha|_W\rangle$ to judge the best fit:

$$\langle|\Delta\alpha|_W\rangle = \frac{1}{\sum_n \sum_j |\overline{F}(R_j, Z_n)|}$$

$$\times \sum_n \sum_j |\overline{F}(R_j, Z_n)| |\alpha_{left}(R_j, Z_n) - \alpha_{right}(R_j, Z_n) - n\pi| \quad (12.15)$$

The amplitude $|\overline{F}|$ used is the average of the left and right side values for each pair of symmetrically related reflections. The phase difference should be zero for perfect helical data. The difference as calculated could also be a multiple of 2π, and thus the actual value used is modulo 2π.

The data on the layer lines are then split into near- and far-side data sets. During this operation one must be aware that what one has is values of $F(R, \Psi, Z)$ from the Fourier transform, but what one needs are values of $G(R, Z)$. One must take note that the left and right sides of the layer lines correspond to different values of Ψ, where R, Ψ, and Z are the cylindrical coordinates of the Fourier transform. If one defines the upper left quadrant of the Fourier transform to be the half-plane $\Psi = -\pi/2$, then according to equation 12.7, $G = F$ in this quadrant. Since the upper right quadrant is related to the upper left quadrant by a rotation of 180° about the meridian, we know that $\Psi = +\pi/2$ and therefore $G = Fe^{in\pi}$ in this quadrant. For odd order layer lines in the upper right quadrant $G = -F$, and for even order layer lines $G = F$. From the two upper quadrants we can thus extract two independent sets of G, one for the near side of the particle and one for the far.

Averaging data sets from many particles having identical selection rules

Each data set from each image could now be used to generate a separate 3-D map, but what is usually done at this stage is to align and average data sets from many images. The alignment consists of adjusting the rotation and axial position of each particle with respect to a reference particle, which may be an average obtained from the previous round of averaging. The criterion for correct alignment is minimization of the phase residual between the particle and the reference. A rotation of $\Delta\phi$ and an axial shift of Δz produce a change in the phases but not the amplitudes of the transform of the particle. If $G'_{n,Z}(R)$ is the transform of the rotated and translated particle and $G_{n,z}(R)$ is the starting value, then the two are related as follows:

$$G'_{n,Z}(R, Z) = G_{n,Z}(R, Z)\, e^{-in\Delta\phi + 2\pi i \Delta_z Z} \tag{12.16}$$

If the structure is polar then both the up and down orientations for the alignment must be tried. The transform, G, of a polar helix in the "up" direction is simply the complex conjugate (G^*) of the transform of the same helix in the "down" direction. To take the complex conjugate, one simply replaces the phase, α, of $G_{n,Z}(R)$ by its negative value, $-\alpha$:

$$G^*_{n,Z}(R, Z) = |G_{n,Z}(R, Z)|\, e^{-i\alpha} \tag{12.17}$$

The search for the best fit is again done by trying values of rotation, shift, and polarity looking for the lowest possible phase residual:

$$\sqrt{\langle \Delta\alpha_W^2 \rangle} = \sqrt{\frac{1}{\sum_n \sum_j |G_{n,Z}(R_j, Z_n)|} \sum_n \sum_j |G_{n,Z}(R_j, Z_n)|\, |\alpha(R_j, Z_n) - \bar{\alpha}(R_j, Z_n)|^2} \tag{12.18}$$

The formula used in equation 12.18 is different from that in equation 12.15 in that the phase differences enter as the square. The justification for doing this is to increase the contribution of the weaker, higher resolution terms, which have smaller amplitudes. The higher resolution terms should in principle produce a more accurate alignment because their phases are more sensitive to rotation and axial shift (Amos, 1975). The values of $|G|$ used are those of the reference particle, which are generally more reliable than those of the individual particles. The quality of each of the particles is assessed from the phase residual and the up/down difference in phase residual if the structure is polar.

Another measure of the quality of a data set is whether the rotational angle and the axial shift that align the near side to the average are the same as those obtained for the far side. In addition, the polarity of the two sides must be the same if the structure is polar. Acceptable ranges of values for phase residual depend on the structure and on how much of the data are included. If only the very strongest values of G are used, then the residuals will be lower than those in which weaker terms are included. Other sources of variation are the degree of defocus, the coherence and wavelength of the electron beam, the stability of the electron microscope stage, the resolution of the data used, the degree of corrugation in the surface of the structure, the thickness of the layer in which the specimen is embedded, and the kind of embedment.

An important change from the method of DeRosier and Moore (1970) is to do the alignment using the entire layer line rather than aligning the near- and far-side data sets. One can both determine tilt out of the plane and locate the helix axis in the same step. By so doing, one is using twice as much data, which increases the signal-to-noise ratio. Moreover, the far-side data of the image are compared only with the far-side data of the reference and the near-side data only to the near-side data of the reference. This alternative procedure gives a more reliable fit. Minimization of the phase residual as given in equation 12.18 provides a good criterion for alignment. During the rounds of alignment and averaging, the weighting of contributing layer lines can be changed to increase the contribution of the higher resolution layer lines to get a more accurate alignment.

Earlier, Morgan et al. (1995) and Unwin (1993) found that upon averaging, one can extract data that previously were hidden by noise. By using suitably prepared specimens and by careful processing of the electron micrographs, they were able to extract high-resolution data from the average that were not evident in the transforms of the individual images. This allowed the extension of resolution from about 3 nm to better than 1 nm. Further improvements have led to resolutions of 0.4 nm for the same structures (Miyazawa et al., 2003; Yonekura et al., 2003).

In merging data from the transforms of individual particles, the real parts and the imaginary parts of G are averaged separately. The variances of these two populations are equal, as expected (Morgan and DeRosier, 1997). From the variance, which is calculated as part of the averaging process, one can test each average amplitude $|\langle G \rangle|$ to see if it is significantly different from that expected for random, uncorrelated data, that is, noise. This is also equivalent to asking whether the reflection is significantly different from 0. If the standard deviation of the population of real or imaginary parts is s and if there are m images in the average, then the amplitude, $|G|$, is significant at the $p\%$ level if:

$$|G| > \frac{s\sqrt{-2\ln\left(1 - \frac{p}{100}\right)}}{\sqrt{m}}. \qquad (12.19)$$

The improvement of signal-to-noise ratio that is achieved from alignment and averaging of many images is evident in the averaging of transforms of the bacterial flagellar filaments. Figure 12.10(a) shows the transform of an image of a single ice-embedded filament, and figure 12.10(b) shows the average generated using 100 filaments. In order to assess the improvement, a number of lines of noise taken between layer lines are carried through the processing. The averaging process is able to extract signal from regions of the transform of a single particle where the noise is greater than the signal. Whereas the transform of a single filament only shows layer lines to a resolution of 2.5 nm, the average clearly shows layer lines at a resolution of about 1 nm. It is not that the individual images did not have the 1 nm data, of course, but that the signal was hidden by the noise.

If one is able to use high-resolution data, one needs to determine the contrast transfer function (CTF) because the phases will go through a shift of 180° whenever the transform crosses one of the CTF nodes. This phase change is relatively straightforward to correct for once the CTF has been determined (Erickson and Klug, 1971). The proper way to combine data from a set of images taken at different degrees of defocus includes more than just changing the phases, however; the average data set is really a weighted sum of the data sets, where the weighting takes into account the loss of

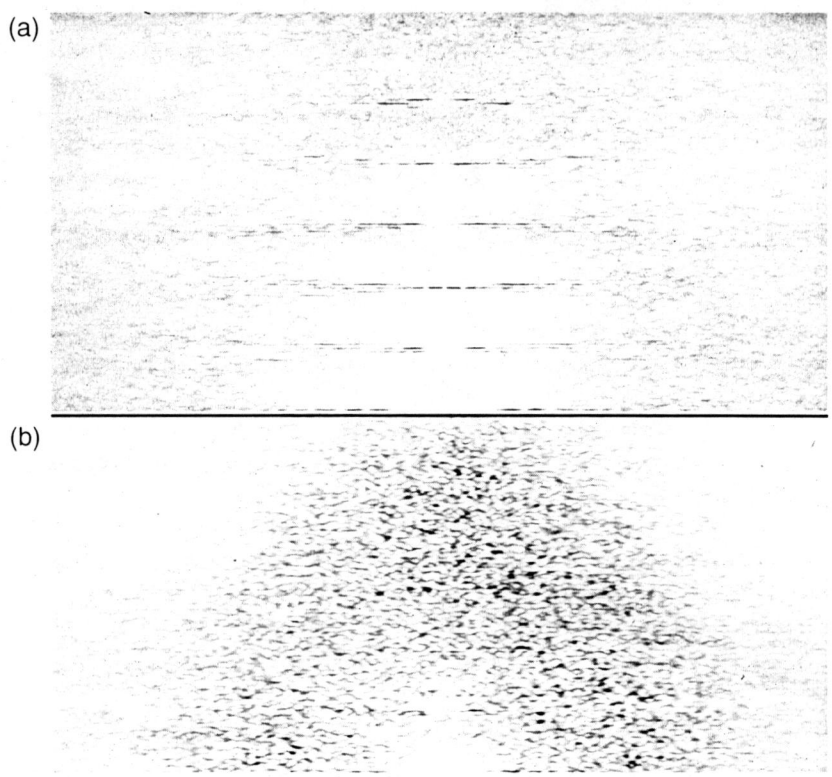

Figure 12.10 The effect of averaging data sets. (a) Fourier transform of 100 filaments averaged. (b) Fourier transform of one filament.

signal at the nodes of the CTF. The appropriate weighting term is given by:

$$W_{n,Z,j}(R) = \frac{CTF_{n,Z,j}(R)}{\sum_{k}(CTF_{n,Z,k}(R))^2 + \frac{N}{S}} \quad (12.20)$$

where $CTF_{n,Z,j}(R)$ is the value of the CTF of the jth image on the layer line n, Z at the radial position, R. N is the noise power and S is the signal power (Saxton, 1978). In practice, if one has a large enough range of defocus in the set of images, the term N/S can be ignored because $\sum_{k}[CTF_{n,Z,k}(R)]^2$ is greater than N/S. The average is then:

$$\langle G_{n,Z}(R)\rangle = \sum_{j} W_{n,Z,j}(R) G_{n,Z,j}(R). \quad (12.21)$$

If the weighting scheme is used, then one cannot compute the variance in a straightforward way. One can instead carry layer lines of background, taken next to the layer lines of data, through the averaging to provide a measure of the noise in the layer line data (Morgan et al., 1995). The local average amplitude, $\langle A \rangle$, on a line of adjacent background can be used to assess the probability that an amplitude on a real layer line is significant at the $p\%$ level:

$$\frac{|G|}{\langle A \rangle} > \sqrt{-\frac{4\ln\left(1 - \frac{p}{100}\right)}{\pi}}. \tag{12.22}$$

The values of $\langle A \rangle$ also provide an estimate of the variance for the values of the real and imaginary parts of $|G|$:

$$s^2 = \frac{2}{\pi}\langle A \rangle^2. \tag{12.23}$$

Averaging data sets from particles that have different selection rules

There are now increasing numbers of techniques by which it is possible to form synthetic, helical arrays of particles. Some membrane proteins, for example, form tubular vesicles as a result of protocols that may have been intended to produce 2-D crystalline sheets. In another example, mentioned previously in section 8.4, his-tagged proteins have been bound to the surface of tubular vesicles in which modified lipids bearing chelated nickel ion (Ni-NTA) have been incorporated as a guest molecule (Kubalek et al., 1998). It is convenient to think of these synthetic helices as being "helical crystals."

Preparations of such helical crystals often contain tubes that have slightly different circumferential vectors and hence different symmetries, a situation not usually encountered in naturally occurring helical structures. This has generated a new type of mathematical problem for merging data, however. $G_{n,Z}(R)$ obtained from structures with different symmetries cannot be directly averaged together, but there are other approaches to merging data. The first, which is the most general but also the most work, is to make a 3-D map from data for each symmetry group, cut out the subunit from the map, and average subunits. This procedure has been used in the case of the acetylcholine receptor (Miyazawa et al., 1999). A second possibility is to average together the functions, $g_{n,Z}(r)$. Consider two helical crystals having different circumferential vectors but deriving from the same 2-D crystal. The functions $g_{n,Z}(r)$ corresponding to the same lattice lines of the 2-D crystal have a special relationship even though they have different orders n and m and occur at different positions, Z and Z'. If the radius of one tube is r and that of the other tube is $r + \Delta r$, then $g_{n,Z}(r) = g_{m,Z'}(r + \Delta r)$ (DeRosier et al., 1999a,b). The details for determining m, n, and Δr are described in these references. This approach was used successfully for the sodium–potassium ATPase at 1.1 nm resolution (Rice et al., 2001). In the case of the acetylcholine receptor however, this second approach failed at about 0.5 nm resolution because of a slight rotation of the receptor relative to the lattice. Thus the first approach was used (Miyazawa et al., 1999).

Automation of data collection using helical cross-correlation

Morgan and DeRosier (1992) have proposed the use of helical cross-correlation to automate the processing of images. This method builds on earlier work by Carragher and others (Bluemke et al., 1988; Carragher et al., 1988). With these automated algorithms, all of the steps that follow the initial rough masking of the images can be done with little or no intervention. These steps include straightening and final masking of the images, determination of the exact positions of the layer lines, collection of data from the Fourier transforms, alignment of the images within the data set, rejection of unsatisfactory images, and averaging of the satisfactory images. The helical cross-correlation defined in equation 12.24 is used to locate each particle's axis and refine the position of the layer lines. Each particle is first cut into segments. Helical cross-correlation allows one to determine the angular and axial position of the segment (denoted by u in equation 12.24) of one image in the set relative to a segment (denoted by ref) of the reference image:

$$c(\omega, \xi) = \frac{2 \sum_j e^{ij\omega} e^{-2\pi i Z_j \xi} \int G_{j,u}(R) G^*_{j,ref}(R) R dR}{\sqrt{\left(\sum_n \int G_{n,u}(R) G^*_{n,u}(R) R dR\right) \left(\sum_m \int G_{m,ref}(R) G^*_{m,ref}(R) R dR\right)}}.$$

(12.24)

The best estimates of angle, ω, and axial position, ξ, are taken at the maximum of the correlation coefficient, c. This assumes that the helix axis has been located at least approximately. Its position can be refined by adjusting Δx (see equation 12.11) until c is maximized. For a perfect helix (but one that has a symmetry slightly different from that of the reference), the observed values of ω and ξ for each succeeding segment in a particle will deviate in a systematic way from the values expected from the reference. The derivations for ω and ξ can be used to determine the exact helical symmetry, which may be somewhat different from the reference particle. The helical symmetry can be used to determine the positions of all the layer lines according to equation 12.13.

For real particles, the observed values of ω and ξ will not vary perfectly systematically. The systematic component can be used to determine the average helical symmetry. The nonsystematic changes provide a measure of the axial and angular disorder present in the image.

As part of the alignment by cross correlation, it is necessary to refine the position of the helical axis of each image segment. If the image is that of a curved particle, then the position of the axis follows a curved path. The cross-correlation by segments thus provides the data necessary to correct automatically for curvature and to determine the positions of the layer lines.

Having determined the axis and knowing the diameter of the particle, one can automatically mask, float, and apodize the images. As part of this process, one has determined for each segment a correlation coefficient, an amount of curvature, and the up/down polarity, and for the particle as a whole, one has determined the difference in twist and stretch (relative to the reference) and the amount of angular and axial disorder. These parameters can be used to assess the quality of the images.

12.6 Calculating and interpreting three-dimensional maps

In papers reporting on the analysis of helical structures (Wagenknecht et al., 1981), one usually finds graphs showing the amplitude and phase distributions for the set of G's.

One often also finds graphs of the amplitudes of the g's. The g's also have phases, but the most informative parts are often just the radial positions of the peaks in $|g|$. The radius at which each $g_{n,Z}(r)$ goes to zero can also be an objective way of determining the outer radius of a structure, as was done for actin (Egelman et al., 1982).

From the values of $g_{n,l}$, the 3-D density distribution follows from equation 12.5. There is one important condition that must be met prior to starting this calculation. Since some particles have slightly different layer line positions, the values, $Z_{n,l}$ must obey the rules of a 2-D lattice. For example $Z_{m,l} + Z_{n,k} = Z_{m+n,l+k}$. If this is not enforced, the density map will not have helical symmetry and the subunit will change shape with position along the helical axis.

There are a variety of ways to view the map (figure 12.11). One way is as an x-projection in which the densities are summed along constant values of x. This projection should look like an averaged electron micrograph. The transverse, or "z," section (figure 12.11(a)), is a convenient way to see the internal features of the structure. The transverse section also preserves the symmetry of the particle. These sections are not helpful, however, if one wants to look in from the side, in which case vertical, or "x," sections (figure 12.11(b)) are useful. The problem with vertical sections is that the contours do not have the symmetry of the particle, and hence it can be hard to figure out where the subunits repeat. Radial sections (figure 12.11(d)) do preserve the symmetry, but they cannot be stacked. Cylindrical sections (figure 12.11(c)) are quite useful since they keep the symmetry of the structure while allowing one to look perpendicular to the axis. Of course, one can produce a transparent outer surface of the structure but, while this is attractive, it does not always give an indication of the internal fluctuations that may indicate the separations between domains of the structure. The two most generally useful representations to start with are z sections and transparent surface representations.

To determine the outer surface of the structure, one can set the volume (nm^3) enclosed to be equal to the subunit mass (daltons) divided by 0.00081. (This value is

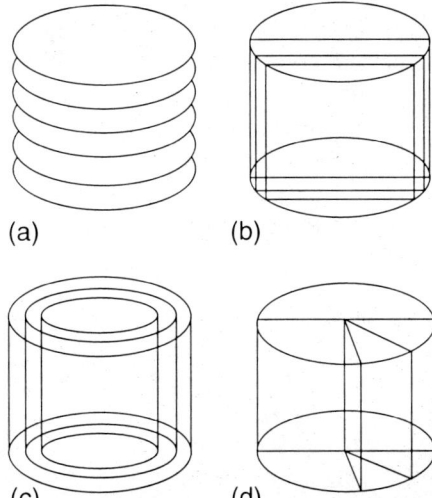

Figure 12.11 Different modes of slicing a three dimensional map in order to display it in two or three dimensions. (a) Transverse sections. (b) Axial sections. (c) Cylindrical sections. (d) Radial sections.

obtained by assuming a *partial specific volume* for proteins of 0.74 cm^3/g converted to cubic nanometers per dalton.) Even smaller volumes are often used because the surface better reveals the channels and knobs within the structure, which can be hidden by a more voluminous representation.

One wants to know if the features in the map are reliable. One possibility is to compute two independent maps and then compare the densities to determine the reliability of the features. This works, but if one combines the two maps, one actually reduces the mean squared error, that is, the variance associated with the averaged density, by a factor of 2 (assuming both maps are derived from equal numbers of equivalent images) (Milligan and Flicker, 1987).

The mean squared error in the map can be determined in other ways that do not sacrifice a factor of 2 in the signal-to-noise ratio. For example, using the methods described in section 12.5, one can determine the error in each of the coefficients **G** used to generate the map. The average of these errors is the average error in density (assuming that the Fourier transform program contains proper normalization). The error determined in this way can be used to assess the reliability of the features. One way is to plot the contours in steps of two standard errors of the mean. Thus, every contour is significantly above the contour below it and significantly less than the contour above it.

In a map, however, the errors may be different in different voxels and therefore the average error might not correctly represent the errors for each voxel. For example, in images of negatively stained specimens, the variance seems to be higher in regions of stain than in regions of protein. To estimate the error on each voxel independently one can use each image in the data set to produce a separate 3-D map. One can then use this set of 3-D maps to produce an average map and, concomitantly, a map of the variances for each voxel.

One can use the measurement of error to assess the reliability of the features within the map. Suppose one chooses ρ_0 to correspond to the outside of the particle. Those values of ρ which are significantly greater than ρ_0 are likely (in the statistical sense) to be part of the structure, while those that are significantly less than ρ_0 are likely to be outside the structure. Those densities that are not significantly different from ρ_0 could be either inside or outside the structure. The *standard t-test* can be used to assess the significance. One can then generate a t-map in which those pixels which are part of the particle are rendered white, those outside black, and those that are not significantly different from the outside contour, ρ_0, are rendered gray. Such an analysis has been used in the study of the bacterial flagellar filament (Trachtenberg and DeRosier, 1987).

In averaging images of a structure, we reduce the "noise" by averaging away variations such as those that might come from extraneous material in the background, from statistical variations in electrons, etc. "Noise" can also arise, however, because different copies of the repeating subunit are structurally different. When actin filaments were decorated with an actin-binding protein, for example, not all sites on actin were occupied by the actin-binding protein. The reduced occupancy lowered the density of the bound subunit in the averaged 3-D map. The resulting reconstruction showed actin clearly but the actin-binding protein appeared to be only a shoulder of density on the actin subunit. The partial occupancy, however, increased the density fluctuations at the site of binding because sometimes there was a protein at the binding site and sometimes not (Hanein et al., 1997). Conformational variations in subunit structure will also result in an increased variation. These variations can be mapped and can allow us to visualize conformational variability (Liu et al., 1995; Liu and Frank, 1995; Rost et al., 1998).

Difference maps are a good way to locate bound ligands. They are particularly helpful when the occupancy of the ligand is less than 100%. The reason is that in the map, the structure to which the ligand is bound appears with unit strength, whereas the component bound to it appears with a strength equal to the fraction of occupancy. In such cases, a large bound protein subunit may appear as a small shoulder on the main structure. As such it is hard to judge its size or shape. In difference maps, however, the size and shape are exhibited correctly (Hanein et al., 1997).

Prior to calculating difference maps, it is important that the maps to be compared are aligned to each other since one wants to reflect actual differences in the density not differences due to misalignment. Two maps can be aligned using the formalism in equation 12.18 in which the r.m.s. phase difference between values of $G_{n,z}(R, Z)$ is minimized. If the two structures, however, are quite dissimilar as is the case when comparing actin to actin decorated with myosin, the alignment procedure may give incorrect results. In such cases, a correct alignment can be obtained by replacing $G_{n,z}(R, Z)$ with $g_{n,z}(r, Z)$ but restricting the radius to those regions where there is only actin (Hanein and DeRosier, 1999). By so doing, one can eliminate the contribution of myosin, which is present in only one of the two structures. Both $G_{n,z}(R, Z)$ and $g_{n,z}(r, Z)$ change phases in the same way upon rotation and translation, and hence the same algorithm can be used for either kind of data set.

A t-map can also be used to determine which differences are significant. The first example of an application of the t-map was in the determination of significant features in a difference map between actin with tropomyosin and plain actin (Milligan and Flicker, 1987). Estimates of errors from the two maps determined which differences were significant. Differences due to the presence of the additional molecule should appear as significantly positive peaks of density. It may also be the case that the binding of a ligand may cause conformational changes, which may appear as pairs of significant positive and negative peaks.

12.7 Helical particles with a seam can be analyzed by extending the method for helical particles

Helical particles such as the 13-protofilament microtubule have a seam (Amos and Klug, 1974) because, in formal terms, the circumferential vector does not extend between two equivalent points in the 2-D net (see section 12.2). Other helical particles, such as the bacterial flagellar filament (Trachtenberg et al., 1998), can have a seam-like feature because of a systematic perturbation in subunit packing. In the case of the flagellar filament, subunit pairing across a five-start helical lattice line, produces a "2.5"-start helical feature, which is inconsistent with the symmetry of the helical lattice; the perturbation thereby generates a superlattice. A superlattice is, however, not a necessary consequence of all perturbations; it is conceivable that, in some cases, the perturbations vary in patches of limited coherence. The Dahlemense strain of tobacco mosaic virus (TMV) may be such a helical structure. The diffraction theory for helical structures with a perturbation such as the Dahlemense strain of TMV has been worked out in terms of ghost spectra (Caspar, 1969); the diffraction theory for helical structures with a seam has been described by Metoz and Wade (1997). Neither of these two theories has led to efficient methods for generating 3-D maps.

The method of Kikkawa (2004) generates a map of the side of the structure away from the seam. Recall that the diffraction pattern can be separated in two parts, one arising essentially from the near side of the particle and the other from the far side. With a nearly vertical seam, Kikkawa has produced a method for selecting the part of the transform coming from the side of the particle without the seam and using those data to generate a 3-D map. The analysis begins as it does for the helical method, that is, with the indexing of the layer lines. In this case, however, nonintegral values are allowed for the orders of the layer lines. In the case of the bacterial flagellar filament, one would have an order of 2.5 for the layer line arising from the pairwise perturbation. The orders and layer line heights obey the selection rule as given in equation 12.13 except that nonintegral values of h and k are allowed. The next step is to collect layer line data for the side of the particle without the seam. The key is to identify the side without the seam. To do so, near- and far-side data sets are collected and used to generate projections of the near and far sides of the particle. In these projections, the seam can be directly visualized as a break along the noninteger order families of helical features; for example, the 2.5-order helical feature of the flagellar filament will show a discontinuity on the side having the seam but will appear continuous on the side without the seam. The final step is to generate a 3-D map using equations 12.8 and 12.5 but with nonintegral values of n. Kikkawa in his paper has an elegant analysis of this seemingly intuitive extension of helical diffraction theory to nonintegral orders. He applied the method to microtubules decorated with kinesin.

12.8 Helical structures can be analyzed using single-particle methods

Egelman has devised a robust single-particle method of generating 3-D maps for helical structures. Figure 12.12 shows the flow chart for his method. Helical particles are cut into segments. The minimum usable length depends on the mass per unit length, which determines the amount of signal in the segment and therefore fixes the accuracy of alignment; the maximum usable length depends on the coherence length of the structure, that is, the length over which the conformation remains constant. The segments are each aligned against a set of 2-D reference images generated from a 3-D model. The model can be a related structure or even be a smooth cylinder. A 3-D map of the aligned particles is generated by *back projection*, and a search is then carried out for helically repeating features in the map. Various combinations of rotation about and translation along the helical axis are tried until the correlation coefficient is maximized. The rotation and translation that best reinforces detail in the map is a first estimate of the helical symmetry operator. The map is symmetrized using this estimated helical symmetry operator, and the symmetrized map is used as a reference for another round of alignment. The rounds of alignment, back projection, and symmetrization are repeated until the symmetry and structure cease to change significantly.

One might ask how such a procedure can work if one begins with a poor reference or worse, a reference that is smooth having only cylindrical symmetry. In the latter case, only the helical axes of the particles are aligned; the rotational and translational

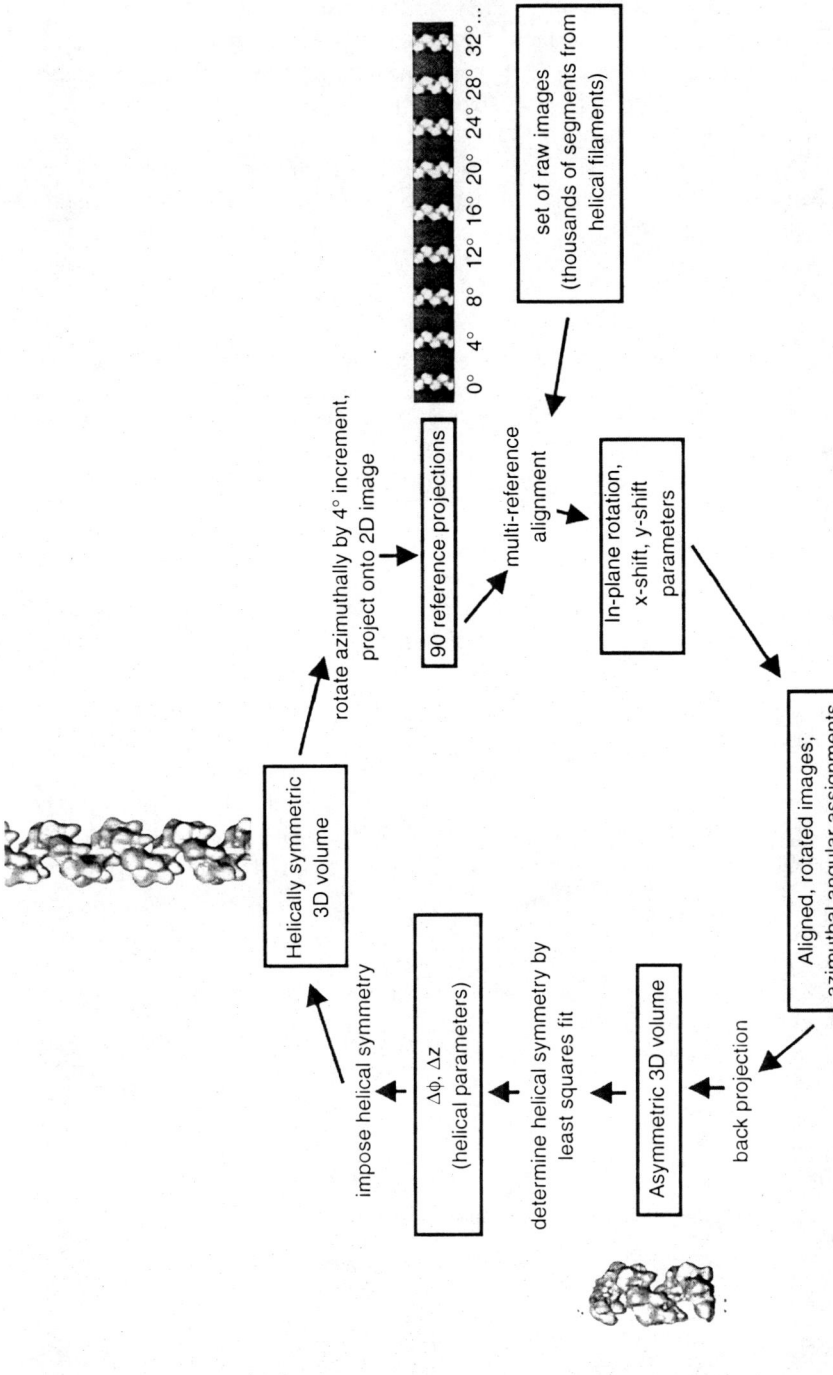

Figure 12.12 The flow chart shown in this figure diagrams the steps in the 3-D reconstruction of helical structures using the single particle approach of Egelman. The figure is reproduced with permission from Egelman (2000).

orientations will be random. Random does not mean the aligned images are uniformly distributed; to see why, consider the simple exercising of flipping a coin. Whereas the probability of heads and tails is 50% for each toss, the absolute value of the difference between the number of heads and tails is expected to increase with the number of coin tosses. In the same way there will always be an excess of particles approximately aligned. The small signal from this set of approximately aligned particles will then be the major aligning feature in the next round. As more particles are aligned to the signal in the ensuing rounds of alignment, the signal in the map will grow. The method has been applied to a number of difficult structures: pili (Mu et al., 2002), actin (Galkin et al., 2002), actin decorated with fragments of actin-binding proteins (Egelman, 2004), dynamin tubes (Chen et al., 2004), and the needles of the type III secretion system (Cordes et al., 2003).

If more than one conformation is present, one can use more than one reference. For example, after the first round of alignment, one can use, as references, both the new map and the initial reference. The images will then sort into two groups, one more like the initial reference and one more like the initial 3-D map. This process can be repeated to generate classes corresponding to different conformations.

12.9 The future looks bright

There seems to be an improvement in the quality of images recorded at higher voltages such as 300 keV electron microscopes that are equipped with field emission guns. An electron energy filter may also provide an improvement in signal-to-noise ratio by reducing the contribution of the inelastically scattered electrons to the image. None of the helical structures solved to date has made use of spot scanning (Downing, 1991), and this too may help. Thus, there is still room for improvement in the images.

The use of electron diffraction has long been important in the analysis of 2-D crystals. It has only begun to be explored for use in the analysis of structures with helical symmetry, however (Ruiz et al., 1994). Electron diffraction will be important in obtaining accurate amplitudes, especially in view of the previously discussed weakness of amplitudes for tubular helical structures.

There are several helical structures solved to better than 1 nm resolution: TMV (Jeng et al., 1989), acetylcholine receptor (Miyazawa et al., 1999), the bacterial flagellar filament (Morgan et al., 1995), and the microtubule (Li et al., 2002). Thus, it seems clear that resolutions of 1 nm or better can be an achievable goal for many different helical structures. The next level worth achieving is a 0.4 nm resolution map. At this resolution, one should be able to resolve the protein chain, since interchain distances are of the order of 0.45 nm. The recent success with the acetylcholine receptor (Miyazawa et al., 2003; Unwin, 2005) and the bacterial flagellar filament (Yonekura et al., 2003; 2005) at 0.4 nm suggests this will be possible for at least some helical structures.

13

Icosahedral Particles

13.1 Introduction

Proteins can self-assemble into highly symmetrical particles that are biologically active and functional. Many viral proteins assemble into icosahedral shells containing a viral genome to become an infectious particle (Casjens, 1997). The icosahedral shell serves as a protective layer for the viral genome and is also involved in various biological processes during the viral infection and replication cycle. The outer shell can interact with immunoresponsive molecules, cellular receptors, and cytoskeletal proteins. In addition, some large enzyme complexes exist as icosahedral particles (Ladenstein et al., 1988; Zhou et al., 2001b; Milne et al., 2002). Structural studies of icosahedral particles thus are highly relevant to understanding their assembly principles and functional mechanisms.

Since *icosahedral particles* generally have a diameter of several tens to hundreds of nanometers, they are well suited for structural determination by single particle electron cryomicroscopy and computer reconstruction (Baker et al., 1999). In theory, the reconstruction algorithm for icosahedral particles could be the same as that used for asymmetric single particles (chapter 14). Historically, however, a very elegant algorithm was developed for icosahedral particle reconstruction which takes advantage of its high symmetry and the fact that there are 60 identical *asymmetric units* per particle (Crowther et al., 1970a). Over the years, the average data-set size for an icosahedral particle reconstruction has expanded from a few tens to thousands and tens of thousands of particle images. Concurrently, the resolution of the reconstruction has gradually extended toward the subnanometer range. Presently, there are a number of algorithms and software packages capable of reconstructing three-dimensional (3-D) structures of icosahedral particles at resolutions ranging from 0.95 to 0.68 nm (see review by

Zhou and Chiu, 2003). The reliability of these density maps is rather high no matter which software package was used. For instance, the cryo-electron microscope density maps of the hepatitis B capsid (Böttcher et al., 1997; Conway et al., 1997) and the rice dwarf virus (Zhou et al., 2001a), match well with their corresponding crystal structures that were subsequently determined (Wynne et al., 1999; Nakagawa et al., 2003). All of the alpha helices, beta sheets and loops overlap well between the two structures of the inner capsid of the rice dwarf virus (Chiu et al., 2005). The current technology has advanced to the stage that solving a structure of an icosahedral particle at subnanometer resolutions by cryo-electron microscope methodology is routine and rapid (Jiang et al., 2001b; Booth et al., 2004).

This chapter begins with a description of the basic structural organization of an icosahedron (section 13.2), and of the *local symmetry* elements (section 13.3) that exist in addition to the *icosahedral symmetry*. It is instructive to understand the symmetry inherent in an icosahedral structure, which forms the basis for 3-D reconstruction and structural analysis. There are numerous refinement approaches for obtaining an icosahedral particle structure. For demonstration purposes, we will describe the classic method of icosahedral particle reconstruction (Crowther et al., 1970a; Crowther, 1971) in section 13.4. Section 13.5 then goes on to review key points that are important in sample preparation and in the experimental procedures used to record images. The subsequent processing of images, including data quality evaluation, *contrast transfer function* and *experimental B factor* estimation and correction, is covered in sections 13.6 and 13.7. Section 13.8 summarizes various approaches used to refine the particles' orientation parameters as more data are added to the process. Section 13.9 describes the criteria used to evaluate the *resolution* of the 3-D reconstruction. Section 13.10 introduces various poststructure analysis and visualization tools to interpret the 3-D density map. The chapter then concludes, in section 13.11, with a perspective on further developments that can be expected within the field of 3-D reconstructions of icosahedral particles.

13.2 Description of an icosahedron

A unit cell of a crystal is made up of one or more asymmetric units whose positions are related by symmetry operations. An asymmetric unit can be as simple as one polypeptide or as complex as several of the same or different kinds of polypeptides. By analogy to the terminology of crystals, an *icosahedron* can be thought of as a single unit cell with 60 asymmetric units related to each other by three types of rotational symmetry axes (5-, 3- and 2-fold axes) (figure 13.1). The icosahedron is composed of 20 equilateral triangles related to each other by 5-fold symmetry operations. Each triangle consists of 3 asymmetric units related by a 3-fold rotational symmetry. In addition to these 5- and 3-fold axes, there are 2-fold symmetry axes at the midpoint of each equilateral triangle edge. These 5-, 3-, and 2-fold symmetry axes are referred to as strict or icosahedral symmetry axes to distinguish them from local, quasi-symmetry axes (see below). All icosahedra, regardless of size, possess the same number of equilateral triangles (20), the number of asymmetric units (60), and the number of 5-fold (6), 3-fold (10), and 2-fold (15) icosahedral symmetry axes (figure 13.1).

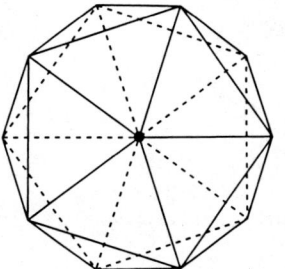 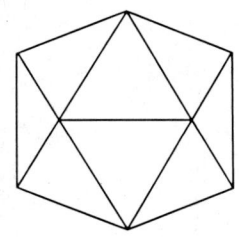 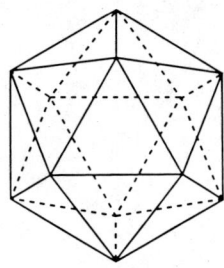

Figure 13.1 A schematic diagram of an icosahedron. It has 20 equilateral triangles related by 12 5-fold symmetry axes. Each triangle has one 3-fold symmetry axis relating three asymmetric units resulting in 20 3-fold symmetry axes and 60 asymmetric units in each icosahedron. There are 30 2-fold symmetry axes at the midpoint of each triangle edge. Projection views of an icosahedron are shown down (a) a 5-fold axis, (b) a 2-fold axis, (c) a 3-fold axis. Courtesy of Wen Jiang at Baylor College of Medicine.

To understand the crystallographic origins of an icosahedron, we can construct it from a hexagonal lattice essentially defined by two vectors (**h, k**) with an included angle of 60° as shown in figure 13.2. The center of each hexagon is a lattice point. The following steps describe how to construct an icosahedron from such a hexagonal lattice (Johnson and Fisher, 1994).

(1) **Define an equilateral triangle.** Set the first vertex of an equilateral triangle at an arbitrary origin of a hexagonal point (figure 13.2(a)). The edge of the equilateral triangle is defined by a vector between the origin and any hexagonal lattice point, which is specified by (h, k), a pair of integer indices of the lattice. The length of this edge determines the size of the equilateral triangle. Due to the symmetry of an equilateral triangle, the third vertex and thus the other two edges are exactly defined on this initial placement.

(2) **Define 20 contiguous equilateral triangles.** Use the edge opposite to the origin of the first equilateral triangle to define the edge of a second equilateral triangle (figure 13.2(b)). Then continue to define a total of 10 contiguous equilateral triangles going in the same direction. These 10 equilateral triangles can be rolled up and the edges between the first and the tenth triangles can be joined together to form the middle portion of the icosahedron. Use one free edge from each previously defined triangle to add a new equilateral triangle, resulting in 20 triangles (figure 13.2(b)).

(3) **Form an icosahedron.** The strip of 20 equilateral triangles can be cut out from the hexagonal lattice as shown in figure 13.2(c). It is then evident that each corner of the equilateral triangle becomes a pentagonal vertex when adjacent free edges are joined with one another.

In the above construction, an icosahedral lattice is characterized by the *triangulation* (T) *number* equal to $(h^2 + k^2 + hk)$ (Caspar and Klug, 1962). In the example of figure 13.2(a), the vector is drawn from the origin to the point $h = 1, k = 1$, so the T number is 3. The T number is related to the edge length of the equilateral triangular edge of an icosahedron expressed in terms of the underlying hexagonal lattice. Because the two unit vectors of the hexagonal lattice are of equal magnitude, there is degeneracy

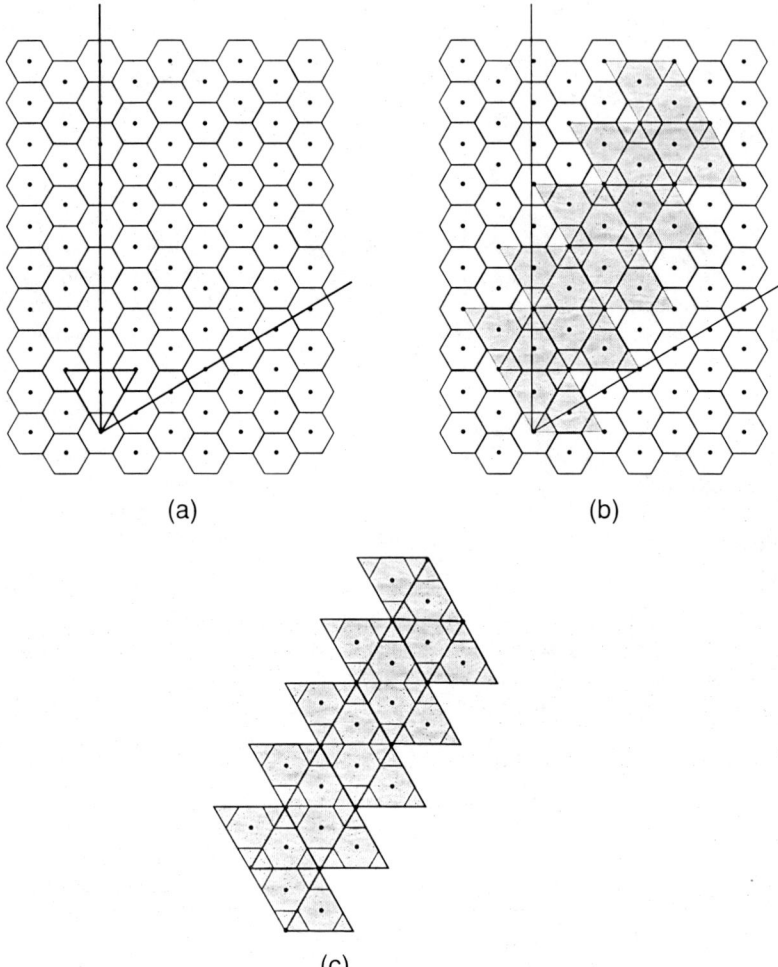

Figure 13.2 Icosahedron derived from a hexagonal lattice. The center of each hexagon is a lattice point. An icosahedron can be built from such a two-dimensional lattice with 12 equilateral triangles related by a 5-fold symmetry operation (Johnson and Fisher, 1994). (a) The first equilateral triangle is defined by one vertex at the center of the hexagonal lattice and another vertex at any lattice point (**h, k**). (b) The middle part of an icosahedron consists of 10 equilateral triangles, which can be built after the first one is established by using one free edge from each previously defined triangle to add a new equilateral triangle. (c) A set of 20 equilateral triangles is cut out of the sheet, and 5 adjacent vertices are merged to forms a pentagon. The strip of 20 equilateral triangles is thus converted into a solid icosahedron. Prepared by Alex Chen at Baylor College of Medicine.

for some *T numbers*. For example, the *T* number is the same both for ($h = 1$, $k = 2$) and ($h = 2$, $k = 1$). Since two such icosahedra are mirror related, they are referred to as icosahedra for which the handedness is either *laevo* or *dextro* (Belnap et al., 1997).

13.3 Local symmetries can be present within an asymmetric unit

As the *T* number increases, the number of lattice points included in an asymmetric unit also increases (Johnson and Fisher, 1994). Because of the hexagonal lattice underlying an icosahedron (see above), the motifs at these lattice points are related by local 2-, 3-, and 6-fold symmetry axes within an asymmetric unit. The number of local symmetry axes in the asymmetric unit varies according to the *T* number. Because the chemical environment around each lattice point is not exactly identical, the local *n*-fold symmetry operation often does not strictly follow the $360°/n$-rotation rule. These non-icosahedral symmetry axes hence have been referred to as local or pseudo symmetry axes. The motifs around each lattice point were referred to as "morphological units or *capsomeres*" when they were first observed as stain-excluding mass-density clusters in electron micrographs of negatively stained virus particles (Horne and Wildy, 1961; Caspar and Klug, 1962).

At low resolution, the morphological units may appear pentameric, hexameric, trimeric, dimeric, or monomeric. The types of morphological units per asymmetric unit vary among particles independent of the *T* number. Given the variability and complexity in structural and compositional arrangement of icosahedral particles, a 3-D structure of the particle derived from electron cryomicroscopy is the most direct and straightforward way to determine the number of subunits in the asymmetric unit (see review by Baker et al., 1999). For instance, the algal virus has only trimers in its $T = 169$ lattice; the herpesvirus capsid has three types of morphological units (hexamers, pentamers, and trimers) in its $T = 16$ lattice; the outer capsid shell of rice dwarf virus is made up of trimers in its $T = 13$ lattice; the capsid has both pentamers and hexamers in the $T = 7$ P22 phage; the papillomavirus contains only pentamers in its $T = 7$ lattice; and the sindbis virus has only trimers arranged on its $T = 4$ outer capsid lattice. The chemical composition in each subunit of the morphological units can be a single polypeptide or multiple polypeptides. In a simple case, such as the P22 phage, only one type of protein makes up both pentamers and hexamers, and each subunit has one protein. In a more complex particle, such as herpesvirus capsid, each of the hexameric subunits is made up of two proteins, and the trimer is a heterotrimer of a dimer of one protein and a monomer of a different protein.

13.4 Theory of icosahedral reconstruction

The procedure described below for calculating a 3-D map of an icosahedron was first introduced in the early 1970s (Crowther et al., 1970a). The algorithm used by Crowther takes advantage of the icosahedral symmetry, so as to reduce the number of projection images required to adequately sample the required 3-D data for a valid reconstruction. Since there are 60 asymmetric units per icosahedron, one could, in principle, reduce the number of particles by a factor of 60 from that required to reconstruct the structure of an asymmetric particle based on the size and resolution relationship. (Crowther et al., 1970b). In practice, the actual number of particle images that are needed is much higher than theoretically estimated, primarily because the images are noisy and an averaging of multiple images is necessary to enhance the statistical definition of fine structural details. In this section, we will first present the mathematical principles of the reconstruction of an icosahedron formulated by Crowther in a cylindrical coordinate system (Crowther, 1971).

The mass density, $\rho(r, \varphi, z)$ of an icosahedral particle can be derived from the computed Fourier transforms of the electron images in the same way that was done for crystalline and helical objects (see chapters 10 and 12). The use of a cylindrical polar coordinate system instead of Cartesian coordinates results in a simple analytical relationship between the mass density function and the image intensity as shown below:

$$\rho(r, \varphi, z) = \sum_{n=-\infty}^{\infty} \int_{-\infty}^{\infty} g_n(r, S_z) e^{in\varphi} e^{2\pi i z S_z} dS_z \qquad (13.1)$$

with

$$g_n(r, S_z) = \int_{0}^{\infty} G_n(S, S_z) J_n(2\pi Sr) 2\pi S dS \qquad (13.2)$$

where g_n is the Fourier–Bessel transform of G_n, J_n is the Bessel function of order n, and (r, φ, z) and (S, Φ, S_z) denote the radii, azimuth angles, and heights in real and Fourier space, respectively. The G_n functions are related to the Fourier transform $F(S, \Phi, S_z)$ of the particle as

$$F(S, \Phi, S_z) = \sum_{n=-\infty}^{\infty} G_n(S, S_z) i^n e^{in\Phi} \qquad (13.3)$$

Note that in equation 13.3, for a given S and S_z, F is a known function of Φ regardless of an object's density function. As shown by the projection theorem, F represents the Fourier coefficients in a central section through 3-D Fourier space. The plane of this central section thus intercepts (i.e., samples) the 3-D structure factor, F, at different S, Φ, and S_z depending upon the orientation of the central section (figure 13.3). The orientation of this Fourier sectional plane is determined by the orientation of the particle with respect to the electron beam direction. For every section, 59 other symmetry-equivalent Fourier sections can be generated to fill up the data sampling in the Fourier space. The upper limits for the summation of n, S, and S_z in the above equations depend on the expected resolution of the reconstruction and the size of the particle. As evident in these three equations, the determination of the set of G_n, followed by that of the set of g_n leads to the final computation of the density (ρ).

Using the above formulation, the first 3-D reconstruction was computed with images of negatively stained tomato bushy stunt virus, a plant virus of moderate size (33 nm in diameter) (Crowther et al., 1970a). This first 3-D reconstruction revealed the overall organization of the virus in terms of its size, triangulation number, and the number of pentameric and hexameric capsomeres. An important turning point in using electron microscopy to study virus structures was when the ice-embedding procedure, rather than negative staining was used in order to better maintain the native structures of the biological specimens (Adrian et al., 1984). In addition, there have been numerous

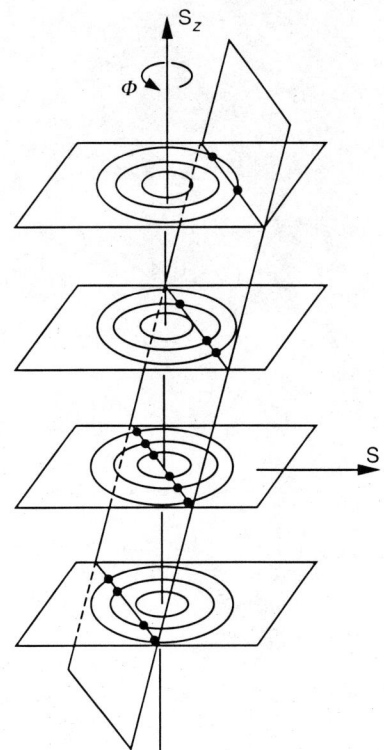

Figure 13.3 A central section of the Fourier transform of an icosahedron as represented by $F(R, \Phi, Z)$ in a cylindrical polar coordinate system. The Fourier transform is sampled at different Z and on rings of constant R within each Z plane. Replicated by Alex Chen after Crowther (1971).

enhancements in the data analysis that were made to compensate for the low contrast of ice-embedded virus particle images and to allow larger data sets for higher resolution reconstruction (see reviews by Baker et al., 1999; Thuman-Commike and Chiu, 2000; Zhou and Chiu, 2003; Jiang and Chiu, 2006).

13.5 Experimental considerations

Ice embedding is the routine technique for preserving icosahedral particles for low dose electron imaging (section 6.6). In general, the particle concentration is around 10^{11}–10^{12} particles/ml; the number of particles per micrograph or CCD frame depends on the particle size and magnification (typically from tens to hundreds). The single particles are suspended in a matrix of vitreous ice across holes either with or without underlying support film, for maximum contrast. The holey grid can be prepared from a triafol solution (Fukami and Adachi, 1965) resulting in different sizes of holes ranging from a fraction of a micrometer to several micrometers in diameter. The plastic net in the holey grid is carbon-coated to enhance its stability under the beam. Alternatively, one may use the commercially available Quantifoil holey grids (Quantifoil Micro Tools GmbH, Jena), where the holes are fabricated in a regular array and with

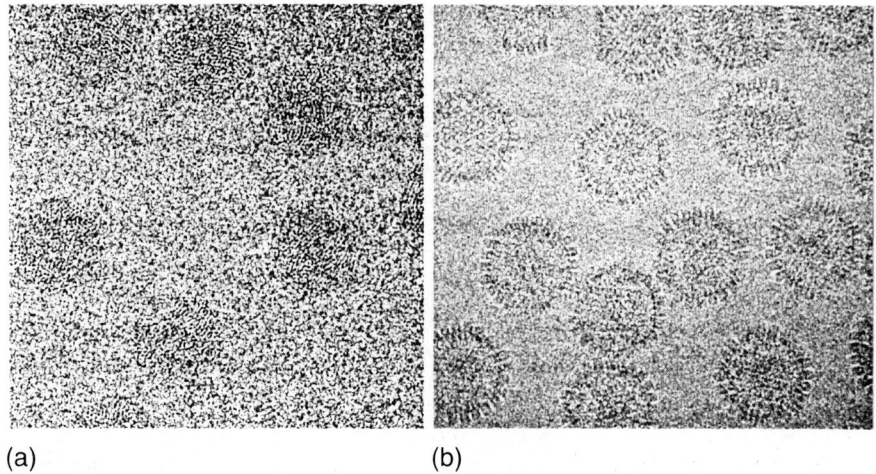

Figure 13.4 Examples of 300 kV images of ice-embedded icosahedral virus particles captured onto a Gatan 4k × 4k CCD camera in a JEM3000 electron cryomicroscope equipped with a liquid helium (4.2 K) cryostage and a field emission gun. (a) Epsilon 15 phage (~700 Å in diameter) provided by Peter Weigele and Jonathan King of MIT. The finger print motif in the particle is the signature feature of the viral dsDNA. (b) Herpes simplex virus type 1 capsid (1250 Å in diameter) provided by Frazer Rixon at MRC, Glasgow. Images were recorded by Joanita Jakana at Baylor College of Medicine.

a constant hole diameter. A glow-discharge treatment of the grid with air can help the particles spread over the holes during the blotting process (sections 6.2, 6.4, and 6.6). With practice, one should be able to routinely obtain grids with a good portion of their area having icosahedral particles embedded in ice of an appropriate thickness. Figures 13.4(a) and (b) show CCD image frames of two different icosahedral virus particles embedded in vitreous ice where the image contrast is relatively high.

If the particle suspensions are in sucrose or glycerol as is the case after centrifugation in many purification schemes, the samples must be dialyzed to remove these viscous solutes. Both sucrose and glycerol not only reduce the image contrast but also often make it difficult to obtain a thin layer of ice-embedded sample. Other problems that have been encountered in preparing and imaging ice-embedded icosahedral particles include aggregation of particles, freezing of particles in preferred rather than random orientations, and local (or global) specimen movement. The first two phenomena cause difficulty in subsequent steps in image processing, which generally require the use of single particles in isolation and in random orientations. Most of these problems have solutions. Aggregation can often be prevented by varying the particle concentration and chemical environment (e.g., buffer, salt, pH, and detergent), or by modifying the particle purification procedure. The preferred orientation of particles, usually at the water–air interface, occurs for some but not all specimens. This problem can be overcome by tilting the specimen at a small angle (e.g., 10–20° depending on the *preferred orientation* pattern) to effectively randomize the orientation distribution of the particles. The third phenomenon is primarily caused by the *beam-induced specimen movement* as described in chapter 5. This problem may be reduced by a variety of methods such as using spot-scan imaging, coating the ice-embedded sample with an additional layer of conducting film, choosing illuminated areas to include adjacent carbon film in the holey net, or using a gold-coated objective aperture and pretreating the grid with extended electron bombardment prior to use. However, these methods have not been generally used in icosahedral particle data collection except in few cases (Zhou et al., 1994; Jakana and Chiu, 2004).

The choice of magnification for imaging icosahedral particles depends on the targeted resolution and the type of recording medium and digitizer. For instance, if using Kodak SO163 film and a scanner, a magnification of $\times 50,000$ is adequate for attaining structural resolution up to 0.7 nm (Zhou and Chiu, 2003). It has been shown that the latest Gatan 4k $\times$ 4k CCD camera is quite suitable for collecting 200 kV images of icosahedral particles at $\times 60,000$ magnification, where 0.9 nm resolution structure can be reconstructed with unambiguous resolution of long alpha helices (Booth et al., 2004). In theory, one can use a CCD camera for even higher resolution data collection at a higher magnification (Booth et al., 2004). A practical drawback in using a CCD camera for data collection is the relatively large size or low concentration of the particles, resulting in too few particles per CCD frame at the chosen magnification for the targeted resolution. This may render such data collection lengthier than using photographic films followed by digitization.

The defocus setting is always a compromise between achieving high image contrast and maximum structural resolution. The use of an electron microscope with a field emission gun has the advantage of allowing a high defocus value while retaining the high-resolution data, due to the high spatial coherence of the electron beam. In the event that sufficiently high *spatial coherence* is not attainable (e.g., with a LaB_6 gun), one can record a focal pair. The first image is taken closer-to-focus in order to record the high-resolution information that is needed for the final reconstruction. The second image is taken further-from-focus to identify the particle positions and to estimate their approximate orientation parameters. This *focal pair* approach has been used successfully to determine a number of subnanometer-resolution structures of icosahedral particles with an electron microscope equipped only with a LaB_6 gun (Zhou et al., 2000; Zhou et al., 2001a; Jiang et al., 2003). However, with the use of a highly coherent field emission gun and an improved reconstruction algorithm, focal pair is no longer necessary even for a subnanometer-resolution structure determination.

13.6 Data evaluation

Similar to other types of specimens described elsewhere in this book, the images have to be evaluated prior to subjecting them to serious image processing. The quality of electron images can be judged initially according to their contrast and lack of specimen movement. If the ice-embedded particles are suspended across holes without a carbon film, it is difficult to detect the *contrast transfer function* (CTF) rings (section 3.8) in an optical diffractometer as is generally done for images of crystalline or helical objects. However, the CTF rings can be seen in the computed Fourier transforms of the digital boxed-out particle images (Zhou et al., 1996). While an image of a single particle does not have sufficient statistical definition to produce the CTF rings, the incoherent average of Fourier intensities of multiple particle images can make the CTF rings visible. This average Fourier spectrum can then be used to characterize the potential resolution of the image data, as well as the amount of defocus, astigmatism, and drift in the micrograph.

The circularly averaged Fourier intensity of the particle images (figure 13.5(a)) can be approximated by

$$F_{obs}^2(S) = F^2(S)\text{CTF}^2(S)\,\text{Env}^2(S) + N^2(S), \qquad (13.4)$$

where $N^2(S)$ is the power spectrum of the noise. In this formulation, the CTF contains the familiar terms for defocus, astigmatism and spherical aberration (as described in chapter 3). In practice, the astigmatism term (section 3.8) can be ignored for high-quality micrographs, at least for reconstruction up to 0.6 nm resolution (Ludtke et al., 2004). For practical simplicity, the overall envelope function (Env) (section 3.9), has been approximated as a Gaussian function, $\exp(-BS^2)$, of width B, which has been referred to as the *experimental B factor* (Saad et al., 2001). Such an approximation has been found to be applicable to image data in many but not all the electron cryo-microscopes. This *experimental B factor* is affected not only by the microscope types and electron optical settings, but also by the specimen movements and the modulation transfer function of the recording medium and/or scanner.

$N(S)$ is a noise function which is dependent upon various experimental factors, including ice thickness and inelastic scattering. It is defined in such as a way that the subtraction of the $N(S)$ from the computed Fourier transform would yield some of the Fourier coefficients to be close to zero as expected from the CTF (figure 13.5(a)). The particle's averaged structure factor (F) is generally unknown but can be derived from *x-ray solution scattering* experiments with a highly concentrated suspension of particles (Thuman-Commike et al., 1999; Saad et al., 2001) as exemplified in figure 13.6. All the parameters in CTF, Env, and N functions can be estimated empirically for each micrograph/frame using a graphical-interface program such as *ctfit* in *EMAN* (Ludtke et al., 1999). It should be pointed out that there are other software packages (e.g., Zhu et al., 1997) that can be used to determine these parameters for single particle reconstruction as discussed in chapter 14.

By fitting all these instrumental and experimental parameters, the signal-to-noise ratio (SNR) of the particle images in a micrograph/frame can be estimated as a function of frequency (figure 13.5(b)). Due to the difficulty in estimating the background function ($N(S)$) at very low and very high frequencies, such an estimate of the SNR is generally not too accurate at those ranges. One should view this type of estimate as a semi-quantitative guide to screen each data set for determining whether they are suited for subsequent analysis at the intended resolution.

$$\text{SNR}(S) = \left(F_{obs}^2(S) - N^2(S)\right)/N^2(S). \qquad (13.5)$$

13.7 Image restoration

Each electron micrograph/CCD frame must be corrected for *phase reversal* and *amplitude modulation* due to the effects of CTF, magnification variation, amplitude contrast contribution, and Fourier amplitude decay (equation 13.4). The Fourier amplitude and

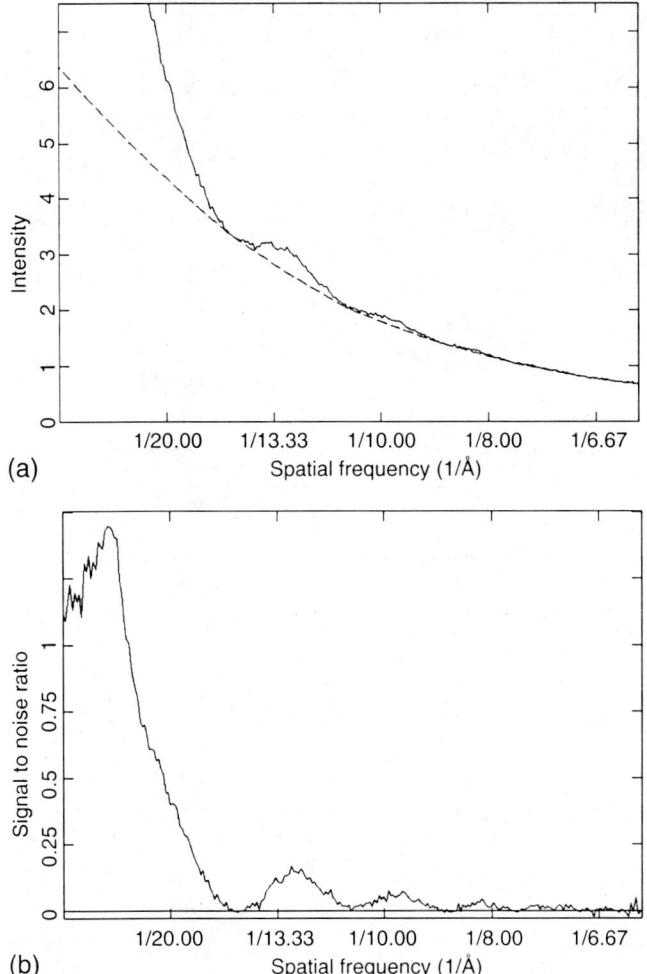

Figure 13.5 Quantitative evaluation of icosahedral particle images. (a) Circular average of the incoherently averaged Fourier transform of ~60 icosahedral particle images without carbon support film from a single CCD frame recorded at 200 kV with the computed background function (Booth et al., 2004). (b) Averaged signal-to-noise ratio of particle images. Provided by Christopher R. Booth at Baylor College of Medicine.

phase corrections for the CTF can be applied either to each individual particle image before computing the 3-D reconstruction (Zhou and Chiu, 2003), or to the individual 3-D reconstructions obtained from each micrograph before they are finally merged (Böttcher et al., 1997). The latter approach is not generally practical when the particle size is large and a single micrograph does not contain a sufficient number of particles even for a moderate resolution reconstruction. In either case, a *Wiener* filter type correction function (section 10.11) is used in order to avoid overscaling of the Fourier amplitudes near the CTF zeros.

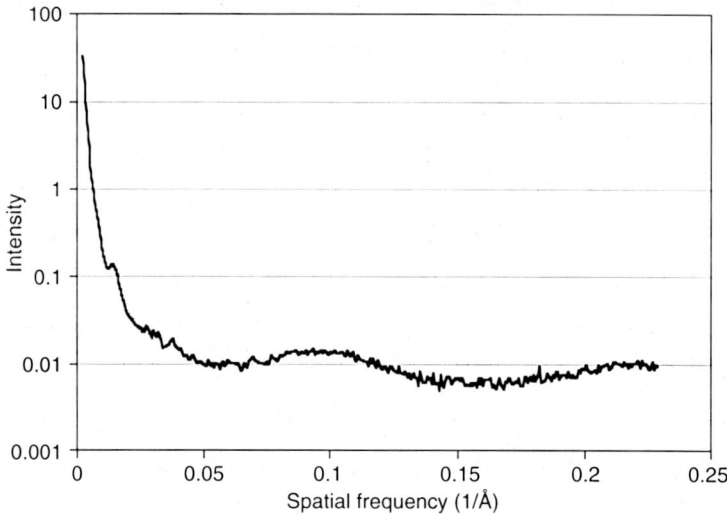

Figure 13.6 X-ray solution scattering intensity of a HSV-1 B capsid suspension as a function of spatial frequency. These data were recorded at the SLAC synchrotron radiation source (courtesy of Hiro Tsuruta, SLAC, Stanford). Reproduced after Saad et al. (2001).

In order to retrieve the structural features properly at subnanometer resolutions, it is necessary to make corrections to the Fourier amplitude fall-off. An *overall B_o factor* analogous to the temperature factor used in crystallography was shown to be an effective form of correction to sharpen the map (Böttcher et al., 1997). This *overall B_o factor* can be larger than the *experimental B factor* defined in section 13.6, because the former factor does not include the Fourier amplitude fall-off due to computational or algorithmic errors. In practice, the computational error for icosahedral particle reconstruction up to 0.7 nm resolution is negligible. It should be cautioned that different literatures defined these "B" factors differently and can be related by a scaling factor of 4 (Böttcher et al., 1997; Saad et al., 2001). There have been different ways of making this map sharpening. For instance, a single *overall B_o factor* was applied in computing the final 3-D map of the hepatitis B core particle (Böttcher et al., 1997) so that the alpha helices became visible. Alternatively, the *experimental B factor* as determined in each micrograph was used to correct each particle image during the refinement process for the subnanometer-resolution structure determination (e.g., Jiang et al., 2003; Booth et al., 2004). Strictly speaking, particles in the same micrograph might have different *experimental B factors* because of possible local specimen movement.

13.8 Initial model building and structure refinement

The 3-D model of an icosahedral particle is reconstructed from images of particles that lie in different orientations. Each particle image is "indexed" by five parameters including the center of the particle (x_o, y_o) and the three *Euler angles* (ϕ, θ, ω) for the *orientation* as defined in box 13.1. Because of the existence of 60 asymmetric units in an icosahedron, the angular range that is used to assign the Euler angles can be reduced from 2π radians to those defining an asymmetric unit (figure 13.7).

Because of the high noise level in the particle images, it is necessary to perform a prefiltering of the image prior to defining its center. *Centering* of the particle can be determined based on the *cross-correlation* of a rotationally averaged particle image followed by a center of gravity search. Once the center of the particle is determined, a circular or Gaussian mask will be applied to the particle before the subsequent steps (see chapter 14 for additional material on the masking step).

The most critical step of the icosahedral particle reconstruction procedure lies in the assignment of the Euler angles for each of the particle images. This assignment is done in a stepwise fashion starting with the low-resolution data and gradually working towards higher resolution. The following sections will describe how to build an initial low-resolution model followed by refinement at higher resolution. There are different algorithms developed for these steps, all of which work almost equally well in the hands of experienced investigators. There have also been attempts to make these steps as automated as possible (Jiang et al., 2001b).

Initial model building

For a given new particle, often the most challenging step is to determine an initial model upon which subsequent refinements can be carried out. One approach is to use the previously determined structure of related particles as the initial model, when that is available. A more general approach is to determine an initial model with a procedure

---

BOX 13.1 Definition of Euler angles relative to the asymmetric unit of an icosahedron

The conventions adopted for the coordinate system of an icosahedral particle are shown in figure B13.1. The three axes are coincident with three 2-fold icosahedral axes, and the origin of the coordinate system is located at the center of the particle. In a randomly oriented sample, the directions of the symmetry axes and thus of the coordinate system axes for each particle would be different from each other. Euler angles (ϕ, θ, ω) are defined to specify the direction of the coordinate system of each particle and the direction of a common coordinate system. Euler angles are three successive angles of rotation performed in a specific sequence, which thereby transforms a given Cartesian coordinate system to another one. ϕ is the first rotational angle around the original z axis. θ is the second angle of rotation around the new y axis and ω is the final angle of rotation around the new z axis. A single asymmetric unit of an icosahedron is bound by two neighboring 5-fold axis and a 3-fold axis which are spanned by the angular ranges of $\phi = \pm 31.72°$, $\theta = 69.09–90.0°$, and $\omega = 0.0–360°$ as illustrated in figure B13.1.

(continued)

BOX 13.1 *(continued)*

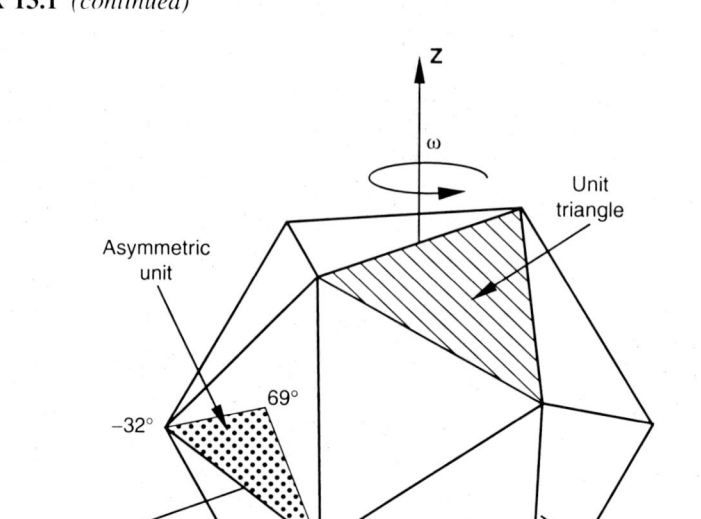

Figure B13.1 Definition of icosahedral particle orientation. All icosahedral particle coordinates (x, y, z) are defined relative to the icosahedral 2-fold axes. An icosahedron is made of 20 triangular units (one being shaded) and the triangular sector representing an asymmetric unit (dotted area) has its vertices at two adjacent 5-fold axes and at the 3-fold axis of the unit triangle. The angles of rotation (θ, ϕ, ω) are depicted which describe the particle orientation with respect to the fixed coordinate system. In the definition of Euler angles, it is essential to note the sequence of rotations about the different axes: the first rotation about the z axis is through an angle ϕ, the second rotation around the new y axis is through an angle θ, and the third rotation around the new z axis is through an angle ω. Note that different software packages may define the Euler angles differently in the sequence of rotation along different axes, but the general concept is identical. Reproduced after Thuman-Commike and Chiu (2000).

known as the *common-line* algorithm (Crowther et al., 1970a). This technique takes advantage of, once again, the existence of 60 redundant asymmetric units in the particle, which are related by symmetry.

The Fourier transform of a single projection image has 59 symmetry-equivalent Fourier sectional planes in other orientations because of icosahedral symmetry. Each of these symmetry-related Fourier sections must pass through the origin, thus intersecting with the original section. The lines along this intersection of any two symmetry-related Fourier sections are called a pair of *self-common lines*, because the corresponding points in these two lines have equal amplitudes and phases (figure 13.7). Although there are 59 *self-common line* pairs in a strict operational sense, some of them overlap. Therefore, there are only a maximum of 37 unique pairs of *self-common lines* in a

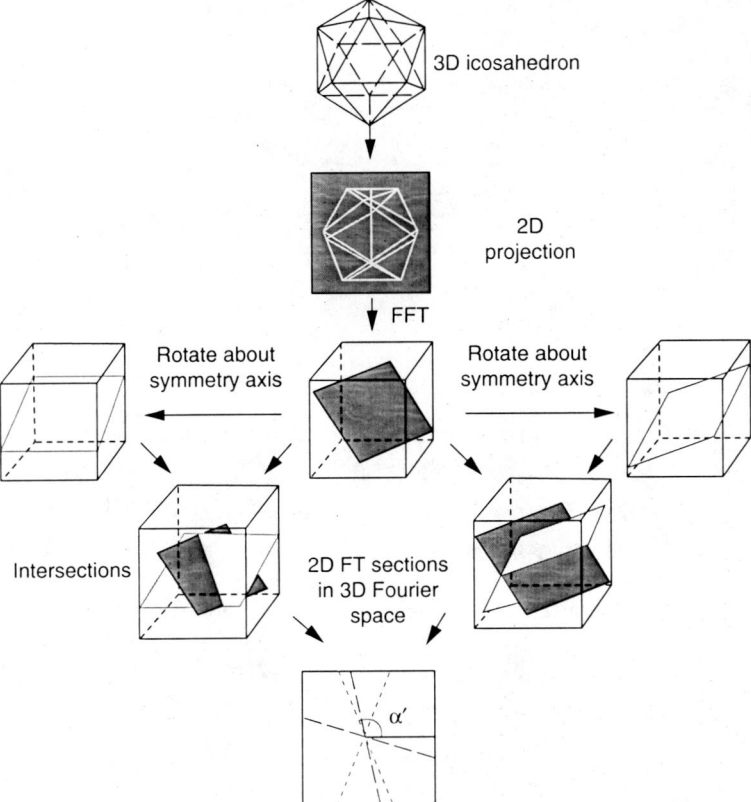

Figure 13.7 Concept of *self-common lines*. A 3-D icosahedral particle in an arbitrary orientation is projected along the z direction to form a 2-D image. The Fourier transform of the projection is shown, intersected by 2 of its 59 symmetry-related central sections. Lines of intersection between the original and each of the 2 symmetry-related central sections are called a pair of self-common lines. The dispositions of the *self-common lines* are uniquely defined in the Fourier transform of the projection image. Provided by Dr. Z. Hong Zhou at the University of Texas Medical School at Houston.

Fourier transform of a projection image of an icosahedron (table 13.1). The locations of these 37 pairs of *self-common lines* in the Fourier transform of a projection image are related to the orientation of the particle and can be used to determine the angular orientation (*Euler angles*) of the particle. Furthermore, if one of the symmetry axes of the particle is along the projection of the image, the number of pairs of *self-common lines* will be further reduced to be fewer than 37, as indicated in table 13.1. The determination of *self-common lines* involves an exhaustive computational search for the minimal averaged *phase residual* along the prospective self-common lines in the Fourier sectional plane (see box 13.2 for definition). It should be pointed out that the *self-common line* search depends on the accurate positioning of the *particle center*, the extent of icosahedral symmetry preserved in the particle image, and the amount of noise in the image.

Table 13.1 Number of self-common lines in the Fourier transforms of 2-D projections parallel to specific symmetry axes of an icosahedron

Symmetry axis	No. of unique common lines per symmetry axis	Total no. of unique symmetry axes	Total no. of self-common lines
5-fold	2 pairs	6	12 pairs
3-fold	1 pair	10	10 pairs
2-fold	1 line[a]	15	15 lines

[a]The pairs of lines are superimposed on each other in this case, and because of Friedel symmetry the result is a single line of real numbers.

BOX 13.2 Definition of phase residual of self- and cross-common lines in the Fourier transforms of icosahedral particle images

Each of the orientations of an icosahedron defines a unique set of self-common lines that exist within the central section (for that particular projection) and cross-common lines that are shared with another central section. The phase residual R_i along the self- and cross common-lines, corresponding to the symmetry-related lines in the ith particle and the symmetry-related lines in each of the jth particles is evaluated according to the following equation (Zhou et al., 1998). The Euler angles (ϕ, θ, ω) and the center (x, y) that minimize this function will be taken as the correct orientation and center parameters of particle i:

$$R_i(\phi, \theta, \omega, x, y) = \frac{\sum_{j=1}^{N} \sum_{k=1}^{k_{max}(j)} \sum_{S=S_{min}}^{S_{max}} \left| \psi_i(S, x_i, y_i, \alpha_{ijk}) - \psi_j(S, x_j, y_j, \alpha_{jik}) \right| w(S, \alpha_{ijk}, \alpha_{jik})}{(S_{max} - S_{min}) \sum_{j=1}^{N} k_{max}(j)}$$

(B13.1)

where ψ is the phase value for particle i and particle j (or its symmetry-related section) along the kth common line; α is the angle of the kth common line in the particle's Fourier transform; S_{min} and S_{max} defines the range of spatial frequency to be evaluated; k_{max} is the maximum number of common lines (a maximum of 37 for self-common lines and 60 for cross-common lines); N is the particle number; w is the weighting function, which is commonly set to be 1 if the Fourier amplitude is above a certain threshold and 0 if the amplitude is below the threshold.

In theory, the *self-common line* procedure is capable of providing the orientation parameters of each single particle. In practice, it is rarely accurate enough because of the high level of noise in the image. A number of improvements using various statistical criteria have been implemented to make the initial estimate more accurate (Thuman-Commike and Chiu, 1997). There are still situations, however, in which it may be difficult to achieve an initial model by this method, especially for particles with a very smooth surface. Empirically, the best option is to take images at a coarser defocus in order to enhance the low-resolution image contrast. Therefore, an initial model is generally obtained at a resolution of 2–3 nm, using only a few tens of particle images.

Structure refinement

Once an initial model is established, all the data can be refined against the initial model, which results in an improved map. Refinement is done in a stepwise fashion in which increasing resolution is achieved with an increasing number of particle images. During the step, both the center and the *Euler angles* for each particle image can be refined simultaneously. For most of the algorithms, the refinement needs an estimate of the *orientation parameters* for each particle image. These estimates can be obtained in a variety of ways. The original way was to use the *self-common line* results. Due to the noisy nature of the particle image, this does not yield a reliable estimate in general, especially when the refinement is done with particle images taken at a smaller defocus. The initial orientation parameters can be derived from the second micrograph of the same particle taken at a higher defocus (Zhou and Chiu, 2003).

There are different algorithms to carry out refinement using icosahedral particle images. Nevertheless, these algorithms follow the same logical steps by comparing each particle image with a set of projection images from a model. The model is continually updated as more data are included. Since each particle image can come from a different micrograph or CCD frame, it would be associated with different microscope parameters, as described in section 13.6. Therefore, the model projection has to be modified with the same amount of defocus and *experimental B factor* prior to making a comparison with the raw particle images.

The number of projection images that should be generated from the model depends on the stage of the refinement and the targeted resolution. The comparison of a given particle image and projections of the model can be made around a narrow range of Euler angles where the initial estimates of Euler angles were made. The better the initial estimates are, of course, the more accurate and the quicker the convergence will be reached.

Different refinement algorithms vary in the criteria used for comparison and in the criteria used to define convergence (Baker and Cheng, 1996; Zhou and Chiu, 2003). The criterion of comparison can be based on the minimization of the *cross-common* lines phase residual (see box 13.2) or simply the best cross-correlation match. Due to the inherent symmetry for the icosahedral particles, the cross-common line criterion is a powerful one because there are many lines to compare and the comparison can thus be statistically well defined. The cross-correlation match, in principle, can be done similarly to any of the single particle reconstruction algorithms (Baker and Cheng, 1996; Frank et al., 1996; van Heel et al., 1996; Ludtke et al., 1999). Any of these mentioned methods for orientation parameter refinement can be used to obtain structures at 0.7–0.9 nm resolution (see review by Zhou and Chiu, 2003). Again, all these refinement procedures are conditioned on first obtaining an initial model.

It should be pointed out that a plausible alternative procedure for structure refinement, without a template model, is to make a comparison among raw particle images based on the cross-common lines phase residual between many pairwise comparisons (see box 13.2). Such an approach is free of model bias, but is computationally a very slow process (Zhou et al., 1998). Therefore, it is hardly used for large data sets.

13.9 Resolution evaluation

Assessment of the resolution of a reconstruction is based on the reproducibility and reliability of structural details shown in the mass density map (also see chapter 14). Three criteria have been employed to make such an assessment. One criterion is based on the phase residual difference determined in the cross-common lines among all raw-particle images. The resolution is defined as being the spatial frequency at which the phase residual difference reaches ~90° (Stewart et al., 1991; Böttcher et al., 1997). Another criterion is based on the Fourier shell correlation of reconstructions from two independent sets of particle images (see box 13.3). Using this formulation, the resolution is commonly defined as being the spatial frequency where the correlation coefficient is 0.5 (Böttcher et al., 1997; Ludtke et al., 2004). A more liberal threshold has also been proposed by Rosenthal and co-workers (appendix in Rosenthal and Henderson, 2003). The third criterion is based on the phase residual difference between two independent reconstructions (see box 13.3), where the resolution is defined as the spatial frequency when the phase residential reaches 45° (Frank et al., 1981; Radermacher et al., 1987; van Heel, 1987). All of these criteria yield a similar resolution cut-off in the same data set of ice-embedded icosahedral particles. Nevertheless, the Fourier shell correlation criterion has become the most commonly used method. It is also common

---

BOX 13.3 Definitions of phase residual difference and Fourier shell correlation

The resolution of an icosahedral reconstruction is defined by comparing two independent data sets using the phase residual difference ($\Delta\psi$) (Frank et al., 1981; Radermacher et al., 1987; van Heel, 1987) or the Fourier shell correlation (FSC) (Harauz and van Heel, 1986; van Heel, 1987), which are defined as:

$$\Delta\psi = \sqrt{\frac{\sum(|F_1|+|F_2|)(\psi_1-\psi_2)^2}{\sum(|F_1|+|F_2|)}} \quad \text{(B13.2)}$$

$$FSC = \frac{\sum(F_1 \cdot F_2^*)}{\sqrt{F_1^2 \cdot F_2^2}} \quad \text{(B13.3)}$$

where F and ψ refer to the Fourier amplitudes and phases at different spatial frequencies, the subscripts refer to the data set 1 and 2, respectively, and the asterisk indicates the complex conjugate of the Fourier coefficient. The sum in these equations refers to all Fourier coefficients at any spatial frequency. In some cases the phrase residual is not weighted by the amplitudes, in which case the numerator in equation B13.2 is the number of terms in the sum.

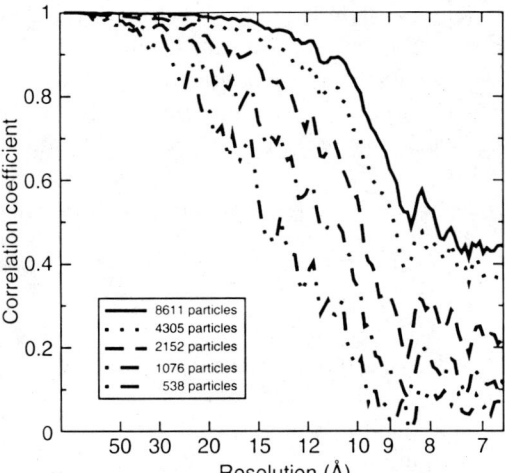

Figure 13.8 Resolution criterion for icosahedral particle reconstruction based on the Fourier shell correlation at 0.5 cutoff. This plot shows the trend for the resolution of the reconstruction to improve as the number of particles increases. Reproduced from Jiang et al. (2001b).

to plot a curve of resolution as a function of the number of particle images used in the reconstruction, as shown in figure 13.8. As expected, the higher the resolution, the more particle images are needed. In practice, more particle images than the simple geometrical relationship between particle number and resolution (DeRosier and Klug, 1968; Crowther et al., 1970b) are needed because of the non-uniform particle quality and the Fourier amplitude decay effects.

In general, an icosahedral particle contains molecular components that do not follow icosahedral symmetry, such as nucleic acids or scaffolding proteins inside a virus particle, or a portal complex at one of the 5-fold vertices. Since the reconstruction is done enforcing icosahedral symmetry, the resolution of the reconstruction defined in any formula above would likely underestimate the true resolution of the icosahedral parts of the reconstruction. As a result, the resolution is sometimes estimated after the nonicosahedral densities are computationally masked off. Furthermore, there are situations where the resolution of the reconstruction is not uniform at different radii due to either biological or experimental/computational reasons (Mancini et al., 2000).

Though the resolution tests are useful to affirm the reproducibility of the 3-D map, they do not guarantee the accuracy because of possible systematic errors introduced in the reconstruction process. As a result, the interpretation of detailed features has to be substantiated by other biochemical/biophysical evidence, and the interpretation should be consistent with the assembly principles of the particle. For instance, a powerful technique to support the claim of the observed features is to use statistically based (e.g., t-test) difference imaging of two different mass density maps that are obtained for biochemically or genetically modified particles and the wild-type particle. Another useful indicator for the correctness of the map of an icosahedral particle is the examination of possible symmetry among subunits around the non-icosahedral symmetry axis. There is no a priori requirement for the subunit conformations to follow the local symmetry. The existence of such local symmetry, as evidenced from the rotational correlation analysis at nonicosahedral symmetry axes, provides a strong argument in support for the reliability of the reconstruction because no such local symmetry is imposed in the reconstruction process (Zhou et al., 2000; He et al., 2001). In this situation, one can then use local symmetry averaging of the subunits to further enhance the statistical definition of the subunit structure before interpreting the detailed features. Lastly, if the density map reaches subnanometer-resolution, a direct identification of alpha helices which match those in some previously determined crystal structure of one of the proteins within the icosahedral particle can strongly corroborate the correctness of the map (Zhou et al., 2001a).

13.10 Poststructure analysis

Visualization of large icosahedral particles with medium-resolution features can be a challenging process because of the large volume of data and the complexity of the structure. For instance, an 0.85 nm map of the herpesvirus capsid occupies 479 × 479 × 239 pixels at a sampling interval of 0.28 nm per pixel and requires a very large computation (section 11.3) to produce a rendering of the isosurface (Zhou et al., 2000). In practice, one needs to visualize only one asymmetric unit of the entire capsid. A shaded surface or wire-frame representation with a single contour level is the most commonly used graphical technique to visualize a molecular structure. For data in the range of 0.7–0.9 nm resolution, it is feasible to detect long alpha helices, as was first shown for hepatitis B core particle (Böttcher et al., 1997; Conway et al., 1997). A proper choice of contour level is important to recognize regions of alpha helices. The characteristic feature of an alpha helix is a rod-shaped density with a diameter of 0.5–0.6 nm, which does not vanish even when contoured at a relatively high threshold value (figure 13.9). Interactive graphical tools are available to highlight the putative alpha helices as cylinders, while beta sheets appear to be a continuous density with variable sizes.

In addition to direct visual identification, it is possible to use computational tools to find regions corresponding to alpha helices and beta sheets. *Helixhunter* is an example of an automatic program with which helices in the density map can be identified computationally (Jiang et al., 2001a). It begins with a six-dimensional correlation search over the structural density with a prototypical helix approximated by a 0.5 nm diameter cylinder of two helix turns. The correlation peaks can be displayed together with the surface density map. It has been found from simulations based on known protein structures that there is a high likelihood of correctly identifying the alpha helix regions and determining the lengths of the helices when the resolution of the density map is in the range 0.7–0.9 nm (Jiang et al., 2001a). A feature extraction step then represents the helices automatically as cylinders (figure 13.9). This automated computational identification of alpha helices makes the map interpretation much easier and can even pick up some of the density regions that may not be obvious by visual inspection of the mass density alone. Automated sheet identification is less trivial but different algorithms have been proposed (Kong and Ma, 2003; Baker et al., 2007). The final verification of the helix and sheet assignment still requires a careful examination of the density map. Finally, because of the structural complexity of large icosahedral viruses, animation of the whole or a portion of the structure has been proven to very effective.

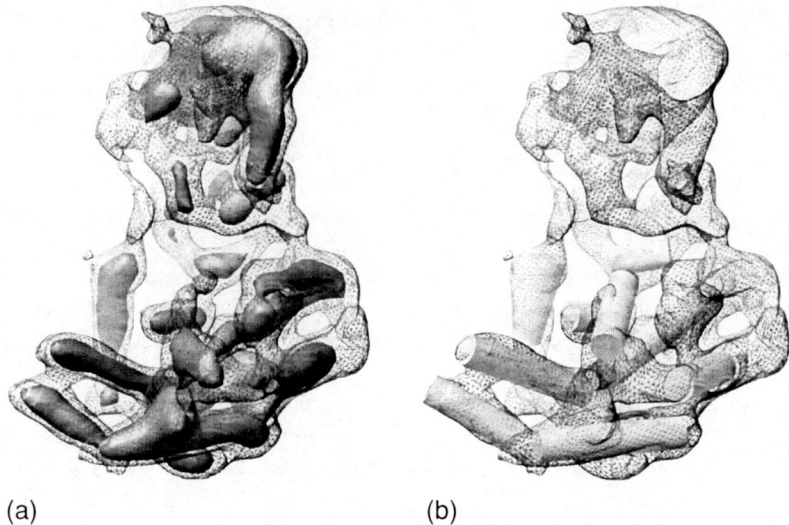

Figure 13.9 Identification of alpha helices of the outer layer capsid shell of the rice dwarf virus annotated as cylinders (right panel) in a subnanometer-resolution density map (left panel displayed in two contour levels using surface rendering and wire frame) (Zhou et al., 2001a) using the *helixhunter* (Jiang et al., 2001a). Prepared by Matthew L. Baker at Baylor College of Medicine.

Animation can be used to communicate a complex structural result to a broad audience as well as to help the investigator to examine the structure in detail.

13.11 Atomic model determination

Folds of individual components in the icosahedral particle can be generated by model building with the structural constraints imposed by the subnanometer-resolution map of the particle or by direct reconstruction of the particle to 0.3–0.4 nm resolution. The increasing power of bioinformatics tools including secondary structure prediction, threading, homology modeling, and de-novo structure prediction has made the first approach rather attractive. Further improvement of the reconstruction software with data recorded from the state of the art instrument will be critical for the second approach.

In the case of rice dwarf virus, it has been possible to obtain models for the folds of its two capsid proteins (Zhou et al., 2001a), which proved to be similar to those found in the crystal structure of the same virus which was subsequently determined (Nakagawa et al., 2003; Chiu et al., 2005). In anticipation of an increase in the number of protein folds that will result from the structural genomics effort, it is likely that a crystal structure of many molecular components or their homologs will be known. Structure-based homology modeling with the constraints of the molecular envelope of the protein components seen in cryo-electron microscope maps would yield a relatively accurate model (Topf et al., 2005). Furthermroe, the use of a density map as a geometrical constraint in identifying the most native-like models from a gallery of models generated by *ab initio* modeling has been shown to be feasible to derive a novel fold of a small protein (70 residues) of the herpesvirus capsid (Baker et al., 2006a). Therefore, fold determination based on medium-resolution density maps can be part of the structural analysis steps.

The current hurdles to reconstructing an icosahedral particle beyond 0.7 nm lie in the acquisition and processing of a large amount of high-quality image data (Jakana and Chiu, 2004). The number of particles needed for a reconstruction is inversely proportional to the intensity of the structure factor (Unwin and Henderson, 1975), and the Fourier intensity fall-off can be approximated as a Gaussian function, $\exp(-2BS^2)$ (Thuman-Commike et al., 1999). Currently, it takes a few thousand icosahedral particles images to reconstruct a subnanometer-resolution map. Based on the *experimental B factor* (50–70 Å^2) of a typical experiment in imaging icosahedral particles with the best type of electron microscope, it has been estimated that one would need 5–10 times more data to reach a resolution where the polypeptide backbone can be directly traced. In making this estimate, the particles are assumed to have their icosahedral symmetry preserved to high resolution.

Since many of the icosahedral particles are relatively large (close to 100 nm in diameter), there is another technical challenge that needs to be overcome due to the limited depth of the field (DeRosier, 2000; Zhou and Chiu, 2003) and other higher order imaging distortions such as astigmatism, magnification variations, and scanning distortions within a single micrograph. A novel refinement schema may be necessary to determine and correct these distortions before a 0.3–0.4 nm resolution structure could be obtained. In addition, the amount of data for such a large particle will be huge. For instance, images of each particle contained in boxes of over 10^6 pixels, for a 100 nm size particle. A data set of 100,000 particles and the necessary intermediate files generated during the data processing steps will occupy terabytes of disk space. The current software and hardware are not adequate to handle this type of computational problem. On the other hand, Moore's law will continue to make the hardware available and affordable in the near future. Electron cryomicroscopy will become the best tool for large icosahedral-particle structural determination, especially in those cases that prove to be difficult to handle by conventional crystallography.

14

Single Particles

14.1 Introduction

The single-particle approach to solving macromolecular structure (see Frank, 1996, 2006) is based on the premise that the specimen is in the form of many isolated *particles* that are randomly oriented, collectively presenting a large range of views. It is further assumed, at least initially, that through application of biochemical methods of isolation and purification, the specimen is *homogeneous*; that is, only macromolecules of one kind are present. Finally, if the specimen is prepared in such a way that anisotropic compression is avoided, then all the molecules on the electron microscopic grid can be assumed to have *identical structure*, and hence the geometrical relationships among the molecules in three-dimensional (3-D) space can be modeled as a set of *rigid body movements*. However, even if this assumption of structural identity is violated, due to some native structural variability, we can still use the single-particle approach for solving the structure at low resolution, and treat the structural variability as a high-resolution perturbation, a topic to be discussed later on.

Embedding the molecule in ice or glucose serves the purpose of ensuring an isotropic close-to-native micro-environment. In contrast, most methods of negative staining followed by air-drying create strongly anisotropic conditions, leading to a unidirectional shrinkage (or flattening; see chapter 6 on specimen preparation). Despite this problem, negative staining is still often used in the beginning phase of a single-particle reconstruction project, when a macromolecule is investigated for the first time with the electron microscope. The reason is that the shrinkage/flattening has only small effects on the image of untilted specimens, since the shrinkage acts in the direction normal to the specimen grid.

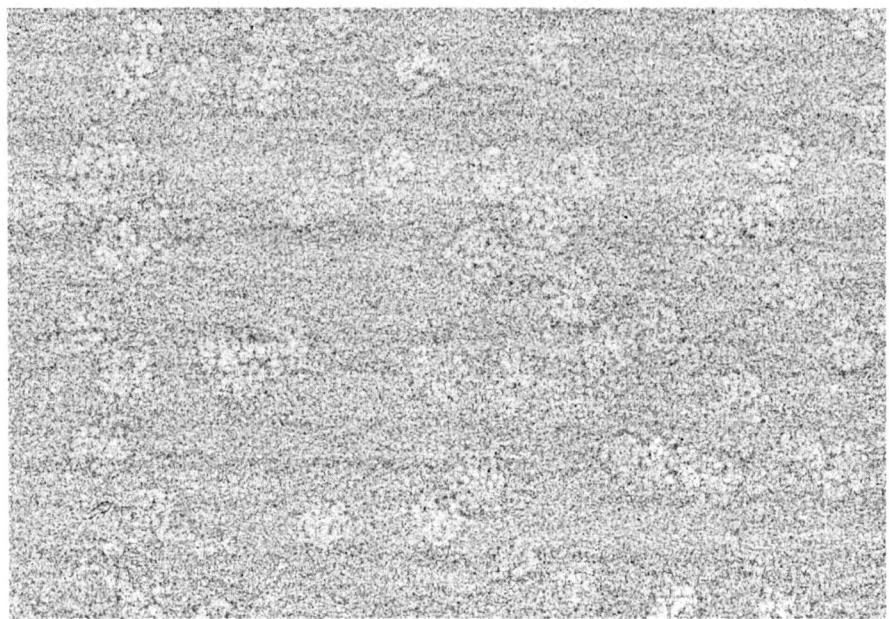

Figure 14.1 Typical cryo-electron micrograph (voltage 200 kV, magnification ×50,000) of 70S ribosomes from *Escherichia coli*.

The main challenge in the single-particle approach to structure research is to infer the angular relationships (i.e., orientations, or viewing directions) from the projections of the particles (an example of a micrograph is shown in figure 14.1). Once this has been accomplished, the molecule can be reconstructed from the electron microscope projections using standard mathematical techniques of 3-D reconstruction, except that these must be modified to account for the *uneven angular spacing*, and to compensate for the effects of the *contrast transfer function*.

As in electron crystallography proper, the molecule reconstructed represents an average over many "repeats" of a structure. For the reconstruction to be valid, however, we have to make sure that the structure itself does not vary; in other words, that the molecule population is homogeneous. In the real case, molecules do exist in different conformations, especially as they are not constrained by crystal packing, and without the benefit of an identical chemical environment that the crystal offers. These differences are particularly prominent for molecules that possess flexible structural components protruding from the overall molecular envelope.

For instance, the ribosome, which will be used throughout this chapter as an example for a totally asymmetric structure, contains a component in its large subunit, the L7/L12 stalk that projects out from the otherwise compact subunit (figure 14.2). This stalk is quite flexible and sometimes it appears bifurcated. Its movements have an important part to play during the process of protein biosynthesis. To discern such a flexible component in an average or reconstruction requires that all molecules be frozen in exactly the same state. Many 3-D reconstructions of the ribosome using single-particle

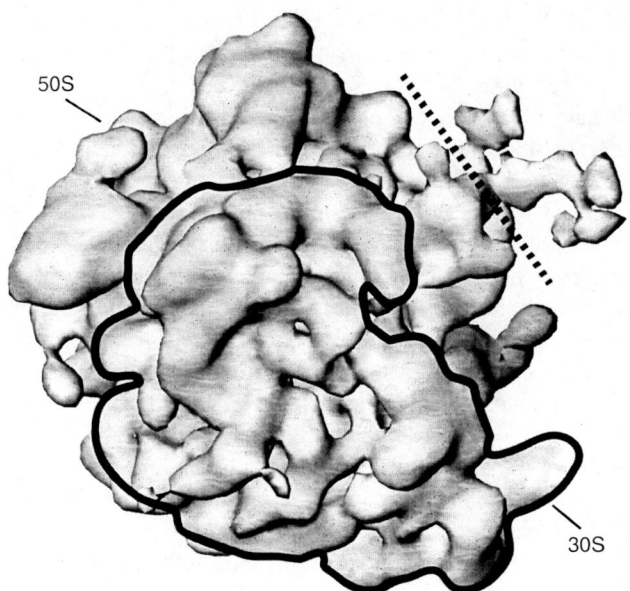

Figure 14.2 Reconstruction of a 70S *E. coli* ribosome to which EF-G (red) is bound. Due to its flexibility, the L7/L12 stalk becomes visible as a bifurcated structure only when the display threshold is lowered (right of the dashed line). From Agrawal et al. (1999); reproduced with permission of Nature Publishing Group.

methods do not show the stalk at all, a fact which indicates that it probably assumes a large range of positions in the individual molecules whose projections contribute to the 3-D map. Sometimes the stalk shows up only when the display threshold is lowered (which is the case with the reconstruction shown in figure 14.2), indicating that it has the same position in at least a subset of the ribosomes. Only in certain ribosome–ligand complexes, which are apparently "locked" in a defined conformational state, does the stalk show up at normal display thresholds.

Single-particle techniques can under certain conditions deal with conformational or compositional heterogeneity of the molecule population. If the conformational changes are large enough, or heterogeneous molecular species are sufficiently dissimilar, then the corresponding images can be separated by *classification*, and molecules in the different conformational states, or of different types, can be separately reconstructed.

The most obvious advantage of the single-particle approach over conventional electron crystallography is that it requires no crystals or otherwise ordered aggregates of any sort, and that the method of data collection is therefore much simplified. Another advantage is that this approach enables visualization of ligand binding states entirely unimpeded by crystal packing forces. Thus, the range of states visualized is likely to be close to the range occurring in the native state. However, among the disadvantages of the approach is the fact that, because of certain requirements of minimum signal-to-noise ratio for the alignment and classification steps, it cannot be used for molecules smaller than a certain minimum size. Furthermore, it is still difficult with present technology — though not impossible on grounds of physical principles (Henderson, 1995) — to achieve near-atomic resolution (see section 14.13).

An unforeseen semantic confusion has arisen recently with the onset of single-molecule techniques of experimentation. The word "single" has two meanings, one referring to the state of spatial isolation and separation from other objects of the same kind, such as the field of ribosome particles shown in figure 14.1, the other to the unique identity of an object. When the single-particle averaging technique was proposed (Frank, 1975), the term was used to distinguish it from the existing approaches characterized by the analysis of specimens containing ordered, closely packed aggregates of molecules. Still, both approaches had in common that the structure is obtained by averaging over images of a large number of "copies" of the molecule. In contrast, many newer experimental techniques obtain observations (such as fluorescence spectra, force measurements, direct imaging by atomic force microscopy, etc.) on individual, single molecules without the need for averaging.

Our description of the single-particle reconstruction approach is structured in the following way. We first define, in section 14.2, the minimum particle size and the minimum number of particles needed for a 3-D reconstruction. These numbers depend in an important way on the maximum electron exposure that the molecule can tolerate (due to radiation damage), and the extent to which the images retain the full contrast that could, in principle, be available on the basis of the atomic scattering cross-sections. In section 14.3 we discuss the further requirement that the reconstruction must include a minimum number of different views of particles; that is, the particles must lie in that many different orientations. The presence of multiple conformational states creates a further complication, in effect increasing the size of the data set that must be collected, as is discussed in section 14.4. Moving on to the tools for merging data from a large number of single-particle images, we discuss alignment (section 14.5) and different methods of classification of the individual particles (sections 14.6, 14.7, and 14.8). A brief summary of alternative methods of 3-D reconstruction is given next, in section 14.9. This is followed by a description of "bootstrapping methods," or approaches to produce an initial reconstruction of an unknown structure (section 14.10) and to refine such a reconstruction toward higher resolution (section 14.11). The chapter continues (section 14.12) with a step-by-step account about the data-processing steps for a particular choice of reconstruction strategy, starting from the raw data and leading to the refined, contrast transfer function (CTF)-corrected reconstruction. Finally, a concluding section (14.13) will address the question how and under what circumstances atomic resolution might be achieved.

For a more comprehensive treatment of single-particle reconstruction, the reader is referred to Frank (1996, 2006), Frank et al. (2000), and van Heel et al. (2000).

14.2 A certain minimum dose is required to align images of single molecules

The basis for imaging in the electron microscope is the scattering of electrons by the specimen, and the ability of magnetic lenses to focus the electron beam. *Elastic scattering*, an interaction without energy loss, leaves the specimen unharmed and provides information on the structure of the specimen. This is always accompanied by *inelastic scattering*, interactions of various kinds accompanied by loss of energy, which result in damage to the specimen and contribute no useful information. Thus, in figuring out how many electrons are required, for statistical reasons, either for imaging molecules or for aligning the resulting images, only the elastically scattered electrons can be counted, whereas, in gauging the extent of radiation damage one has to go by the number of inelastically scattered electrons (for details on the ratio of inelastic-to-elastic scattering events and its voltage dependence, see chapter 15).

For a crystal, it is possible to distribute the radiation load among a large number of repeats of the molecule — in principle even below the level of one electron per repeat. In contrast, single molecules must be exposed to a certain minimum number of electrons so as to ensure that their projections contain sufficient features to allow alignment. According to an expression derived by Saxton and Frank (1977), the minimum dose p_{min} (electrons/unit area) required to detect a particle of diameter D and resolution d using cross-correlation is given by

$$p_{min} \geq \frac{3}{c^2 Dd} \tag{14.1}$$

where c is the contrast. If we now stipulate that the dose should be smaller than p_{crit}, the critical dose for which unacceptable damage occurs, we obtain

$$D_{min} = \frac{3}{c^2 d p_{crit}} \tag{14.2}$$

for the minimum particle size D_{min}. From this expression it is clear — since the contrast cannot be controlled unless staining is used — that *the success of the approach depends critically on the size of the particle*. That means if there is a choice between studying a molecule separately and studying it as part of a larger assembly, the latter is always preferable. Equation 14.2 also shows the reciprocal relationship, akin to the uncertainty principle, between particle size and resolution: smaller particles can be studied only at lower resolution if the radiation level is to be kept constant.

14.3 Due to the lack of symmetries, 3-D imaging requires coverage of the entire angular space

Three-dimensional reconstruction requires 3-D Fourier space to be sampled out to the desired resolution, each projection providing one central section at a time. Even though the reconstruction is often carried out using a real-space algorithm — for example, by weighted back-projection (box 14.4) or variants of the algebraic reconstruction technique (ART) (box 14.5) — the requirements and limitations of the technique are most easily analyzed in Fourier space.

Reconstruction of an object from a finite number of projections is possible because each Fourier coefficient of the object is surrounded by a "region of influence," a region within which the complex values of the Fourier transform are related to one another. The shape and size of this region is determined by the object's *shape transform*, that is, the Fourier transform of a 3-D mask function describing the shape of the object, whose values are 1 inside the object's boundary and 0 outside. The number of projections needed follows from the requirement that in Fourier space, out to the desired resolution, sample points must not be separated by more than the region within which the shape transform has appreciable nonzero values. This stipulation, when applied to a tilt series of N equispaced projections of an object that has no symmetries, leads directly to the known formula (DeRosier and Klug, 1968; Crowther et al., 1970) for the number of projections required to reconstruct an object of diameter D to the resolution d:

$$N = \frac{\pi D}{d}. \tag{14.3}$$

In other words, the larger the object, the smaller the extent of its shape transform, hence the more projections are required overall. However, there are three reasons why the number of projections has to be much larger in practice than N given by equation 14.3. One reason is that the projections are not equispaced as assumed in the derivation of this equation, but statistically distributed over the whole angular range. This means that a gap larger than expected with equal angular spacings will be encountered with high probability. The second reason is the low signal-to-noise ratio of the data, a consequence of the use of low electron dose, which makes it necessary to collect data redundantly. The third reason (which might or might not apply) is due to structural heterogeneity, as will be detailed below.

14.4 Conformational variability increases the total number of images needed to achieve higher resolution

A complication is introduced by the fact that the molecule may be present in different conformations. Thus the data collection may initially produce a conformational mixture that has to be sorted in the course of the processing. The level of scrutiny required to perform this sorting must necessarily increase with the resolution sought, since a large number of conformational changes involve rearrangements or flexing of subdomains and secondary structure elements (helices, β-sheets, etc.) over short distances only. Leaving aside for the moment the exact means of achieving a division of the data into homogeneous subsets, it is clear that the mathematical and statistical requirements spelled out in the preceding section will pertain to each of these separate subsets. Thus, simply put, if the reconstruction of a molecule to a given resolution requires 30,000 particles, and it exists in 3 different conformations in equal proportions, then the total data set obviously must comprise a minimum of 90,000 particles. We can speak of a successive resolution-dependent "conformational bifurcation" in the treatment of a set of single molecules, since each subset with distinctly different structure must be processed separately, and since the number of subsets increases with increasing resolution.

An important question to be answered at the outset in the practical execution of an experiment is whether all conformational states are subject to interest, or perhaps only a subset of these states, or even only a single state. This decision would determine the way the data are efficiently screened from the very start. With electron microscopes that have integrated functions for image readout, instrument control, and preprocessing, it is now possible to write software that screens for data that are compatible with pre-existing knowledge regarding particle shapes or density statistics (Potter et al., 2004).

14.5 Alignment of particles is required for averaging and image reconstruction, and its principal tool is the cross-correlation function

Alignment is used to establish a common frame of reference for the particle projections. Images need to be rotated and translated with respect to one another, so that any two projections showing the molecule in the same view come to a precise pixel-by-pixel overlap. Moreover, we also want to make sure that even for images of particles exhibiting *different* views, some type of consistent "alignment" can be achieved, in the sense that if we align any image showing view number 1 with any image showing view number 2, we expect to always get the same relative position.

The basis for the alignment is that the two images have a common "signal" — here the molecule projection. The two versions of the signal come into register only for a particular choice of translation and rotation that is applied to one image relative to the other. When the two versions of the signal exactly overlap, the cross-correlation function (CCF) between the two images is maximized. Conversely, the CCF can be used to find the exact overlap position, and thus the CCF in its various forms has an important role to play in particle alignment.

The cross-correlation function between two images, $f_1(\mathbf{r}_j)$ and $f_2(\mathbf{r}_j)$, represented by measurements on a regular grid with coordinates $\mathbf{r}_j = (x_j, y_j)$, can be derived by considering the different terms of the *generalized Euclidean distance*, a measure of similarity (see box 14.1). Most convenient for numerical purposes is the *translational CCF*,

$$\text{CCF}(\mathbf{r}_k') = \sum_j f_1(\mathbf{r}_j) f_2(\mathbf{r}_j + \mathbf{r}_k') \tag{14.4}$$

which is a function of two variables (x_k', y_k'; that is, the components of the probing vector $\mathbf{r}_k'$) only. This function can easily be computed by using the fast Fourier algorithm: both images are Fourier-transformed, the conjugate product of their Fourier transforms is computed, and the result is finally inversely transformed.

The translational CCF can be used only when the images lie in the same or closely similar orientation, otherwise it gives a very weak signal. Hence, in practice, the translational search must be combined with an orientational search. One way to accomplish this is by centering the particles first, by aligning them with a centered "blob" (e.g., a low-pass filtered disc) of the same size, then rotating them with respect to one another around the new origin while maximizing their cross-product, then again using translational search, etc. Orientational search employs an expression quite similar to the CCF defined in equation 14.4, except that now the translation in the argument is replaced by a rotation (expressed by a 2-D rotation matrix $\mathbf{R}_\phi$ with angle ϕ):

$$\text{RCCF}(\phi) = \sum_j f_1(\mathbf{r}_j) f_2(\mathbf{R}_\phi \mathbf{r}_j). \tag{14.5}$$

It is practical to introduce a polar coordinate system at this stage. The images, originally represented by samples on a Cartesian system, are re-sampled by interpolation on a set of concentric circles, leading to new arrays $f_1(r_j, \phi_j)$, $f_2(r_j, \phi_j)$ indexed

BOX 14.1 The generalized Euclidean distance

The generalized Euclidean distance is a concept that is quite useful in understanding the meaning of the cross-correlation function, as well as providing the basis for multivariate data analysis of images. Two images, $f_1(\mathbf{r}_j)$ and $f_2(\mathbf{r}_j)$, represented by samples ("pixels") on a discrete grid $\{\mathbf{r}_j, j = 1\ldots J\}$, may be considered vectors in a J-dimensional space $\mathbf{R}^J$. Thus, each grid point is a component of the high-dimensional vector. In that space, a distance E can be defined, in analogy to the distance defined in the 3-D Euclidean space, so that its square is

$$E^2 = \sum_j |f_1(\mathbf{r}_j) - f_2(\mathbf{r}_j)|^2 \tag{B14.1}$$

Images that are dissimilar have a large Euclidean distance, since they normally differ in the values of all pixels. This distance definition is impractical, however, since it depends on the — often arbitrary — relative geometric position of the two images. For example, two images that are identical, but are rotated and shifted with respect to each other would be considered quite dissimilar, since E^2 would have a large value. It is useful, therefore, to regard E^2 as a function of the relative translation (expressed by a vector $\mathbf{r}'$) and rotation (angle α, expressed by a rotation matrix $\mathbf{R}_\alpha$) between the two images,

$$E^2(\mathbf{R}_\alpha, \mathbf{r}') = \sum_j |f_1(\mathbf{r}_j) - f_2(\mathbf{R}_\alpha \mathbf{r}_j + \mathbf{r}')|^2 = \sum_j |f_1(\mathbf{r}_j)|^2 + \sum_j f_2(\mathbf{R}_\alpha \mathbf{r}_j + \mathbf{r}')|^2$$

$$-2 \sum_j f_1(\mathbf{r}_j) f_2(\mathbf{R}_\alpha \mathbf{r}_j + \mathbf{r}')| \tag{B14.2}$$

Our goal is to pay special attention to the minimum that the expression assumes as $\mathbf{r}'$ and $\mathbf{R}_\alpha$ are varied over their entire value range. The principle on which this choice relies is that the degree of similarity of two images should be a measure unaffected by their relative positions, and that one should go with the *highest* similarity that is attained for *any* choice of relative positions.

It is important to see that the first two terms in the above expression are invariant under relative movements (translations, rotations) of the images. These two terms are, in fact, related to the variance of the images, which is a quantity that is obviously invariant under translation and rotation. In contrast, the most important, third term (called the *cross-correlation function,* CCF) critically depends on the values of the arguments $\mathbf{r}'$ and α. If the two images $f_1(\mathbf{r}_j)$ and $f_2(\mathbf{r}_j)$ are noise-corrupted versions of the same underlying signal, then the CCF is *maximized* for a position that brings the signal component of the two images into perfect register, leading to the global *minimum* of E^2.

The CCF so introduced is a function of three variables: two translational components x, y and one angle, α. Alignment of two images is achieved by computing their CCF and searching for the combination (x_m, y_m, α_m) that leads to a maximum. For practical reasons, a simplified version of the CCF is used, which depends on two variables (x, y) only: the translational CCF.

by radius r_j and azimuthal angle ϕ_j, so that the rotational cross-correlation function RCCF becomes

$$\text{RCCF}(\phi') = \sum_j f_1(r_j, \phi) f_2(r_j, \phi'). \qquad (14.6)$$

Alignment of an entire image set can be achieved by sequentially aligning each image of the set to one or several reference images. Many different strategies have been devised; however, since alignment is always followed by some type of classification, the only issue that counts, in judging the relative merits of different algorithms, is to what extent similar images will be aligned consistently.

In order to show why the CCF (or, following similar arguments, the RCCF) can be used for alignment, we need to introduce the autocorrelation function (ACF), which is simply the CCF of an image with itself. The ACF has a sharp peak at the origin, reflecting the fact that for $\mathbf{r}' = \mathbf{0}$, all elements of the image come into a precise overlap. Now let us consider the CCF of two images $i_1(\mathbf{r}_j)$, $i_2(\mathbf{r}_j)$ that contain the same particle projection $p(\mathbf{r})$ in two different positions separated by a translation vector $\Delta \mathbf{r}$ and superposed with two different realizations of noise, $n_1(\mathbf{r}_j)$, $n_2(\mathbf{r}_j)$:

$$\begin{aligned} i_1(\mathbf{r}_j) &= p(\mathbf{r}_j) + n_1(\mathbf{r}_j); \\ i_2(\mathbf{r}_j) &= p(\mathbf{r}_j + \Delta \mathbf{r}) + n_2(\mathbf{r}_j) \end{aligned} \qquad (14.7)$$

The only term in the CCF of the images in equation (14.7) that does not involve a noise term is

$$\text{CCF}(\mathbf{r}_k') = \sum_j p(\mathbf{r}_j) p(\mathbf{r}_j + \Delta \mathbf{r} + \mathbf{r}_k') = \text{ACF}(\mathbf{r}_k' + \Delta \mathbf{r}), \qquad (14.8)$$

which means that the ACF of the particle appears in a position that is shifted by $-\Delta \mathbf{r}$, the negative of the translation vector. The CCF therefore has a sharp peak at $-\Delta \mathbf{r}$. Hence, conversely, the appearance of a peak at that location signifies that the two identical motifs are shifted by $\Delta \mathbf{r}$.

Since the height of the CCF peak above noise determines the feasibility and accuracy of particle alignment, it is important to understand how the CCF is affected by the CTF. In Fourier space,

$$\mathcal{F}\{\text{CCF}\} = \mathcal{F}\{i_1\} \mathcal{F}^*\{i_2\} \qquad (14.9)$$

where symbol $\mathcal{F}$ denotes Fourier transformation and $*$ denotes complex conjugation. If the images contain the same projection p but they were obtained with different defocus, then $\mathcal{F}\{i_{1/2}\} = \mathcal{F}\{p\} \, \text{CTF}_{1/2}$, where $\text{CTF}_{1/2}$ are the contrast transfer functions associated with the two defocus settings. This means that

$$\mathcal{F}\{\text{CCF}\} = |\mathcal{F}\{p\}|^2 \text{CTF}_1 \text{CTF}_2. \qquad (14.10)$$

From this formula one can gauge the effect of the CTFs on the value of the CCF peak (which is obtained by integration of equation 14.10 over the Fourier domain). The best result is evidently achieved when the two CTFs are precisely matched, since in that case their product and hence the entire integrand is positive-definite. When the CTFs are different, as a consequence of different defoci, then they tend to get increasingly out of phase with increasing spatial frequency, leading in the worst case to spatial frequency intervals where a positive lobe of one CTF coincides with the negative lobe of the other. Thus the cross-product $CTF_1 \times CTF_2$ in equation 14.10 is no longer positive everywhere in the spatial frequency range. This has the consequence that the CCF peak resulting from an integration of equation 14.10 will be small and blurred, and thus more likely to disappear in the noise background.

The sensitivity of the CCF to a CTF mismatch has an important implication in the reference-based orientation determination (section 14.11): it will become clear that, to optimize the conditions for alignment and classification of experimental projections, one should always multiply the 3D reference with the CTF of the current experimental data.

14.6 Classification may be used to divide the projection set according to viewing directions, conformations, and ligand-binding states

An electron micrograph presents particles in different viewing directions, and often also in different conformations and ligand-binding states ("occupied" or "unoccupied"). Various tools of classification have been developed to divide the data into homogeneous groups that can be processed separately. Classification uses either data-intrinsic features (e.g., the existence of a division into clusters, requiring an analysis of the image set as a whole), or comparison of each image with some kind of template or reference. The first kind of classification is called "*unsupervised*," the second "*supervised*." Current methods of single-particle analysis makes use of both kinds of classification at different stages, and as part of different strategies of reconstruction.

An example for classification of the first kind occurs at the very beginning of a project: dealing with an unknown structure, one wishes to have an inventory of existing views, to develop a strategy for data collection and 3-D reconstruction. The second kind of classification, involving comparison with a 3-D reference, is exemplified by the orientation determination through 3-D projection alignment, a technique that is now routine for the refinement stage of each project (section 14.11). The remainder of the present section, as well as the following two sections (14.7 and 14.8), will provide an introduction into techniques of classification of the unsupervised kind. (At a later stage in this chapter (section 14.12), the topic of classification will be revisited in the context of sample heterogeneity, as a technique intrinsically linked to the reconstruction procedure itself. At the same place the 3-D variance will be introduced, as a diagnostic tool helping to determine if conformational heterogeneity exists, and wherein the molecule it is located.)

Classification entails a multiple comparison of the images in the space used for their discrete representation. For example, a set of images represented by arrays of 100×100 pixels can in principle vary in any of the 10,000 pixels, or any combination thereof. Thus the space of representation in this case has a dimension of 10,000, making multiple comparisons virtually impossible. However, classification is greatly facilitated by the use of a special kind of reduced data representation, achieved through multivariate data analysis[1] (MDA; see section 14.7). The idea here is that any significant variability among images of a molecule never affects a single pixel only, but entire groups of pixels, in a correlated way. For example, a change in the position of a flexible stalk (the example used in section 14.1) would affect many pixels along the length of the stalk: the pixels at the front of such a movement would gain density, the ones in the back would lose density in a correlated fashion. Similarly, the presence or absence of a ligand would affect pixels within an entire region of the molecule as seen in projection.

The fact that variations affect extended regions comprising many pixels has the important consequence that the investigation of the conformational changes in the molecule population and their clustering can be done in a *subspace of much lower dimensionality*. The next section provides an introduction into the technique that is used to obtain such a low-dimensional representation. The section after that will explain how classification may be done efficiently in this reduced space.

14.7 Variational patterns among images of macromolecules can be found by using multivariate data analysis or self-organized maps

Let us imagine a mixed set of aligned images, for example, a mixture of images containing molecules with a ligand bound, and those without. (We confine the analysis to molecules lying in the same orientation.) This is a special, "binary" case of a more general phenomenon of variability, which is often continuous: a ligand could be bound to each molecule, but it could be positioned in any of a continuous range of orientations. There are two approaches that have been taken to analyze images from such mixed objects: one is driven by an analysis of the problem in a high-dimensional space, the other by using a self-organizational principle.

Multivariate data analysis

In the high-dimensional space of the original representation, each image is represented by a vector going from the origin to the vector end point (see box 14.2). Thus the entire image set is represented by a "cloud" of points. The grouping of images according to the presence or absence of the bound ligand will be reflected by the distribution of points within the data cloud. Depending on the amount of noise present, the cloud may be sharply divided into two subclouds, or there could be a fuzzy transition between two portions of the cloud, each having a high density of "points."

MDA starts out with the following rationale: we wish to design a new, low-dimensional representation of these data by using a set of orthogonal basis vectors of unit length that are tailored to the distribution: we want the vector pointing into the direction of the largest data variability, that is, the one pointing from the center of subcloud #1 to that of subcloud #2, to be the most important basis vector. Next we want to determine that vector which has the following properties: (1) it goes into the

---

[1] Since "multivariate statistical analysis" is used in the general literature for a wide range of techniques, some of which utilize statistical models, the term "multivariate data analysis" is preferable here.

BOX 14.2 Multivariate data analysis

We wish to analyze an image set for the existence of clusters or other variational patterns. For the purpose of the analysis, each image of the set is initially represented by a vector in the space $\mathbf{R}^J$ already introduced in the definition of the generalized Euclidean distance (box 14.1). As pointed out in the text, images typically do not vary in single pixels, but in combinations of many pixels. Such combinations correspond to oblique vectors in $\mathbf{R}^J$. For this reason, $\mathbf{R}^J$ is not well suited for analyzing the problem, but a space should be used whose axes point along the most prominent directions of variability. Figure B14.1 shows how such a space is constructed, based on the desired property of being well tailored to the data.

The space $\mathbf{R}^J$ is populated with vectors $\mathbf{x}_i = \overline{OP_i}$ representing the images, where P_i are the vector end points and O is the origin. We now introduce a new vector $\mathbf{u}_1$ that we wish to position optimally with respect to the existing data vectors: we want it to point in the direction that best reflects the shape of the cloud of vector end points. To this end, we specifically stipulate that the projections of the data vectors onto the new vector, $\overline{OP'_i}$, be maximal. This leads to the condition, in the construction of $\mathbf{u}_1$, that the sum of the squared projections of $\mathbf{x}_i$ onto $\mathbf{u}_1$ is a maximum. Having found this vector, we next

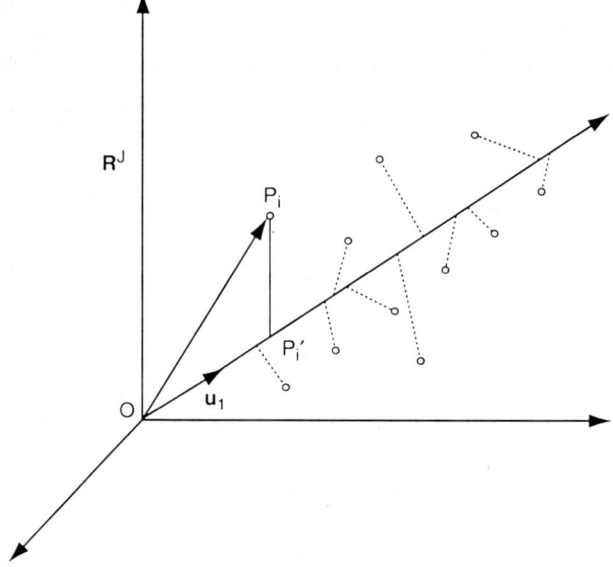

Figure B14.1 Construction of a vector that optimally represents the direction of a data cloud in a high-dimensional space. Each point is the end point of a vector that represents an image in $\mathbf{R}^J$. For details, see text. From Frank (2006); reproduced with permission of Oxford University Press.

(continued)

wish to find a vector $\mathbf{u}_2$ that is orthogonal to $\mathbf{u}_1$ and points into the direction of maximal extension of the data cloud, etc.

The mathematical analysis that follows the above concept (Lebart et al., 1984) leads to the common eigenvector-eigenvalue problem of the form

$$\mathbf{Du} = \lambda \mathbf{u} \qquad (B14.3)$$

where **D** is a so-called symmetric variance/covariance matrix with the general element

$$d_{ii'} = \sum_j x_{ij} x_{i'j}, \qquad (B14.4)$$

with x_{ij} standing for "jth element of image i." (That means, in this notation, that the image is considered a 1-D array of pixels arranged in "lexicographic order" — with no explicit reference to the arrangement of the pixels into rows and columns.) In correspondence analysis (Frank and van Heel, 1980; Lebart et al., 1984; Borland and van Heel, 1990), a variant of principle component analysis, the element of the matrix is normalized in a particular way, but the analysis proceeds as in principal component analysis.

This matrix equation has a series of solutions $\mathbf{u}_i$ (eigenvectors of the problem formulated in equation B14.1) with associated eigenvalues λ_i. These solutions are in fact the basis vectors (often called "factors") that we were looking for, and the associated eigenvalues measure the share of variational energy that each factor accounts for.

In the following, for illustration, we use a set of images that are generated from 8 distinct prototypes (figure B14.2) by the addition of random noise (figure B14.3). A histogram of eigenvalues makes the drop-off of variational energy immediately apparent (figure B14.4). On the basis of such a histogram, a decision can be reached on which number, M, of highest ranking factors express the most important variations. In the example, M is evidently equal to 3. Thus, each image is now represented by a set of $M = 3$ coordinates, each coordinate

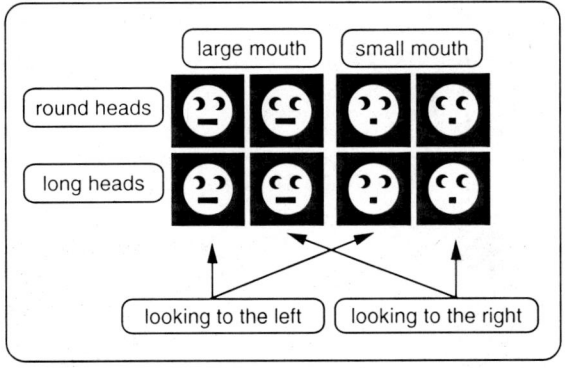

Figure B14.2 Motif to demonstrate multivariate data analysis. The motif varies in three features (shape of head; width of mouth; position of eyes), each of which exists in two variants (round/oval; wide/narrow; left-/right-facing). (Kindly provided by N. Boisset).

(continued)

BOX 14.2 *(continued)*

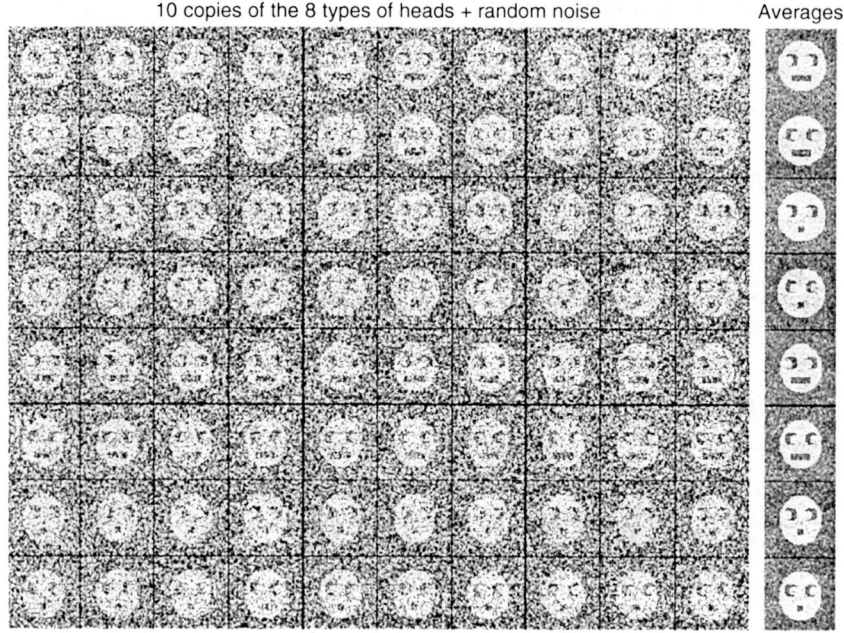

Figure B14.3 Noisy set of 80 images created by superimposing noise on all 8 variants. Kindly provided by N. Boisset.

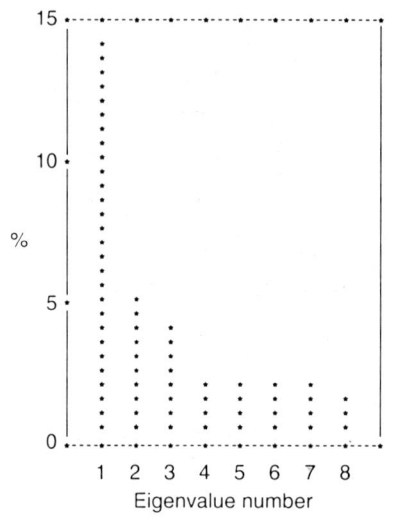

Figure B14.4 Eigenvalue histogram, showing that three factors are quite prominent. Kindly provided by N. Boisset.

(continued)

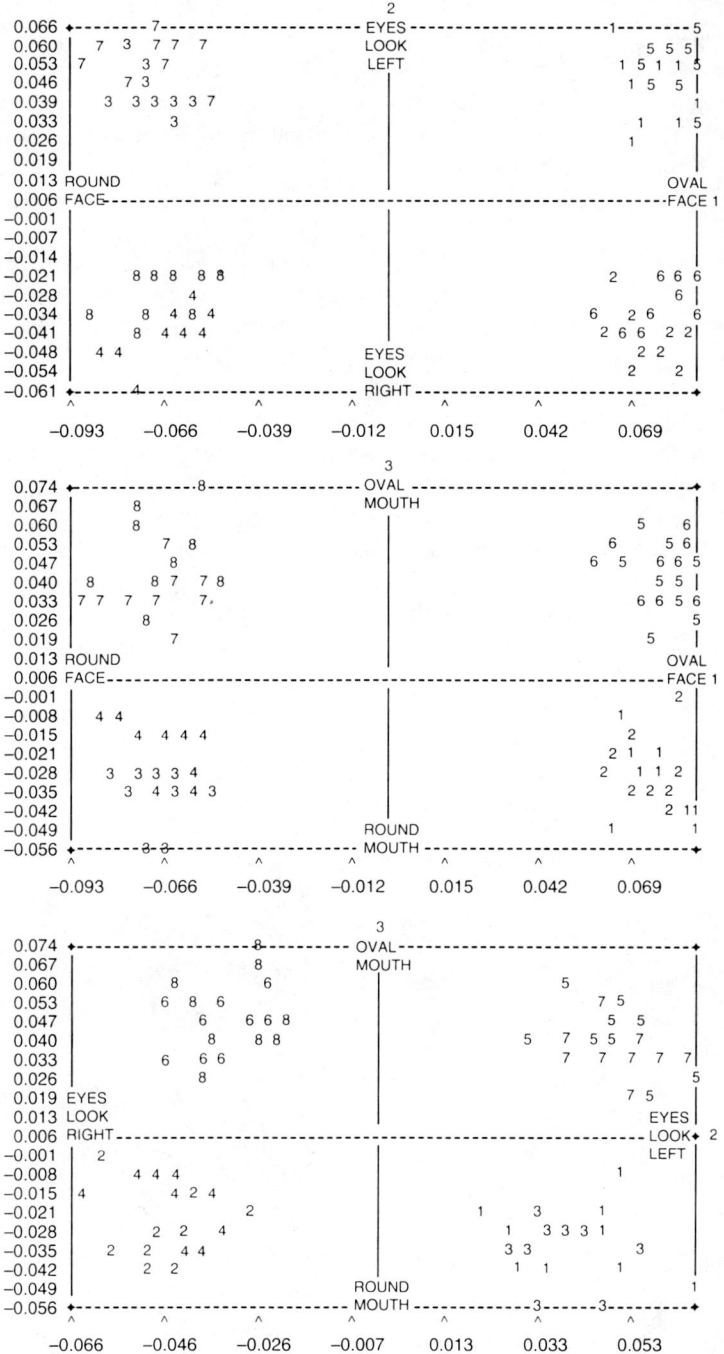

Figure B14.5 Factor maps (1 versus 2, 1 versus 3, and 2 versus 3) showing the sharp clustering of the points representing the images into the four corners of a cube. Kindly provided by N. Boisset.

(continued)

BOX 14.2 *(continued)*

being the projection of the data vector, representing the image, onto one of the factors. To find out the grouping of images and possible clustering, we make factor maps — these are maps of any combination of two factors on which the images are shown as points or symbols, according to their coordinates. Since we have found 3 factors to be significant, we need to look at 3 factor maps (figure B14.5). Here we see sharp clustering — the 8 subsets of images originating from the 8 prototypes evidently cluster in the 8 corners of the 3-D factor space. To see how the differences in the images are expressed in the eigenvector expansion, we can look at the eigenvectors in pictorial format (hence the term "eigenimages") (figure B14.6).

Any image of the image set can be recovered by summing over terms that describe the contributions from each factor. Geometrically, this recovery (which is termed "reconstitution"; Lebart et al., 1984) can be visualized as a vector summation, starting from the vector representing the average image, gradually adding vectors of different lengths that point in the directions of the eigenvectors, and eventually leading to the vector representing the image that is being considered. Since variational energy is strongly concentrated in a relatively small set of highest ranking factors, one is able to obtain a reasonable estimate of the image when breaking up the summation after a few terms ("partial reconstitution"; see figure B14.7).

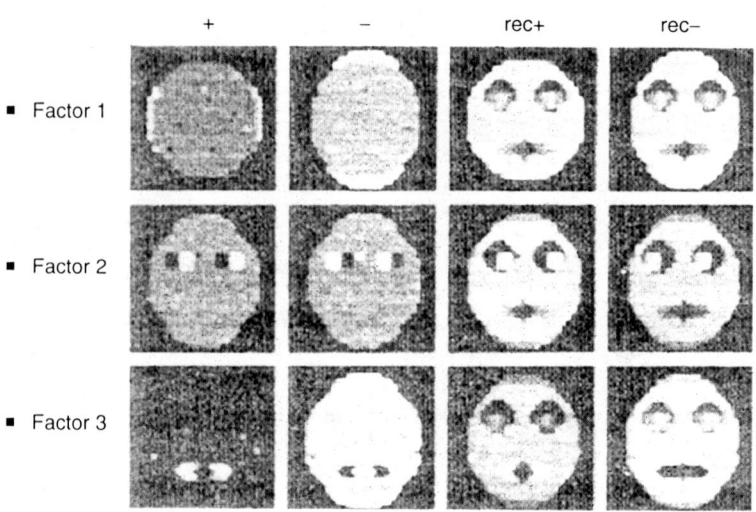

Figure B14.6 Columns 1 and 2: positive and negative parts of the eigenimages related to the three varying features. Columns 3 and 4: eigenimages added to, or subtracted from, the average image. Note that in each case, the result is an image that exhibits one feature sharply but is blurred with respect to the other two. For example, the two images on the top have distinct shapes (round, oval) but blurred eyes and mouth portions. Kindly provided by N. Boisset.

(continued)

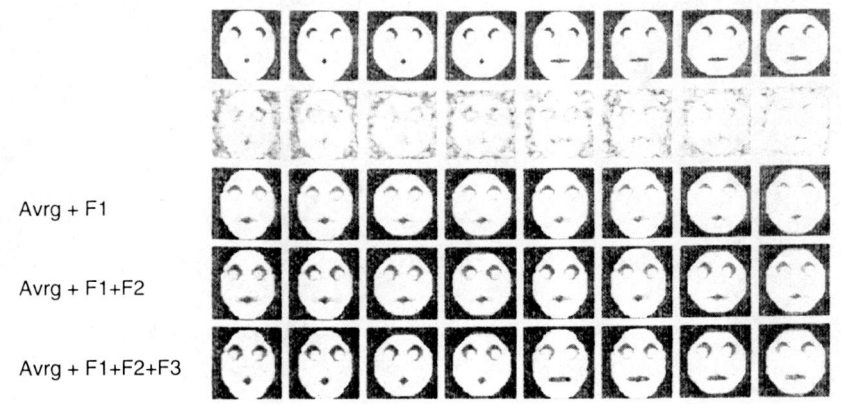

Figure B14.7 Stepwise recovery of the motif underlying some noisy images of the original set of 80 images, by partial reconstitution. First row: original motifs; second row: noise-corrupted versions of these motifs. Third through fifth row: successive restoration of the motif by adding contributions due to the first, second, then third factor to the average image. Kindly provided by N. Boisset.

direction of the next-largest extension of the cloud; (2) it runs orthogonal to the first; and (3) it is normalized to unit length. In this way, we proceed until all basis vectors have been found.

The mathematical analysis of the problem presented in the foregoing shows that the basis vectors are the eigenvectors of the variance/covariance matrix associated with the image data (box 14.2). It is common to refer to these eigenvectors as *factors*. The corresponding eigenvalues, ranked by size, are a measure for the importance of each factor in expressing the fraction of the total "variational energy" that they account for. Variational energy is the term often used for the total variance of a data set.

The dimensionality of the space of this representation, that is, the maximum number of factors found when we proceed in this way, is either equal to the number of pixels or the number of images, whatever number is smaller. For example, if the data set consists of 2000 particle images that are each represented by 75×75 pixels, or a total of 5625, then the dimensionality of the space is 2000, and hence only that number of factors are needed for a full representation. As a result of the construction principle and the ranking implicit in it, the variational energy associated with each factor in this expansion drops off very rapidly for images encountered in electron microscopy. What that means is that a relatively small number of factors are normally sufficient to represent the entire pattern of significant variations, while the remaining bulk portion of the basis vectors represent variations due to noise. Figure 14.3 illustrates the example of a cloud consisting of two subclouds in a space spanned by three basis vectors.

The approach of MDA described thus far is known as principal component analysis (see Lebart et al., 1984). Mainly for historic reasons, the first applications of MDA in electron microscopy (e.g., van Heel and Frank, 1981; Frank et al., 1982) made use of a variant known as correspondence analysis (Frank and van Heel, 1980; Lebart et al., 1984). Here the elements of the matrix subject to diagonalization in the

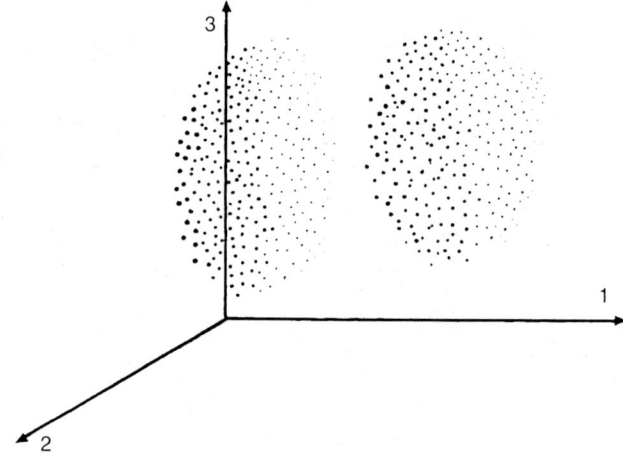

Figure 14.3 Schematic diagram depicting the distribution of data on "clouds" in factor space. From Frank (1985); reproduced with permission of Van Nostrand Reinhold.

eigenvector-eigenvalue equation (equation B14.1) are normalized in a particular way. Experience has shown, however, that the results of the two approaches when applied to single-particle images are quite similar (see also Borland and van Heel, 1990).

Most important for the practical application is the fact that instead of images, in their original, exhaustive representation, we now deal with points in a low-dimensional space. Thus each image is expressed by a set of coordinates, one for each eigenvector deemed significant. Since the eigenvectors are called *factors*, a map that shows the locations of the images as points in a 2-D subspace is called a *factor map*. A factor map can be regarded as a projection of the data cloud, initially represented in a high-dimensional space, onto a plane that is spanned by two selected eigenvectors. Such factor maps are particularly useful in those cases where variational energy is concentrated in very few factors. Images of particles prepared by the negative staining technique often show such behavior (see early applications of correspondence analysis: van Heel and Frank, 1981; Frank et al., 1982). Inspection of factor maps has been very useful in examining data clustering or continuous trends that reflect orientational preferences or "rocking" behavior of molecules on the support film that is apparent in stained, air-dried specimens.

Variability among projections of single molecules embedded in ice is dominated by shape variations due to changes in orientation. Even when we look at a subset of such molecules presenting the same general view, they prove to be much more variable than molecules prepared by negative staining, as reflected by a much broader range of factors. However, more prominent changes such as the presence or absence of a large ligand may still lead to a division of the data along one of the highest-ranking factors.

For illustration of MDA, an artificial projection set was created (figure 14.4). Projections were computed from a $1/1.15$ nm^{-1} reconstruction of the *E. coli* ribosome (described in Gabashvili et al., 2000) in three entirely different directions. For each

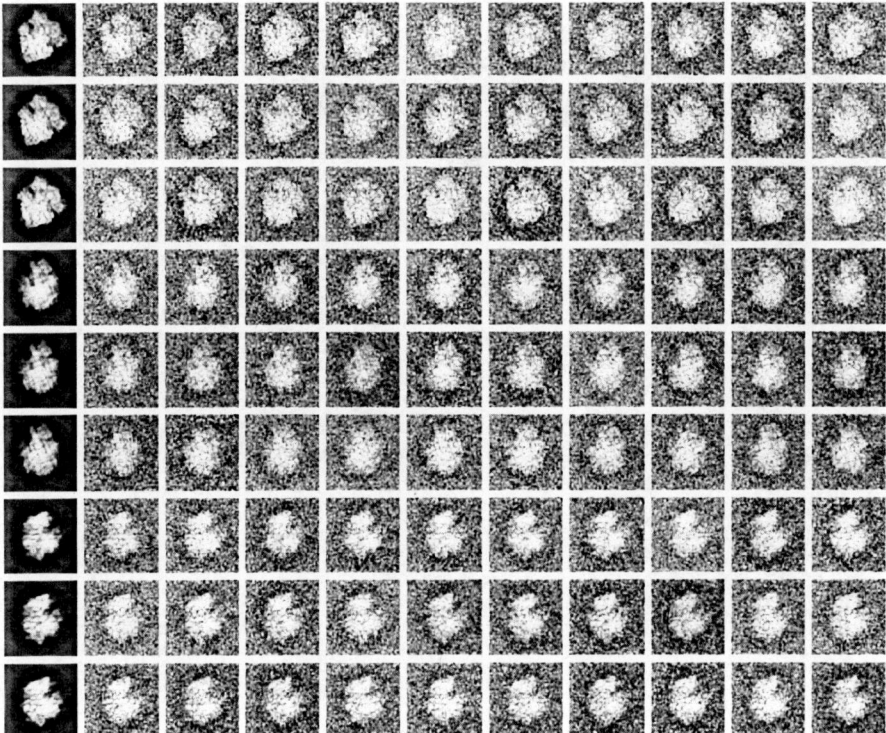

Figure 14.4 Test data set to demonstrate correspondence analysis of particle images. Projections of a density map of the *E. coli* ribosome (Gabashvili et al., 2000) were created in 3 different viewing directions (grouped in rows of 3). In each direction, the angle was wobbled by ±5°, and noise (SNR = 1) was added. From Frank (2006); reproduced with permission of Oxford University Press.

direction, the angle was in addition slightly wobbled (±5°) to create three distinct yet very similar views. The nine projections thus created are seen in the left-most column of figure 14.4. For each of these projections, a set of 10 noisy projections was created by adding random noise in the ratio of 1:1. The resulting 90 "raw" images were subjected to correspondence analysis.

In the following discovery process, we first inspect the histogram of eigenvalues (figure 14.5) and find that two eigenvalues stand out well above the rest, which tells us that the first two factors capture the most important aspects of the data grouping. This means that a single factor map, 1 versus 2, will be most informative, and that we can ignore higher factors. Based on this information, we proceed to make a map of factors 1 versus 2 (figure 14.6), which indeed shows complete separation of the data into three clusters. There is no evidence, however, that the slight changes in angle leading to a further subdivision of the data are reflected in the map (nor in other factor maps, not shown here). The changes in features due to the slight "wobbling" are evidently obscured by the noise.

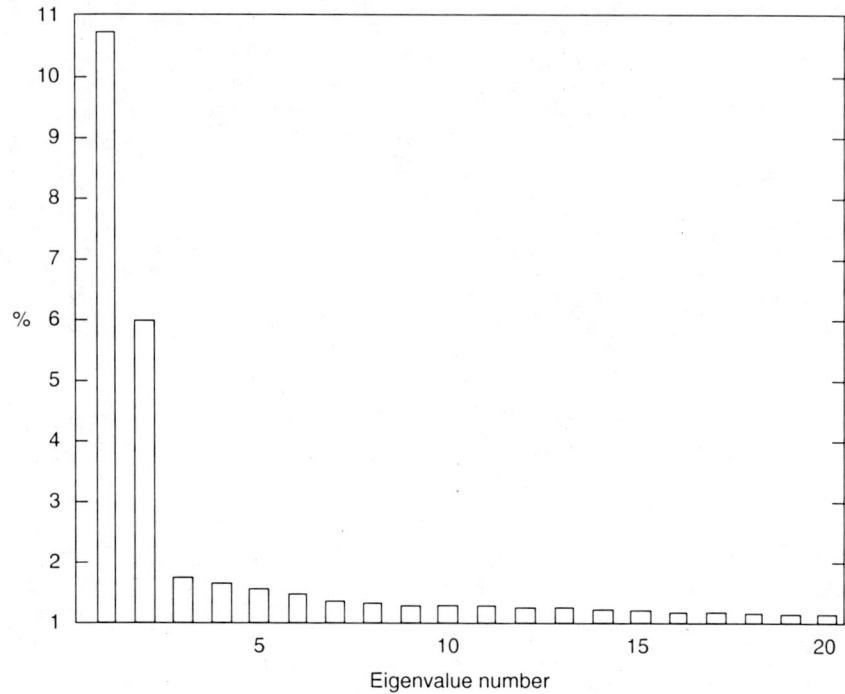

Figure 14.5 Eigenvalue histogram obtained by correspondence analysis of test data shown in figure 14.4. The fact that only two eigenvalues stand out means that two factors are sufficient to represent the shape variations associated with the change of viewing directions. From Frank (2006); reproduced with permission of Oxford University Press.

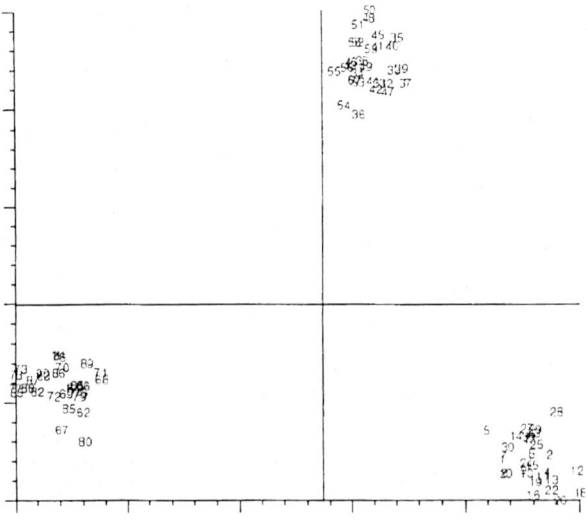

Figure 14.6 Factor map (1 versus 2) showing the segregation of the data set into 3 sharp clusters, which correspond to the three viewing directions. Although the wobbling by ±5° contributes to the spread of the clusters, it is drowned in noise. From Frank (2006); reproduced with permission of Oxford University Press.

14.8 Two useful methods of classification in single particle analysis are hierarchical ascendant classification and K-means clustering

Computational approaches to classification of single molecules use as input the low-dimensional representation of the data obtained by principal component analysis or correspondence analysis. In other words, these techniques operate on points (each representing an image) in an M-dimensional space, where M is a number much smaller than the number of pixels or images. Sometimes, M can be as small as 2 or 3. The objective of the classification is to divide the points into groups, or classes, based on the way they are clustered. (We must keep in mind that "closeness" of the points in factor space reflects the degree of similarity of the images they represent.) Hence, the output of classification is simply a list of classes and their members. In the following, the two most useful classification techniques will be discussed, *hierarchical ascendant classification* (HAC) and the *K-means clustering method*.

HAC is a classification method that proceeds by merging the points successively. In this manner, a pattern of hierarchical relationships evolves, which is symbolically represented by a *classification dendrogram* (see example in figure 14.7). The dendrogram can be likened to a family tree turned upside down: each branch of the tree represents a grouping of similar objects. The farther down we come in the tree, the more groups we have, the smaller each group, and the closer the similarity among the objects contained within it.

Programs that perform HAC start by computing a distance table that contains the $N(N-1)/2$ distances among the N points representing N images. The two points lying closest together are merged into a "cluster," represented henceforth by a single point. This step reduces the number of points by 1. The analysis proceeds in this way until only two points remain to be merged into one, which finally represents a single, all-encompassing cluster.

Obviously, the choice of merging criterion, or of what constitutes closeness between two clusters whose merger is considered next, is pivotal for the construction of the hierarchy of relationships. One choice that is often used is called *Ward's criterion*. According to this criterion, at each stage of the construction, *those two clusters are merged whose merger results in the smallest increase in intra-class variance*.

As is the case with all computational classification methods, the decision on how many classes exist cannot be made by the computer, but involves visual judgment or some other discretionary choice based on the properties of the dendrogram. The ordered hierarchy constructed by HAC can obviously be cut on many different levels, each time resulting in a different number of classes. However, HAC provides a plausible way for gauging the cutting level, and hence for obtaining the number of classes that best represents the clustering of the data: it should be done at that level which is least sensitive to "noise"; i.e., to fluctuations of branching levels.

While HAC achieves a classification by merging images successively, K-means algorithms proceed in the opposite direction, by splitting the data set into a predefined number, K, of groups. This process works in an iterative way, starting with a set of K "seeds" serving as organizing centers. The first partition is obtained by dividing the data set (again conveniently visualized as a set of points in a high-dimensional space, each representing an image) according to the proximity of the individual points to these

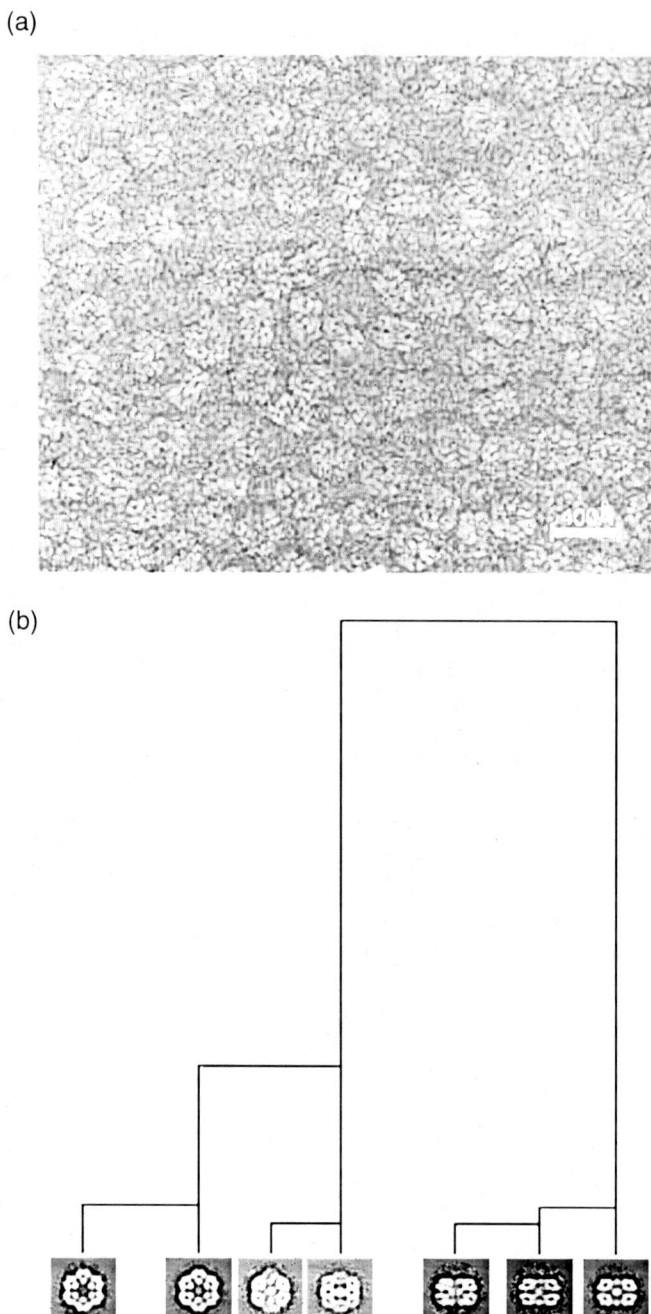

Figure 14.7 Example for the application of hierarchical ascendant classification: classification of 2450 images of *Lumbricus terrestris* (earthworm) hemocyanin (Mouche et al., 2001). (a) typical micrograph field; (b) HAC dendrogram, with class averages inserted. It is seen that on the uppermost level, the two branches split all top views from all side views. More subtle differentiations are visible at the lowest level shown. Kindly provided by N. Boisset (unpublished).

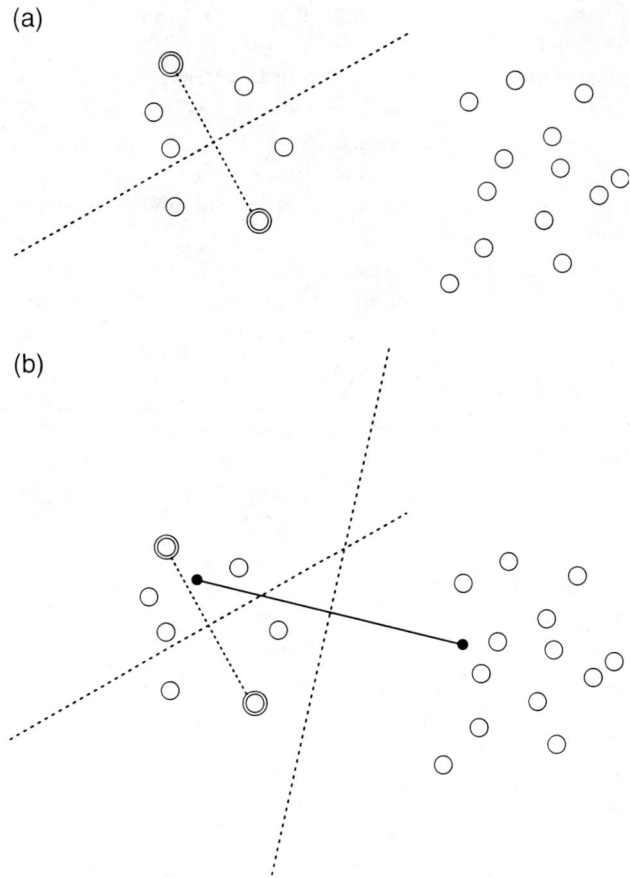

Figure 14.8 Principle of the method of K-means clustering, shown for the case of $K = 2$. In (a), two seeds (circles surrounded by stipples) are arbitrarily selected, defining the initial partition (indicated by a dashed line). In (b), two new points (fat dots) are constructed, each by computing the center of gravity of the objects in the current partition. These two newly constructed points define a new partition, indicated again by a dashed line. Repeated application of this "moving center" construction leads to a stable partition. In the case shown, two iterations were already sufficient. From Frank (1990); reproduced with permission of Cambridge University Press.

seeds (figure 14.8 shows this method of construction for the case of $K = 2$). In the next round, a new set of K centers is obtained by computing the centers of gravity for all groups of the partition. These centers are now used to determine a new partition of the data set, etc. In this way the algorithm proceeds until convergence is achieved, as indicated by the fact that the group centers and associated constituencies stabilize. In contrast to HAC, which gives a clue on what number of groups best describe the diversity among the data, K-means offers no such clue, and one is normally left with a trial-and-error approach.

Instead of the images themselves, any invariant function (with respect to rotation and translation) derived from them can be used for the approach of multivariate data analysis and classification. Examples of useful invariants are the *rotational power spectrum* and the so-called *double-autocorrelation function (DACF)* (Schatz and van Heel, 1990). In this way, the images can be directly classified without the need for their pre-alignment. Of course, the problem of alignment is in this case only deferred to a later stage, but then it only needs to be carried out among the images *within* each of the classes found.

Classification of rotational power spectra results in groups of images that have the same ratios of rotational powers. This allows the selection of data with high preservation of certain symmetries. The DACF is derived from the autocorrelation function (ACF) by expressing the ACF in a polar coordinate system and computing the rotational autocorrelation for each ring (Schatz and van Heel, 1990). Thus, like the rotational power spectrum, the DACF of an image is invariant under both translation and rotation. Classification on the basis of the appearance of their DACF will result in groups of molecule images that have the same distribution of intramolecular distances, which normally means that they present similar views. However, the loss of information in each of the two autocorrelation steps has the undesired consequence that the analysis will fail to distinguish between motifs that are related by a 180° rotation or a mirror reflection, or between motifs that have the same diffraction pattern in common but differ in the phases of their Fourier transforms. Thus, for instance, opposite views of the same structure will not be distinguished by this method of classification.

14.9 Real-space reconstruction techniques can deal with the general 3-D projection geometries encountered in single-particle reconstruction

One of the challenges posed by the single-particle reconstruction is the irregularity of the projection geometry, since particle orientations cannot be controlled in the experiment, as is the case in electron tomography. This lack of regularity and the fact that usually no symmetry can be utilized are the main reasons why real-space reconstruction techniques are preferred.

Before going into the details of reconstruction techniques and strategies, a word is in order regarding the description and definition of projection geometry. Early on in the development of 3-D electron microscopy, Eulerian angles were adopted to express the relationship between projections and the object they originate from (Radermacher, 1980). The conventional definition of the three Eulerian angles that goes back to Sommerfeld (1964) is described in box 14.3.

One explicit approach to the solution of the reconstruction problem is by *matrix inversion*. In this approach (Crowther et al., 1970; Klug and Crowther, 1972), the relationship between the projections and the object they originate from is formulated as a set of equations in real space, with as many unknowns as the number of voxels used in a discrete representation of the object. However, the large size of the matrices occurring in such a problem makes this approach impractical: an object represented by 64 × 64 × 64 voxels, for which, say, 5000 projections *in general directions* are available, would lead to a matrix with 250,000 columns and 5000 rows that has to be solved by a least-squares method. For comparison, an asymmetric single particle reconstruction at 1/1.15 nm^{-1} resolution made use of 73,000 projections, and a representation of the 3-D object by 140 × 140 × 140 voxels (Gabashvili et al., 2000). It should be noted that the key complication in the matrix inversion approach is introduced by the necessity, in single particle reconstruction, to allow a general projection geometry, that is, to allow arbitrary intervals in the entire angular space. If this stipulation can be dropped, as in single-axis electron tomography, then the numerical problem involves much smaller matrices, each dealing with the reconstruction of a single 2-D slice from its 1-D projections.

A common reconstruction method, often preferred in the case of a large number of projections because of its computational speed, *is weighted back-projection* (see box 14.4). As an introduction, it is helpful to first understand the properties of a reconstruction by back-projection *without* weighting. Here the basic operation is the back-projection of a single projection, or the creation of an "elementary volume" by translating (and "smearing out") the selected projection in the direction opposite to

---

BOX 14.3 Definition of Eulerian angles

The angular relationship between two differently oriented particles is given by a series of three rotations, which are specified as in celestial mechanics (figure B14.8), by three Eulerian angles, ψ, θ, and ϕ. Each of these rotations is expressed by a rotation matrix; thus, the relationship between equivalent vectors **r**′ and **r** in two differently oriented versions of a molecule is given by

$$\mathbf{r}' = \mathbf{R}_\psi \mathbf{R}_\theta \mathbf{R}_\phi \mathbf{r}. \quad (B14.5)$$

In Radermacher's (1991) convention, which follows Sommerfeld's (1964), the rotation matrices (indexed by i, to identify the molecule) are defined as follows:

$$\mathbf{R}_{\psi_i} = \begin{pmatrix} \cos\psi_i & \sin\psi_i & 0 \\ -\sin\psi_i & \cos\psi_i & 0 \\ 0 & 0 & 1 \end{pmatrix} \quad (B14.6)$$

$$\mathbf{R}_{\theta_i} = \begin{pmatrix} \cos\theta_i & 0 & -\sin\theta_i \\ 0 & 1 & 0 \\ \sin\theta & 0 & \cos\theta_i \end{pmatrix} \quad (B14.7)$$

$$\mathbf{R}_{\phi_i} = \begin{pmatrix} \cos\phi_i & \sin\phi_i & 0 \\ -\sin\phi_i & \cos\phi_i & 0 \\ 0 & 0 & 1 \end{pmatrix}. \quad (B14.8)$$

With reference to figure B14.8, these rotations are defined as follows: the molecule is first rotated by ϕ ("azimuth") in the positive direction around the z axis, then by θ ("tilt") in the negative direction around the y axis, and finally by ψ in the positive direction around the z axis.

(continued)

BOX 14.3 *(continued)*

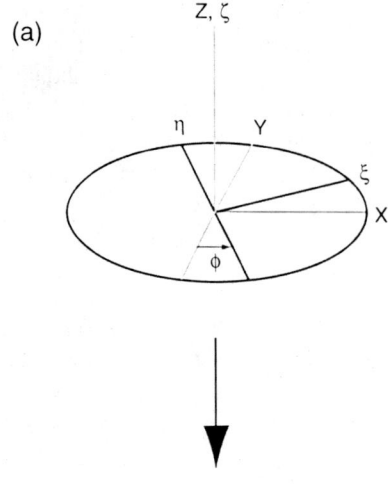

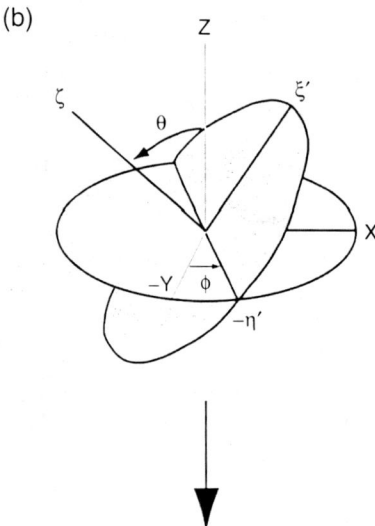

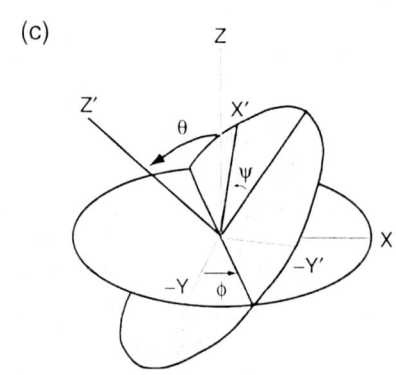

Figure B14.8 Definition of Eulerian angles. The three angles ϕ, θ, ψ represent three successive coordinate transformations. (a) Coordinate system (x, y, z) is rotated by ϕ, leading to the new system (ξ, η, ζ); (b) coordinate system (ξ, η, ζ) is tilted by θ around the "line of nodes," η, leading to coordinate system (ξ', η', ζ'); (c) coordinate system (ξ', η', ζ') is rotated in tilted plane by ψ around axis ζ', yielding the final coordinate system (x', y', z'). From Frank (2006); reproduced with permission of Oxford University Press.

BOX 14.4 Weighted back-projection

Weighted back-projection is a fast real-space method of 3-D reconstruction. It is best explained by reference to the much cruder method of (unweighted) back-projection (see Radermacher, 1992).

Back-projection is a simple operation whereby a volume is created from each projection, by "translating" and "smearing out" the projection along its associated viewing direction (figure B14.9). All volumes created by this operation, termed "back-projection bodies," are then summed. The mathematical analysis shows that the object recovered by this summation is a strongly blurred version of the original, as a result of the implicit over-representation of low spatial frequencies. The fact that Fourier components with lower spatial frequencies are strongly over-represented follows from a closer analysis of the geometry in Fourier space.

According to the projection theorem (see section 4.3), the projection of an object is represented by a central section of the object's 3-D Fourier transform. Translating a projection by a distance D (which should be large enough to fully contain the object) has the effect of convoluting its Fourier transform with a sinc function, essentially spreading it out into a modulated "thick section" of the object's Fourier transform. So, for instance, if we consider a projection in the direction of the z-axis of the object, then its contribution to the 3-D Fourier

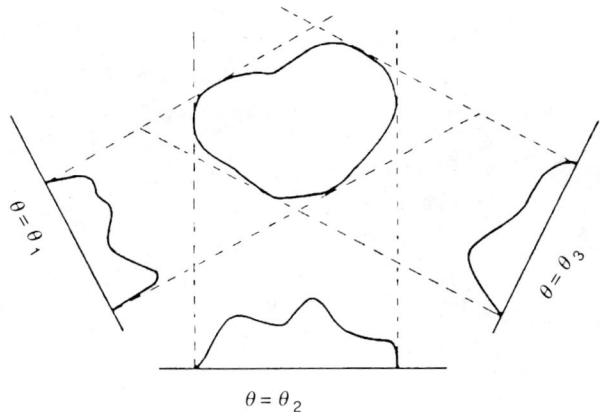

Figure B14.9 Back-projection method of reconstruction. Projections (indicated as profiles arranged in their proper geometric relationships) are "back-projected," that is, for each projection, a volume is created by translating the 2-D density distribution of the projection (or a weighted form of it) along the direction of projection. In the end, all those "projection volumes" are summed to obtain the reconstruction. From Frank et al. (1985); reproduced with permission of Van Nostrand-Reinhold, New York.

(continued)

BOX 14.4 *(continued)*

transform, $P(s_x, s_y)$, is initially confined to the plane (S_x, S_y). The contribution, to the 3-D Fourier transform, by the 3-D object that is formed by translating and smearing-out the projection along the z-axis by the distance D is $D \operatorname{sinc}(D\pi s_z)$. Thus the 3-D Fourier transform created by superimposing all those thick sections associated with different projections (indexed by i) becomes in the end

$$\mathbf{F}(\mathbf{S}) = \sum_i \mathbf{P}(s_x^{(i)} s_y^{(i)}) D \operatorname{sinc}(D\pi s_z^{(i)}). \tag{B14.9}$$

(Here, for simplicity, we use a notation in which each projection and the volume generated by its translation is represented in its original coordinate system, but with the understanding that the summation in equation B14.10 implies the application of appropriate coordinate transformations; see Radermacher, 1988.)

By assuming the simplest object, a point represented by a delta function at the origin, we obtain $\mathbf{P}(s_x^{(i)}, s_x^{(i)}) = 1$, so that equation B14.9 collapses into

$$H(\mathbf{S}) = \sum_i D \operatorname{sinc}(D\pi s_z^{(i)}), \tag{B14.10}$$

whereas the Fourier transform of the point we wished to recover is $\mathbf{F}(\mathbf{S}) = 1$. This consideration leads us to conclude that $W(\mathbf{S}) = 1/H(\mathbf{S})$ is the appropriate choice of weighting function in Fourier space that restores the equal weighting necessary to retrieve the object without distortion. The particular case of equispaced tilting around a single axis — the situation most common in other applications of weighted back-projection — leads to the simple weighting by

$$W(\mathbf{s}) = W(s_y, s_z) = [s_y^2 + s_z^2]^{\frac{1}{2}}, \tag{B14.11}$$

if the tilt axis is assumed to coincide with the x-axis. (This is sometimes referred to as "r* weighting," following the convention in x-ray crystallography to denote reciprocal vectors, here denoted as $\mathbf{s}$ in the 2-D case, by attaching an asterisk to the real-space vector.) Multiplication of the Fourier transform with a factor that is proportional to the absolute spatial frequency has the effect of applying a high-pass filter.

In the current application, however, where we are dealing with arbitrarily spaced projection directions, the weighting function must be derived numerically, by carrying out the summation in equation B14.10 for the particular geometry (in terms of the Eulerian angles of all projections) found in the dataset, and forming the inverse in equation B14.11.

the direction of the parallel beam that created it in the experiment (see figure B14.9). However, the reconstruction obtained by summing over all these elementary volumes results in a misrepresentation of the original object, for a reason that is clear from the analysis of the sampling geometry in Fourier space: samples are increasingly sparse as the radius in Fourier space increases. As a result, the reconstruction has a strongly blurred appearance due to the preponderance of low spatial frequencies. This misrepresentation calls for an analytical correction by application of a weighting function in Fourier space.

In the usual application of the technique, the experimental images are first Fourier-transformed, and each Fourier transform is multiplied with a *weighting function*, then the inverse Fourier transform is computed. The resulting *weighted projections*, which have a ghost-like appearance because of the strong high-pass filtration effect of the weighting function, are then back-projected, each in its proper geometrical orientation. Only in those cases where the distribution of angles is regular, the weighting function is the same for all projections, and simply equal to s, the spatial frequency radius (so-called $r*$ weighting, in the notation of x-ray crystallography, where an argument in reciprocal space is denoted by adding an asterisk to the argument in real space). However, the generally irregular nature of the projection geometry encountered in the single-particle reconstruction has the consequence that the weighting function is different for each projection, hence the set of weighting functions has to be customized for each data set, as outlined in box 14.4.

Algebraic reconstruction techniques (see box 14.5) work by an iterative process of back-projection, designed to correct the current volume for each projection direction

---

BOX 14.5 Algebraic reconstruction techniques

Algebraic reconstruction techniques (ART; Gordon et al., 1970; SIRT: Gilbert, 1972; general: Herman, 1980) are based on an algebraic statement of the projection relationship between the unknown object and its projections, both represented by arrays of discrete density values. The algebraic relationship can be visualized by dividing the object evenly into parallel strips running in the direction of projection (figure B14.10): any pixel of the projection is the sum over all the pixels of the corresponding strip of the object. Such a relationship holds for each projection. The set of equations that link the pixels of the object to those of the projections can be approximately solved by an iterative method that is at the heart of all ART. First, an arbitrary density distribution is used for the object at the start. The object is then projected, by forming the sums along the strips, into the direction of the first experimental projection. There will be a discrepancy between the projection of the object and the experimental projection it is supposed to match. The trick of ART methods is to apply a correction to the object along the strips, by adding (i.e., back-projecting, as explained in box B14.4) a term to the object's pixels along each strip such that the discrepancy is removed. Now, moving on to the next projection, we will again discover that the object's current projection disagrees with its experimental counterpart; hence an appropriate correction is applied, etc. At the very end, when all projections have been similarly used in updating the object, we move back to projection number 1. It is no surprise that the exact agreement, brought about by the correction in the first iteration, is now destroyed: this is on account

(continued)

BOX 14.5 *(continued)*

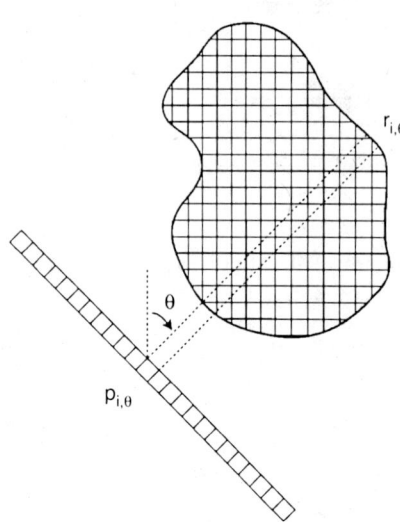

Figure B14.10 Relationship between projection and object in ART, SIRT, and related real-space reconstruction methods. For each pixel p (indexed by i and projection angle θ) of the array representing the experimentally measured projection, a ray with finite width is drawn in the direction of projection. Along that ray lie all the voxels (i.e., the discrete elements representing the 3-D object) that contribute to that pixel $p_{i,\theta}$. Since the ray normally intersects voxels partially, geometry-dependent weights have to be used in the summation. ART uses this relationship in two different ways: (i) to calculate the current value of the projection along the ray and (ii) to distribute, to all voxels along the ray, an adjustment whose purpose is to enforce the equality between experimental value $p_{i,\theta}$ and the projection sum ("back-projection").

of all the intervening corrections for other projection directions. However, the discrepancy will now be smaller than at the start. As we move on through successive iterations, we find that the discrepancy decreases for all projections, until it stabilizes at some finite positive value, signifying that an approximate solution has been reached.

Many variants of ART exist, often distinguished by differences in the numerical algorithm, but also by the introduction of constraints.

such that it agrees with the corresponding experimental projection. By their design, these algorithms do not require a regular geometry and hence are perfectly suited for single-particle reconstruction, but they have the shortcoming of comparatively low computational speed since they work in an iterative way. An advantage of these techniques is that they lend themselves readily to the application of real-space constraints (e.g., Penczek et al., 1992; Carazo, 1992).

Fourier reconstruction techniques (see implementations by Lanzavecchia and Bellon, 1996; Grigorieff, 1998; Penczek et al., 2004) make use of interpolation as an approximate method to find the values of the 3-D Fourier transform on the 3-D Cartesian grid. As stated before, the measured values of the Fourier transform lie on central section planes, which are derived from experimental projections by virtue of the projection theorem. Thus, experimental projections are first Fourier-transformed, and all their 2-D Fourier transforms are placed into the 3-D Fourier geometry according to the known particle orientations. It is easy to see that most of the samples do not fall on the points of the Cartesian grid, but somewhere in between. For each grid point of the Cartesian 3-D Fourier transform, the program has to find those experimental points lying in the vicinity that give rise to a significant contribution to the 3-D Fourier transform. Schemes as simple and crude as bilinear interpolation may be used if the objective of the exercise is to have a fast look at the reconstruction, or for computing intermediate, lower-quality reconstructions that are to be eventually refined. It is clear from the foregoing, however, that bilinear interpolation can be used only within the core range of the shape transform of the particle; that is, within the region where the shape transform changes very little in amplitude and phase. Lanzavecchia and Bellon's (1998) and Penczek et al.'s (2004) methods are designed to minimize numerical errors resulting from the limited range of existing Fourier samples considered in computing the Fourier value on the "new" grid point. Grigorieff's (1998) algorithm is noteworthy for the fact that it approximately corrects for the CTF simultaneously with the Fourier interpolation.

14.10 Random-conical and common-lines methods can provide angular relationships among the molecule projections, as a way to jump-start a reconstruction project

In each single-particle reconstruction project, we have to distinguish between two stages: the starting stage and the subsequent refinement stage. Initially, the structure may be completely unknown, and the first objective is then to obtain a crude reconstruction. In the second stage, this crude reconstruction is refined toward higher resolution, often by inclusion of a much larger dataset. The starting stage is usually realized with one of two different techniques: the *random-conical data collection method* or the *method of common lines* (also *"angular reconstitution"* or the *Radon transform method*). While the former establishes the angular relationship among a certain subset of the projections by a tilt experiment, the latter interrelates the projections by making use of their common lines in Fourier space. It will become clear from considering the design of these techniques that the random-conical method is able to furnish the geometry in an unambiguous way. The common-lines method, in contrast, allows two solutions with opposite handedness, from which the correct one can only be picked by resorting to extraneous information, for example, an additional tilting experiment (e.g., Bellon et al., 1998) or, when the resolution is sufficiently high, by recognizing the known chirality of naturally occurring helices.

Random-conical data collection

The idea underlying the random-conical data-collection method (Radermacher et al., 1987; Radermacher, 1988, 1991) is to establish defined geometrical relationships at least among a subset of the molecule projections. Once these relationships are known, and the angular space is sufficiently closely covered, an approximate reconstruction of the molecule can be obtained. Such a reconstruction can then serve as the basis for 3-D projection refinement (to be described in section 14.11) as more and more data are added.

To obtain defined angular relationships, we focus on a subset of the molecules that all present the same view, that is, molecules that face the specimen grid in a particular orientation but still have random azimuths ϕ_i (figure 14.9(a)). When the grid is tilted by a fixed angle θ (figure 14.9(b)), the analysis of the geometry in a coordinate system affixed to the molecule shows that all directions of projection come to lie on a cone

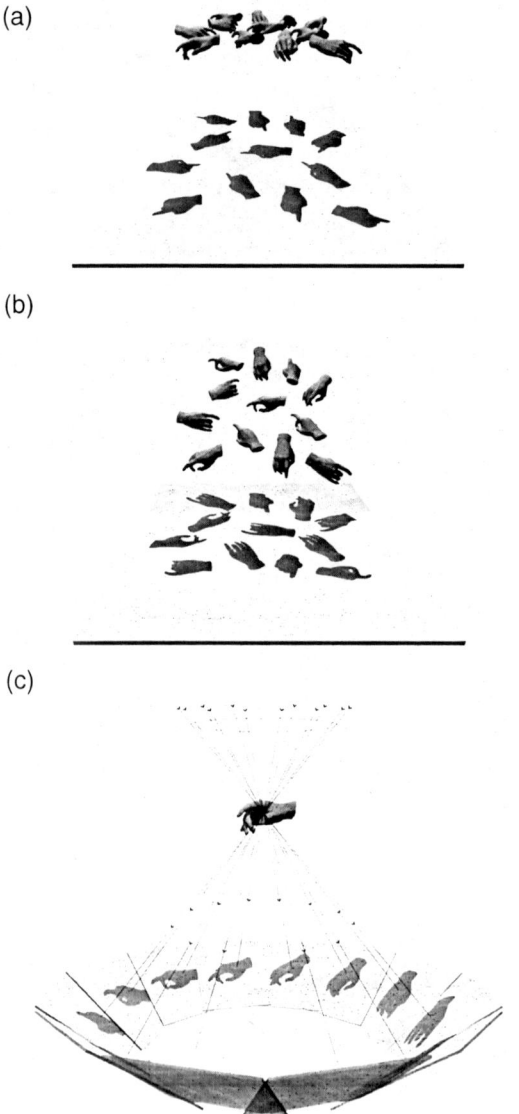

Figure 14.9 Principle of the random-conical data collection method. (a) Assume multiple copies of an object (here symbolized by a hand) lie on the support in the same orientation — the only freedom is a rotation in the plane of the grid. A transmission electron microscope image of the untilted grid shows identical projections, only distinguished by rotation. (b) An image of the tilted grid shows a set of different projections that can all be arranged in a conical geometry, as shown in (c). In the actual data collection, the experiment (b) is carried out first and that described in (a) second, to reduce the amount of radiation the molecules experience when the tilted-grid data set is collected.

with half-angle θ at the azimuths ϕ_i (figure 14.9(c)). These azimuths can be easily determined by analyzing the same particles in the image of the untilted specimen: all relative rotation angles are obtained when these control projections are aligned. If the tilt angle θ is large enough, and the azimuths are spaced closely enough (a condition that can always be satisfied by pooling particles from several micrographs), then the entire volume of Fourier space is covered, out to the limiting resolution, except for a cone-shaped gap. It is easy to see that the angle of this empty cone is $90° - \theta$. Evidently, the closer θ is to $90°$, the better the quality of the reconstruction. Yet the increasing thickness of the support film as θ is increased sets a practical limit which is between $50°$ and $60°$. Experience has shown that some usable reconstructions can be obtained for as little as $\theta = 45°$, although fairly strong artifacts (smearing out of features in the direction of the missing cone) will occur. To eliminate the gap in the angular range and create a density map that is bias-free for use in subsequent 3-D projection alignment steps (see section 14.11), it is necessary to extend the random-conical data collection to at least one additional view of the molecule (Penczek et al., 1994). This second view should be sufficiently different from the view initially selected, so that the entire angular space may be covered with projections.

Common-lines methods

Common-lines approaches to the determination of relative orientations make use of the fact (see also section 14.3) that each projection is represented by a central plane in the object's Fourier transform. The two central planes from any two projections intersect along a line (the common line) where their Fourier transforms must match. This principle was first used by Crowther and co-workers (1970) to interrelate different projections of an icosahedral virus. Searching for this line fixes one azimuthal angle, but the two planes are free to rotate around the common line. However, addition of a third projection, from a third unique direction, produces two additional common lines which fix the orientations of the first two projections as well as the orientation of the third one (van Heel, 1987; Goncharov et al., 1987). Generalization to any number of projections is straightforward, since a new-coming $(N + 1)$th projection will intersect the already imported N projections in Fourier space along N common lines, and will have to be oriented in such a way that the cross-correlation or some other measure of overall consistency is maximized. Instead of proceeding successively, one projection at a time, we can make use of a steepest-descent algorithm that seeks a solution by taking into account the entire projection set simultaneously (Penczek et al., 1996).

In the absence of symmetry, the presence of noise makes it difficult to find the common lines for raw, unaveraged particle images. This is the reason why the common-lines approach is always based on class averages obtained after some type of (unsupervised) classification; see sections 14.6–14.8.

In the method of "angular reconstitution" (van Heel, 1987; Goncharov et al., 1987), one of the variants of the common-lines methods, the equivalent real-space relationships among three 2-D projections of a 3-D object are exploited: for any two 2-D projections, the common line in Fourier space corresponds in real space to a common 1-D projection (the so-called common-line projection). To search for this common-line projection, it is convenient to compute a 2-D map that contains all possible 1-D projections (with a small angular increment) — the so-called *sinogram* (see figure 14.10). The search

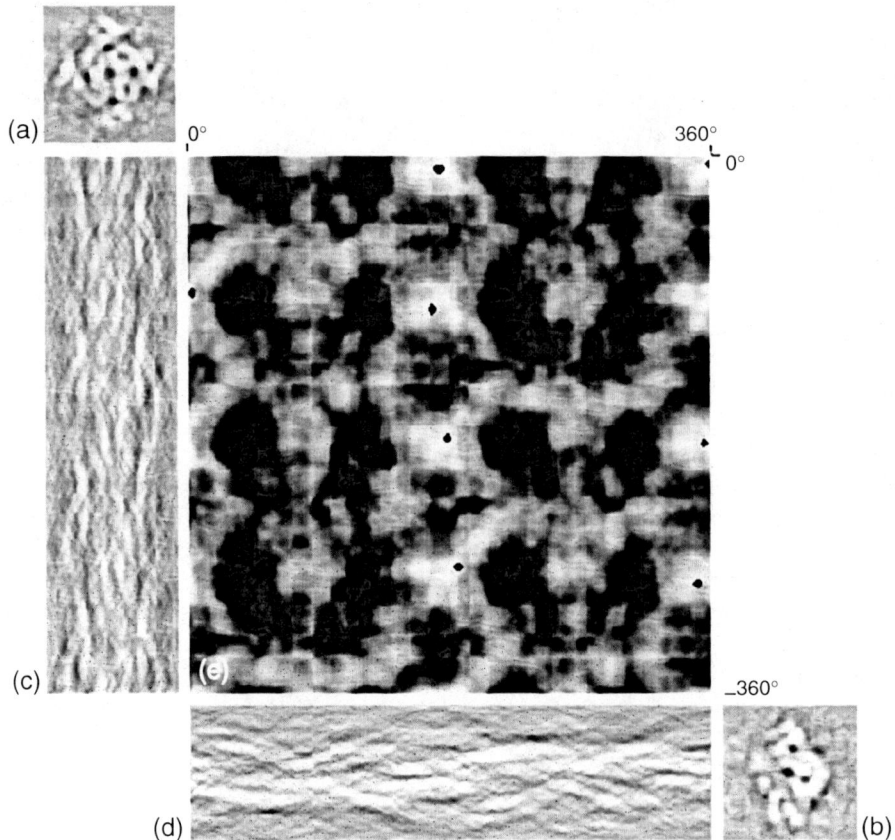

Figure 14.10 Determination of common-line projections (the real-space equivalent of the common lines), used in angular reconstitution, by the sinogram correlation function (SCF). (a, b) Two class averages (top view and a ~45° view) of the ryanodine receptor/calcium release channel and (c, d) their respective sinograms. Each line in the sinogram corresponds to a 1-D projection of the 2-D molecule in a particular direction. Lines are ordered according to increasing angles, such that the complete sinogram covers the entire 360° range. (e) SCF between the sinograms (c) and (d). This function is obtained as follows: the cross-correlation coefficient is computed between the first row of sinogram 1 and each row of sinogram 2, and the resulting array of values are placed into the first row of the SCF. Similarly, the second row of the SCF is obtained by cross-correlating the second row of sinogram 1 with each row of sonogram 2, and so forth. The position of each maximum in (e) indicates the angular relationship between the projections. From Serysheva et al. (1995); reproduced with permission of Nature Publishing Group.

for the orientation at which the 1-D projections match is done by cross-correlation of these sinograms (van Heel, 1987; Schatz and van Heel, 1990; Serycheva et al., 1995). The same relationships can also be cast into the formalism of 1-D and 2-D Radon transforms (Lanzavecchia et al., 1996; Lanzavecchia and Bellon, 1998).

14.11 Angular refinement methods are used to proceed from the initial reconstruction to the final reconstruction

Neither the random-conical nor the common-lines methods lead to a reconstruction that represents the most effective exploitation of the information present in the data. In the case of the random-conical method, selection of a view class on the basis of the appearance of particles in the untilted specimen field is bound to include particles with a certain range of orientations. In the case of the common-lines methods, the need to work with averages instead of raw images entails a certain amount of resolution loss because of the inclusion of particles with a finite angular range in each class average. Thus, in both methods the angles assigned to the individual particles can only be approximate.

To seek a better solution, which produces a better overall agreement between the experimental projections and the reconstruction, some type of refinement technique must be used. All existing techniques — various forms of the 3-D projection alignment method (bootstrap: Harauz and Ottensmeyer, 1984b; van Heel, 1984; reference-based, for viruses: Cheng et al., 1994; reference-based, for asymmetric particles: Penczek et al., 1994) and those utilizing the 3-D Radon transform (Radermacher, 1994) — are mathematically equivalent, though they differ in the practical details and the numerical behavior. In the following, the reference-based 3-D projection alignment method (Penczek et al., 1994) will be described (figure 14.11).

The method starts out with an existing density map as reference. This density map might be the result of one of the starting strategies outlined above, or it might have been derived from other experimental techniques. This density map is projected in virtually "equidistant" directions (i.e., in directions that intersect the surface of a sphere on a close-to-regular grid). The resulting library of "reference projections" are stored in computer memory, to be then compared with every single experimental projection by a cross-correlation matching procedure.

This matching yields a list of all particles with their Eulerian angles (i.e., χ, θ, and ϕ for each particle; see box 14.3 for definition) that give the best match, along with their cross-correlation coefficients indicating the quality of the agreement. The value of this cross-correlation coefficient can be used as a basis for deciding whether the particle should be included in the reconstruction. (Note that 3-D projection matching can be seen as a variant of supervised classification, since each experimental projection is compared with a large number of 2-D "templates," i.e., the projections of the 3-D reference.) In a situation where multiple conformations are known to coexist, each described by a 3-D reference density map, the cross-correlation coefficients can also be used in a multiple comparison to decide which conformation the particle most likely belongs to (Valle et al., 2002; Heymann et al., 2004; Gao et al., 2004).

The process of angular refinement is an iterative procedure. In each iteration, the new assignments of Eulerian angles are used to compute a new density map from the

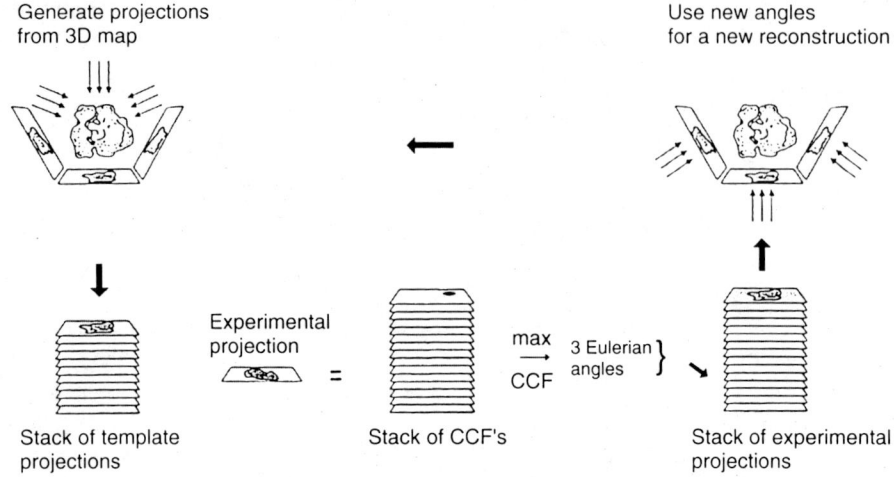

Figure 14.11 Principle of the 3-D projection alignment (and refinement) method. Top left: projections of an existing reconstruction of the object are computed for all possible directions, and are stored in memory for fast access, to serve as references. Bottom row: reference projections generated in the first step are matched with each of the experimental projections by using a fast correlation algorithm. Each comparison results in the determination of three Eulerian angles and a quality score (usually the maximum value of the correlation function). The set of Eulerian angles giving the highest score are then assigned to the current experimental projection. For expediency, both positions related by a 180° rotation are always tried. In the next step of the procedure, a new reconstruction (top right) is computed from the projection data set, which is again used for generating a reference projection set following the scheme indicated at the start. Adapted from Penczek et al. (1994).

experimental data set. Each iteration results in a reassignment of projection directions and a gradual improvement of the reconstruction, until convergence is achieved, as indicated by a stabilization of the angles. It is most efficient to start the first iterations with a coarse angular grid and then gradually make the grid finer toward the end. For a homogeneous dataset, as the refinement proceeds, the chance of the angles changing by a drastic amount quickly becomes small, so it is possible after a few iterations to confine the fine-grid search for each particle to a small conical neighborhood of the angular direction assigned in the previous iteration.

It is sufficient initially to use a grid with approximately 15° spacing on the surface of the sphere, a choice which results in 83 directions on the hemisphere (note that the projections resulting from placing the 83 directions on the other hemisphere are simply mirrors of the initial set). Experience with the ribosome has shown that, to realize resolutions in the neighborhood of 1 nm with an asymmetric molecule, the angular grid must be tightened to a spacing of 2.0° (resulting in 5088 directions), or even to 1.5° (9076 directions). The rapid increase in the number of reference projections, as the angular increment is being reduced, means that the achievement of resolutions approaching atomic resolution will pose a formidable computational challenge.

It should be noted that angular refinement and reconstruction do not have to be organized in an alternate iterative way as presented here. In a new approach

(Yang et al., 2005), the entire system of projection equations describing the relationship between the experimental data and the 3-D object (in which the values of all voxels and all parameters describing positions are angles of projections appear as unknowns) is solved simultaneously following a conjugate gradient method.

14.12 Single-particle reconstruction in practice

By now it will be clear that the single-particle approach faces a much larger range of variations and problems than do approaches in electron crystallography that make use of strict order in the specimen. Over the past 25 years, a whole new set of tools has been developed to deal with these problems, and the aim of the preceding sections was to introduce these tools and describe their applications. It is difficult to formulate a general "recipe" for how to address a new reconstruction project, since this will depend on the answers to a number of questions related to the size of the molecule, its inherent flexibility, the degree of conformational heterogeneity, and the extent of orientational preferences. In the following, a recipe of some sort is presented to deal with the situation where a rough reconstruction is already available that can serve as reference. Step-by-step introductions on how to obtain such a preliminary reconstruction can be found in the literature (random-conical reconstruction: Radermacher, 1988; angular reconstitution: Serysheva et al., 1995; Schatz et al., 1997; van Heel et al., 2000; multiple common lines: Penczek et al., 1996).

Defocus estimation is based on a computation of a power spectrum for each micrograph

For each micrograph, the power spectrum must be computed, so that the defocus of the micrograph can be estimated. To appreciate the importance of this estimation, recall that for a reconstruction to be valid and useful, data from a wide range of defocus values must be merged with proper accounting for the CTF of each projection. CTF correction follows two alternative strategies: according to one strategy, the entire micrograph (or the set of particles selected from it) is CTF-corrected right at the start, so that all particles can be processed in a single batch from there on (Ludtke et al., 2001). Following the second strategy, particles are categorized into "defocus groups" and, accordingly, placed in separate batches that are separately reconstructed and refined, to be merged and CTF-corrected at the very end by Wiener filtering (Penczek et al., 1997; Frank et al., 2000) (see below). Strategy I is evidently much simpler, but it may not result in an optimal use of the data. The reason is that raw data have a very small signal-to-noise ratio, hence the application of Wiener filtering at that point must exclude data within a substantial region around the zeros of the CTF to avoid numerical instabilities. In the following, the second strategy is sketched out.

In grouping the particles into defocus groups, the size of the defocus range must be small enough to avoid resolution loss as the data are being combined in the course of the reconstruction. Such resolution loss may occur because a defocus difference δz will affect the transfer function progressively as the spatial frequency is increased, shifting the zeros of the CTF. Mixing data whose defocus varies by δz has the same effect on the CTF as a lens whose current fluctuates during the time of the exposure.

Both will produce a sharp band limitation, expressed by an envelope function (Wade and Frank, 1977; see chapter 3) that falls off with $\exp[-(s/s_0)^4]$, where s is the spatial frequency radius and s_0 is proportional to $(\delta z)^{1/2}$. On the other hand, the defocus range into which data are binned must be large enough for practical reasons, to accommodate a sufficient number of particles necessary to obtain a well-behaved reconstruction, in which features of the molecule can be readily recognized. From this it follows that, as the resolution is pushed higher and higher, the defocus range δz must be made progressively narrower, implying that the total amount of data to be collected must necessarily increase.

Because of the importance of knowing the power spectrum for each micrograph, it is necessary to make sure that enough background scattering is provided in the specimen. To ensure that this is the case, a thin carbon film can be added in the preparation of the specimen grid, which covers the holes of the thick carbon (Wagenknecht et al., 1988). Alternatively, the power spectrum can be retrieved from the selected particle images, as commonly done for virus particles (see chapter 12).

The power spectrum is defined as the expectation value of $|\mathbf{F}(\mathbf{s})|^2$, where $\mathbf{F}(\mathbf{s})$ is the Fourier transform of the image. For power spectrum estimation (Zhu et al., 1997; Fernandez et al., 1997), we can divide the micrograph into partially (50%) overlapping patches of size $\sim 500 \times 500$. For each patch, the Fourier transform, and from this $|\mathbf{F}(\mathbf{s})|^2$ (the "periodogram") is computed. All patterns $|\mathbf{F}(\mathbf{s})|^2$ are then averaged, to give the estimate of the power spectrum, $\langle|\mathbf{F}(\mathbf{s})|^2\rangle$. This averaged pattern displays the characteristic Thon rings (section 10.11), and serves as a valuable diagnostic. Instead of using the patches of the whole micrograph, one can also compute $|F(\mathbf{s})|^2$ for each particle selected, and then average over all contributions from all particles of this micrograph (Sander et al., 2003).

At this stage, micrographs should be rejected whose representative power spectrum shows any of the following deficiencies:

- rings do not extend far enough (indicative for instability of the instrument, as this leads to absence of information at higher spatial frequencies)
- rings are elliptic or hyperbolic (indicative for the presence of axial astigmatism)
- ring pattern is unidirectionally limited, as though seen through a rectangular window (indicative of the presence of unidirectional lateral drift or vibrations during the exposure time).

For micrographs that pass this test, the power spectrum is next rotationally averaged, leading to a 1-D profile in which zeroes of the CTF show up as minima (figure 14.12). Because of the general falloff of power toward higher spatial frequencies, caused by the combined effect of instrument instabilities and factors described by envelope functions, the curve always declines overall. The decline is quite steep at lowest spatial frequencies, in the range where contributions not following the CTF are found, as the result of inelastic scattering being confined to low angles and the presence of overall intensity variations, e.g., due to uneven illumination across the image field.

To estimate the defocus, it is practical to use an interactive tool that enables the simultaneous display of the power spectrum profile "on the fly" along with the CTF that is calculated for the estimated defocus values (CTFFIND2 in the MRC package: Crowther et al., 1996; CTF tool in Web/SPIDER: Frank et al., 1996) (figure 14.12). The objective of this interactive display is simply to make the minima of the power spectrum

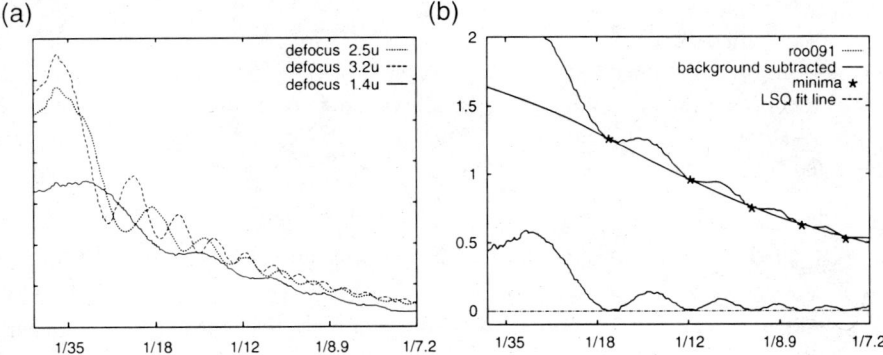

Figure 14.12 CTF determination. (a) Examples of a rotationally averaged power spectra for different defocus values. (b) Background subtraction, by using a smooth curve following the minima of the power spectrum, produces a curve that can be matched with the square of the CTF. From Frank (2006); reproduced with permission of Oxford University Press.

coincide with the zeroes of the CTF. Automated techniques for CTF estimation also increasingly come into use (see Huang et al., 2003; Mindell and Grigorieff, 2003; Sander et al., 2003). Either way, a defocus is assigned to each micrograph and the corresponding set of single molecules. On this basis, all particles are then pooled into one of the "defocus groups." As pointed out at the beginning of this section, the rationale here is to create subsets of the data set, one for each defocus group, that are each large enough (~3000 particles) to support computation of a statistically well-defined 3-D reconstruction.

Particle picking and normalization

Particles may be picked by using a fast and simple computer procedure such as the following. The micrograph is first subjected to low-pass filtration, with a filter bandwidth corresponding to the inverse of the particle size. Elimination of very low spatial frequencies (through high-pass filtration) may also be beneficial, as a means to suppress long-range density variations (e.g., due to changes in illumination intensity across the field). On the resulting bandpass-filtered micrograph, particles in the expected size range show up as "bumps." A simple peak search on this smoothed version of the image produces a list of positions (x, y coordinates) of particle candidates, and, on the basis of this list, windows of appropriate size are extracted from the original micrograph and stored as individual files, or as part of an image stack. Next these particle candidates are displayed in a sequence of galleries, to allow rejection of candidates that are clearly misclassified.

Hand-picking is both tedious and subjective, and the need for ever-larger numbers of particles has led to the development of automated particle-picking methods. A variety of these is now available, as documented in Potter et al. (2004). Of these, a technique based on locally normalized cross-correlation (Roseman, 2000, 2003; Rath and Frank, 2004) is arguably most robust, and can be easily implemented in general image processing software packages without the need for special routines. Following this technique, the CCF is computed between the micrograph and a reference. This reference can be a simple disc, or a rotationally averaged, representative projection, or — if a previous reconstruction is available — the average of computed projections whose orientations coarsely sample the full angular range. The local variance of the micrograph is now computed for all positions of a circular mask whose size is large enough to contain the particle fully. In this way, a locally normalized cross-correlation coefficient is obtained for each point of the micrograph field.

Next, all the images selected are normalized. To this end, from within each micrograph, a small portion of the image that is free of particles is analyzed by computing the density histogram. This histogram is used to normalize each of the particle images belonging to the corresponding micrograph. The normalization procedure ensures that data from different micrographs are statistically comparable, not only in terms of their variance, but also their higher-order statistics (Boisset et al., 1993).

Orientational classification and alignment

The objective of this step is to determine the particle orientation by reference to a set of projections of an existing reconstruction. For an angular spacing of 15°, one obtains $N = 83$ reference projections. Each particle of the data set is aligned, in two dimensions, with each of these reference projections, producing a set of N cross-correlation values and an associated set of Eulerian angles. The angles that produce the highest cross-correlation value are taken to be an estimate of the actual particle orientation.

If the existing 3-D reference — normally the outcome of a bootstrap reconstruction via random-conical or common lines, or the result of a previous project — has been CTF-corrected, theory predicts that the size of the cross-correlation peak in the 3-D projection alignment is adversely affected by the oscillations of the CTF in the experimental data; see section 14.5 and, in particular, equation 14.10. To circumvent this problem, and to maximize the signal-to-noise ratio in the cross-correlation detection, the reference volume must be first multiplied with the CTF of the relevant defocus group. This CTF should be computed by using the average defocus of the group, which is usually quite close to the point halfway between minimum and maximum defocus values of the micrograph associated with the defocus group.

Within each defocus group, the procedure described above creates orientational classes of particles. Particles falling within a given class are expected to be similar, and their average should resemble the corresponding reference projection quite closely. It is useful at this point to perform a spot check, by displaying a gallery of images falling into particular orientation classes (figure 14.13(a)), and a gallery of averages side by side with a gallery of the reference projections (figure 14.13(b)). It is also interesting to have a look at the images that have been classified as belonging to a particular view (figure 14.13(c)).

These comparisons show the extent to which individual particles conform to the reference they have been attributed to according to the CCF. It is quite common to find outliers with poor resemblance to the reference, since the CCF can be influenced by extraneous data. The degree of existing heterogeneity also can be assessed in this way.

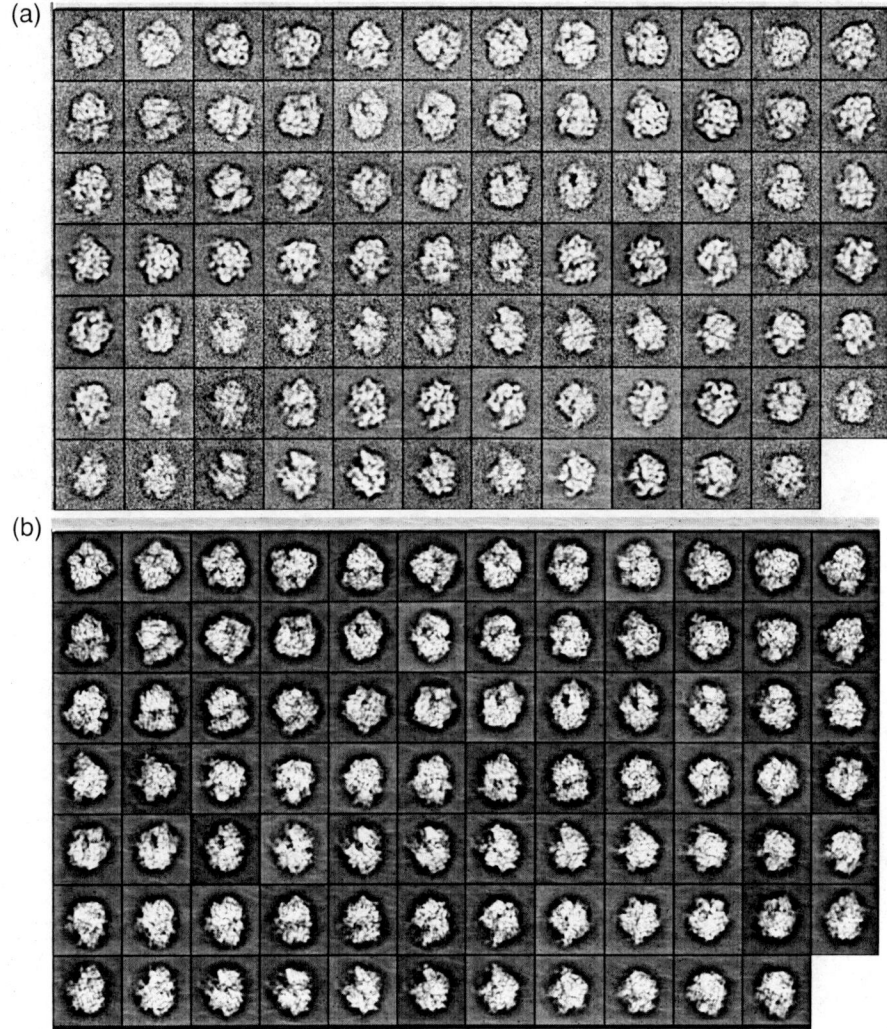

Figure 14.13 Example for 3-D projection alignment. (a) Gallery of 83 reference projections, corresponding to a coarse angular grid with 15° increments; (b) gallery showing averages of experimental projections classified by a cross-correlation match with the reference projections in (a); (c) the 100 highest ranking (in terms of cross-correlation) experimental projections assigned to the top left reference projection in (a); (d) example for angular statistics: for each of the 83 viewing directions, the position of the circle indicates the angular position, while its area is proportional to the number of particles encountered. Clusters of large circles indicate the existence of some orientational preferences in this data set.

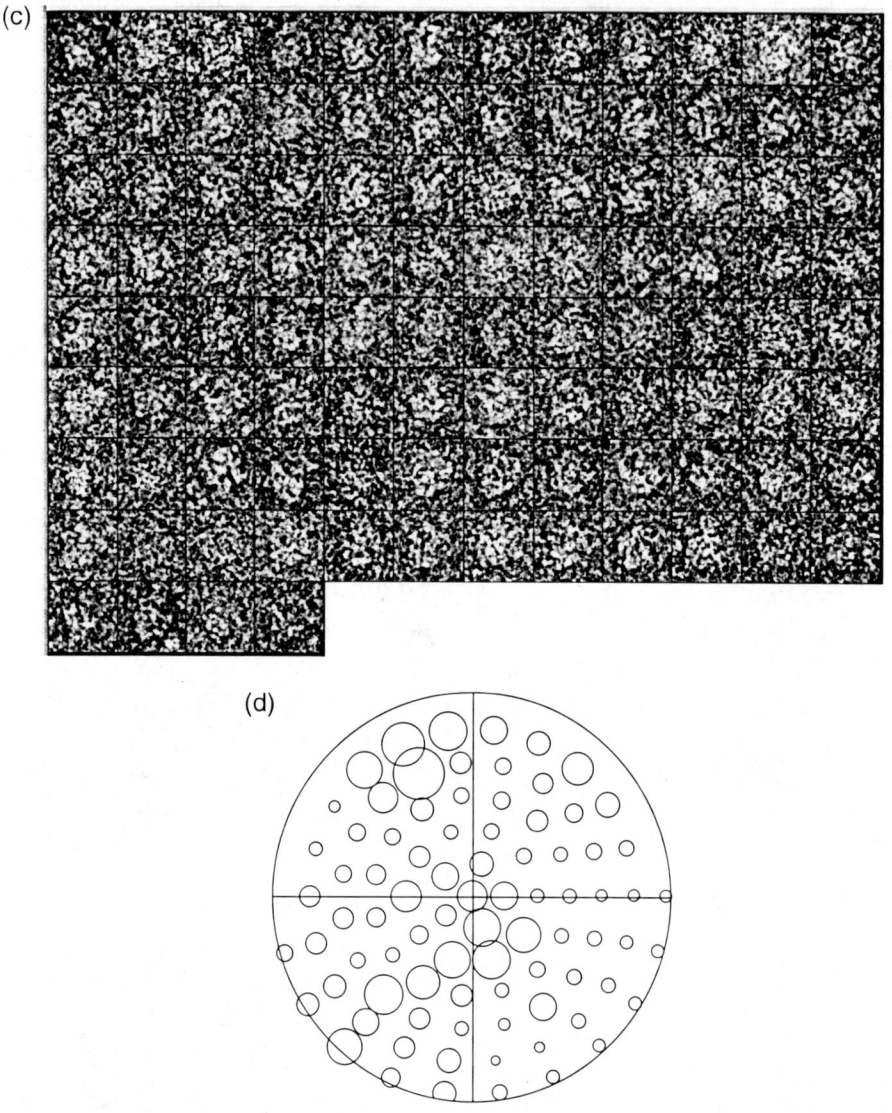

Figure 14.13 (*continued*)

The second comparison is a check on the overall consistency of the program. Any deviation from randomness in angles is immediately apparent, by the different degrees of statistical definition of the averages. (The existence of orientational preferences can, of course, also be detected by looking at a table, or histogram, of particles falling into the N directions.) Another, quick way to display orientational preferences is by means of a 2-D plot in which the number of particles falling into each of the 83 directions is symbolized by the size of a circle placed into the corresponding viewing direction (figure 14.13(d)). It is important to look for such preferences as they can have an effect on the reconstruction quality, as will be discussed in the following section.

Within each view group, as identified by the orientational classification, there can be residual heterogeneity. The reasons can be manifold: the most obvious is the failure of the 3-D projection alignment algorithm to find the correct assignments for a certain subset of the particles. The second, more important reason is that there might be genuine conformational heterogeneity among the particles. Strategies and algorithms to deal with this situation are still subject to further methodological development.

Three-dimensional reconstruction of each defocus group

Data from each defocus group are used for separate reconstructions. At this stage of the project, it is possible to use any reconstruction algorithms that are fast, even at the expense of accuracy, such as a direct Fourier reconstruction based on bilinear interpolation. Strong orientational preferences (i.e., large fractions of particles falling into a few directions) may present a problem in the reconstruction, because of possible directional artifacts (Boisset et al., 1998). In such cases the number of particles should be limited to create more directional uniformity within the particle set, or certain parameters of the reconstruction algorithm should be adjusted (e.g., iteration count, relaxation parameter) (Sorzano et al., 2001).

Resolution estimation

For each reconstruction, the resolution is estimated by the usual method of comparing two half-set reconstructions in Fourier space (see box 14.6 for definition of these measures). Here the estimation fulfills two purposes: first, it gives an idea about the quality of the data sets prior to refinement, allowing data sets to be expanded, decimated, or eliminated, depending on the outcome of the test; second, the resolution estimation produces a set of curves, one for each defocus group, that allows serious gaps in the data collection to be spotted and appropriate remedial experiments to be designed targeting specific defocus settings (see figure B14.11, which shows an example of the most frequently used *Fourier shell correlation*).

Angular refinement, and computation of a merged, defocus-corrected final reconstruction

Refinement involves the repeated application of the 3-D projection alignment followed by a round of 3-D reconstruction. Each cycle leads to a reassignment of projection directions until the angles stabilize. In this phase of the procedure, a highly accurate 3-D reconstruction algorithm is used; for example, iterative back-projection (see Penczek et al., 1992). Refinement can be thought of as a way to seek an object that closely explains the observed projections. Because of the presence of noise and inconsistencies among the data, this problem has many approximate solutions, which all prove quite similar in the detailed analysis.

The progress toward a stable solution is monitored by the program, by keeping track of the sum of the squared angle reassignments in each iteration. When this quantity

BOX 14.6 Resolution criteria

In electron crystallography, where the 3-D density map is synthesized from Fourier amplitudes and phases, the resolution is defined as the spatial frequency corresponding to the highest diffraction order that contributes significantly to the map (see section 4.6). Because information associated with the repeating structure is concentrated in the reciprocal lattice spots or lines, there is no ambiguity in this definition. Since such a demarcation between signal and noise is missing in the Fourier transform of a single-particle reconstruction, resolution must be determined in a different way: by comparing two reconstructions from independent data sets. These data sets are obtained by splitting the entire projection set randomly in half. In Fourier space, the reproducibility of the two reconstructions is then measured as a function of spatial frequency.

Two measures of agreement that have gained practical importance are the Fourier shell correlation (FSC) (introduced in the form of the Fourier ring correlation for 2-D averages; see Saxton and Baumeister, 1982; van Heel et al., 1982) and the differential phase residual (DPR) (Frank et al., 1981). In both, the 3-D Fourier transforms are analyzed over shells (with radius $s = |\mathbf{s}|$ and width Δs). The definitions are as follows:

$$\text{Fourier shell correlation: FSC}(s, \Delta s) = \frac{\text{Re}\left\{\sum_{[s,\Delta s]} \mathbf{F}_1(\mathbf{s})\mathbf{F}_2^*(\mathbf{s})\right\}}{\left\{\sum_{[s,\Delta s]} |\mathbf{F}_1(\mathbf{s})|^2 \sum_{[s,\Delta s]} |\mathbf{F}_2(\mathbf{s})|^2\right\}^{1/2}}. \quad (B14.12)$$

The FSC should be above a critical value which assures that the signal-to-noise ratio (SNR) stays above a certain minimum. The threshold for FSC that is often adopted (Böttcher et al., 1997; Conway et al., 1997; Malhotra et al., 1998) is 0.5 (an example is shown in figure B14.11). Below it is shown that for this value of FSC, the SNR falls to 1.

Differential phase residual:

$$\text{DPR}(s, \Delta s) = \left| \frac{\sum_{[s,\Delta s]} [\Delta\phi(\mathbf{s})]^2 [|\mathbf{F}_1(\mathbf{s})| + |\mathbf{F}_2(\mathbf{s})|]}{\sum_{[s,\Delta s]} [|\mathbf{F}_1(\mathbf{s})| + |\mathbf{F}_2(\mathbf{s})|]} \right|^{1/2}. \quad (B14.13)$$

Here the stipulation is that the DPR$(s, \Delta s)$ should be above 45°.

Both DPR and FSC have the disadvantage that they are based on a splitting of the data set into two, leading to increased statistical uncertainty. In the case of the averaging of images, this consideration led Unser et al. (1987) to propose a measure that is based on a multiple comparison, the spectral signal-to-noise ratio, SSNR:

$$\text{SSNR}(s, \Delta s) = \frac{\sigma_s^2(s, \Delta s)}{\sigma_n^2(s, \Delta s)/N}, \quad (B14.14)$$

(continued)

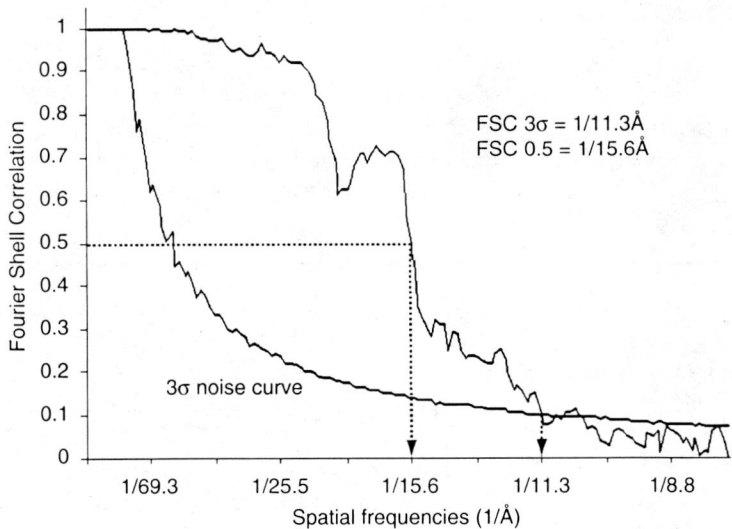

Figure B14.11 Example for resolution determination by Fourier shell correlation, using different cutoff criteria. The molecule studied here is the keyhole limpet hemocyanin type I. The FSC = 0.5 criterion yields $1/1.56$ nm^{-1}, while the FSC = 3σ criterion (the point at which the FSC curve crosses the 3σ noise curve) yields $1/1.13$ nm^{-1}. A factor of $\sqrt{10}$ was applied to the noise curve due to the D5 point-group symmetry imposed in the course of the reconstruction. From Mouche et al. (2003); reproduced with permission of Elsevier Science.

where N is the number of images, and the following definitions apply:

$$\sigma_s^2(s, \Delta s) = \frac{1}{n_{s,\Delta s}} \sum_{s,\Delta s} |\mathbf{F}(\mathbf{s})|^2 \text{ (signal variance on the shell)}$$

$$\sigma_n^2(s, \Delta s) = \frac{\sum_{(s,\Delta s)} \sum_{i=1}^{N} |\mathbf{F}^{(i)}(\mathbf{s}) - \mathbf{F}(\mathbf{s})|^2}{(N-1)n_{s,\Delta s}} \text{ (noise variance on the shell).}$$

(B14.15)

It has proved more difficult to find a similar measure for the 3-D case. However, if the Fourier interpolation method is used for the reconstruction, an estimate of the 3D SSNR can be obtained (Grigorieff, 1998; Penczek, 2002).

The spectral signal-to-noise ratio can be tied to the FSC on the corresponding Fourier shell, since

$$\text{SSNR}(s, \Delta s) = \text{FSC}(s, \Delta s)/[1 - \text{FSC}(s, \Delta s)]$$

or

$$\text{FSC}(s, \Delta s) = \text{SSNR}(s, \Delta s)/[1 + \text{SSNR}(s, \Delta s)] \quad \text{(B14.16)}$$

(continued)

> **BOX 14.6** *(continued)*
>
> following a similar relationship between the cross-correlation coefficient and SNR (Frank and Al-Ali, 1975). This means that for FSC = 0.5, SSNR = 1 (Penczek, appendix in Malhotra et al., 1998). The FSC has a desirable property that goes beyond the estimation of resolution: it shows the local drop in the SNR at the zeros of the CTF quite sharply, so that the positions of the FSC minima can be used to refine the CTF estimation (Mouche et al., 2001).

falls below a certain minimum value, or the iteration count exceeds a certain allowed maximum, the refinement is stopped.

At the end of the refinement, the individual reconstructions of each defocus group are simultaneously merged and CTF-corrected. The need for CTF correction has been explained in section 3.8. The projection images are altered by the effects of the CTF in two ways: in their Fourier transforms, phases are incorrectly measured (changed by 180° in certain zones), and amplitudes are reduced. The basis of the CTF correction and merging is the *Wiener filter* (Schiske, 1968; Frank and Penczek, 1995; Penczek et al., 1997). If the CTF is rotationally symmetric, that is, in the absence of axial astigmatism, both CTF (denoted by $H_i(s)$ for the ith defocus group reconstruction) and the Wiener filter $W_i(s)$ are functions solely of the spatial frequency radius $s = |\mathbf{s}|$:

$$W_i(s) = \frac{H_i(s)}{\sum_i |H_i(s)|^2 + \frac{1}{\text{SNR}}}, \qquad (14.11)$$

where SNR is the signal-to-noise ratio, defined as the ratio of signal and noise variances. Even though it is strictly true that the SNR is a function of s, as well, it has proved sufficient in the medium resolution range (up to $s = 1$ nm^{-1}) to work with a single empirical constant. The filter is tailored to the analytical behavior of the CTF, to both correct the phases and restore the amplitudes of the Fourier transform in the merged reconstruction (figure 14.14). This Fourier transform is finally computed as

$$\mathbf{F}_{merge}(s) = \sum_i W_i(s)\mathbf{F}_i(s). \qquad (14.12)$$

It should be noted that the CTF-correction approach implied in equations 14.11 and 14.12 entails a number of assumptions and approximations. It is strictly valid only for the part of the Fourier transform that represents the linear portion of the elastic bright-field image, but not for those portions that arise from the quadratic term or from inelastically scattered electrons. A detailed analysis of the error made in these approximations has yet to be done.

An example for a merged, CTF-corrected reconstruction at 1/9 Å^{-1} resolution (80S ribosome from yeast; Halic et al., 2005) is presented in figure 14.15.

Final resolution estimation and low-pass filtration

At this point, the resolution is estimated for the merged, CTF-corrected reconstruction. Beyond the point where the Fourier shell correlation drops to 0.5, the SNR falls to 1

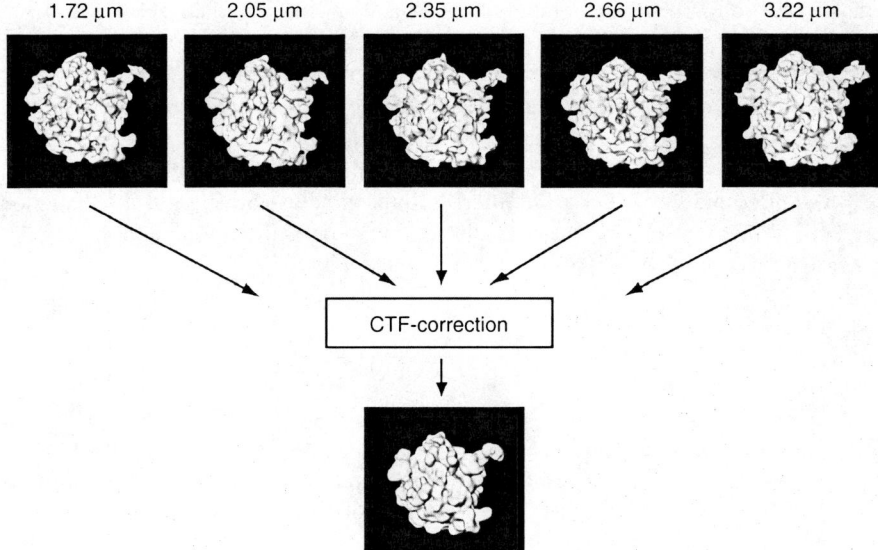

Figure 14.14 Merging of reconstructions from different defocus groups. In this example, five separate ribosome reconstructions were obtained from five different defocus groups, each characterized by a mean defocus. The reconstructions bear the signatures of the CTF: the lower the defocus, the smaller the structural features that are pronounced. Some of these features are distorted because of sign reversals of the CTF. Merging and CTF correction is accomplished by using a Wiener filter, which restores the correct phases and amplitudes within the resolution range. From Frank (2006); reproduced with permission of Oxford University Press.

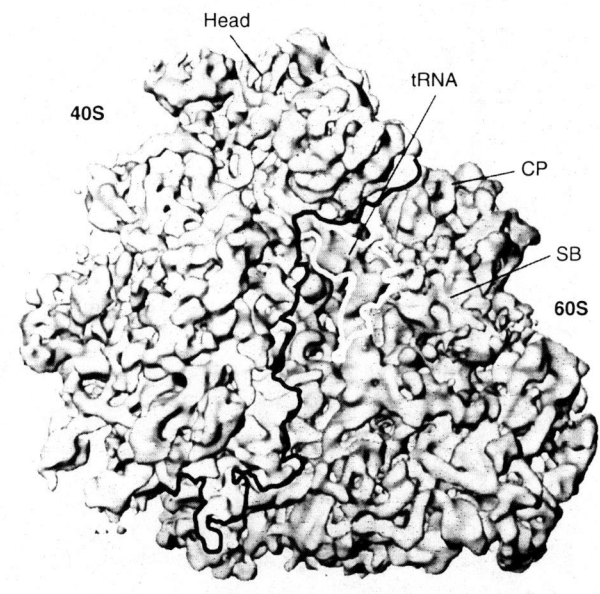

Figure 14.15 80S ribosome from yeast at 1/9 nm^{-1} resolution (Halic et al., 2005), shown in a side view. 40S, small subunit (light gray); 60S, large subunit (dark gray). Other landmarks: Head, head of the small subunit; tRNA, P-site tRNA; CP, central protuberance; SB, stalk base. (tRNA is outlined in white to make up for lack of color in reproduction.) From Halic et al. (2005); reproduced with permission of Nature Publishing Group.

(Frank and Al-Ali, 1975; Malhotra et al., 1998 and appendix by Penczek therein), and inclusion of such data is difficult to justify. A low-pass filter is therefore used to limit the Fourier transform of the reconstruction to a sphere of this radius. The shape of such a function has to be carefully designed, so as to avoid *truncation artifacts* (see Frank, 1996, 2006). The idea here is to have a smooth transition between the "pass" and the "no pass" region. There are two filters often used in electron microscope applications: the *Fermi filter* and the *Butterworth filter*. Both have adjustable parameters that allow the width of the transition zone to be adjusted to the particular problem at hand.

The Fermi filter (Frank, 1985) is a filter defined as

$$W(s) = \frac{1}{1 + [1 + (\exp(s - s_0)/T)]}, \quad (14.13)$$

where s_0 is the cutoff frequency, and the "temperature" T controls the width of the transition zone, with $T = 0$ representing zero width. (The Fermi filter is borrowed from physics, where it represents the distribution of electrons at a given temperature T according to the Fermi–Dirac statistics.)

The Butterworth filter is defined as (Pratt, 1991)

$$W(s) = \frac{1}{1 + (s/s_0)^{2n}}, \quad (14.14)$$

where s_0 is again the "cutoff frequency" at which the filter value falls to 0.5. Here the parameter n controls the steepness of falloff, or, in other words, the width of the transition zone. In contrast to the Fermi filter, the Butterworth filter is specially designed to minimize ("apodize") the unwanted side ripples.

Relationship between a low-resolution reconstruction and a low-pass filtered high-resolution reconstruction

In the course of a project, as more data are collected, it is normal to accumulate several versions of a 3-D density map, distinguished by different nominal resolutions as determined by the Fourier shell correlation or other measures. Works by different groups of researchers on the same molecule might also generate reconstructions with different resolutions. The question is, how are these versions related to one another? Superficial analysis might suggest that, as in x-ray crystallography, one should be able to regenerate the low-resolution versions by suitably low-pass filtering any of the higher resolution density maps. According to this line of reasoning, failure to obtain the former from the latter would indicate inconsistency between the two results. However, due to the way angular refinement works, the angular definition at any stage of resolution benefits from the definition at the previous stage. Thus, the relationship among the versions of particle positions at different stages of refinement is dynamic, not static as it is in the relationship among unit cells of a crystal. This means that a low-resolution version of a high-resolution reconstruction is not necessarily identical with a reconstruction obtained at an early stage of angular refinement.

3-D variance estimation

It was earlier pointed out that for the reconstruction to be valid, the molecules in the sample must all have the same conformation and ligand binding state. Whether or not heterogeneity exists, and where the varying features are located in the molecule can be determined from the 3-D variance map. Depending on the outcome of this analysis, it may be necessary to apply classification to the whole projection data set such that homogeneous subsets of the particles can be identified for separate reconstructions. (The 3-D variance is also useful in determining whether a difference found in the comparison between the density map for a complex resulting from an experiment and the control is statistically significant.)

A way to estimate the 3-D variance has been developed by Penczek and coworkers (2006a). Multiple reconstructions are done by repeatedly drawing (with replacement) a sample of projections from the whole dataset. The authors show that the 3-D variance map computed from these subset reconstructions is related to the desired 3-D variance estimate of the whole dataset by a simple factor. Applications of this "bootstrap variance analysis" are found in the work of Penczek et al. (2006b) and Grob et al. (2006).

Classification of a heterogeneous projection dataset

Indications for the presence of heterogeneity are (i) failure to reach a reasonable resolution in the course of refinement; (ii) the density obtained for a ligand molecule in the reconstruction is smaller than the density obtained for a protein of the host molecule ("partial occupancy"); or (iii) the 3-D variance map shows highlights of localized variability. In the simplest case, two different molecule populations may co-exist in the sample. In this case, if density maps or models for the two pure populations are available, then it is possible to separate the two subpopulations in the sample by using 3-D projection matching with the two 3-D references (Valle et al., 2002; Gao et al., 2004). A histogram displaying the difference in cross-correlation values for each particle when matched with reference #1 versus reference #2 is helpful in separating the subpopulations (Valle et al., 2002).

Reference-based (supervised) classification has the obvious disadvantage that the 3-D references must be known beforehand, and that incorrect choices of references based on an incorrect hypothesis may yield meaningless results. A way out of this problem is a newly developed method of maximum-likelihood classification (Scheres et al., 2007), which requires a minimum of advance knowledge, such as the number of subpopulations anticipated.

14.13 What are the prospects of achieving atomic resolution?

Routine achievement of resolutions in the range of $1/0.8$ to $1/1.2$ nm^{-1}, which is now possible for a number of molecules including GroEL and the ribosome, has opened up studies of ligand binding and dynamics of such molecules under a variety of conditions. In these studies, the lack of atomic information can be made up by fitting and docking of structural components known from x-ray crystallography or NMR (section 11.6).

Nevertheless, atomic resolution is the ultimate goal, and there is no physical reason why it could not be reached for molecules beyond a certain size.

Henderson (1995) investigated the range of conditions for which single-particle imaging could lead to high — potentially atomic — resolution, and came to the following conclusions: theoretically, the information contained in images of molecules with molecular weights above 10^5 Da should be sufficient to obtain alignment accuracies required for attainment of 1/0.3 nm^{-1} resolution. Somewhat surprisingly, the number of particles required comes out in these calculations to be independent of the particle size, at around 13,000. In practice, even the ribosome with 2.7 MDa molecular mass and a diameter of 25 nm poses a formidable problem, indicated by the fact that it required 29,000 single particles for achieving 1/1.5 nm^{-1} (Malhotra et al., 1998) and 73,000 for 1/1.15 nm^{-1} resolution (Gabashvili et al., 2000). More recent reconstructions of the ribosome have benefited from improvements in instrumentation and image processing, and have brought the number required for 1 nm^{-1} down to ~35,000 (see Valle et al., 2003). Indeed, Henderson (1995) pointed out that the real imaging situation might produce much more unfavorable conditions than assumed in the calculations, due to a rapid falloff of information toward higher resolution. Recent estimates, guided by experience, put the number of particles needed to achieve 1/0.3 nm^{-1} resolution at a million (Frank, 2002; Sali et al., 2003). Another factor, the multiplicity of conformational states that may prevail under most native conditions (see section 14.4 above), could in fact lead to a further explosion in the number of images required for achieving atomic resolution.

15

Special Considerations Encountered with Thick Specimens

15.1 Introduction

A number of idealized assumptions formed the basis of the majority of what was said about electron microscopy of biological macromolecules in the previous chapters. The most important of these assumptions were: (1) elastically scattered electrons undergo no more than one scattering event, (2) curvature of the Ewald sphere can be neglected, and (3) inelastic scattering can be ignored. While these represent satisfactory approximations for suitably thin specimens, each one becomes progressively less good as the specimen thickness increases.

If deviations from these idealized assumptions become large enough, the use of electron crystallographic structure analysis becomes more difficult than what we have discussed so far. It is important, therefore, that failure of any of the approximations should be small enough to cause only small effects during the subsequent structure analysis. The purpose of this chapter is to describe the complications that occur when any of the three principal assumptions mentioned above start to fail.

Section 15.2 provides a general overview of different theoretical treatments that are used to describe multiple elastic scattering (dynamical diffraction), while section 15.3 goes on to give practical guidance as to when the single scattering (kinematic scattering) approximation is likely to fail in a significant way. Section 15.4 then discusses the possibility that dynamical effects can still be accommodated during refinement of a structural model.

The fact that Fresnel diffraction can occur within the finite thickness of the sample itself is closely connected to the amount of curvature of the Ewald sphere. Section 15.5 explains why these interrelated concepts are more important for thick specimens than they are for thin objects. Section 15.6 then goes on to explain how curvature of the

Ewald sphere, as well as multiple elastic scattering, can destroy the approximation that the scattered electron wave possesses Friedel symmetry.

Inelastic scattering becomes increasingly significant as the sample thickness increases, as is explained in section 15.7. The most serious consequences of inelastic scattering occur when the sample thickness is so great that an appreciable fraction of the electrons experience both inelastic and elastic scattering events. At this stage the useful signal decreases rather than increases as the sample thickness continues to increase.

Each of the thick-specimen effects described in this chapter — inelastic scattering, dynamical diffraction, and sensitivity to curvature of the Ewald sphere — will cause the scattered wave to deviate from Friedel symmetry. Section 15.8 concludes the chapter by summarizing how failure of the diffracted wave to satisfy Friedel symmetry can determine whether images and diffraction patterns can be used to complete a structure analysis.

15.2 Dynamical diffraction can be described by a number of different, but equivalent mathematical formalisms

The term "*dynamical diffraction*" is commonly used to refer to situations in which the single-scattering approximation is not valid. The single-scattering approximation, on the other hand, gives rise to the theory of *kinematic diffraction*. The terms "kinematic" and "dynamical" are used extensively in crystallography, and the book by James (1982) presents a clear account of the historical origin of this terminology.

The Born series formulation of scattering theory

In quantum mechanics, the theory of electron scattering is traditionally developed with the mathematical approach known as perturbation theory. The scattered wave is represented as a small perturbation that is added to a plane wave, the latter being the solution of Schrödinger's equation when the scattering object is not present. The perturbed wave function can be represented with increasing accuracy by an iterative series of integrals, called the *Born series*, which is described in Cowley (1995) and most textbooks on quantum mechanics. The first-order term in this series is simply the Fourier transform of the scattering potential (evaluated on the surface of the Ewald sphere). The *first Born approximation* is therefore identical to the kinematic approximation.

The second Born approximation adds a new term, which is just the scattered wave (produced from the first Born approximation) convoluted with itself. In other words, every beam that is produced by single scattering gives rise, in turn, to another full scattering pattern, centered about the direction of the first, scattered beam. Continuing in this way, triple scattering is described by the third Born approximation, quadruple scattering by the fourth Born approximation, and so on.

The dynamically scattered wave function can be represented to arbitrary accuracy by including a sufficiently large number of terms in the Born series. It is evident, however, that the simple Fourier transform relationship between the object and the scattered wave is lost as soon as double scattering and higher order terms begin to be important. Crystallographic theory, as we normally know it, therefore becomes invalid when the kinematic (single scattering) approximation begins to fail.

Although the infinite Born series does not generally sum to an analytic, closed-form result, Schiff (1956) was able to show that a very simple result is obtained at small scattering angles, in the high-energy limit. According to the closed-form solution obtained by Schiff, the dynamically scattered wave is given by

$$\psi_{scattered}(\mathbf{s}) = \mathcal{F}\{e^{-i2\pi \frac{e}{hv} V'(\mathbf{x})}\} \tag{15.1}$$

where h is Planck's constant, v is the electron velocity, and $V'(\mathbf{x})$ is the 2-D projection of the scattering potential. Equation 15.1 is identical to saying that the wave function transmitted through the object is $T(\mathbf{x}) = e^{-i2\pi \frac{e}{hv} V'(\mathbf{x})}$. This, in turn, is simply the (strong) phase object approximation, which we have mentioned in section 4.2. Equation 15.1 is also the result obtained by the semiclassical Wenzel–Kramers–Brillouin (WKB) approximation, described in most books on quantum mechanics. Others, including Moliere (1947), have also derived this result by equivalent methods.

It is instructive to point out some internal consistencies that are associated with the closed-form solution given in equation 15.1. In the high-energy limit, the electron wavelength goes to zero and curvature of the Ewald sphere therefore becomes negligible. The central section theorem (section 4.3) tells us that the Fourier transform of a projection in real space is a central section of the 3-D Fourier transform of the object. It is therefore natural that the scattered wave function should involve the projection of the scattering potential if the Ewald sphere is a plane, as it is in the high-energy (or the small-angle scattering) approximation. Expansion of the complex exponential in equation 15.1 as a power series in $V'(\mathbf{x})$ gives the weak phase object approximation, as mentioned previously in section 4.2, plus terms involving higher products of $V'(\mathbf{x})$. These latter terms give rise to the successive convolutions that make up the Born series, as indicated above, when the Fourier transform is used to obtain the scattered wave function.

The path-integral formulation of scattering theory

Feynman developed another mathematical approach that is widely used in scattering physics. Referred to as the *path integral* formulation of quantum mechanics, the final wave function is obtained as the superposition of wave functions for all possible scattering paths. Classically forbidden (i.e., tunneling) paths are included along with classically allowed paths in the superposition (Feynman and Hibbs, 1965). When the path integral is limited to single scattering events, Feynman's formulation of quantum mechanics results in the kinematic (or first Born) approximation. As shown by Jap and Glaeser (1978) in a real-space derivation and earlier by van Dyck (1975) in a Fourier-space derivation, the Feynman path integral yields the (strong) phase object approximation, equation 15.1, or the Cowley–Moodie multislice approximation (see below) when additional paths of an appropriate type are included in the path-length integral.

The Cowley–Moodie formulation of scattering theory

The so-called multislice method developed by Cowley and Moodie (1957) is a formulation of dynamical diffraction theory that is even more widely applicable than the phase object approximation. The Cowley–Moodie multislice formulation is based on an intuitive picture of wave propagation that is used in physical optics. Simply stated, the *Cowley–Moodie multislice approximation* first describes the propagation of an incident wave through a very thin slice of the scattering object in terms of a transmittance function for that slice. The transmitted wave is then propagated by Fresnel diffraction to the position of the next slice, where once again it is modified by the transmittance function of that slice. The process of alternating transmission and Fresnel propagation continues until the wave exits the specimen.

Because the slice thicknesses used in the multislice method must be very small, the weak phase object approximation may often be an acceptable description of the transmittance function of a slice. However, there is no computational advantage in using the weak phase object approximation rather than the more accurate, strong phase object transmittance function described in equation 15.1. In either case, it is important to keep each slice at a small enough thickness that the projection approximation remains valid. The thickness required for this to occur depends upon the electron voltage (wavelength), and on the resolution (highest scattering angle used in the calculation). More is said on this particular point in section 15.3.

The Cowley–Moodie multislice formulation of dynamical diffraction theory can be applied equally well to single molecules, amorphous materials or crystals. Unlike the Bloch wave formulation described in the following paragraph, the Cowley–Moodie formulation requires no special boundary conditions or other restrictions that need to be satisfied by the specimen. As for any formulation of dynamical theory, however, the diffracted wave can be calculated only if the structure — which can be a model, of course — is specified in advance. Computer programs for doing the Cowley–Moodie multislice calculations are available as commercial software as well as from many research groups who do high-resolution electron microscopy of inorganic materials.

The Bloch wave formulation of scattering theory

Dynamical diffraction in perfect crystals can be elegantly described by expanding the electron wave function as a sum of *Bloch waves* (Cowley, 1995), just as is done in describing the electronic properties of solid-state materials. The Bloch wave formulation becomes highly impractical as a description of dynamical diffraction in materials with very large unit cells, such as protein crystals, however. As a result, it is never used for work that is directly relevant to the scope of this chapter, and it is mentioned here only for completeness.

Additional perspective is contributed by each of these equivalent formulations

The different formulations of dynamical diffraction theory mentioned above each contribute a different perspective on the effects caused by multiple scattering. The Born series and the Feynman path integral formulations both lend themselves well to the physical picture of single scattering, double scattering, etc. By including enough terms they are able to describe dynamical diffraction to arbitrary accuracy. The Born series is never used as such; rather, its importance lies in providing the formalism for the derivation of the phase object approximation. The phase object approximation, on the other hand, is an easily used, closed-form approximation, but it is valid only in the limit of short electron wavelength, or in the limit of small-angle scattering. Similarly, the Feynman path integral formalism is never used as such; its importance lies in providing insight into the Cowley–Moodie approximation, as well as the physical effects that are neglected in that formalism. The Cowley–Moodie formulation, on the other hand, achieves arbitrary accuracy in the limit that infinitely thin slices are used in the calculation. For our purposes, which involve single macromolecules as well as crystals with very large unit cells, the Cowley–Moodie formalism provides the only practical method that can be used to get an accurate idea of the magnitude of dynamical diffraction effects.

15.3 Conditions when kinematic diffraction theory fails

Many factors determine when dynamical effects become important

In order for the kinematic, or single-scattering approximation to be valid, the fraction of electrons that are scattered even once must be very small. This condition ensures that the number of electrons that are scattered twice, three times, etc., will be too small to have an important effect on the data.

A first estimate of the sample thickness for which the kinematic approximation is valid can be made on the basis of calculated atomic scattering cross-sections. For example, the cross section for elastic scattering of 100 keV electrons by a single carbon (nitrogen or oxygen) atom is approximately 70 pm^2 (Langmore et al., 1973). In a hypothetical sample made up solely of carbon, nitrogen, and oxygen, with a mass density of 1.5 g/cm^3, the mean free path for elastic scattering would then be 190 nm. In order for the total number of scattered electrons to be no more than 10% of the incident electrons, the sample thickness would have to be about one-tenth of the mean free path. By this semiquantitative argument alone, one can see that dynamical effects might begin to be significant for 100 keV electrons at a specimen thickness greater than about 20 nm.

The chemical composition of the specimen and the incident electron energy are also important factors in determining the validity of the single-scattering approximation. As might be intuitively clear, dynamical effects are smaller for light elements, which scatter more weakly, and for higher energy electrons. In addition to these two points, however, the unit cell size and the alignment of atoms either within the unit cell or from one unit cell to the next can also have a relatively large effect on the validity of the kinematic approximation, as will be discussed below.

Even scattering from single atoms shows dynamical effects

Arguments based on the calculated atomic scattering cross-section require that the kinematic approximation is valid at least at the level of single atoms. We make this point with some emphasis because — surprisingly — it is not always true that the kinematic approximation is valid even for scattering from single atoms! The scattering of 40 keV electrons from single molecules of UF_6, in the gas phase represents a famous and historically significant example in which multiple elastic scattering within the shielded Coulomb potential of a single uranium atom was found to have a major effect (Glauber and Schomaker, 1953). Another measure of the failure of the first Born approximation is given by the ratio of elastic scattering cross-sections as estimated by the phase object (i.e., Moliere) approximation and by the first Born approximation (Langmore et al., 1973). At high atomic number, the first Born approximation can overestimate the scattering produced by a single atom by as much as ~30% at 100 keV, and even by ~20% at 300 keV.

Second- and higher order terms in the Born series will necessarily be present in every scattering situation, even for single carbon atoms. As a result, the atomic scattering factor, usually approximated as a real-valued function, will have both real and imaginary parts:

$$f(s) = f_1(s) + i f_2(s). \tag{15.2}$$

When the kinematic approximation is valid (for single atoms), the imaginary part, $f_2(s)$, is too small to be significant, and thus it is normally ignored. The imaginary part is nevertheless always present, and, according to figure 5.5 in Reimer (1989), $f_1(0) \simeq 0.2$ nm while $f_2 \simeq 0.01$ nm for 100 keV electrons scattered by a single carbon atom.

Alignment of several atoms aggravates the dynamical effect

Dynamical scattering becomes much more significant when individual atoms in the specimen are precisely aligned in the direction of the electron beam. An example that illustrates the effect of close atomic alignment is given by Cowley–Moodie calculations of diffraction by the 0.34 nm Bragg spacing in graphite (Jefferson et al., 1976). According to this calculation, dynamical effects can become very important within a thickness of carbon as small as 5 nm.

Important insight into what happens when atoms are aligned directly above each other can be obtained by considering the expression for the phase object approximation in equation 15.1. The superposition of two or more atomic potentials has an exponential effect on the transmitted wave, making it clear that the scattering intensity for a column of "N" atoms can become far greater than N times the scattering intensity for one atom. When the atoms are shifted laterally from one another, however, the exponential buildup does not occur and the amount of scattering remains additive.

Even when atoms are aligned vertically (i.e., in the direction of the incident electron beam), the resulting dynamical effect will not be quite as severe as is suggested by the phase object approximation, because Fresnel propagation from one atomic position to the next tends to flatten out the phase perturbation caused by the first atom before the transmitted wave passes through the second atom. Expressed in another way, Fresnel propagation tends to invalidate the projection approximation, inherent in equation 15.1, and thus weakens the dynamical effect from superimposed atoms.

The nonlinear rate of increase in the dynamical effect can also be appreciated by considering, for the sake of clarity, just the contribution of double scattering. While single scattering increases linearly with any increase in potential, double scattering increases as the square of the potential. A twofold increase in the projected potential will lead to a fourfold increase in the contribution made by double scattering, while a tenfold increase in the projected potential will lead to a 100-fold increase in the contribution made by double scattering. The effect of atomic alignment will be brought up again at the end of this section, when we discuss the results of multislice calculations for some real specimens.

Experimental tests for dynamical effects are rarely simple

One experimental test of the validity of the kinematic approximation that is relatively easy to perform — but only for a favorable type of specimen — is to measure the fraction of incident electrons that are scattered into individual diffraction spots. This measurement can be made by defocusing the electron diffraction pattern somewhat, such that each spot forms a miniature dark-field image of the crystal, but still the spots do not overlap one another. By using a very short exposure, one can record an image of the central, unscattered beam, which produces a miniature bright-field image of the crystal. Densitometry of the two types of images, allowing for the differences in exposure time, gives an accurate measurement of the unscattered beam intensity and the diffracted beam intensities.

Crystalline plates of $C_{44}H_{90}$ (a paraffin wax) were observed in this way to produce first-order diffraction spots whose intensities are each about 2% of the incident intensity (Henderson and Glaeser, 1985). Since there are six such spots, and there is additional diffraction intensity in spots at higher resolution, the elastically scattered electrons represent far more than 10% of the incident electrons. Diffraction from paraffin thus cannot be expected to behave according to the kinematic theory, even though this organic crystal is less than 5 nm in thickness. The early advent of dynamical effects in this case is not unexpected, since the carbon atoms are aligned above one another in columns, corresponding to the crystalline packing of the long, aliphatic chains. Dorset, in section 5.1 of his book (Dorset, 1995), provides a detailed discussion of other experiments that have systematically varied the specimen thickness and the accelerating voltage, in order to demonstrate the dynamical nature of diffraction from thin paraffin crystals.

There are only a few other experimental techniques that can be used to test whether the kinematic scattering theory is valid for a given specimen. One test is provided by the requirement that kinematic diffraction intensities must satisfy Friedel symmetry (box 3.3). Since the transmittance function for the phase object approximation is a complex-valued function, its Fourier transform is no longer required to obey Friedel symmetry. In applying the test of Friedel symmetry to measured diffraction intensities, however, one must be careful to show that an apparent failure of Friedel's law is not just a trivial inequivalence caused by curvature of the Ewald sphere (see section 15.6). If curvature of the Ewald sphere is too large, of course, the intensities that are recorded are no longer at approximately symmetric vector positions S and $-S$, and thus the measured intensities cannot be expected to be identical.

The failure of Friedel's law has been used to show that electron diffraction intensities obtained from purple membrane are measurably dynamical at electron energies of 100 keV or less (Glaeser and Ceska, 1989). At 100 keV the effect seen with this

~4.5 nm thick, 2-D protein crystal is small, however, and comparable to the error inherent in measuring the diffraction intensities. As a result, the failure of Friedel's law was measurable only as a statistical bias. At 20 keV, on the other hand, the failure of Friedel's law for the (1,2) family of reflections is so pronounced — 40% — that one can see the unequal intensities of Friedel mates by eye.

The appearance of diffraction intensity in symmetry-forbidden reflections is another measurement that has been used to test whether the kinematic approximation has failed, as is discussed in section 5.2 of Dorset (1995). The absence of measurable intensity in symmetry-forbidden reflections is actually a rather weak test of the importance of dynamical diffraction, however. Even a nonlinear function of the projection of a structure will have the same symmetry as does the projection itself. Thus, the phase object transmittance function, which itself is highly dynamical, will have the same symmetry — and the same symmetry-forbidden reflections in its Fourier transform — as does the weak phase object transmittance function. Thus, to summarize, the appearance of measurable diffraction intensity in reflections that are symmetry-forbidden is clear evidence for failure of the kinematic approximation, while the absence of measurable intensity in symmetry-forbidden reflections only establishes that the projection approximation is still valid.

Model calculations provide our best understanding of dynamical effects

Because experimental investigation of dynamical diffraction is technically difficult, computational simulations for appropriate model structures have been used to augment our understanding of conditions under which the kinematic approximation may fail. As an example, an atomic model of the structure of bacteriorhodopsin has been used for multislice calculations at various specimen thicknesses, equal to multiples of the protein thickness (Glaeser and Downing, 1993). At an energy of 100 keV the R-factor between Friedel mates, R_{sym}, was calculated to be 0.11 for a monolayer crystal, similar to the value estimated in the experimental measurements (Glaeser and Ceska, 1989).

Two additional effects have been discovered from the multislice calculations on bacteriorhodopsin, which would be very difficult to learn about from experimental measurements. (1) The intensity-weighted phase error between the multislice dynamical-diffraction spots and the corresponding weak-phase-object diffraction spots was only 5.3° for the monolayer protein crystal (Glaeser and Downing, 1993). This demonstrates that a small failure of the kinematic approximation, as reflected by the 10% intensity differences in Friedel pairs, would not have a significant effect on the actual data analysis. (2) In addition, the calculations showed that the average diffraction intensity of a Friedel pair is almost the same as the kinematic intensity, even though their difference in intensity may be as large as 10%. This must mean that the Friedel difference is initially split between an increase in intensity for one spot and a corresponding decrease for its Friedel mate. As a result, the average intensity for Friedel mates can be used for normal structure analysis, even though dynamical effects have produced a significant Friedel difference.

An analytic derivation can, in fact, be given within the phase object approximation, to explain why the average intensity of Friedel mates should be close to the kinematic intensity, as long as the Friedel difference is itself small. If one first expands the phase object approximation to second order, we obtain

$$T(\mathbf{x}) = e^{-i2\pi \frac{e}{h v} V'(\mathbf{x})} \simeq 1 - i2\pi \frac{e}{hv} V'(\mathbf{x}) + \frac{1}{2}\left(-i2\pi \frac{e}{hv} V'(\mathbf{x})\right)^2$$

$$\simeq 1 - i2\pi \frac{e}{hv} V'(\mathbf{x}) - 2\pi^2 \left(\frac{e}{hv}\right)^2 V'^2(\mathbf{x}). \tag{15.3}$$

As a result, the scattered wave will be of the form

$$\psi_{scattered}(\mathbf{s}) = \delta(\mathbf{s}) - i\mathbf{F}(\mathbf{s}) - \frac{1}{2}\{\mathbf{F} * \mathbf{F}\}(\mathbf{s}). \tag{15.4}$$

Since the projected Coulomb potential, $V'(\mathbf{x})$, is a real function, the middle term is actually Friedel antisymmetric because of the factor $i = \sqrt{-1}$ (although the antisymmetric character is normally ignored, since it is not apparent in the corresponding diffraction intensities), while the third term is properly Friedel symmetric. Squaring the scattered wave function, we obtain the results:

$$I(\mathbf{s}) = F^2(\mathbf{s}) + \mathfrak{Im}\mathbf{F}(\mathbf{s})\{\mathbf{F} * \mathbf{F}\}^*(\mathbf{s}) + \frac{1}{4}\{\mathbf{F} * \mathbf{F}\}^2(\mathbf{s}), \tag{15.5}$$

$$I(-\mathbf{s}) = F^2(\mathbf{s}) - \mathfrak{Im}\mathbf{F}(\mathbf{s})\{\mathbf{F} * \mathbf{F}\}^*(\mathbf{s}) + \frac{1}{4}\{\mathbf{F} * \mathbf{F}\}^2(\mathbf{s}), \text{ and}$$

$$\frac{I(\mathbf{s}) + I(-\mathbf{s})}{2} = F^2(\mathbf{s}) + \frac{1}{4}\{\mathbf{F} * \mathbf{F}\}^2(\mathbf{s}) \simeq F^2(\mathbf{s}),$$

where $\mathfrak{Im}$ denotes the imaginary part of the expression. In other words, as long as one can neglect terms of order F^4 relative to terms of order F^2, the average values of Friedel mates will closely approximate the kinematic diffraction intensities.

Naturally, the conclusion that measurable dynamical effects do not initially lead to significant errors in the structure analysis cannot hold indefinitely as the specimen thickness increases. Results found in the multislice calculations at 100 keV for bacteriorhodopsin (Glaeser and Downing, 1993), which are reproduced in table 15.1,

Table 15.1 Estimate of the errors that occur because of dynamical effects, when the weak phase object approximation is used to interpret electron diffraction data for thin specimens

Thickness (nm) (no. of unit cells)	$R^a_{Friedel}$	$R^b_{Dynamical/WeakPhase}$	Phase residual[c] (degrees)
4 (1)	11.1	1.0	5.3
8 (2)	20.9	3.4	10.0
12 (3)	29.5	7.4	14.9
16 (4)	36.8	12.7	19.9
20 (5)	42.5	19.0	25.2

[a]$R_{Friedel} = 100 \frac{\sum ||F(\mathbf{g})| - |F(-\mathbf{g})||}{\sum \frac{1}{2}(|F(\mathbf{g})| + |F(-\mathbf{g})|)}$, where the structure factors, F, are calculated with the multislice algorithm. The summation goes over all Friedel mates.
[b]The R-factor is again calculated with the modulus of the structure factors, but in this case the comparison is between (1) the average value of Friedel mates that are calculated with the multislice algorithm and (2) those calculated with the weak phase object approximation. The two sets of structure factors were scaled such that the integrated diffraction intensity was the same in both cases.
[c]Intensity-weighted phase difference between the phases calculated with the multislice algorithm and those calculated with the weak phase object approximation.

demonstrate how the kinematic approximation becomes progressively less satisfactory as the sample thickness increases. By the time that the sample thickness increases to 20 nm, for example, the intensity-weighted phase error has increased to 25° and the R-factor between the weak phase approximation and the average value of the Friedel mates (in the multislice approximation) has risen to 0.19. The contribution made to the R-factor from dynamical diffraction alone, at a thickness of 20 nm, would no doubt bring the value of R_{free} to a large enough value that one would have limited confidence that the structure was really correct. The large value (about 42%) of the R-factor between Friedel mates provides an unambiguous warning that the kinematic approximation is not well obeyed. The use of specimen thicknesses significantly greater than 20 nm is therefore likely to be problematic for high-resolution structure determinations with 100 keV electrons. For any given degree of failure of the kinematic approximation, one can use higher energy electrons (e.g., 300 keV) to roughly double the sample thickness before the error is expected to become similar to the estimates given in table 15.1.

The multislice formulation has also been used to map out a fairly detailed "domain of validity" of the weak phase object approximation. These calculations have been done for a relatively small protein structure, cytochrome b5, as a model (Ho et al., 1988). In interpreting the data that were obtained from those calculations, one must recognize that there is no absolute threshold for the validity (or failure) of the kinematic approximation. Instead, there is a progressive worsening of the kinematic approximation as the specimen thickness increases, as the resolution increases, and as the electron energy decreases.

As is shown in figure 15.1, the multislice calculations again suggest that the weak phase object approximation appears to be quite acceptable for proteins, at 100 keV, for a specimen thickness of at least 15 to 20 nm, even at high resolution. At greater thickness than this the R-factor begins to exceed the value that one expects to have in a properly refined, high-resolution structure. For simpler organic molecules, however, which have a much smaller unit-cell repeat in the direction of the beam, the limiting specimen thickness was found to be only a half or a third of that which can be tolerated with large, complex structures (Jap and Glaeser, 1980). As we have indicated previously, an earlier onset of dynamical effects is expected in crystals of small molecules because of the greater degree of superposition of atomic potentials that occurs in crystals with smaller unit cell size.

15.4 Strong dynamical diffraction effects need not interfere with subsequent refinement of an atomic-resolution model of the structure

The chief disadvantage introduced by dynamical diffraction lies in the fact that the Fourier transform relationship is lost between the structure of the specimen and the scattered wave function. Both the amplitudes and the phases of the scattered wave are affected, of course. While small deviations from kinematic values can only result in small errors in the density map, at some point the failure of the kinematic approximation will become so severe that the Fourier transform produced from dynamical image data cannot be interpreted in terms of the correct structure.

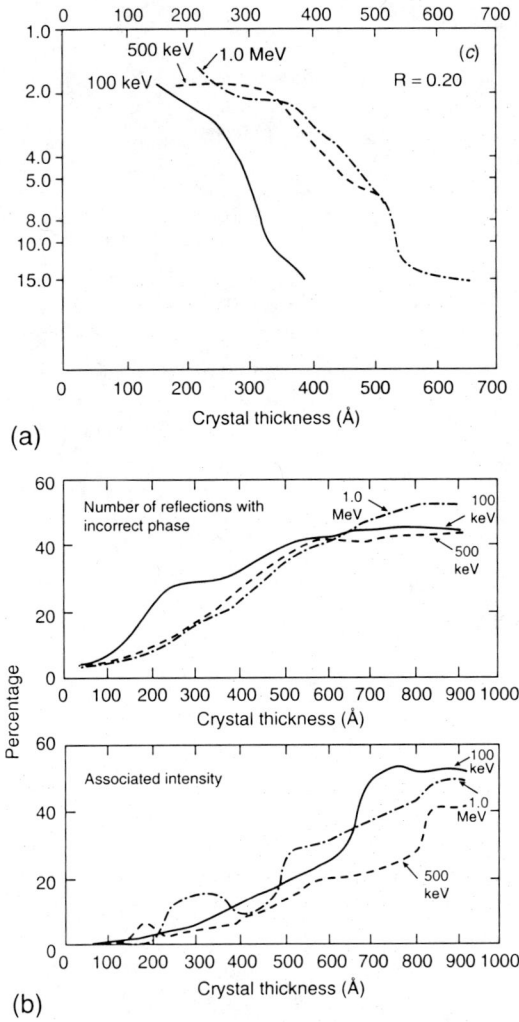

Figure 15.1 Representative estimates of the domain of validity of the weak phase object approximation, obtained by comparing structure factors calculated (for the same object) with the weak phase object approximation and with the "exact" Cowley–Moodie formulation of dynamical diffraction theory. The model specimen used for these calculations was the crystal structure of cytochrome b5. Panel (a) shows boundary lines in the 2-D plane of resolution and crystal thickness, to the left of which the R-factor between the two calculated structure factors is less than 20%, and to the right of which the R-factor is greater than 20%. The domain of validity depends, of course, on the degree of error that one is willing to accept. The weak phase object approximation is valid up to greater thickness at low resolution than it is at high resolution. The domain of validity improves at high electron energy, but this improvement saturates at an energy of 500 keV. Panel (b) shows the percentage of diffraction spots that have incorrect phases (top) and the percentage of the total diffracted intensity that is associated with reflections that have incorrect phase (bottom). The model specimen is centrosymmetric in projection for the crystal orientation chosen in this work, and thus one can assign a value of the phase as being either correct or incorrect, depending upon whether the dynamical value is closer to the "kinematic" value or to the "kinematic" value plus π. The data in both panels indicate that the weak phase object approximation is sufficiently accurate to be used for specimen thicknesses of at least 20 nm for 100 keV electrons, and for specimen thicknesses of at least 40 nm in the high-energy limit.

On the other hand, there is no difficulty in employing dynamical diffraction data to distinguish which of two (or more) models of a structure is correct. This can be accomplished by using an appropriate version of dynamical diffraction theory (e.g., the Cowley–Moodie formulation) to calculate either the image intensities or the diffraction pattern intensities. The model structure that gives (dynamical) calculations that agree best with the experimental measurements is taken to be the best model of the real structure. Image-matching calculations of this type are a common element of the interpretation process for high-resolution electron microscopy in materials science research. Refinement of a model structure in which the multislice formulation is used to compute F_{calc} would be just as valid, of course. Model-based comparisons can even work well as a method of structure determination, if the structures being studied are simple, the number of models to be distinguished are few, and the number of parameters to be varied is small. Model-based comparisons cannot be expected to work as an ab initio method with complex structures such as biological macromolecules, however.

15.5 Fresnel diffraction alone can become significant in thick specimens

If electrons are scattered from layers that are separated by a distance "Δz" parallel to the beam axis, as is shown schematically in figure 15.2, the diffracted beams from the first layer will differ in phase from those scattered by the second layer due to the path length difference between the two scattered waves. The geometric construction that is presented in figure 15.2 shows how one can calculate the difference in pathlength, which, in the example shown there, ignores the difference in "refractive index" inside and outside the specimen. The phase difference that is associated with the difference in path length is $2\pi \frac{\text{path difference}}{\text{wavelength}}$, which, for small scattering angles, is

$$\varphi = 2\pi \frac{\Delta z \lambda}{2} S^2, \qquad (15.6)$$

where Δz is the separation between the two layers, λ is the electron wavelength, and S is the spatial frequency associated with the angle at which the electron beam has been scattered. This phase difference is nothing else, of course, than the phase shift associated with the amount of defocus, Δz, between the two layers.

The effect that defocus has on images of thin samples has been covered extensively in chapter 10, and defocus poses no special difficulties in that case. Thicker samples, on the other hand, present severe difficulties whenever the sample thickness exceeds the *depth of field* for a given resolution. The depth of field is defined as the range of defocus, Δz, over which features of a given resolution are not significantly affected. To illustrate this, we might say that Fourier components of the object that have a wavelength of $d = 1/S$, and which are located in thin slabs of the specimen at different depths, Δz, will be imaged with a relative phase error, φ, as is given in equation 15.6. If we allow this phase error, due to defocus, to be as large as $\pm 45°$ relative to the midpoint, we then find that

$$\Delta z = \frac{d^2}{2\lambda}. \qquad (15.7)$$

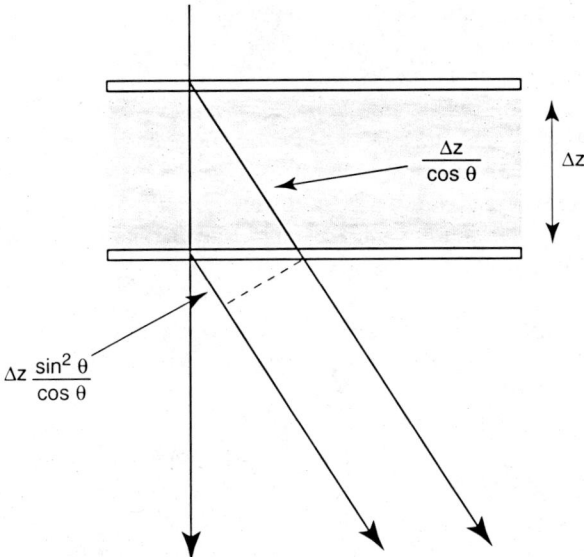

Figure 15.2 Rays scattered at an angle θ from the top of a specimen travel a shorter distance than those scattered at the bottom, as is evident by inspection of the diagram shown here. In the former case the distance is $\Delta z/\cos\theta$, where Δz is the total thickness of the specimen, while in the latter case the path length is $\Delta z(1 + \sin^2\theta/\cos\theta)$. The difference in path length, when θ is small, is $\frac{\Delta z}{2}\theta^2$. The wave scattered from the top of the specimen and that scattered from the bottom will have a phase difference of $2\pi \frac{\text{path difference}}{\text{electron wavelength}}$. When the angle θ satisfies Bragg's law for a spatial frequency S, and the angle is again small, the phase difference for rays scattered at the top of the sample and those scattered at the bottom is given by equation 15.6.

If we imagine processing images of many identical objects, each of which are much thinner than the depth of field, but which are distributed at different levels of a "thick" specimen, it is clear that each would have to be corrected separately for its own contrast transfer function before the data were merged. As pointed out by Jensen (2001), this is likely to be an important consideration when processing high-resolution images of particles distributed at various depths in a film of vitreous ice.

Even if samples embedded in vitreous ice can be confined to a sufficiently thin slab, the problem would again become important if the sample were, for any reason, tilted at a fairly high angle. It is for this reason that the defocus ramp must be removed when processing images of tilted, two-dimensional crystals, for example (see section 10.11). An experimental solution to this problem, in the case of tilted samples, is to simply use spot-scan imaging along with dynamic defocusing (see section 5.7).

A more intractable situation arises, however, when the biological object itself is thicker than the depth of field. If an object is more than 20 nm thick, the phase-contrast transfer function for high-resolution Fourier components of the Coulomb potential will not have the same value at all levels within the specimen. If the specimen is 100 nm thick, for example, the finite depth of field becomes a limitation at $\sim$0.8 nm resolution for 100 keV electrons, and at $\sim$0.6 nm resolution for 300 keV electrons. The maximum specimen thickness that is allowed for a resolution of 0.3 nm, in order to meet the depth of field criterion in equation 15.7, is only 12 nm at 100 keV, and 22 nm at 300 keV.

Ignoring the phase shift due to defocus and simply summing the transmitted waves for each slice is, of course, what happens when one makes the projection approximation. Appropriately enough, one can see from equation 15.6 that the phase shift is zero — and the projection approximation is exact — if the wavelength, λ, is zero. When the wavelength is not zero, however, the Ewald sphere has finite curvature and the projection approximation must eventually fail as the thickness increases.

15.6 Curvature of the Ewald sphere destroys the appearance of Friedel symmetry at high resolution and at high tilt angles

Curvature of the Ewald sphere is determined only by the wavelength; it is the same for a thin sample as it is for a thick sample. Why then should the curvature of the Ewald sphere have a greater effect for a thick sample than for a thin one? The answer to this question is seen from a comparison of the Fourier transforms of thick versus thin samples. As is easily seen for a rectangle function, the Fourier transform of a thin sample cannot change appreciably over distances in z^* that are small compared to 1/thickness. The Fourier transform of a thin sample, evaluated on the Ewald sphere, will be almost identical to the value found on a central section, if the distance in z^* between the Ewald sphere and the central section is much smaller than 1/thickness of the specimen. The important conclusion here is that the projection approximation will always be valid, regardless of the electron wavelength, if the specimen is thin enough. The same conclusion is easily drawn from equation 15.6, where one can see that the phase shift due to defocus goes to zero for any electron wavelength, as Δz goes to zero. Conversely, as the sample thickness increases, the projection approximation begins to fail and one must take curvature of the Ewald sphere into account.

Curvature of the Ewald sphere does not, by itself, influence whether the scattered electron wave obeys Friedel symmetry, if the measurement is actually made at (vector) spatial frequencies that are related by inversion symmetry. Friedel symmetry is simply a mathematical property of the Fourier transform of real-valued functions (see box 3.3). What we do want to discuss here, however, is the fact that curvature of the Ewald sphere will cause the Fourier transform of an object to be sampled at pairs of vectors in reciprocal space that may appear to be, but are not true Friedel mates. As is shown in figure 15.3, two points on the Ewald sphere may lie at equal and opposite distances from the origin, but S_2 is not equal to $-S_1$ as required for the Friedel symmetry, and as would be the case if the Ewald sphere were a plane.

The situation is especially easy to describe in the case of an untilted sample, that is, when the incident beam is perpendicular to the plane of a 2-D crystal. In this case the two diffraction spots that correspond to the intersection of the Ewald sphere with the (h, k) and $(-h, -k)$ reciprocal lattice rods, respectively, appear to be Friedel mates. While there is nothing about the geometric position of these spots in the diffraction pattern that would show that they are not Friedel mates, the sketch of the Ewald sphere presented in figure 15.3(a) shows that S_2 has a positive z^* component while the true Friedel mate, $-S_1$ has a negative z^* component. As a result, neither the amplitude nor the phase of the two reflections need to be related by Friedel's law. If the sample is

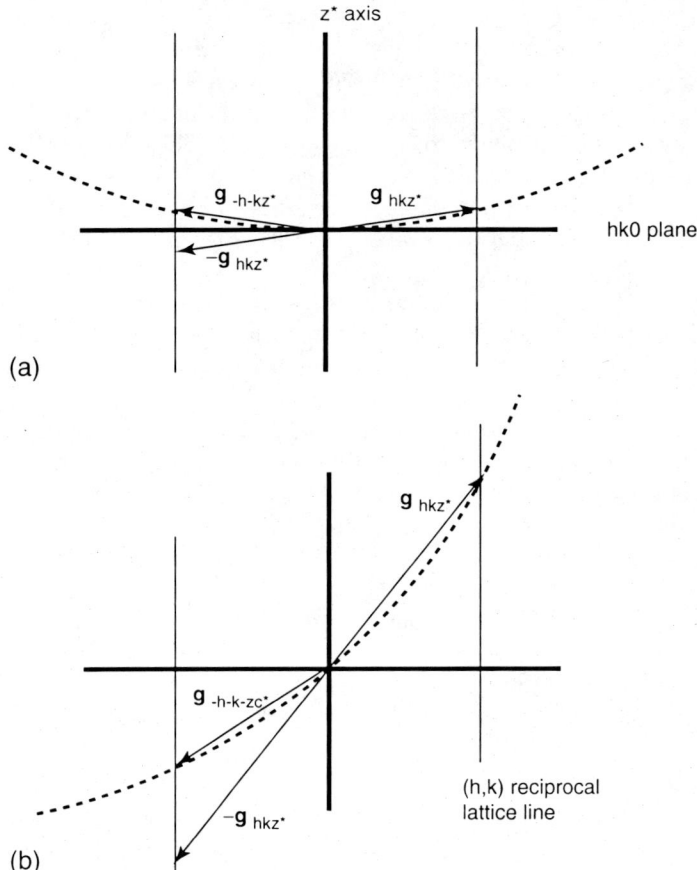

Figure 15.3 Curvature of the Ewald sphere causes the 3-D Fourier transform to be sampled at positions that are not precisely related by inversion symmetry. As a result, the diffraction data may not satisfy Friedel's law, as it is required to do when the weak phase object approximation is valid. The axes of the reciprocal-space coordinate system are shown with thick lines, two reciprocal lattice lines are shown with thin lines, and a central portion of the Ewald sphere is shown with a dashed line. (a) When the incident beam is perpendicular to the plane of a 2-D crystal, the Ewald sphere intersects pairs of reciprocal lattice lines at vector positions S_1 and S_2, which appear to be precise Friedel mates because they lie at equal and opposite distances from the origin, but which are not true Friedel mates (i.e., $S_2 \neq -S_1$), as can be seen in the diagram. (b) When the specimen is tilted, even the lengths of the vectors S_1 and S_2 are different.

thin enough, of course, then the value of the Fourier transform at S_2 and at $-S_1$ must still be very similar, and Friedel symmetry will still be very nearly preserved.

When the specimen is tilted, as is shown in figure 15.3, then even the vector lengths of S_1 and S_2 will no longer be equal. The unequal lengths of S_1 and S_2 can have important effects in the analysis of experimental diffraction patterns and in the analysis of experimental images. The condition of Friedel symmetry no longer holds in this case, of course, because the scattered waves at the "Friedel" vector positions $-S_1$

and $-S_2$ will be zero when the vector lengths of S_1 and S_2 are unequal. Images that are obtained under these conditions must therefore be described as *single side band images* (see section 3.11). Since the lengths of the two vectors S_1 and S_2 will be close to each other, however, one expects that the two single side band contributions will beat against each other, thereby producing a superlattice effect, or a "Moire pattern" effect in images of 2-D crystals. Since the consequences of curvature of the Ewald sphere are not important for thin samples, for modest tilt angles and for modest resolution, the effects discussed in this paragraph have received little attention in the literature.

In the case of noncrystalline specimens, whose Fourier transforms are continuous rather than being limited to discrete points in reciprocal space, the scattered wave function will normally be nonzero for both of the two spatial frequencies that appear to be Friedel mates, whether the specimen is tilted or untilted. When specimens of this type are thick enough that one must take curvature of the Ewald sphere into consideration, there is more than one way in which that can be done (DeRosier, 2000). The most general and the most straightforward approach described by DeRosier is to record two images at different values of defocus. Any image can be described as the superposition of single-sideband image intensities (see box 3.3), whether the diffracted wave has Friedel symmetry or not. When the diffracted wave is Friedel-symmetric, one image is sufficient to determine the contributions that are made by the two Friedel mates, since the contribution from one can be known exactly from the other. An alternative way to view this situation is to say that Friedel symmetry is a constraint (i.e., another equation), which reduces the number of unknowns that must be determined from the experimental observations. When the diffracted wave in not Friedel-symmetric, however, then a second image (at a different value of defocus) can be used in order to provide as many experimental observations as there are experimental unknowns.

15.7 Inelastic scattering becomes an important consideration in thick specimens

According to data presented in figures 5 and 6 of Wall et al. (1974), the cross-section for inelastic electron scattering in organic materials is about two times greater than the elastic scattering cross-section, although we note that earlier authors felt that the inelastic cross-section for carbon atoms would be about three times the elastic scattering cross-section. In the following we will assume that the inelastic scattering cross-section is ~ 140 pm^2, twice the value quoted previously for the elastic scattering cross-section. The *mean free path* for *inelastic scattering* of 100 keV electrons in organic materials, with a mass density of 1.5 g/cm^3, is therefore estimated to be about 100 nm, while in ice (with a mass density of ~ 1 g/cm^3) the mean free path for inelastic scattering would be about 150 nm. Similar, but slightly larger values of the mean free path were estimated by Grimm et al. (1996).

On the basis of Poisson statistics, 63% of the incident electrons suffer one or more inelastic scattering events within a thickness that corresponds to one mean free path, while only 37% are transmitted as either unscattered electrons or elastically scattered electrons. The mean free path increases at higher energies, however, asymptotically reaching a value about 2.5 times greater than that at 100 keV. At so-called

Table 15.2 Representative estimates of the fraction of incident electrons that are transmitted through different specimen thicknesses without suffering an inelastic scattering event

Electron energy (keV)	10 nm	20 nm	30 nm	50 nm	100 nm	200 nm	300 nm
100	0.90	0.82	0.74	0.61	0.37	0.14	0.05
300	0.95	0.90	0.86	0.78	0.61	0.37	0.22

"intermediate" energies, about 300 or 400 keV, the mean free path for scattering is about twice the value at 100 keV.

Many biological specimens that are used for electron crystallography have a total specimen thickness of ∼100 nm. The actual biological object may be much thinner than that, but it may be supported by a carbon film (to reduce specimen charging and to confer mechanical support), and there may be additional thickness associated with embedment in glucose or vitreous ice. In any case, it is instructive to note that more than half of the electrons will be scattered inelastically in a 100 nm sample of ice at 100 keV, and more than a third will be scattered inelastically at 300 keV. Thus, even at 300 keV, the inelastically scattered electrons are not a negligible fraction of the total. A more complete set of "representative estimates" of inelastic scattering, as a function of thickness, is presented in table 15.2.

When an electron is inelastically scattered, it no longer contributes to the image intensity in the way that is described by Fourier optics (chapter 3). Abbe's theoretical picture, in which the scattered wave interferes coherently with the unscattered wave, clearly is not applicable to an inelastically scattered electron. Instead, inelastic scattering leads to two separate effects. (1) The transmitted intensity of electrons that have not suffered an inelastic event — we will now call these the *no-loss electrons* — is reduced in proportion to the intensity of inelastic scattering. (2) A complementary electron-intensity function is formed, which is associated with the inelastically scattered electrons.

The fact that the amount of inelastic scattering can vary at different points over the specimen will clearly affect the transmission function that is used to describe the "no-loss" electrons, that is, those which are either unscattered or elastically scattered. The extreme example of there being spatially varying amounts of inelastic scattering would be given by a totally absorbing aperture, in which the intensity transmitted through the open hole is equal to the incident intensity, while it is zero everywhere else.

The effect on the transmission function (that describes the no-loss electrons) is actually the same as if the inelastically scattered electrons had been absorbed in the specimen. This point can be fully appreciated by considering an experiment in which an energy filter is used to transmit only the no-loss electrons to the final image plane. Inelastic scattering thus causes the transmission function of the object to modulate the amplitude as well as the phase of the incident wave.

The inelastically scattered electrons are not actually absorbed in the specimen, of course. High-energy electrons lose only a minute fraction of their incident energy in each inelastic scattering event, and as a result they are still transmitted through the specimen. Thus, unless one uses a microscope that is equipped with an energy filter, it is important to understand how the inelastically scattered electrons may affect the measured data when the image that they form is superimposed on the image formed by the no-loss electrons.

Previously unaccounted for image intensity is produced by inelastically scattered electrons

The inelastically scattered electrons form what is, in effect, a self-luminous image of the specimen. For every point in the object, the inelastic electron image intensity will increase in direct proportion to the number of inelastic scattering events. In the absence of chromatic aberration, the image intensity can therefore be thought of as a projection of the inelastic scattering cross-section of the object. Like a dark-field image, this image will be brighter where the scattering cross-section is greater, thus it is reasonable to refer to the image formed by inelastically scattered electrons as a dark-field image. The image formed by a monochromatic subset of inelastically scattered electrons will additionally exhibit phase contrast to the extent that the spherical wave representing an inelastically scattered electron must experience spatially varying phase shifts (i.e., elastic scattering) as it continues to propagate through the remainder of the specimen.

Even though the inelastic scattering cross-section is about two (or possibly three) times greater than the elastic scattering cross-section, the increment of contrast that is associated with the inelastic "dark-field" image of a thin specimen remains much smaller than the phase contrast that is generated in the bright-field image of a thin specimen. The explanation, of course, is that the contrast in the bright-field image arises from the interference of the weak, scattered wave and the strong, unscattered wave. The magnitude of the phase contrast is thus linear in the weakly scattered wave amplitude, while the dark-field image intensity is effectively proportional to the square of the amplitude rather than the scattering amplitude itself. For thin objects, intensity terms that are proportional to the square of the amplitude can be ignored. As the specimen thickness increases, however, all types of nonlinear terms will inevitably become more important. These will initially include the double-scattering (dynamical) terms discussed above, the quadratic term produced when squaring the image wave function (see equation 3.18), and the image intensity formed by the inelastically scattered electrons.

At relatively high resolution, better than ~2 nm, one must also understand the fact that inelastic scattering results from a rather long-range interaction between the incident electron and the target atom. An extremely short electric-field pulse is generated by the passing electron, effectively bathing the atom in a spectrum of virtual photons, which correspond to the various Fourier components of the pulse of electric field strength (Fano, 1963; Christophorou, 1971). As a result, inelastic scattering events are delocalized, in the sense that they can occur even when the electron passes at quite large distances from the target atom (Isaacson et al., 1974). As is shown in figure 15.4, about 98% of the inelastic scattering events involve energy loses in the range ~10 to ~300 eV (Isaacson, 1977), and the majority of these low-energy excitations are associated with a transverse delocalization in the range of 1 to 2 nm. The net outcome is that the inelastic dark-field image must always be (for low energy losses) a rather low-resolution representation of the scattering density (mass density) of the object.

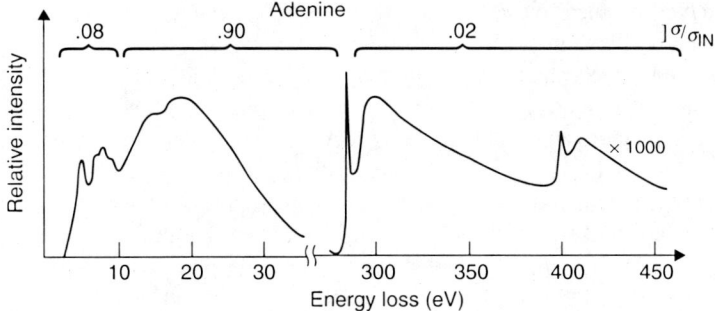

Figure 15.4 Experimental energy loss spectrum for a film of adenine that was approximately 50 nm thick, supported on a 3 nm thick carbon film. As is indicated in the figure, about 8% of the inelastic scattering events occur with an energy loss between zero and 10 eV, about 90% with an energy loss between 10 and the threshold for ionization of the 1S (core) electrons, and only 2% for energy losses above that threshold. The energy of the incident electrons was only 25 keV in this experiment, but the energy-loss spectrum for higher energy electrons would be the same. Reproduced with permission from Isaacson (1977).

The resolution of the inelastic dark-field image intensity will be further degraded by the effect of chromatic aberration, although this can be compensated in any energy window by an appropriate amount of defocus. In real space, chromatic aberration will blur a point into a disc with radius

$$\Delta r = C_c \frac{\Delta E}{E} \theta, \quad (15.8)$$

where C_c = coefficient of chromatic aberration, ΔE = energy loss, E = incident energy, and θ = maximum scattering angle. The expression in equation 15.8 is identical to that for blurring of a point due to defocus of the objective lens, with $\Delta z = C_c \cdot \Delta E / E$. To illustrate with a specific example, a 50 eV energy loss, at 100 keV, leads to a defocus of 1 μm for a chromatic aberration coefficient of 2 mm.

Even with an energy loss as large as 50 eV (and a defocus of 1 μm as calculated above), the amount of image blurring will be relatively small because of the fact that inelastic scattering is normally distributed over relatively small angles. Chromatic aberration becomes a more severe effect, however, when electrons experience at least one elastic event as well as one or more inelastic events. Elastic events typically scatter the electrons to relatively wide angles, in the range of 10^{-2} radians. Again taking the defocus caused by chromatic aberration to be 1 μm, the radius of the disc of confusion that is produced in the image is approximately 10 nm, much greater than the extent of the delocalization of the initial scattering event. As the specimen thickness increases, the blurring due to chromatic aberration will increase even further because of the larger magnitude of energy losses accumulated through multiple inelastic events, as well as the larger scattering angles, θ, produced as the result of multiple scattering events.

Two extreme limits can thus be discussed when describing the contribution of the inelastically scattered electrons. In the limit of a very thin samples and small defocus, the inelastic dark-field image will almost match, and thus compensate for, the removal of no-loss electrons from the transmitted beam. Stated in another way, the image intensity added by the inelastically scattered electrons cancels the amplitude contrast term that is introduced in the image formed by the no-loss electrons. The fact that the two terms cancel each other simply reflects the fact that the total number of electrons must be constant if there is no absorption and no phase contrast. The final result is that there will be almost no indication that there has even been any inelastic scattering, unless an energy filter is used to isolate just the no-loss electrons. In the limit of rather thick specimens, on the other hand, multiple elastic and inelastic events cause the inelastically scattered electrons to be blurred out over a much broader area of the image than that corresponding to the point of origin. To a first approximation, the inelastically scattered electrons will then form a background distribution of uniform intensity. In this case the inelastically scattered electrons only contribute to the average electron intensity and to the shot-noise, and they do not "cancel out" the amplitude contrast variations in the image formed by the no-loss electrons, as they do in the limit of thin specimens.

Inelastic scattering invariably degrades the image quality of thick specimens

The thick-specimen limit involves two effects that are very damaging to the image contrast. (1) A serious problem arises because multiple elastic–inelastic scattering depletes (effectively "absorbs") the no-loss electrons that would otherwise have been used to produce a phase-contrast image of the specimen. The absolute amount of phase contrast (and amplitude contrast, for that matter) will therefore begin to decrease rather than to increase as the specimen thickness continues to increase. (2) In addition, the diminishing amount of contrast that remains begins to be overwhelmed by the increasing amount of uniform-background intensity that is contributed by the inelastically scattered electrons.

An energy filter lens can be used to remove the background of inelastically scattered electrons, and doing so does improve our ability to see the contrast that is present in the image formed by the remaining no-loss electrons, as is shown in panel (a) and panel (b) of figure 15.5. Even so, the image intensity associated with the remaining, no-loss electrons will become very low, and the signal-to-noise ratio at a given incident flux (and thus a corresponding amount of radiation damage) will become worse, the thicker the specimen. Panel (c) of figure 15.5 shows how the relative proportions of electrons that are unscattered, that are elastically scattered, and that are inelastically scattered varies with the specimen thickness. As can be appreciated from these curves, there is rather little electron intensity left when the specimen is greater than two mean free pathlengths in thickness, if the energy filter is set to reject essentially all of the inelastically scattered electrons.

Inelastically scattered electrons also add a strong, unwanted background to electron diffraction patterns

Inelastic scattering also has important effects in degrading the quality of electron diffraction patterns. In thin specimens the principal effect of inelastic scattering is to produce an intense background at small angles. In much thicker specimens, multiple scattering events again deplete the elastically scattered electrons, and at the same time

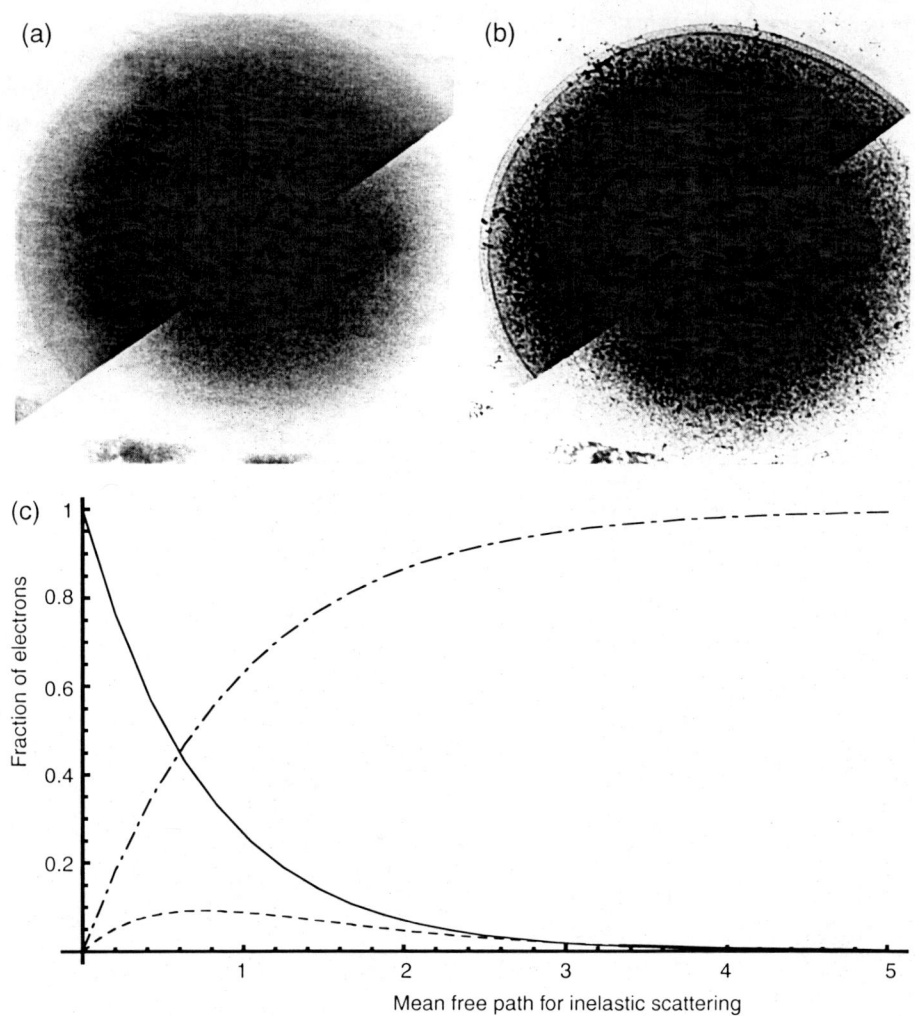

Figure 15.5 When an energy filter is used to remove inelastically scattered electrons, the image contrast of thick specimens is improved, but the overall image intensity is decreased. Images are shown for the identical area of a thick, ice-embedded specimen (a) recorded with an energy filter, to isolate the no-loss electrons, and (b) without the use of the energy filter (Koster et al., 1997). The specimen is an ice-embedded microbial cell, and the specimen thickness was estimated to be 600 nm. The image intensities in the top left and bottom right portions of the panels are scaled differently, in order to display a greater portion of the dynamic range of intensities. Panel (c) shows calculations of the fraction of electrons that are transmitted through a specimen as zero-loss electrons (solid line), inelastically scattered electrons (dot-dash line) and as elastic, singly scattered electrons (dashed line), all as a function of the specimen thickness (expressed in units of the mean free path for inelastic scattering).

the background becomes so dominant that one may not even be able to tell whether the specimen is crystalline or not. The effect that inelastic scattering has on diffraction patterns of very thick samples is thus similar to what it has on images.

The ratio of elastic and inelastic scattering has been calculated as a function of angle for organic materials by Wall et al. (1974). The two curves for the differential scattering cross-section are compared in figure 15.6. The shielded Coulomb potential of isolated

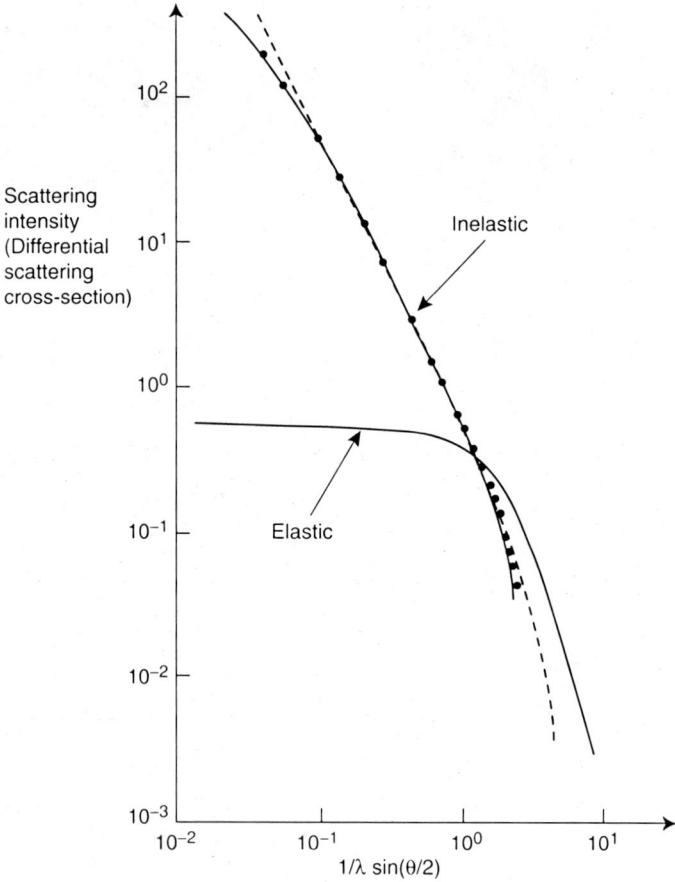

Figure 15.6 Calculated differential scattering cross-sections for the nucleic acid base, adenine, redrawn from figure 2 in Wall et al. (1974). The differential scattering cross-sections are in units of nm^2/steradian, and the values of the abscissa must be multiplied by 2 in order to get the spatial frequency in nm^{-1}. The dashed line is the inelastic scattering curve calculated for 1 MeV electrons, while the solid line represents the scattering curve for 17.5 keV electrons, corresponding to the energy of incident electrons used for the measurements (represented by black dots). Note that the inelastic scattering cross-section extrapolates to a value more than 1000 times greater than the elastic scattering cross-section at small angle, but the inelastic scattering cross-section falls below that for elastic scattering at a spatial frequency higher than about 2.5 nm^{-1} (i.e., a resolution of $\sim$0.4 nm). The total cross-section for inelastic scattering in biological specimens is variously stated to be 2 to 3 times that for elastic scattering.

atoms, calculated from accurate atomic wave functions, was used to calculate the elastic scattering cross-section, while experimental measurements of the energy loss spectrum of a thin, organic film were used to calculate the inelastic scattering cross-section. As shown in figure 15.6, the differential scattering cross-section for inelastic scattering is about 1000 times greater than that for elastic scattering at very small angle, but the inelastic scattering cross-section falls rapidly with scattering angle, and becomes less than the elastic scattering cross-section at spatial frequencies $2/\lambda \cdot \sin\theta/2 \simeq 2.5$ nm^{-1}, corresponding to Bragg spacings of 0.4 nm.

In the case of a thin sample, where the number of elastic and inelastic scattering events is relatively small, the electron diffraction pattern from a 2-D crystal will be superimposed upon a continuous background of inelastically scattered electrons. As we know from the theoretical distribution shown in figure 15.6, however, the inelastic background will be strongly concentrated around the unscattered, central spot. The relative intensity of the diffraction spots and the inelastic background depends upon (1) the thickness of the support film, embedment and other noncrystalline material relative to (2) the thickness of the crystalline specimen. The electron diffraction spots are normally stronger than the background, however, in spite of the fact that the atomic differential scattering cross-section for the inelastically scattered electrons may be much greater than the atomic differential cross-section for elastic scattering. The explanation is that the elastic electrons are concentrated into sharp diffraction spots while the inelastically scattered electrons are spread continuously over the background. Nevertheless, the background of inelastically scattered electrons can be so intense that it quickly causes a photographic film, and even a CCD camera, to be saturated in the small-angle region.

In practical circumstances, exposures that are long enough to record the high-resolution part of a diffraction pattern may result in saturation of the detector (especially when recording data on film) at spacings larger than ~ 2 nm, as is illustrated in panel (a) of figure 15.7. Many of these lower resolution diffraction spots can still be recorded by using an exposure that is 5 or 10 times shorter than that used to record the high-resolution reflections. Thus, when recording diffraction data on film, two or more diffraction patterns, taken with different exposure times, may be needed to record low-resolution diffraction spots as well as the high-resolution part of the pattern. It is even more effective, of course, to use an energy filter lens to remove the background of inelastically scattered electrons, so that the full pattern of diffraction spots can be recorded in a single exposure, as is illustrated in panels (b) through (e) of figure 15.7.

15.8 A final caution: failure of Friedel symmetry for thick specimens can be due to curvature of the Ewald sphere, dynamical diffraction, or inelastic scattering

The expression "satisfaction of *Friedel symmetry*" is usually taken to mean that diffraction intensities are identical at pairs of points in reciprocal space that are related by inversion through the origin. As was written in equation 3.23, however, a more complete definition of Friedel symmetry states that the value of the diffracted wave function at a spatial frequency $-\mathbf{S}$ is the complex conjugate of the value at a spatial frequency $\mathbf{S}$. As a result, one should keep in mind that the more commonly assumed meaning of "Friedel symmetry" is really just a consequence of the more complete definition.

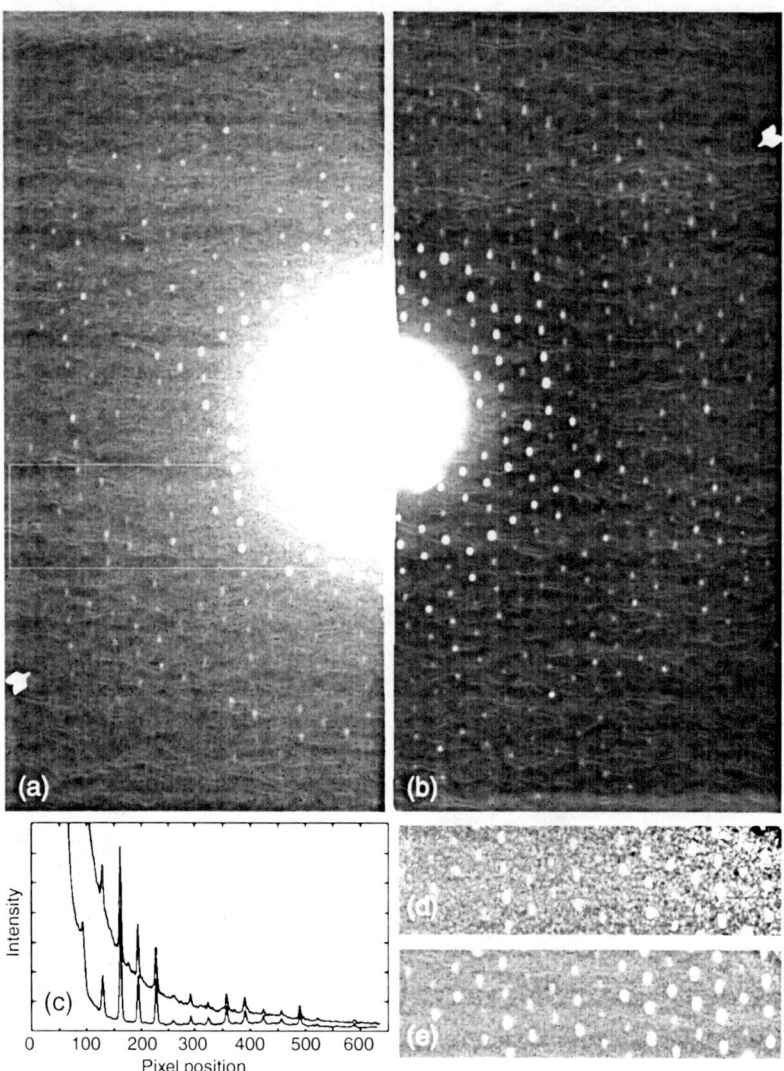

Figure 15.7 Representative example of the high background intensity of inelastic scattering that occurs in the center of electron diffraction patterns, that is, at small angles. (a) When the electron exposure is high enough to record the high-resolution diffraction spots, the image is nearly saturated by the inelastic background at Bragg spacings below ~2 nm. The low-resolution part of the diffraction pattern is easily recorded with an exposure that is 5 to 10 times shorter than what is used in this image, but only at the expense of losing the weaker, high-resolution part of the diffraction pattern. (b) When the background intensity of inelastically scattered electrons is removed with an energy filter, however, one can record diffraction spots over nearly the entire range of spatial frequencies with a single expose. (c) A more quantitative comparison of filtered and unfiltered diffraction patterns is shown by the pair of line tracings shown here. (d) and (e) show the same portion of one area of the electron diffraction pattern, recorded without (d) and with (e) the use of the energy filter. The diffraction pattern in the example shown here was obtained from a glucose-embedded crystal of bacteriorhodopsin, using the CCD camera in a Gatan (postcolumn) energy filter.

Satisfaction of Friedel symmetry is a basic requirement for the rigorously valid application of many standard tools, both practical and theoretical, in electron crystallography. As an example, the observed differences in the intensities measured for Friedel mates are commonly used to calculate an R-factor, $R_{Friedel}$, which is one of the standard ways to evaluate the quality of the diffraction data. Even more importantly, the usual theory of image formation, and in particular the theory of the phase contrast transfer function (section 3.8), requires that the scattered electron wave function obeys Friedel symmetry. As the sample thickness increases, however, the degree to which Friedel symmetry fails to hold is likely to become more and more noticeable. Indeed, there is more than one reason why Friedel symmetry may not hold, as the heading of this section points out.

The requirement that diffraction intensities should be identical for two diffraction spots whose positions are related by inversion symmetry can appear to be violated because of curvature of the Ewald sphere, as was pointed out in section 15.6. This apparent failure of Friedel's law is not a real violation, of course, because two points on the surface of the Ewald sphere are never related by inversion through the origin. Although apparent failure of Friedel symmetry occurs for a trivial reason in this case, its effect on data processing for a thick sample can become pronounced. The point to remember in this case is that the value of the Fourier transform of the object is expected to change more rapidly, at nearby points in reciprocal space, as the specimen thickness increases. As a result, the value of the Fourier transform can become quite different at two points on the surface of the Ewald sphere that appear to be Friedel mates in a 2-D diffraction pattern, but actually are not Friedel mates in 3-D reciprocal space.

There is a simple solution that can be used when one must account for curvature of the Ewald sphere, which is to separately index the two members of a pair of diffraction spots that appear — naively — to be Friedel mates, rather than to average the two measurements. Failure to use separate indexing of reflections that are not, in fact, Friedel mates could result in unwanted error in the structure-factor magnitudes, high values of $R_{Friedel}$, and correspondingly high values in R_{merge}. Separate indexing of reflections that appear to lie at S and $-S$, but do not do so because of curvature of the Ewald sphere, is not currently provided for in the commonly used software discussed in chapter 9, however, even though it is an essential feature of x-ray diffraction software.

The correct way to handle images, when Friedel symmetry cannot be assumed due to curvature of the Ewald sphere, is somewhat more difficult. If the apparent Friedel mates are at slightly different spatial frequencies, for example at S and $-(S + \delta S)$, then the two reflections will give separate, single-sideband contributions to the image intensity. The formalism described in section 3.11, rather than the conventional contrast transfer function formalism, would then have to be used in order to recover structure factor phases from the images. If, on the other hand, pairs of reflections appear to be true Friedel mates but their structure factors do not obey Friedel symmetry, one can use a pair of images recorded at two different values of defocus, as was described in section 15.6.

The true violation of Friedel symmetry that is due to dynamical diffraction has been discussed in section 15.3. When dynamical effects become large enough, neither images nor diffraction intensities can be interpreted in terms of a projection of the structure. In addition, the concepts of phase-contrast and amplitude-contrast transfer functions must also be abandoned. In fact, the use of linear, Fourier transform methods of structure determination are no longer a useful way to analyze the data. As was discussed in section 15.4, however, dynamical theory can be used to compare computed results to the experimental data, in order to rule out incorrect models and even to refine an initial model of the structure.

A true violation of Friedel symmetry can also be due to the fact that inelastic scattering causes the transmittance function of an object to have both real and imaginary parts. This effect of inelastic scattering is perhaps most easily explained from a phenomenological point of view. The spatial variation in the modulus of the wave function describing the transmitted, no-loss electrons can be written as an exponential term, $e^{-\frac{1}{2}\mu(\mathbf{x})}$, that modulates the transmittance function of a pure phase object. The modified transmittance function will then be written

$$T(\mathbf{x}) = e^{-i2\pi \frac{e}{h v} V'(\mathbf{x})} e^{-\frac{1}{2}\mu(\mathbf{x})} = e^{-i\left(2\pi \frac{e}{h v} V'(\mathbf{x}) - i\frac{1}{2}\mu(\mathbf{x})\right)}, \quad (15.9)$$

where $\mu(\mathbf{x})$ describes the spatial variation in the "absorption" of electron intensity.

In the special case that a linear expansion of both exponential functions is a reasonably accurate approximation, the transmittance function simplifies to

$$T(\mathbf{x}) \simeq 1 - i\left(2\pi \frac{e}{h v} V'(\mathbf{x}) - i\frac{1}{2}\mu(\mathbf{x})\right). \quad (15.10)$$

This linearized version of the transmittance function is the same one that was used in equation 3.34 to describe a mixed weak-phase and weak-amplitude object, but it is rewritten here in a form that emphasizes how the absorption term, $i\frac{1}{2}\mu(\mathbf{x})$, can be thought of as an imaginary part of the Coulomb potential. The point being made in this case is that the transmitted wave for even a weak-phase, weak-amplitude object is no longer a real function, and as a result the diffracted electron wave will not obey Friedel's law.

References

Abbe E. 1873. Beitrage zur theorie des mikroskops und der mikroskopischen wahrnehmung. Archiv fuer mikroskopische. Anatomie 9:413–468.
Adrian M, Dubochet J, Lepault J, McDowall AW. 1984. Cryo-electron microscopy of viruses. Nature 308(5954):32–36.
Adrian M, Dubochet J, Fuller SD, Harris JR. 1998. Cryo-negative staining. Micron 29(2–3): 145–160.
Aebi U, Fowler WE, Isenberg G, Pollard TD, Smith PR. 1981. Crystalline actin sheets: their structure and polymorphism. Journal of Cell Biology 91(2):340–351.
Agrawal RK, Heagle AB, Penczek P, Grassucci RA, Frank J. 1999a. EF-G-dependent GTP hydrolysis induces translocation accompanied by large conformational changes in the 70S ribosome. Nature Structural Biology 6(7):643–647.
Agrawal RK, Penczek P, Grassucci RA, Burkhardt N, Nierhaus KH, Frank J. 1999b. Effect of buffer conditions on the position of tRNA on the 70 S ribosome as visualized by cryoelectron microscopy. Journal of Biological Chemistry 274(13):8723–8729.
Amos LA. 1975. Combination of data from helical particles: correlation and selection. Journal of Molecular Biology 99(1):65–73.
Amos LA, Klug A. 1974. Arrangement of subunits in flagellar microtubules. Journal of Cell Science 14(3):523–549.
Amos LA, Henderson R, Unwin PNT. 1982. 3-Dimensional structure determination by electron-microscopy of two-dimensional crystals. Progress in Biophysics & Molecular Biology 39(3):183–231.
Anderson TF. 1952. A technique for retaining three-dimensional structures in drying specimens for the electron microscope. Journal of Applied Physics 23(1):158–158.
Aoyama K, Ogawa K, Kimura Y, Fujiyoshi Y. 1995. A method for 2D crystallization of soluble-proteins at liquid–liquid interface. Ultramicroscopy 57(4):345–354.
Asturias FJ, Kornberg RD. 1995. A novel method for transfer of 2-dimensional crystals from the air–water interface to specimen grids: sample preparation lipid-layer crystallization. Journal of Structural Biology 114(1):60–66.
Asturias FJ, Kornberg RD. 1999. Protein crystallization on lipid layers and structure determination of the RNA polymerase II transcription initiation complex. Journal of Biological Chemistry 274(11):6813–6816.

Auer M, Scarborough GA, Kuhlbrandt W. 1998. Three-dimensional map of the plasma membrane H^+-ATPase in the open conformation. Nature 392(6678):840–843.

Auer M, Scarborough GA, Kuhlbrandt W. 1999. Surface crystallisation of the plasma membrane H^+-ATPase on a carbon support film for electron crystallography. Journal of Molecular Biology 287(5):961–968.

AvilaSakar AJ, Chiu W. 1996. Visualization of beta-sheets and side-chain clusters in two-dimensional periodic arrays of streptavidin on phospholipid monolayers by electron crystallography. Biophysical Journal 70(1):57–68.

AvilaSakar AJ, Guan TL, Arad T, Schmid MF, Loke TW, Yonath A, Piefke J, Franceschi F, Chiu W. 1994. Electron cryomicroscopy of bacillus-stearothermophilus 50-S ribosomal-subunits crystallized on phospholipid monolayers. Journal of Molecular Biology 239(5): 689–697.

Baker TS, Caspar DLD. 1984. Computer image modeling of pentamer packing in polyoma virus "hexamer" tubes. Ultramicroscopy 13(1–2):137–151.

Baker TS, Cheng RH. 1996. A model-based approach for determining orientations of biological macromolecules imaged by cryoelectron microscopy. Journal of Structural Biology 116(1):120–130.

Baker TS, Olson NH, Fuller SD. 1999. Adding the third dimension to virus life cycles: three-dimensional reconstruction of icosahedral viruses from cryo-electron micrographs. Microbiology and Molecular Biology Reviews 63(4):862.

Baker ML, Jiang W, Wedemeyer WJ, Rixon FJ, Baker D, Chiu, W. 2006a. Ab initio modeling of the herpesvirus VP26 core domain assessed by cryoEM density. PLoS Comput Biol 2(10):e146.

Baker ML, Ju T, Chiu W. 2007. Identification of secondary structure elements in intermediate resolution density maps. Structure, 15(1):7–19.

Baldwin J, Henderson R. 1984. Measurement and evaluation of electron-diffraction patterns from 2-dimensional crystals. Ultramicroscopy 14(4):319–335.

Baldwin JM, Henderson R, Beckman E, Zemlin F. 1988. Images of purple membrane at 2.8 A resolution obtained by cryo-electron microscopy. Journal of Molecular Biology 202(3): 585–591.

Barrett AN, Leigh JB, Holmes KC, Leverman R, Mandelkow E, Sengbusch PV. 1972. An electron-density map of tobacco mosaic virus at 10 angstrom resolution. Cold Spring Harbor Symposium Quantitative Biology 36:433–448.

Bassi R, Magaldi AG, Tognon G, Giacometti GM, Miller KR. 1989. Two-dimensional crystals of the photosystem-II reaction center complex from higher plants. European Journal of Cell Biology 50(1):84–93.

Batson PE, Dellby N, Krivanek OL. 2002. Sub-angstrom resolution using aberration corrected electron optics. Nature 418(6898):617–620.

Battersby BJ, Sharp JCW, Webb RI, Barnes GT. 1994. Vitrification of aqueous suspensions from a controlled environment for electron microscopy: an improved plunge cooling device. Journal of Microscopy (Oxford) 176:110–120.

Baumeister W. 2004. Mapping molecular landscapes inside cells. Biological Chemistry 385(10):865–872.

Baumeister W, Barth M, Hegerl R, Guckenberger R, Hahn M, Saxton WO. 1986. 3-Dimensional structure of the regular surface-layer (hpi layer) of *Deinococcus radiodurans*. Journal of Molecular Biology 187(2):241–253.

Bellon PL, Lanzavecchia S, Scatturin V. 1998. A two exposures technique of electron tomography from projections with random orientations and a quasi-Boolean angular reconstitution. Ultramicroscopy 72(3–4):177–186.

Belnap DM, Olson NH, Baker TS. 1997. A method for establishing the handedness of biological macromolecules. Journal of Structural Biology 120(1):44–51.

Berger MJ, Selzer SM. 1964. Tables of Energy Losses and Ranges of Electrons and Positrons. Washington, DC: National Academy of Sciences–National Research Council, pp. 205–268.

Berman H, Henrick K, Nakamura H. 2003. Announcing the worldwide protein data bank. Nature Structural Biology 10(12):980.

Bernstein FC, Koetzle TF, Williams GJB, Meyer EF, Brice MD, Rodgers JR, Kennard O, Shimanouchi T, Tasumi M. 1977. Protein data bank: computer-based archival file for macromolecular structures. Journal of Molecular Biology 112(3):535–542.

Beroukhim R, Unwin N. 1997. Distortion correction of tubular crystals: improvements in the acetylcholine receptor structure. Ultramicroscopy 70(1–2):57–81.

Berriman J, Unwin N. 1994. Analysis of transient structures by cryomicroscopy combined with rapid mixing of spray droplets. Ultramicroscopy 56(4):241–252.

Bluemke DA, Carragher B, Potel MJ, Josephs R. 1988. Structural analysis of polymers of sickle-cell hemoglobin: 2. Sickle hemoglobin macrofibers. Journal of Molecular Biology 199(2):333–348.

Blundell TL, Johnson LN. 1976. Protein Crystallography. New York: Academic Press.

Boisset N, Penczek P, Pochon F, Frank J, Lamy J. 1993. 3-Dimensional architecture of human alpha-2-macroglobulin transformed with methylamine. Journal of Molecular Biology 232(2):522–529.

Boisset N, Penczek PA, Taveau JC, You V, de Haas F, Lamy J. 1998. Overabundant single-particle electron microscope views induce a three-dimensional reconstruction artifact. Ultramicroscopy 74(4):201–207.

Booth CR, Jiang W, Baker ML, Zhou ZH, Ludtke SJ, Chiu W. 2004. A 9 angstrom single particle reconstruction from CCD captured images on a 200 kV electron cryomicroscope. Journal of Structural Biology 147(2):116–127.

Booy FP, Pawley JB. 1993. Cryo-crinkling: what happens to carbon-films on copper grids at low-temperature. Ultramicroscopy 48(3):273–280.

Borland L, van Heel M. 1990. Classification of image data in conjugate representation spaces. Journal of the Optical Society of America A: Optics Image Science and Vision 7(4):601–610.

Born M, Wolf E. 1997. Principles of Optics: Electromagnetic Theory of Propagation, Interference and Diffraction of Light. Cambridge, UK: Cambridge University Press.

Bottcher B. 1995. Electron cryomicroscopy of graphite in amorphous ice. Ultramicroscopy 58(3–4):417–424.

Bottcher B, Wynne SA, Crowther RA. 1997. Determination of the fold of the core protein of hepatitis B virus by electron cryomicroscopy. Nature 386(6620):88–91.

Box HC. 1975. Free radicals in radiation biology and enzymology. Bulletin of the American Physical Society 20(3):336–336.

Bozzola JJ, Russell LD. 1999. Electron Microscopy: Principles and Techniques for Biologists. Sudbury, MA: Jones and Bartlett.

Branden CI, Jones TA. 1990. Between objectivity and subjectivity. Nature 343(6260):687–689.

Braun N, Meining W, Hars U, Fischer M, Ladenstein R, Huber R, Bacher A, Weinkauf S, Bachmann L. 2002. Formation of metal nanoclusters on specific surface sites of protein molecules. Journal of Molecular Biology 321(2):341–353.

Bravais A. 1969. On the Systems Formed by Points Regularly Distributed on a Plane or in Space. Shaler AJ, translator. Pittsburgh, PA: Polycrystal Book Service, American Crystallographic Association.

Breedlove JR, Trammell GT. 1970. Molecular microscopy: fundamental limitations. Science 170(3964):1310–1313.

Brenner S, Horne RW. 1959. A negative staining method for high resolution electron microscopy of viruses. Biochimica et Biophysica Acta 34(1):103–110.

Brillinger DR, Downing KH, Glaeser RM. 1990. Some statistical aspects of low-dose electron imaging of crystals. Journal of Statistical Planning and Inference 25:235–259.

Brink J, Chiu W. 1991. Contrast analysis of cryo-images of n-paraffin recorded at 400 kV out to 2.1 A resolution. Journal of Microscopy (Oxford) 161:279–295.

Brink J, Chiu W. 1994. Applications of a slow-scan CCD camera in protein electron crystallography. Journal of Structural Biology 113(1):23–34.

Brink J, Chiu W, Dougherty M. 1992. Computer-controlled spot-scan imaging of crotoxin complex crystals with 400 keV electrons at near atomic resolution. Ultramicroscopy 46(1–4):229–240.

Brink J, Gross H, Tittmann P, Sherman MB, Chiu W. 1998a. Reduction of charging in protein electron cryomicroscopy. Journal of Microscopy (Oxford) 191:67–73.

Brink J, Sherman MB, Berriman J, Chiu W. 1998b. Evaluation of charging on macromolecules in electron cryomicroscopy. Ultramicroscopy 72(1–2):41–52.

Brisson A, Unwin PNT. 1985. Quaternary structure of the acetylcholine receptor. Nature 315(6019):474–477.

Brisson A, Bergsma-Schutter W, Oling F, Lambert O, Reviakine I. 1999. Two-dimensional crystallization of proteins on lipid monolayers at the air–water interface and transfer to an electron microscopy grid. Journal of Crystal Growth 196(2–4):456–470.

Brunger AT. 1992. Free R-value: a novel statistical quantity for assessing the accuracy of crystal structures. Nature 355(6359):472–475.

Brunger AT. 1997. Free R value: cross-validation in crystallography. Macromolecular Crystallography, Pt B. pp. 366–396.

Brunger AT, Huber R, Karplus M. 1987. Trypsinogen trypsin transition: a molecular-dynamics study of induced conformational change in the activation domain. Biochemistry 26(16):5153–5162.

Bullivant S. 1973. Freeze-etching and freeze-fracturing. In: Koehler JK, editor. Techniques in Biological Electron Microscopy. New York: Springer-Verlag.

Bullough P, Henderson R. 1987. Use of spot-scan procedure for recording low-dose micrographs of beam-sensitive specimens. Ultramicroscopy 21(3):223–229.

Carazo JM. 1992. The fidelity of 3D reconstruction from incomplete data and the use of restoration methods. In: Frank J, editor. Electron Tomography. New York: Plenum, pp. 88–91.

Carragher B, Bluemke DA, Gabriel B, Potel MJ, Josephs R. 1988. Structural analysis of polymers of sickle-cell hemoglobin: 1. Sickle hemoglobin fibers. Journal of Molecular Biology 199(2):315–331.

Casjens S. 1997. Principles of virion structure, function, and assembly. In: Chiu W, Burnett RM, Garcea RL, editors. Structural Biology of Viruses. New York: Oxford University Press.

Caspar DLD, Holmes KC. 1969. Structure of Dahlemense strain of tobacco mosaic virus — a periodically deformed helix. Journal of Molecular Biology 46(1):99.

Caspar DLD, Klug A. 1962. Physical principles in the construction of regular viruses. Cold Spring Harbor Symposium Quantitative Biology 27:1–24.

Cejka Z, Magaldi AG, Tognon G, Zanotti G, Baumeister W. 1989. The molecular architecture of extracellular hemoglobin of eophila-tellinii. Journal of Ultrastructure and Molecular Structure Research 102(1):71–81.

Celia H, Hoermann L, Schultz P, Lebeau L, Mallouh V, Wigley DB, Wang JC, Mioskowski C, Oudet P. 1994. 3-Dimensional model of Escherichia coli gyrase-b subunit crystallized in 2-dimensions on novobiocin-linked phospholipid films. Journal of Molecular Biology 236(2):618–628.

Celia H, Wilson-Kubalek E, Milligan RA, Teyton L. 1999. Structure and function of a membrane-bound murine MHC class I molecule. Proceedings of the National Academy of Sciences of the United States of America 96(10):5634–5639.

Ceska TA, Henderson R. 1990. Analysis of high-resolution electron diffraction patterns from purple membrane labeled with heavy atoms. Journal of Molecular Biology 213(3):539–560.

Chalcroft JP, Davey CL. 1984. A simply constructed extreme-tilt holder for the Philips eucentric goniometer stage. Journal of Microscopy (Oxford) 134:41–48.

Chami M, Pehau-Arnaudet G, Lambert O, Ranck JL, Levy D, Rigaud JL. 2001. Use of octyl beta-thioglucopyranoside in two-dimensional crystallization of membrane proteins. Journal of Structural Biology 133(1):64–74.

Chang S, Head-Gordon T, Glaeser RM, Downing KH. 1999. Chemical bonding effects in the determination of protein structures by electron crystallography. Acta Crystallographica A 55(2):305–313.

Chen YJ, Zhang PJ, Egelman EH, Hinshaw JE. 2004. The stalk region of dynamin drives the constriction of dynamin tubes. Nature Structural & Molecular Biology 11(6):574–575.

Cheng AC, vanHoek AN, Yeager M, Verkman AS, Mitra AK. 1997. Three-dimensional organization of a human water channel. Nature 387(6633):627–630.

Cheng RH, Reddy VS, Olson NH, Fisher AJ, Baker TS, Johnson JE. 1994. Functional implications of quasi-equivalence in a T = 3 icosahedral animal virus established by cryoelectron microscopy and x-ray crystallography. Structure 2(4):271–282.

Cheong GW, Guckenberger R, Fuchs KH, Gross H, Baumeister W. 1993. The structure of the surface-layer of Methanoplanus limicola obtained by a combined electron-microscopy and scanning-tunneling-microscopy approach. Journal of Structural Biology 111(2): 125–134.
Cheong GW, Typke D, Baumeister W. 1996. Projection structure of the surface layer of Methanoplanus limicola at 10 angstrom resolution obtained by electron cryomicroscopy. Journal of Structural Biology 117(2):138–144.
Chiu W, Hosoda J. 1978. Crystallization and preliminary electron-diffraction study to 3.7 Å of DNA helix-destabilizing protein-Gp32-Star-I. Journal of Molecular Biology 122(1): 103–107.
Chiu W, Jeng TW, Degn LL, Prasad BVV. 1986. Potential for high-resolution electron crystallography at intermediate high-voltage. Annals of the New York Academy of Sciences 483:149–156.
Chiu W, Schmid MF, Prasad BVV. 1993. Teaching electron diffraction and imaging of macromolecules. Biophysical Journal 64(5):1610–1625.
Chiu W, Avila-Sakar AJ, Schmid MF. 1997. Electron crystallography of macromolecular periodic arrays on phospholipid monolayers. Advances in Biophysics 34:161–172.
Chiu W, Baker ML, Jiang W, Dougherty M, Schmid MF. 2005. Electron cryomicroscopy of biological machines at subnanometer resolution. Structure 13(3):363–372.
Christophorou LG. 1971. Atomic and Molecular Radiation Physics. London: Wiley-Interscience.
Cochran W, Crick FHC, Vand V. 1952. The structure of synthetic polypeptides: 1. The transform of atoms on a helix. Acta Crystallographica 5(5):581–586.
Cohen H, Chiu W, Hosoda J. 1983. Structural analysis of T4 DNA helix destabilizing protein (Gp32-Star-I) crystal by electron microscopy. Journal of Molecular Biology 169(1): 235–248.
Cohen HA, Jeng TW, Grant RA, Chiu W. 1984. Specimen preparative methods for electron crystallography of soluble proteins. Ultramicroscopy 13(1–2):19–25.
Colovos C, Yeates TO. 1993. Verification of protein structures: patterns of nonbonded atomic interactions. Protein Science 2(9):1511–1519.
Conway JF, Cheng N, Zlotnick A, Wingfield PT, Stahl SJ, Steven AC. 1997. Visualization of a 4-helix bundle in the hepatitis B virus capsid by cryo-electron microscopy. Nature 386(6620):91–94.
Cordes FS, Komoriya K, Larquet E, Yang SX, Egelman EH, Blocker A, Lea SM. 2003. Helical structure of the needle of the type III secretion system of Shigella flexneri. Journal of Biological Chemistry 278(19):17103–17107.
Corless JM, McCaslin DR, Scott BL. 1982. Two-dimensional rhodopsin crystals from disk membranes of frog retinal rod outer segments. Proceedings of the National Academy of Sciences of the United States of America 79(4):1116–1120.
Cowley JM. 1995. Diffraction Physics. Amsterdam: Elsevier.
Cowley JM, Moodie AF. 1957. The scattering of electrons by atoms and crystals: 1. A new theoretical approach. Acta Crystallographica 10(10):609–619.
Crepeau RH. 1980. Determination of the degree of elliptical flattening for helical fibers from tilted views. Journal of Molecular Biology 139(2):144–145.
Crick FHC, Magdoff BS. 1956. The theory of the method of isomorphous replacement for protein crystals: 1. Acta Crystallographica 9(10):901–908.
Crowe JH, Hoekstra FA, Crowe LM. 1992. Anhydrobiosis. Annual Review of Physiology 54:579–599.
Crowther RA. 1971. Procedures for 3-dimensional reconstruction of spherical viruses by Fourier synthesis from electron micrographs. Philosophical Transactions of the Royal Society of London B: Biological Sciences 261(837):221.
Crowther RA, Amos LA, Finch JT, Derosier DJ, Klug A. 1970a. 3 Dimensional reconstructions of spherical viruses by fourier synthesis from electron micrographs. Nature 226(5244):421.
Crowther RA, DeRosier DJ, Klug A. 1970b. Reconstruction of 3 dimensional structure from projections and its application to electron microscopy. Proceedings of the Royal Society of London A: Mathematical and Physical Sciences 317(1530):319.

Crowther RA, Padron R, Craig R. 1985. Arrangement of the heads of myosin in relaxed thick filaments from tarantula muscle. Journal of Molecular Biology 184(3):429–439.

Crowther RA, Henderson R, Smith JM. 1996. MRC image processing programs. Journal of Structural Biology 116(1):9–16.

Cyrklaff M, Kuhlbrandt W. 1994. High-resolution electron-microscopy of biological specimens in cubic ice. Ultramicroscopy 55(2):141–153.

Cyrklaff M, Roos N, Gross H, Dubochet J. 1994. Particle–surface interaction in thin vitrified films for cryoelectron microscopy. Journal of Microscopy (Oxford) 175:135–142.

Cyrklaff M, Auer M, Kuhlbrandt W, Scarborough GA. 1995. 2-D structure of the Neurospora crassa plasma-membrane ATPase as determined by electron cryomicroscopy. EMBO Journal 14(9):1854–1857.

Daberkow I, Herrmann KH, Liu L, Rau WD, Tietz H. 1996. Development and performance of a fast fibre-plate coupled CCD camera at medium energy and image processing system for electron holography. Ultramicroscopy 64(1–4):35–48.

Daillant J, Bosio L, Benattar JJ, Meunier J. 1989. Capillary waves and bending elasticity of monolayers on water studied by x-ray reflectivity as a function of surface pressure. Europhysics Letters 8(5):453–458.

Darst SA, Ribi HO, Pierce DW, Kornberg RD. 1988. Two-dimensional crystals of *Escherichia coli* RNA polymerase holoenzyme on positively charged lipid layers. Journal of Molecular Biology 203(1):269–273.

Darst SA, Ahlers M, Meller PH, Kubalek EW, Blankenburg R, Ribi HO, Ringsdorf H, Kornberg RD. 1991a. 2-Dimensional crystals of streptavidin on biotinylated lipid layers and their interactions with biotinylated macromolecules. Biophysical Journal 59(2):387–396.

Darst SA, Kubalek EW, Edwards AM, Kornberg RD. 1991b. 2-Dimensional and epitaxial crystallization of a mutant form of yeast RNA polymerase-II. Journal of Molecular Biology 221(1):347–357.

de Groot BL, Engel A, Grubmuller H. 2003. The structure of the aquaporin-1 water channel: a comparison between cryo-electron microscopy and x-ray crystallography. Journal of Molecular Biology 325(3):485–493.

Dekker JP, Betts SD, Yocum CF, Boekema EJ. 1990. Characterization by electron-microscopy of isolated particles and 2-dimensional crystals of the Cp47-D1-D2-cytochrome-B-559 complex of photosystem-II. Biochemistry 29(13):3220–3225.

DeRosier DJ. 2000. Correction of high-resolution data for curvature of the Ewald sphere. Ultramicroscopy 81(2):83–98.

DeRosier D, Stokes DL, Darst SA. 1999. Averaging data derived from images of helical structures with different symmetries. Journal of Molecular Biology 289(1):159–165.

DeRosier DJ, Klug A. 1968. Reconstruction of 3 dimensional structures from electron micrographs. Nature 217(5124):130.

DeRosier DJ, Moore PB. 1970. Reconstruction of 3-dimensional images from electron micrographs of structures with helical symmetry. Journal of Molecular Biology 52(2):355.

DeRuijter WJ. 1995. Imaging properties and applications of slow-scan charge-coupled-device cameras suitable for electron microscopy. Micron 26(3):247–275.

Dietrich I, Fox F, Knapek E, Lefranc G, Nachtrieb K, Weyl R, Zerbst H. 1977. Improvements in electron microscopy by application of superconductivity. Ultramicroscopy 2(2–3):241–249.

Dorset DL. 1995. Structural Electron Crystallography. New York: Plenum Press.

Dorset DL, Engel A, Haner M, Massalski A, Rosenbusch JP. 1983. Two-dimensional crystal packing of matrix porin: a channel forming protein in Escherichia coli outer membranes. Journal of Molecular Biology 165(4):701–710.

Downing KH. 1979. Possibilities of heavy-atom discrimination using single-sideband techniques. Ultramicroscopy 4(1):13–31.

Downing KH. 1984. Electron crystallographic studies of DNA. Ultramicroscopy 13(1–2):35–46.

Downing KH. 1988. Observations of restricted beam-induced specimen motion with small-spot illumination. Ultramicroscopy 24(4):387–398.

Downing KH. 1991. Spot-scan imaging in transmission electron microscopy. Science 251(4989):53–59.

Downing KH. 1992. Automatic focus correction for spot-scan imaging of tilted specimens. Ultramicroscopy 46(1–4):199–206.
Downing KH, Glaeser RM. 1980. Electron diffraction from single crystals of DNA. Biophysical Journal 32(2):851–856.
Downing KH, Glaeser RM. 1986. Improvement in high-resolution image quality of radiation-sensitive specimens achieved with reduced spot size of the electron beam. Ultramicroscopy 20(3):269–278.
Downing KH, Grano DA. 1982. Analysis of photographic emulsions for electron microscopy of two-dimensional crystalline specimens. Ultramicroscopy 7(4):381–403.
Downing KH, Hendrickson FM. 1999. Performance of a 2k CCD camera designed for electron crystallography at 400 kV. Ultramicroscopy 75(4):215–233.
Downing KH, Jontes J. 1992. Projection map of tubulin in zinc-induced sheets at 4-angstrom resolution. Journal of Structural Biology 109(2):152–159.
Downing KH, Li HL. 2001. Accurate recording and measurement of electron diffraction data in structural and difference Fourier studies of proteins. Microscopy and Microanalysis 7(5):407–417.
Downing KH, Siegel BM. 1973. Phase-shift determination in single-sideband holography. Optik 38(1):21–28.
Downing KH, Siegel BM. 1975. Discrimination of heavy and light components in electron-microscopy using single-sideband holographic techniques. Optik 42(2):155–175.
Drenth J. 1994. Principles of Protein X-ray Crystallography. New York: Springer-Verlag.
Dubochet J, McDowall AW. 1981. Vitrification of pure water for electron microscopy. Journal of Microscopy (Oxford) 124:RP3–RP4.
Dubochet J, Chang JJ, Freeman R, Lepault J, McDowall AW. 1982a. Frozen aqueous suspensions. Ultramicroscopy 10(1–2):55–61.
Dubochet J, Lepault J, Freeman R, Berriman JA, Homo JC. 1982b. Electron microscopy of frozen water and aqueous solutions. Journal of Microscopy (Oxford) 128:219–237.
Dubochet J, Adrian M, Chang JJ, Homo JC, Lepault J, McDowall AW, Schultz P. 1988. Cryo-electron microscopy of vitrified specimens. Quarterly Reviews of Biophysics 21(2):129–228.
Durchschlag H, Zipper P. 1997. Calculation of partial specific volumes and other volumetric properties of small molecules and polymers. Journal of Applied Crystallography 30(2):803–807.
Egelman EH. 1986. An algorithm for straightening images of curved filamentous structures. Ultramicroscopy 19(4):367–373.
Egelman EH. 2000. A robust algorithm for the reconstruction of helical filaments using single-particle methods. Ultramicroscopy 85(4):225–234.
Egelman EH. 2004. More insights into structural plasticity of actin binding proteins. Structure 12(6):909–910.
Egelman EH, Francis N, DeRosier DJ. 1982. F-actin is a helix with a random variable twist. Nature 298(5870):131–135.
Egerton RF. 1982. Organic mass-loss at 100-K and 300-K. Journal of Microscopy (Oxford) 126:95–100.
Eisenberg D, Luthy R, Bowie JU. 1997. VERIFY3D: assessment of protein models with three-dimensional profiles. Macromolecular Crystallography Pt B:396–404.
Engel A, Hoenger A, Hefti A, Henn C, Ford RC, Kistler J, Zulauf M. 1992. Assembly of 2-D membrane-protein crystals: dynamics, crystal order, and fidelity of structure analysis by electron microscopy. Journal of Structural Biology 109(3):219–234.
Erickson HP, Klug A. 1971. Measurement and compensation of defocusing and aberrations by Fourier processing of electron micrographs. Philosophical Transactions of the Royal Society of London B: Biological Sciences 261(837):105.
Ermantraut E, Wohlfart K, Tichelaar W. 1998. Perforated support foils with pre-defined hole size, shape and arrangement. Ultramicroscopy 74(1–2):75–81.
Eusemann R, Rose H, Dubochet J. 1982. Electron scattering in ice and organic materials. Journal of Microscopy (Oxford) 128:239–249.
Fano U. 1963. Penetration of protons, alpha particles, and mesons. Annual Reviews of Nuclear Science, 1–66.

Faruqi AR, Andrews HN, Henderson R. 1995. A high sensitivity imaging detector for electron microscopy. Nuclear Instruments & Methods in Physics Research A: Accelerators Spectrometers Detectors and Associated Equipment 367(1–3):408–412.

Faruqi AR, Henderson R, Subramaniam S. 1999. Cooled CCD detector with tapered fibre optics for recording electron diffraction patterns. Ultramicroscopy 75(4):235–250.

Ferrier RP, Murray RT. 1966. Low-angle electron diffraction. Journal of the Royal Microscopical Society 85:323–335.

Feynman RP, Hibbs AR. 1965. Quantum Mechanics and Path Integrals. New York: McGraw-Hill.

Finch JT. 1972. Hand of helix of tobacco mosaic virus. Journal of Molecular Biology 66(2):291.

Ford RC, Hefti A, Engel A. 1990. Ordered arrays of the photosystem-I reaction center after reconstitution: projections and surface reliefs of the complex at 2 nm resolution. EMBO Journal 9(10):3067–3075.

Frank J. 1973. The envelope of electron microscopic transfer-functions for partially coherent illumination. Optik 38(5):519–536.

Frank J. 1985. Computer processing of electron microscopic images. In: Wu TT, editor. New Methodologies in Studies of Protein Configuration. New York: Van Nostrand Reinhold.

Frank J. 1990. Classification of macromolecular assemblies studied as single particles. Quarterly Reviews of Biophysics 23(3):281–329.

Frank J. 1996. Three-Dimensional Electron Microscopy of Macromolecular Assemblies. San Diego, CA: Academic Press.

Frank J. 2002. Single-particle imaging of macromolecules by cryo-electron microscopy. Annual Review of Biophysics and Biomolecular Structure 31:303–319.

Frank J. 2006. Three-Dimensional Electron Microscopy of Macromolecular Assemblies: Visualization of Biological Molecules in Their Native State. New York: Oxford University Press.

Frank J, Al-Ali L. 1975. Signal-to-noise ratio of electron micrographs obtained by cross-correlation. Nature 256(5516):376–379.

Frank J, Penczek P. 1995. On the correction of the contrast transfer function in biological electron microscopy. Optik 98(3):125–129.

Frank J, van Heel M. 1980. Intelligent averaging of single molecule using computer alignment and correspondence analysis: I. The basic method. In: Brederoo P, de Priestser W, editors. Seventh European Congress on Electron Microscopy Foundation, pp. 690–691.

Frank J, Verschoor A, Boublik M. 1981. Computer averaging of electron micrographs of 40S ribosomal subunits. Science 214(4527):1353–1355.

Frank J, Verschoor A, Boublik M. 1982. Multivariate statistical analysis of ribosome electron micrographs: L and R lateral views of the 40-S-subunit from Hela cells. Journal of Molecular Biology 161(1):107–133.

Frank J, Radermacher M, Penczek P, Zhu J, Li YH, Ladjadj M, Leith A. 1996. SPIDER and WEB: processing and visualization of images in 3D electron microscopy and related fields. Journal of Structural Biology 116(1):190–199.

Frank J, Penczek P, Agrawal RK, Grassucci RA, Heagle AB. 2000. Three-dimensional cryoelectron microscopy of ribosomes. Methods Enzymology 317:276–291.

Frey TG, Murray JM. 1994. Electron microscopy of cytochrome-C-oxidase crystals: monomer–dimer relationship and cytochrome-C binding site. Journal of Molecular Biology 237(3):275–297.

Fromherz P. 1971. Electron microscopic studies of lipid protein films. Nature 231(5300):267.

Fujiyoshi Y, Uyeda N, Morikawa K, Yamagishi H. 1984. Electron microscopy of transfer-RNA crystals: 2.4-Å resolution diffraction pattern and substantial stability to radiation damage. Journal of Molecular Biology 172(3):347–354.

Fujiyoshi Y, Mizusaki T, Morikawa K, Yamagishi H, Aoki Y, Kihara H, Harada Y. 1991. Development of a superfluid-helium stage for high-resolution electron microscopy. Ultramicroscopy 38(3–4):241–251.

Fujiyoshi Y, Mitsuoka K, de Groot BL, Philippsen A, Grubmuller H, Agre P, Engel A. 2002. Structure and function of water channels. Current Opinion in Structural Biology 12(4):509–515.

Fukami A, Adachi K. 1965. A new method of preparation of a self-perforated micro plastic grid and its application. Journal of Electron Microscopy 14:112–118.

Gabashvili IS, Agrawal RK, Spahn CMT, Grassucci RA, Svergun DI, Frank J, Penczek P. 2000. Solution structure of the *E. coli* 70S ribosome at 11.5 angstrom resolution. Cell 100(5): 537–549.
Galkin VE, VanLoock MS, Orlova A, Egelman EH. 2002. A new internal mode in F-actin helps explain the remarkable evolutionary conservation of actin's sequence and structure. Current Biology 12(7):570–575.
Gao HX, Valle M, Ehrenberg M, Frank J. 2004. Dynamics of EF-G interaction with the ribosome explored by classification of a heterogeneous cryo-EM dataset. Journal of Structural Biology 147(3):283–290. [Corrigendum 148(3):375–375].
Garrison WM, Jayko ME, Rodgers MAJ, Sokol HA, Bennett W. 1968. Ionization and excitation in peptide radiolysis. Advances in Chemistry Series 81:384–396.
Gilbert PFC. 1972. Reconstruction of a 3-dimensional structure from projections and its application to electron microscopy: 2. Direct methods. Proceedings of the Royal Society of London B: Biological Sciences 182(1066):89.
Glaeser RM. 1971. Limitations to significant information in biological electron microscopy as a result of radiation damage. Journal of Ultrastructure Research 36(3–4): 466–482.
Glaeser RM. 1975. Radiation damage and biological electron microscopy. In: Siegel BM, Beaman DR, editors. Physical Aspects of Electron Microscopy and Microanalysis. New York: Wiley. pp. 205–229.
Glaeser RM. 1992. Specimen flatness of thin crystalline arrays: influence of the substrate. Ultramicroscopy 46(1–4):33–43.
Glaeser RM. 1999. Review. Electron crystallography: present excitement, a nod to the past, anticipating the future. Journal of Structural Biology 128(1):3–14.
Glaeser RM, Ceska TA. 1989. High-voltage electron diffraction from bacteriorhodopsin (purple membrane) is measurably dynamical. Acta Crystallographica A 45:620–628.
Glaeser RM, Downing KH. 1993. High-resolution electron crystallography of protein molecules. Ultramicroscopy 52(3–4):478–486.
Glaeser RM, Downing KH. 2004. Specimen charging on thin films with one conducting layer: discussion of physical principles. Microscopy and Microanalysis 10(6):790–796.
Glaeser RM, Hobbs LW. 1975. Radiation damage in stained catalase at low-temperature. Journal of Microscopy (Oxford) 103:209–214.
Glaeser RM, Taylor KA. 1978. Radiation damage relative to transmission electron microscopy of biological specimens at low temperature: review. Journal of Microscopy (Oxford) 112: 127–138.
Glaeser RM, Thomas G. 1969. Application of electron diffraction to biological electron microscopy. Biophysical Journal 9(9):1073.
Glaeser RM, Kuo I, Budinger TF. 1971. Method for processing of periodic images at reduced levels of electron irradiation. In: Areceneaux CJ, editor. Boston.
Glaeser RM, Tong L, Kim SH. 1989. 3-Dimensional reconstructions from incomplete data: interpretability of density maps at atomic resolution. Ultramicroscopy 27(3):307–318.
Glaeser RM, Zilker A, Radermacher M, Gaub HE, Hartmann T, Baumeister W. 1991. Interfacial energies and surface-tension forces involved in the preparation of thin, flat crystals of biological macromolecules for high-resolution electron microscopy. Journal of Microscopy (Oxford) 161:21–45.
Glauber R, Schomaker V. 1953. The theory of electron diffraction. Physical Review 89(4): 667–671.
Glusker JP, Lewis M, M. R. 1994. Crystal Structure Analysis for Chemists and Biologists. New York: Wiley-VCH.
Gogol E, Unwin N. 1988. Organization of connexons in isolated rat-liver Gap-junctions. Biophysical Journal 54(1):105–112.
Golas MM, Sander B, Will CL, Luhrmann R, Stark H. 2003. Molecular architecture of the multiprotein splicing factor SF3b. Science 300(5621):980–984.
Goncharov AB, Vainshtein BK, Ryskin AI, Vagin AA. 1987. Three-dimensional reconstruction of arbitrarily oriented particles from their electron photomicrographs. Soviet Physics, Crystallography 32:504–509.

Gonen T, Sliz P, Kistler J, Cheng YF, Walz T. 2004. Aquaporin-0 membrane junctions reveal the structure of a closed water pore. Nature 429(6988):193–197.

Gonen T, Cheng YF, Sliz P, Hiroaki Y, Fujiyoshi Y, Harrison SC, Walz T. 2005. Lipid–protein interactions in double-layered two-dimensional AQPO crystals. Nature 438(7068): 633–638.

Goodman JW. 1968. Introduction to Fourier Optics. San Francisco: McGraw-Hill.

Gordon R, Bender R, Herman GT. 1970. Algebraic reconstruction techniques (ART) for 3-dimensional electron microscopy and x-ray photography. Journal of Theoretical Biology 29(3):471–481.

Grigorieff N. 1998. Three-dimensional structure of bovine NADH: ubiquinone oxidoreductase (complex I) at 22 angstrom in ice. Journal of Molecular Biology 277(5):1033–1046.

Grigorieff N, Ceska TA, Downing KH, Baldwin JM, Henderson R. 1996. Electron crystallographic refinement of the structure of bacteriorhodopsin. Journal of Molecular Biology 259(3):393–421.

Grigorieff N, Henderson R. 1995. Diffuse scattering in electron diffraction data from protein crystals. Ultramicroscopy 60(2):295–309.

Grimm R, Typke D, Barmann M, Baumeister W. 1996. Determination of the inelastic mean free path in ice by examination of tilted vesicles and automated most probable loss imaging. Ultramicroscopy 63(3–4):169–179.

Grob P, Cruse MJ, Inouye C, Peris M, Penczek PA, Tjian R, Nogales E. 2006. Cryo-electron microscopy studies of human TFIID: conformational breathing in the integration of gene regulatory cues. Structure 14(3):511–520.

Guckenberger R. 1985. Surface reliefs derived from heavy-metal-shadowed specimens: Fourier space techniques applied to periodic objects. Ultramicroscopy 16(3–4):357–370.

Gyobu N, Tani K, Hiroaki Y, Kamegawa A, Mitsuoka K, Fujiyoshi Y. 2004. Improved specimen preparation for cryo-electron microscopy using a symmetric carbon sandwich technique. Journal of Structural Biology 146(3):325–333.

Hahn T, editor. 2002. International Tables for Crystallography, Volume A: Space Group Symmetry, 5th edn. Dordrecht: Kluwer.

Halic M, Becker T, Frank J, Spahn CMT, Beckmann R. 2005. Localization and dynamic behavior of ribosomal protein L30e. Nature Structural & Molecular Biology 12(5):467–468.

Hall CE. 1955. Electron densitometry of stained virus particles. Journal of Biophysical and Biochemical Cytology 1(1):1–12.

Hall CE. 1966. Introduction to Electron Microscopy. New York: McGraw-Hill.

Hamilton JF, Marchant JC. 1967. Image recording in electron microscopy. Journal of the Optical Society of America 57(2):232–239.

Han BG, Vonck J, Glaeser RM. 1994a. The bacteriorhodopsin photocycle: direct structural study of 2 substrates of the M-intermediate. Biophysical Journal 67(3):1179–1186.

Han BG, Wolf SG, Vonck J, Glaeser RM. 1994b. Specimen flatness of glucose-embedded biological materials for electron crystallography is affected significantly by the choice of carbon evaporation stock. Ultramicroscopy 55(1):1–5.

Hanein D, DeRosier D. 1999. A new algorithm to align three-dimensional maps of helical structures. Ultramicroscopy 76(4):233–238.

Hanein D, Matsudaira P, DeRosier DJ. 1997. Evidence for a conformational change in actin induced by fimbrin (N375) binding. Journal of Cell Biology 139(2):387–396.

Hanssen KJ. 1971. The optical transfer theory of the electron microscope: fundamental principles and applications. In: Barer R, Cosslett VE, editors. Advances in Optical and Electron Microscopy. New York: Academic Press, pp. 1–84.

Hanssen KJ, Trepte L. 1971. Contrast transfer of electron microscope with partial coherent illumination: B. Disk-shaped source. Optik 33(2):182–198.

Harauz G, Ottensmeyer FP. 1984a. Direct 3-dimensional reconstruction for macromolecular complexes from electron micrographs. Ultramicroscopy 12(4):309–320.

Harauz G, Ottensmeyer FP. 1984b. Nucleosome reconstruction via phosphorus mapping. Science 226(4677):936–940.

Harauz G, van Heel M. 1986. Exact filters for general geometry three dimensional reconstruction. Optik 73:146–156.

Harris JR. 1997. Negative staining and cryoelectron microscopy: the thin film techniques. Oxford: BIOS Scientific.
Harris JR, Horne RW. 1994. Negative staining: a brief assessment of current technical benefits, limitations and future possibilities. Micron 25(1):5–13.
Harris JR, Cejka Z, Wegenerstrake A, Gebauer W, Markl J. 1992. 2-Dimensional crystallization, transmission electron-microscopy and image-processing of keyhole limpet hemocyanin (Klh). Micron and Microscopica Acta 23(3):287–301.
Hasler L, Walz T, Tittmann P, Gross H, Kistler J, Engel A. 1998. Purified lens major intrinsic protein (MIP) forms highly ordered tetragonal two-dimensional arrays by reconstitution. Journal of Molecular Biology 279(4):855–864.
Hayat MA. 2000. Principles and Techniques of Electron Microscopy: Biological Applications. Cambridge, UK: Cambridge University Press.
Hayward SB, Glaeser RM. 1979. Radiation damage of purple membrane at low temperature. Ultramicroscopy 4(2):201–210.
Hayward SB, Glaeser RM. 1980. High-resolution cold stage for the Jeol 100B and 100C electron microscopes. Ultramicroscopy 5(1):3–8.
Hayward SB, Stroud RM. 1981. Projected structure of purple membrane determined to 3.7 Å resolution by low-temperature electron microscopy. Journal of Molecular Biology 151(3):491–517.
He J, Schmid VF, Zhou ZH, Rixon F, Chiu W. 2001. Finding and using local symmetry in identifying lower domain movements in hexon subunits of the herpes simplex virus type 1 B capsid. Journal of Molecular Biology 309(4):903–914.
Hebert H, Schmidt-Krey I, Morgenstern R, Murata K, Hirai T, Mitsuoka K, Fujiyoshi Y. 1997. The 3.0 angstrom projection structure of microsomal glutathione transferase as determined by electron crystallography of p2(1)2(1)2 two-dimensional crystals. Journal of Molecular Biology 271(5):751–758.
Hemming SA, Bochkarev A, Darst SA, Kornberg RD, Ala P, Yang DSC, Edwards AM. 1995. The mechanism of protein crystal growth from lipid layers. Journal of Molecular Biology 246(2):308–316.
Henderson R. 1990. Cryoprotection of protein crystals against radiation damage in electron and x-ray diffraction. Proceedings of the Royal Society of London B: Biological Sciences 241(1300):6–8.
Henderson R. 1992. Image contrast in high-resolution electron microscopy of biological macromolecules: TMV in ice. Ultramicroscopy 46:1–18.
Henderson R. 1995. The potential and limitations of neutrons, electrons and x-rays for atomic-resolution microscopy of unstained biological molecules. Quarterly Reviews of Biophysics 28(2):171–193.
Henderson R, Glaeser RM. 1985. Quantitative analysis of image contrast in electron micrographs of beam-sensitive crystals. Ultramicroscopy 16(2):139–150.
Henderson R, Moffat JK. 1971. Difference Fourier technique in protein crystallography: errors and their treatment. Acta Crystallographica B: Structural Crystallography and Crystal Chemistry B 27:1414–1420.
Henderson R, Unwin PNT. 1975. 3-Dimensional model of purple membrane obtained by electron microscopy. Nature 257(5521):28–32.
Henderson R, Baldwin JM, Downing KH, Lepault J, Zemlin F. 1986. Structure of purple membrane from Halobacterium halobium: recording, measurement and evaluation of electron micrographs at 3.5 A resolution. Ultramicroscopy 19(2):147–178.
Henderson R, Baldwin JM, Ceska TA, Zemlin F, Beckmann E, Downing KH. 1990. Model for the structure of bacteriorhodopsin based on high-resolution electron cryomicroscopy. Journal of Molecular Biology 213(4):899–929.
Henderson R, Raeburn C, Vigers G. 1991. A side-entry cold holder for cryoelectron microscopy. Ultramicroscopy 35(1):45–53.
Hendrickson WA, Ogata CM. 1997. Phase determination from multiwavelength anomalous diffraction measurements. Macromolecular Crystallography Pt A:494–523.
Herman GT. 1980. Image Reconstruction from Projections: The Fundamentals of Computerized Tomography. New York: Academic Press.

Heuser J. 1980. 3-Dimensional visualization of coated vesicle formation in fibroblasts. Journal of Cell Biology 84(3):560–583.
Heuser J. 1989. Effects of cytoplasmic acidification on clathrin lattice morphology. Journal of Cell Biology 108(2):401–411.
Heymann JAW, Hirai T, Shi D, Subramaniam S. 2003. Projection structure of the bacterial oxalate transporter OxlT at 3.4 angstrom resolution. Journal of Structural Biology 144(3):320–326.
Heymann JB, Conway JF, Steven AC. 2004. Molecular dynamics of protein complexes from four-dimensional cryo-electron microscopy. Journal of Structural Biology 147(3):291–301.
Hinsen K, Reuter N, Navaza J, Stokes DL, Lacapere JJ. 2005. Normal mode-based fitting of atomic structure into electron density maps: application to sarcoplasmic reticulum Ca-ATPase. Biophysical Journal 88(2):818–827.
Hirai T, Heymann JAW, Shi D, Sarker R, Maloney PC, Subramaniam S. 2002. Three-dimensional structure of a bacterial oxalate transporter. Nature Structural Biology 9(8):597–600.
Hirai T, Heymann JAW, Maloney PC, Subramaniam S. 2003. Structural model for 12-helix transporters belonging to the major facilitator superfamily. Journal of Bacteriology 185(5):1712–1718.
Hirsch P, Howie A, Nicholson RB, Pashley DW, Whelan MJ. 1965. Electron Microscopy of Thin Crystals. Washington, DC: Butterworths.
Ho MH, Jap BK, Glaeser RM. 1988. Validity domain of the weak-phase-object approximation for electron diffraction of thin protein crystals. Acta Crystallographica A 44:878–884.
Hoenger A, Aebi U. 1996. 3-D reconstructions from ice-embedded and negatively stained biomacromolecular assemblies: a critical comparison. Journal of Structural Biology 117(2):99–116.
Holmes KC, Mandelkow E, Leigh JB. 1972. Determination of heavy atom positions in tobacco mosaic virus from double heavy atom derivatives. Naturwissenschaften 59(6):247–254.
Holser WT. 1958. Point groups and plane groups in a two-sided plane and their subgroups. Zeitschrift für Kristallographie 110:266–281.
Hooft RWW, Vriend G, Sander C, Abola EE. 1996. Errors in protein structures. Nature 381(6580):272.
Hope H. 1988. Cryocrystallography of biological macromolecules: a generally applicable method. Acta Crystallographica B: Structural Science 44:22–26.
Hope H. 1990. Crystallography of biological macromolecules at ultra-low temperature. Annual Review of Biophysics and Biophysical Chemistry 19:107–126.
Hoppe W. 1970. Principles of structure analysis at high resolution using conventional electron microscopes and computers. Berichte der Bunsen-Gesellschaft für Physikalische Chemie 74(11):1090–1100.
Hoppe W, Langer R, Knesch G, Poppe C. 1968. Protein crystal structure analysis with electron rays. Naturwissenschaften 55(7):333–336.
Hoppe W, Gassmann J, Hunsmann N, Schramm HJ, Sturm M. 1974. 3-Dimensional reconstruction of individual negatively stained yeast fatty-acid synthetase molecules from tilt series in electron microscope. Hoppe-Seylers Zeitschrift Fur Physiologische Chemie 355(11):1483–1487.
Horne RW, Wildy P. 1961. Symmetry in virus architecture. Virology 15(3):348–373.
Horne RW, Wildy P. 1979. Historical account of the development and applications of the negative staining technique to the electron-microscopy of viruses. Journal of Microscopy (Oxford) 117:103–122.
Hosokawa F, Tomita T, Naruse M, Honda T, Hartel P, Haider M. 2003. A spherical aberration-corrected 200 kV TEM. Journal of Electron Microscopy 52(1):3–10.
Hovmoller S, Slaughter M, Berriman J, Karlsson B, Weiss H, Leonard K. 1983. Structural studies of cytochrome reductase: improved membrane crystals of the enzyme complex and crystallization of a subcomplex. Journal of Molecular Biology 165(2):401–406.
Huang Z, Baldwin PR, Mullapudi S, Penczek PA. 2003. Automated determination of parameters describing power spectra of micrograph images in electron microscopy. Journal of Structural Biology 144(1–2):79–94.
Huxley HE. 1963. Electron microscope studies on structure of natural and synthetic protein filaments from striated muscle. Journal of Molecular Biology 7(3):281–308.

Huxley HE. 1967. Actin is polar by S1 decoration. Journal of Molecular Biology 7:281–308.
Isaacson M. 1977. Specimen damage in the electron microscope. In: Hayat MA, editor. Principles and Techniques of Electron Microscopy: Biological Applications. New York: Van Nostrand Reinhold. pp. 1–78.
Isaacson M, Langmore JP, Rose H. 1974. Determination of nonlocalization of inelastic scattering of electrons by electron microscopy. Optik 41(1):92–96.
Ishizuka K. 1993. Analysis of electron image detection efficiency of slow-scan CCD cameras. Ultramicroscopy 52(1):7–20.
Jaffe JS, Glaeser RM. 1984. Preparation of frozen-hydrated specimens for high-resolution electron microscopy. Ultramicroscopy 13(4):373–377.
Jahn W. 1995. Easily prepared holey films for use in cryoelectron microscopy. Journal of Microscopy (Oxford) 179:333–334.
Jakana J, Chiu W. 2004. Subnanometer imaging of spherical viruses in a JEOL3000SF liquid helium electron cryomicroscope. Microscopy and Microanalysis 10(Supplement 2):1504–1505.
Jakubowski U, Baumeister W, Glaeser RM. 1989. Evaporated carbon stabilizes thin, frozen-hydrated specimens. Ultramicroscopy 31(4):351–356.
James RW. 1982. The Optical Principles of the Diffraction of X-rays. Woodbridge, CT: Ox Bow Press.
Jancarik J, Kim SH. 1991. Sparse-matrix sampling: a screening method for crystallization of proteins. Journal of Applied Crystallography 24:409–411.
Jap BK. 1988. High-resolution electron-diffraction of reconstituted PhoE porin. Journal of Molecular Biology 199(1):229–231.
Jap BK, Glaeser RM. 1978. Scattering of high-energy electrons: 1. Feynman path-integral formulation. Acta Crystallographica A 34:94–102.
Jap BK, Glaeser RM. 1980. Scattering of high-energy electrons: 2. Quantitative validity domains of the single-scattering approximations for organic crystals. Acta Crystallographica A 36:57–67.
Jap BK, Li HL. 1995. Structure of the osmo-regulated H2O-channel, Aqp-Chip, in projection at 3.5 angstrom resolution. Journal of Molecular Biology 251(3):413–420.
Jap BK, Walian PJ, Gehring K. 1991. Structural architecture of an outer-membrane channel as determined by electron crystallography. Nature 350(6314):167–170.
Jap BK, Zulauf M, Scheybani T, Hefti A, Baumeister W, Aebi U, Engel A. 1992. 2D crystallization: from art to science. Ultramicroscopy 46(1–4):45–84.
Jefferson DA, Millward GR, Thomas JM. 1976. Role of multiple-scattering in study of lattice images of graphitic carbons. Acta Crystallographica A 32:823–828.
Jeng TW, Chiu W. 1983. Low-dose electron microscopy of the crotoxin complex thin crystal. Journal of Molecular Biology 164(2):329–346.
Jeng TW, Crowther RA, Stubbs G, Chiu W. 1989. Visualization of alpha-helices in tobacco mosaic virus by cryo-electron microscopy. Journal of Molecular Biology 205(1):251–257.
Jensen GJ. 2001. Alignment error envelopes for single particle analysis. Journal of Structural Biology 133(2–3):143–155.
Jiang W, Baker ML, Ludtke SJ, Chiu W. 2001a. Bridging the information gap: computational tools for intermediate resolution structure interpretation. Journal of Molecular Biology 308(5):1033–1044.
Jiang W, Chiu W. 2006. Cryoelectron microscopy of icosahedral virus particles. In: Kuo J, editor. Methods in Molecular Biology. Totowa, NJ: The Humana Press.
Jiang W, Li ZL, Zhang ZX, Booth CR, Baker ML, Chiu W. 2001b. Semi-automated icosahedral particle reconstruction at sub-nanometer resolution. Journal of Structural Biology 136(3):214–225.
Jiang W, Li ZL, Zhang ZX, Baker ML, Prevelige PE, Chiu W. 2003. Coat protein fold and maturation transition of bacteriophage P22 seen at subnanometer resolutions. Nature Structural Biology 10(2):131–135.
Johnson JE, Fisher AJ. 1994. Principles of virus structure. In: Webster RG, Granoff A, editors. Encyclopedia of Virology. London: Academic Press, pp. 1573–1586.

Jones TA, Kjeldgaard M. 1997. Electron-density map interpretation. Macromolecular Crystallography Pt B:173–208.
Jones TA, Kleywegt GJ, Brunger AT. 1996. Storing diffraction data. Nature 383(6595):18–19.
Jontes JD, Milligan RA, Pollard TD, Ostap EM. 1997. Kinetic characterization of brush border myosin-I ATPase. Proceedings of the National Academy of Sciences of the United States of America 94(26):14332–14337.
Kabius B, Haider M, Uhlemann S, Schwan E, Urban K, Rose H. 2002. First application of a spherical aberration corrected transmission electron microscope in materials science. Journal of Electron Microscopy 51:S51–S58.
Karplus M, Petsko GA. 1990. Molecular dynamics simulations in biology. Nature 347(6294):631–639.
Karrasch S, Typke D, Walz T, Miller M, Tsiotis G, Engel A. 1996. Highly ordered two-dimensional crystals of photosystem I reaction center from Synechococcus sp: functional and structural analyses. Journal of Molecular Biology 262(3):336–348.
Kellenberger E, Haner M, Wurtz M. 1982. The wrapping phenomenon in air-dried and negatively stained preparations. Ultramicroscopy 9(1–2):139–150.
Kikkawa M. 2004. A new theory and algorithm for reconstructing helical structures with a seam. Journal of Molecular Biology 343(4):943–955.
Kimura Y, Vassylyev DG, Miyazawa A, Kidera A, Matsushima M, Mitsuoka K, Murata K, Hirai T, Fujiyoshi Y. 1997. Surface of bacteriorhodopsin revealed by high-resolution electron crystallography. Nature 389(6647):206–211.
Kiselev NA, Sherman MB, Tsuprun VL. 1990. Negative staining of proteins. Electron Microscopy Reviews 3(1):43–72.
Kistler J, Kellenberger E. 1977. Collapse phenomena in freeze-drying. Journal of Ultrastructure Research 59(1):70–75.
Kleywegt GJ, Jones TA. 2002. Homo crystallographicus: quo vadis? Structure 10(4):465–472.
Klug A, Crowther RA. 1972. 3-Dimensional image reconstruction from viewpoint of information theory. Nature 238(5365):435–440.
Klug A, DeRosier DJ. 1966. Optical filtering of electron micrographs: reconstruction of one-sided images. Nature 212(5057):29–32.
Klug A, Crick FHC, Wyckoff HW. 1958. Diffraction by helical structures. Acta Crystallographica 11(3):199–213.
Kong YF, Ma JP. 2003. A structural-informatics approach for mining beta-sheets: locating sheets in intermediate-resolution density maps. Journal of Molecular Biology 332(2):399–413.
Kong YF, Zhang X, Baker TS, Ma JP. 2004. A structural-informatics approach for tracing beta-sheets: building pseudo-C-alpha traces for beta-strands in intermediate-resolution density maps. Journal of Molecular Biology 339(1):117–130.
Koster AJ, Grimm R, Typke D, Hegerl R, Stoschek A, Walz J, Baumeister W. 1997. Perspectives of molecular and cellular electron tomography. Journal of Structural Biology 120(3):276–308.
Krebs A, Edwards PC, Villa C, Li JD, Schertler GFX. 2003. The three-dimensional structure of bovine rhodopsin determined by electron cryomicroscopy. Journal of Biological Chemistry 278(50):50217–50225.
Krivanek OL, Nellist PD, Dellby N, Murfitt MF, Szilagyi Z. 2003. Towards sub-0.5 angstrom electron beams. Ultramicroscopy 96(3–4):229–237.
Kubalek EW, Kornberg RD, Darst SA. 1991. Improved transfer of 2-dimensional crystals from the air–water interface to specimen support grids for high-resolution analysis by electron microscopy. Ultramicroscopy 35(3–4):295–304.
Kubalek EW, Legrice SFJ, Brown PO. 1994. 2-Dimensional crystallization of histidine-tagged, HIV-1 reverse-transcriptase promoted by a novel nickel-chelating lipid. Journal of Structural Biology 113(2):117–123.
Kuhlbrandt W. 1982. Discrimination of protein and nucleic-acids by electron-microscopy using contrast variation. Ultramicroscopy 7(3):221–232.
Kuhlbrandt W. 1992. 2-Dimensional crystallization of membrane proteins. Quarterly Reviews of Biophysics 25(1):1–49.

Kuhlbrandt W, Unwin PNT. 1982. Distribution of RNA and protein in crystalline eukaryotic ribosomes. Journal of Molecular Biology 156(3):431–448.

Kuhlbrandt W, Wang DN. 1991. 3-Dimensional structure of plant light-harvesting complex determined by electron crystallography. Nature 350(6314):130–134.

Kuhlbrandt W, Wang DN, Fujiyoshi Y. 1994. Atomic model of plant light-harvesting complex by electron crystallography. Nature 367(6464):614–621.

Kunji ERS, Harding M. 2003. Projection structure of the atractyloside-inhibited mitochondrial ADP/ATP carrier of Saccharomyces cerevisiae. Journal of Biological Chemistry 278(39):36985–36988.

Lacapere JJ, Stokes DL, Olofsson A, Rigaud JL. 1998. Two-dimensional crystallization of Ca-ATPase by detergent removal. Biophysical Journal 75(3):1319–1329.

Ladenstein R, Schneider M, Huber R, Bartunik HD, Wilson K, Schott K, Bacher A. 1988. Heavy riboflavin synthase from *Bacillus subtilis*: crystal-structure analysis of the icosahedral-beta-60 capsid at 3.3-A resolution. Journal of Molecular Biology 203(4):1045–1070.

Lamvik MK. 1991. Radiation damage in dry and frozen hydrated organic material. Journal of Microscopy (Oxford) 161:171–181.

Landau EM, Rosenbusch JP. 1996. Lipidic cubic phases: a novel concept for the crystallization of membrane proteins. Proceedings of the National Academy of Sciences of the United States of America 93(25):14532–14535.

Langmore JP, Wall J, Isaacson MS. 1973. Collection of scattered electrons in dark field electron microscopy: 1. Elastic scattering. Optik 38(4):335–350.

Lanio S, Rose H, Krahl D. 1986. Test and improved design of a corrected imaging magnetic energy filter. Optik 73(2):56–68.

Lanyi JK, Schobert B. 2002. Crystallographic structure of the retinal and the protein after deprotonation of the Schiff base: the switch in the bacteriorhodopsin photocycle. Journal of Molecular Biology 321(4):727–737.

Lanzavecchia S, Bellon PL. 1996. Electron tomography in conical tilt geometry. The accuracy of a direct Fourier method (DFM) and the suppression of non-tomographic noise. Ultramicroscopy 63(3–4):247–261.

Lanzavecchia S, Bellon PL. 1998. Fast computation of 3D Radon transform via a direct Fourier method. Bioinformatics 14(2):212–216.

Lanzavecchia S, Bellon PL, Scatturin V. 1993. Spark, a kernel of software programs for spatial reconstruction in electron microscopy. Journal of Microscopy (Oxford) 171:255–266.

Larquet E, Boisset N, Pochon F, Lamy J. 1994. Architecture of native human alpha(2)-macroglobulin studied by cryoelectron microscopy and 3-dimensional reconstruction. Journal of Structural Biology 113(1):87–98.

Larsson H, Wallin M, Edstrom A. 1976. Induction of a sheet polymer of tubulin by Zn^{2+}. Experimental Cell Research 100(1):104–110.

Leapman RD, Sun SQ. 1995. Cryoelectron energy-loss spectroscopy: observations on vitrified hydrated specimens and radiation damage. Ultramicroscopy 59(1–4):71–79.

Leapman RD, Brink J, Chiu W. 1993. Low-dose thickness measurement of glucose-embedded protein crystals by electron-energy-loss spectroscopy and stem dark-field imaging. Ultramicroscopy 52(2):157–166.

Lebart L, Maurineau A, Warwick KM. 1984. Multivariate Descriptive Statistical Analysis. New York: Wiley.

Lebeau L, Lach F, Venien-Bryan C, Renault A, Dietrich J, Jahn T, Palmgren MG, Kuhlbrandt W, Mioskowski C. 2001. Two-dimensional crystallization of a membrane protein on a detergent-resistant lipid monolayer. Journal of Molecular Biology 308(4):639–647.

Lembcke G, Durr R, Hegerl R, Baumeister W. 1991. Image analysis and processing of an imperfect 2-dimensional crystal: the surface layer of the archaebacterium *Sulfolobus acidocaldarius* re-investigated. Journal of Microscopy (Oxford) 161:263–278.

Lenz F. 1971. In: Valdre U, editor. Electron Microscopy in Material Science. New York: Academic Press.

Leonard K, Haiker H, Weiss H. 1987. 3-Dimensional structure of NADH–ubiquinone reductase (complex-I) from neurospora mitochondria determined by electron microscopy of membrane crystals. Journal of Molecular Biology 194(2):277–286.

Lepault J, Dubochet J. 1986. Electron microscopy of frozen hydrated specimens: preparation and characteristics. Methods in Enzymology 127:719–730.

Lepault J, Booy FP, Dubochet J. 1983. Electron microscopy of frozen biological suspensions. Journal of Microscopy (Oxford) 129:89–102.

Lepault J, Dargent B, Tichelaar W, Rosenbusch JP, Leonard K, Pattus F. 1988. 3-Dimensional reconstruction of maltoporin from electron-microscopy and image-processing. EMBO Journal 7(1):261–268.

Levy D, Mosser G, Lambert O, Moeck GS, Bald D, Rigaud JL. 1999. Two-dimensional crystallization on lipid layer: a successful approach for membrane proteins. Journal of Structural Biology 127(1):44–52.

Levy D, Chami M, Rigaud JL. 2001. Two-dimensional crystallization of membrane proteins: the lipid layer strategy. Febs Letters 504(3):187–193.

Li HL, Lee S, Jap BK. 1997. Molecular design of aquaporin-1 water channel as revealed by electron crystallography. Nature Structural Biology 4(4):263–265.

Li HL, DeRosier DJ, Nicholson WV, Nogales E, Downing KH. 2002. Microtubule structure at 8 angstrom resolution. Structure 10(10):1317–1328.

Lindahl M, Henderson R. 1997. Structure of the bacteriorhodopsin D85N/D96N double mutant showing substantial structural changes and a highly twinned, disordered lattice. Ultramicroscopy 70(1–2):95–106.

Lipson SG, Lipson H, Tannhauser DS. 1995 Optical Physics. Cambridge: Cambridge University Press.

Liu J, Wendt T, Taylor D, Taylor K. 2003. Refined model of the 10S conformation of smooth muscle myosin by cryo-electron microscopy 3D image reconstruction. Journal of Molecular Biology 329(5):963–972.

Liu WP, Frank J. 1995. Estimation of variance distribution in 3-dimensional reconstruction: 1. Theory. Journal of the Optical Society of America A: Optics Image Science and Vision 12(12):2615–2627.

Liu WP, Boisset N, Frank J. 1995. Estimation of variance distribution in 3-dimensional reconstruction: 2. Applications. Journal of the Optical Society of America A: Optics Image Science and Vision 12(12):2628–2635.

Lovell SC, Davis IW, Adrendall WB, de Bakker PIW, Word JM, Prisant MG, Richardson JS, Richardson DC. 2003. Structure validation by C alpha geometry: phi,psi and C beta deviation. Proteins-Structure Function and Genetics 50(3):437–450.

Lowe J, Li H, Downing KH, Nogales E. 2001. Refined structure of alpha beta-tubulin at 3.5 Å resolution. Journal of Molecular Biology 313(5):1045–1057.

Lucic V, Forster F, Baumeister W. 2005. Structural studies by electron tomography: From cells to molecules. Annual Review of Biochemistry 74:833–865.

Ludtke SJ, Baldwin PR, Chiu W. 1999. EMAN: semiautomated software for high-resolution single-particle reconstructions. Journal of Structural Biology 128(1):82–97.

Ludtke SJ, Jakana J, Song JL, Chuang DT, Chiu W. 2001. A 11.5 angstrom single particle reconstruction of GroEL using EMAN. Journal of Molecular Biology 314(2):253–262.

Ludtke SJ, Chen DH, Song JL, Chuang DT, Chiu W. 2004. Seeing GroEL at 6 angstrom resolution by single particle electron cryomicroscopy. Structure 12(7):1129–1136.

Luecke H, Schobert B, Richter HT, Cartailler JP, Lanyi JK. 1999. Structure of bacteriorhodopsin at 1.55 angstrom resolution. Journal of Molecular Biology 291(4):899–911.

Malhotra A, Penczek P, Agrawal RK, Gabashvili IS, Grassucci RA, Jünemann R, Burkhardt N, Nierhaus KH, Frank J. 1998. *Escherichia coli* 70 S ribosome at 15 angstrom resolution by cryo-electron microscopy: Localization of fMet-tRNA(f)(Met) and fitting of L1 protein. Journal of Molecular Biology 280(1):103–116.

Mancini EJ, Clarke M, Gowen BE, Rutten T, Fuller SD. 2000. Cryo-electron microscopy reveals the functional organization of an enveloped virus, Semliki Forest virus. Molecular Cell 5(2):255–266.

Mannella CA. 1984. Phospholipase-induced crystallization of channels in mitochondrial outer membranes. Science 224(4645):165–166.

Martin-Benito J, Gavilanes F, de los Rios V, Mancheno JM, Fernandez JJ, Gavilanes JG. 2000. Two-dimensional crystallization on lipid monolayers and three-dimensional structure of

sticholysin II, a cytolysin from the sea anemone Stichodactyla helianthus. Biophysical Journal 78(6):3186–3194.
Massover WH, Marsh P. 1997. Unconventional negative stains: heavy metals are not required for negative staining. Ultramicroscopy 69(2):139–150.
Massover WH, Marsh P. 2000. Light atom derivatives of structure-preserving sugars are unconventional negative stains. Ultramicroscopy 85(2):107–121.
Matricardi VR, Moretz RC, Parsons DF. 1972. Electron diffraction of wet proteins: catalase. Science 177(4045):268–270.
McBride JM, Segmuller BE, Hollingsworth MD, Mills DE, Weber BA. 1986. Mechanical stress and reactivity in organic solids. Science 234(4778):830–835.
McPherson A. 1990. Current approaches to macromolecular crystallization. European Journal of Biochemistry 189(1):1–23.
McRee DE. 1993. Practical Protein Crystallography. San Diego, CA: Academic Press.
Meier T, Matthey U, von Ballmoos C, Vonck J, von Nidda TK, Kuhlbrandt W, Dimroth P. 2003. Evidence for structural integrity in the undecameric c-rings isolated from sodium ATP synthases. Journal of Molecular Biology 325(2):389–397.
Metoz F, Wade RH. 1997. Diffraction by helical structures with seams: microtubules. Journal of Structural Biology 118(2):128–139.
Milligan RA, Flicker PF. 1987. Structural relationships of actin, myosin, and tropomyosin revealed by cryoelectron microscopy. Journal of Cell Biology 105(1):29–39.
Milne JLS, Shi D, Rosenthal PB, Sunshine JS, Domingo GJ, Wu XW, Brooks BR, Perham RN, Henderson R, Subramaniam S. 2002. Molecular architecture and mechanism of an icosahedral pyruvate dehydrogenase complex: a multifunctional catalytic machine. EMBO Journal 21(21):5587–5598.
Min GW, Zhou G, Schapira M, Sun TT, Kong XP. 2003. Structural basis of urothelial permeability barrier function as revealed by Cryo-EM studies of the 16 nm uroplakin particle. Journal of Cell Science 116(20):4087–4094.
Mindell JA, Grigorieff N. 2003. Accurate determination of local defocus and specimen tilt in electron microscopy. Journal of Structural Biology 142(3):334–347.
Mindell JA, Maduke M, Miller C, Grigorieff N. 2001. Projection structure of a ClC-type chloride channel at 6.5 angstrom resolution. Nature 409(6817):219–223.
Mitra AK, Vanhoek AN, Wiener MC, Verkman AS, Yeager M. 1995. The chip28 water channel visualized in ice by electron crystallography. Nature Structural Biology 2(9):726–729.
Mitsuoka K, Hirai T, Murata K, Miyazawa A, Kidera A, Kimura Y, Fujiyoshi Y. 1999. The structure of bacteriorhodopsin at 3.0 angstrom resolution based on electron crystallography: implication of the charge distribution. Journal of Molecular Biology 286(3):861–882.
Miyazawa A, Fujiyoshi Y, Stowell M, Unwin N. 1999. Nicotinic acetylcholine receptor at 4.6 angstrom resolution: transverse tunnels in the channel wall. Journal of Molecular Biology 288(4):765–786.
Miyazawa A, Fujiyoshi Y, Unwin N. 2003. Structure and gating mechanism of the acetylcholine receptor pore. Nature 423(6943):949–955.
Mohraz M, Simpson MV, Smith PR. 1987. The 3-dimensional structure of the Na,K-Atpase from electron microscopy. Journal of Cell Biology 105(1):1–8.
Moliere G. 1947. Theorie Der Streuung Schneller Geladener Teilchen: 1. Zeitschrift Fur Naturforschung: Journal of Physical Sciences 2(3):133–145.
Moody MF. 1967a. Structure of sheath of bacteriophage T4: 2. Rearrangement of sheath subunits during contraction. Journal of Molecular Biology 25(2):201.
Moody MF. 1967b. Structure of sheath of bacteriophage T4: 1. Structure of contracted sheath and polysheath. Journal of Molecular Biology 25(2):167.
Moody MF. 1971. Application of optical diffraction to helical structures in bacteriophage tail. Philosophical Transactions of the Royal Society of London B: Biological Sciences 261(837):181.
Moody MF. 1991. Image Analysis of Electron Micrographs. Biophysical Electron Microscopy. New York: Academic Press, pp. 145–287.
Moore PB, Huxley HE, DeRosier DJ. 1970. 3-Dimensional reconstruction of F-actin, thin filaments and decorated thin filaments. Journal of Molecular Biology 50(2):279–295.

Morais MC, Tao YZ, Olson NH, Grimes S, Jardine PJ, Anderson DL, Baker TS, Rossmann MG. 2001. Cryoelectron-microscopy image reconstruction of symmetry mismatches in bacteriophage phi 29. Journal of Structural Biology 135(1):38–46.

Morgan DG, DeRosier D. 1992. Processing images of helical structures: a new twist. Ultramicroscopy 46(1–4):263–285.

Morgan DG, DeRosier DJ. 1997. An evaluation of noisy images of helical structures and the detection of high resolution data. Scanning Microscopy Supplement 11:1–21.

Morgan DG, Owen C, Melanson LA, DeRosier DJ. 1995. Structure of bacterial flagellar filaments at 11 angstrom resolution: packing of the alpha-helices. Journal of Molecular Biology 249(1):88–110.

Mosser G, Mallouh V, Brisson A. 1992. A 9 angstrom 2-dimensional projected structure of cholera-toxin B-subunit-G(M1) complexes determined by electron crystallography. Journal of Molecular Biology 226(1):23–28.

Mouche F, Boisset N, Penczek PA. 2001. Lumbricus terrestris hemoglobin: the architecture of linker chains and structural variation of the central toroid. Journal of Structural Biology 133(2–3):176–192.

Mouche F, Zhu YX, Pulokas J, Potter CS, Carragher B. 2003. Automated three-dimensional reconstruction of keyhole limpet hemocyanin type 1. Journal of Structural Biology 144(3):301–312.

Mu XQ, Egelman EH, Bullitt E. 2002. Structure and function of Hib pili from Haemophilus influenzae type b. Journal of Bacteriology 184(17):4868–4874.

Murata K, Mitsuoka K, Hirai T, Walz T, Agre P, Heymann JB, Engel A, Fujiyoshi Y. 2000. Structural determinants of water permeation through aquaporin-1. Nature 407(6804): 599–605.

Murray JM, Ward R. 1987a. Preparation of holey carbon films suitable for cryoelectron microscopy. Journal of Electron Microscopy Technique 5(3):285–290.

Murray JM, Ward R. 1987b. Principles for the construction and operation of a device for rapidly freezing suspensions for cryoelectron microscopy. Journal of Electron Microscopy Technique 5(3):279–284.

Murray RT, Ferrier RP. 1967. Biological applications of electron diffraction. Journal of Ultrastructure Research 21(5–6):361.

Nabarro FRN. 1987. Theory of Crystal Dislocations. New York: Dover Publications.

Nakagawa A, Miyazaki N, Taka J, Naitow H, Ogawa A, Fujimoto Z, Mizuno H, Higashi T, Watanabe Y, Omura T and others. 2003. The atomic structure of rice dwarf virus reveals the self-assembly mechanism of component proteins. Structure 11(10):1227–1238.

Nakazato K, Toyoshima C, Enami I, Inoue Y. 1996. Two-dimensional crystallization and cryoelectron microscopy of photosystem II. Journal of Molecular Biology 257(2):225–232.

Namba K, Stubbs G. 1986. Structure of tobacco mosaic virus at 3.6-A resolution: implications for assembly. Science 231(4744):1401–1406.

Nettles JH, Li HL, Cornett B, Krahn JM, Snyder JP, Downing KH. 2004. The binding mode of epothilone A on alpha,beta-tubulin by electron crystallography. Science 305(5685):866–869.

Newman RH. 1991. 2-Dimensional crystallization of proteins on lipid monolayers. Electron Microscopy Reviews 4(2):197–203.

Nogales E, Wolf SG, Zhang SX, Downing KH. 1995. Preservation of 2-D crystals of tubulin for electron crystallography. Journal of Structural Biology 115(2):199–208.

Nogales E, Wolf SG, Downing KH. 1997. Visualizing the secondary structure of tubulin: three-dimensional map at 4 angstrom. Journal of Structural Biology 118(2):119–127.

Nogales E, Wolf SG, Downing KH. 1998. Structure of the alpha beta tubulin dimer by electron crystallography. Nature 393(6663):199–203.

Oling F, Bergsma-Schutter W, Brisson A. 2001. Trimers, dimers of trimers, and trimers of trimers are common building blocks of annexin A5 two-dimensional crystals. Journal of Structural Biology 133(1):55–63.

Olofsson A, Mallouh V, Brisson A. 1994. 2-Dimensional structure of membrane-bound annexin-V at 8-angstrom resolution. Journal of Structural Biology 113(3):199–205.

Oostergetel GT, Keegstra W, Brisson A. 1998. Automation of specimen selection and data acquisition for protein electron crystallography. Ultramicroscopy 74(1–2):47–59.

Ostermeier C, Harrenga A, Ernier U, Michel H. 1997. Structure at 2.7 Å resolution of the *Paracoccus denitrificans* two-subunit cytochrome c oxidase complexed with an antibody F_v fragment. Proceedings of the National Academy of Sciences of the United States of America 94(20):10547–10553.
Owen C, DeRosier D. 1993. A 13-angstrom map of the actin scruin filament from the limulus acrosomal process. Journal of Cell Biology 123(2):337–344.
Parsons DF. 1974. Structure of wet specimens in electron microscopy. Science 186(4162): 407–414.
Parsons DF, Martius U. 1964. Determination of alpha-helix configuration of poly-gamma-benzyl-L-glutamate by electron diffraction. Journal of Molecular Biology 10(3):530–533.
Penczek P, Radermacher M, Frank J. 1992. 3-Dimensional reconstruction of single particles embedded in ice. Ultramicroscopy 40(1):33–53.
Penczek P, Zhu J, Schoder R, Frank J. 1997. Three-dimensional reconstruction with contrast transfer compensation from defocus series. Scanning Microscopy. O'Hare (Chicago), IL: Scanning International, pp. 147–154.
Penczek PA. 2002. Three-dimensional spectral signal-to-noise ratio for a class of reconstruction algorithms. Journal of Structural Biology 138(1–2):34–46.
Penczek PA, Grassucci RA, Frank J. 1994. The ribosome at improved resolution: new techniques for merging and orientation refinement in 3D cryoelectron microscopy of biological particles. Ultramicroscopy 53(3):251–270.
Penczek PA, Zhu J, Frank J. 1996. A common-lines based method for determining orientations for N>3 particle projections simultaneously. Ultramicroscopy 63(3–4):205–218.
Penczek PA, Renka R, Schomberg H. 2004. Gridding-based direct Fourier inversion of the three-dimensional ray transform. Journal of the Optical Society of America A: Optics Image Science and Vision 21(4):499–509.
Penczek PA, Yang C, Frank J, Spahn CMT. 2006a. Estimation of variance in single-particle reconstruction using the bootstrap technique. Journal of Structural Biology 154(2):168–183.
Penczek PA, Frank J, Spahn CMT. 2006b. A method of focused classification, based on the bootstrap 3D variance analysis, and its application to EF-G-dependent translocation. Journal of Structural Biology 154(2):184–194.
Perkins GA, Burkard F, Liu E, Glaeser RM. 1993. Glucose alone does not completely hydrate bacteriorhodopsin in glucose-embedded purple membrane. Journal of Microscopy (Oxford) 169:61–65.
Pollard TD. 1986. Rate constants for the reactions of ATP-actin and ADP-actin with the ends of actin filaments. Journal of Cell Biology 103(6):2747–2754.
Porter AB. 1906. On the diffraction theory of microscopic vision. Philosophical Magazine 11:154–166.
Potter CS, Chu H, Frey B, Green C, Kisseberth N, Madden TJ, Miller KL, Nahrstedt K, Pulokas J, Reilein A and others. 1999. Leginon: a system for fully automated acquisition of 1000 electron micrographs a day. Ultramicroscopy 77(3–4):153–161.
Potter CS, Zhu Y, Carragher B. 2004. Journal of Structural Biology (special issue).
Pratt WK. 1991. Digital Image Processing. New York: John Wiley.
Rachel R, Jakubowski U, Tietz H, Hegerl R, Baumeister W. 1986. Projected structure of the surface protein of *Deinococcus radiodurans* determined to 8 Å resolution by cryomicroscopy. Ultramicroscopy 20(3):305–316.
Radermacher M. 1980. Dreidimensionale Rekonstruktion bei kagelförmiger Kippung im Elektronenmikroskop. Munich: Technical University.
Radermacher M. 1988. 3-Dimensional reconstruction of single particles from random and nonrandom tilt series. Journal of Electron Microscopy Technique 9(4):359–394.
Radermacher M. 1991. Three-dimensional reconstruction of single particles in electron microscopy. In: Haeder D-P, editor. Image analysis in biology. Boca Raton, FL: CRC Press.
Radermacher M. 1992. Weighted back-projection methods. In: Frank J, editor. Electron Tomography. New York: Plenum.
Radermacher M. 1994. 3-Dimensional reconstruction from random projections: orientational alignment via Radon transforms. Ultramicroscopy 53(2):121–136.

Radermacher M, Wagenknecht T, Verschoor A, Frank J. 1987. 3-Dimensional reconstruction from a single-exposure, random conical tilt series applied to the 50S ribosomal subunit of *Escherichia coli*. Journal of Microscopy (Oxford) 146:113–136.
Rash JE, Hudson CS. 1979. Freeze Fracture: Methods, Artifacts, and Interpretations. New York: Raven Press.
Rath BK, Frank J. 2004. Fast automatic particle picking from cryo-electron micrographs using a locally normalized cross-correlation function: a case study. Journal of Structural Biology 145(1–2):84–90.
Rayment I, Holden HM, Whittaker M, Yohn CB, Lorenz M, Holmes KC, Milligan RA. 1993. Structure of the actin–myosin complex and its implications for muscle contraction. Science 261(5117):58–65.
Reimer LR. 1989. Transmission Electron Microscopy. Berlin: Springer-Verlag.
Ren G, Cheng A, Reddy V, Melnyk P, Mitra AK. 2000. Three-dimensional fold of the human AQP1 water channel determined at 4 angstrom resolution by electron crystallography of two-dimensional crystals embedded in ice. Journal of Molecular Biology 301(2):369–387.
Rhodes G. 1993. Crystallography Made Crystal Clear: A Guide for Users of Macromolecular Models. San Diego, CA: Academic Press.
Ribi HO, Reichard P, Kornberg RD. 1987. Two-dimensional crystals of enzyme effector complexes: ribonucleotide reductase at 18 Å resolution. Biochemistry 26(24):7974–7979.
Ribi HO, Ludwig DS, Mercer KL, Schoolnik GK, Kornberg RD. 1988. 3-Dimensional structure of cholera-toxin penetrating a lipid membrane. Science 239(4845):1272–1276.
Rice WJ, Young HS, Martin DW, Sachs JR, Stokes DL. 2001. Structure of Na^+,K^+-ATPase at 11-angstrom resolution: comparison with Ca^{2+}-ATPase in E-1 and E-2 states. Biophysical Journal 80(5):2187–2197.
Rigaud JL, Mosser G, Lacapere JJ, Olofsson A, Levy D, Ranck JL. 1997. Bio-beads: an efficient strategy for two-dimensional crystallization of membrane proteins. Journal of Structural Biology 118(3):226–235.
Ringler P, Muller W, Ringsdorf H, Brisson A. 1997. Functionalized lipid tubules as tools for helical crystallization of proteins. Chemistry: a European Journal 3(4):620–625.
Ringler P, Heymann B, Engel A. 2000. Two-dimensional crystallization of membrane proteins. In: Baldwin SA, editor. Membrane Transport: A Practical Approach. Oxford: Oxford University Press.
Robinson JP, Schmid MF, Morgan DG, Chiu W. 1988. 3-Dimensional structural-analysis of tetanus toxin by electron crystallography. Journal of Molecular Biology 200(2):367–375.
Rodgers DW. 1997. Practical cryocrystallography. Macromolecular Crystallography Pt A: 183–203.
Rose A. 1948. Television Camera Tubes and the Problem of Vision. Advances in Electronics and Electron Physics. New York: Academic Press, pp. 131–166.
Roseman AM. 2000. Docking structures of domains into maps from cryo-electron microscopy using local correlation. Acta Crystallographica D: Biological Crystallography 56: 1332–1340.
Roseman AM. 2003. Particle finding in electron micrographs using a fast local correlation algorithm. Ultramicroscopy 94(3–4):225–236.
Rosenthal PB, Henderson R. 2003. Optimal determination of particle orientation, absolute hand, and contrast loss in single-particle electron cryomicroscopy. Journal of Molecular Biology 333(4):721–745.
Rossmann MG, Tao YZ. 1999. Cryo-electron microscopy reconstruction of partially symmetric objects. Journal of Structural Biology 125(2–3):196–208.
Rost LE, Hanein D, DeRosier DJ. 1998. Reconstruction of symmetry deviations: a procedure to analyze partially decorated F-actin and other incomplete structures. Ultramicroscopy 72(3–4):187–197.
Ruiz T, Ranck JL, Diazavalos R, Caspar DLD, DeRosier DJ. 1994. Electron diffraction of helical particles. Ultramicroscopy 55(4):383–395.
Saad A, Ludtke SJ, Jakana J, Rixon FJ, Tsuruta H, Chiu W. 2001. Fourier amplitude decay of electron cryomicroscopic images of single particles and effects on structure determination. Journal of Structural Biology 133(1):32–42.

Sali A, Glaeser R, Earnest T, Baumeister W. 2003. From words to literature in structural proteomics. Nature 422(6928):216–225.
Samudzi CT, Fivash MJ, Rosenberg JM. 1992. Cluster analysis of the biological macromolecule crystallization database. Journal of Crystal Growth 123(1–2):47–58.
Sander B, Golas MM, Stark H. 2003. Automatic CTF correction for single particles based upon multivariate statistical analysis of individual power spectra. Journal of Structural Biology 142(3):392–401.
Saxton WO. 1978. Computer Techniques for Image Processing in Electron Microscopy. New York: Academic Press.
Saxton WO, Frank J. 1977. Motif detection in quantum noise-limited electron micrographs by cross-correlation. Ultramicroscopy 2(2–3):219–227.
Saxton WO, Baumeister W. 1982. The correlation averaging of a regularly arranged bacterial-cell envelope protein. Journal of Microscopy (Oxford) 127:127–138.
Schatz M, van Heel M. 1990. Invariant classification of molecular views in electron micrographs. Ultramicroscopy 32(3):255–264.
Schatz M, Orlova A, Dube P, Stark H, Zemlin F, van Heel M. 1997. Angular reconstitution in three-dimensional electron microscopy: practical and technical aspects. Scanning Microscopy. O'Hare (Chicago), IL: Scanning International, pp. 179–193.
Scheres SHW, Gao H, Valle M, Herman GT, Eggermont PPB, Frank J, Carazo JM. 2007. Disentangling conformational states of macromolecules in 3D cryo-EM through likelihood optimization. Nature Methods 4(1):20–21.
Scherzer O. 1947. Sphärische und chromatische Korrektur von Elektronen-Linsen. Optik 2(2):114–132.
Scherzer O. 1949. The theoretical resolution limit of the electron microscope. Journal of Applied Physics 20(1):20–29.
Schiff LI. 1956. Approximation method for high-energy potential scattering. Physical Review 103(2):443–453.
Schiske P. 1968. Zur Frage der Bildrekonstruktion durch Fokusreihen [Regarding the question of image reconstruction using focus series]; Rome, pp. 145–146.
Schmid MF, Dargahi R, Tam MW. 1993a. Spectra: a system for processing electron images of crystals. Ultramicroscopy 48(3):251–264.
Schmid MF, Robinson JP, Dasgupta BR. 1993b. Direct visualization of botulinum neurotoxin-induced channels in phospholipid vesicles. Nature 364(6440):827–830.
Schmidt-Krey I, Murata K, Hirai T, Mitsuoka K, Cheng YF, Morgenstern R, Fujiyoshi Y, Hebert H. 1999. The projection structure of the membrane protein microsomal glutathione transferase at 3 angstrom resolution as determined from two-dimensional hexagonal crystals. Journal of Molecular Biology 288(2):243–253.
Schmidt-Krey I, Mitsuoka K, Hirai T, Murata K, Cheng Y, Fujiyoshi Y, Morgenstern R, Hebert H. 2000. The three-dimensional map of microsomal glutathione transferase 1 at 6 angstrom resolution. EMBO Journal 19(23):6311–6316.
Schmitt L, Dietrich C, Tampe R. 1994. Synthesis and characterization of chelator-lipids for reversible immobilization of engineered proteins at self-assembled lipid interfaces. Journal of the American Chemical Society 116(19):8485–8491.
Serysheva II, Orlova EV, Chiu W, Sherman MB, Hamilton SL, van Heel M. 1995. Electron cryomicroscopy and angular reconstitution used to visualize the skeletal-muscle calcium-release channel. Nature Structural Biology 2(1):18–24.
Seymour J, DeRosier DJ. 1987. The projection of a negatively stained filamentous object down its central axis as revealed by image-reconstruction from tilt series. Journal of Microscopy (Oxford) 148:195–210.
Shaw PJ, Hills GJ. 1981. Tilted specimen in the electron microscope: a simple specimen holder and the calculation of tilt angles for crystalline specimens. Micron 12(3):279–282.
Sherman MB, Brink J, Chiu W. 1996. Performance of a slow-scan CCD camera for macromolecular imaging in a 400 kV electron cryomicroscope. Micron 27(2):129–139.
Shi D, Hsiung HH, Pace RC, Stokes DL. 1995. Preparation and analysis of large, flat crystals of Ca^{2+}-Atpase for electron crystallography. Biophysical Journal 68(3):1152–1162.

Skriver E, Maunsbach AB, Jorgensen PL. 1981. Formation of two-dimensional crystals in pure membrane-bound Na^+,K^+-Atpase. Febs Letters 131(2):219–222.

Smith DJ, Saxton WO, O'Keefe MA, Wood GJ, Stobbs WM. 1983. The importance of beam alignment and crystal tilt in high-resolution electron microscopy. Ultramicroscopy 11(4):263–281.

Smith PR, Aebi U. 1974. Computer-generated Fourier transforms of helical particles. Journal of Physics A: Mathematical and General 7(13):1627–1633.

Smith PR, Ivanov IE. 1980. Surface reliefs computed from micrographs of isolated heavy-metal shadowed particles. Journal of Ultrastructure Research 71(1):25–36.

Smith PR, Aebi U, Josephs R, Kessel M. 1976. Studies of structure of T4-bacteriophage tail sheath: 1. Recovery of 3-dimensional structural information from extended sheath. Journal of Molecular Biology 106(2):243–271.

Solodukhin AS, Caldwell HL, Sando JJ, Kretsinger RH. 2002. Two-dimensional crystal structures of protein kinase C-delta, its regulatory domain, and the enzyme complexed with myelin basic protein. Biophysical Journal 82(5):2700–2708.

Sommerfeld A. 1964. Vorlesungen fuer Theoretische Physik: Mechanik. Leipzig: Akademische Verlagsgesellschaft, Geest & Partig KG.

Sorzano COS, Marabini R, Boisset N, Rietzel E, Schröder R, Herman GT, Carazo JM. 2001. The effect of overabundant projection directions on 3D reconstruction algorithms. Journal of Structural Biology 133(2–3):108–118.

Sosinsky G, Schekman R, Glaeser RM. 1986. Morphological observations on the formation and stability of the crystalline arrays in the plasma membrane of *Saccharomyces cerevisiae*. Journal of Ultrastructure and Molecular Structure Research 94(1):37–51.

Spence JCH. 1981. Experimental High-Resolution Electron Microscopy. Oxford: Clarendon Press.

Stauffer KA, Hoenger A, Engel A. 1992. 2-Dimensional crystals of *Escherichia coli* maltoporin and their interaction with the maltose-binding protein. Journal of Molecular Biology 223(4):1155–1165.

Stewart M. 1988. Computer image-processing of electron micrographs of biological structures with helical symmetry. Journal of Electron Microscopy Technique 9(4):325–358.

Stewart M, Kensler RW, Levine RJC. 1981. Structure of limulus telson muscle thick filaments. Journal of Molecular Biology 153(3):781–790.

Stewart PL, Burnett RM, Cyrklaff M, Fuller SD. 1991. Image reconstruction reveals the complex molecular organization of adenovirus. Cell 67(1):145–154.

Stokes DL, Green NM. 1990. 3-Dimensional crystals of CaATPase from sarcoplasmic reticulum: symmetry and molecular packing. Biophysical Journal 57(1):1–14.

Stout GH, Jensen LH. 1989. X-ray Structure Determination. New York: John Wiley.

Subramaniam S, Henderson R. 1999. Electron crystallography of bacteriorhodopsin with millisecond time resolution. Journal of Structural Biology 128(1):19–25.

Subramaniam S, Gerstein M, Oesterhelt D, Henderson R. 1993. Electron diffraction analysis of structural changes in the photocycle of bacteriorhodopsin. EMBO Journal 12(1):1–8.

Sun WQ, Leopold AC, Crowe LM, Crowe JH. 1996. Stability of dry liposomes in sugar glasses. Biophysical Journal 70(4):1769–1776.

Tahara Y, Ohnishi S, Fujiyoshi Y, Kimura Y, Hayashi Y. 1993. A pH induced 2-dimensional crystal of membrane-bound Na^+,K^+-Atpase of dog kidney. Febs Letters 320(1):17–22.

Tate CG, Ubarretxena-Belandia I, Baldwin JM. 2003. Conformational changes in the multidrug transporter EmrE associated with substrate binding. Journal of Molecular Biology 332(1):229–242.

Taylor KA. 1978. Structure determination of frozen, hydrated, crystalline biological specimens. Journal of Microscopy (Oxford) 112:115–125.

Taylor KA, Glaeser RM. 1974. Electron diffraction of frozen, hydrated protein crystals. Science 186(4168):1036–1037.

Taylor KA, Glaeser RM. 1976. Electron microscopy of frozen hydrated biological specimens. Journal of Ultrastructure Research 55(3):448–456.

Taylor KA, Taylor DW. 1993. Projection image of smooth-muscle alpha-actinin from 2-dimensional crystals formed on positively charged lipid layers. Journal of Molecular Biology 230(1):196–205.

Taylor KA, Varga S. 1994. Similarity of 3-dimensional microcrystals of detergent-solubilized (Na^+,K^+)-Atpase from pig kidney and Ca^{2+}-ATPase from skeletal muscle sarcoplasmic reticulum. Journal of Biological Chemistry 269(13):10107–10111.

Taylor KA, Deatherage JF, Amos LA. 1982. Structure of the S-layer of *Sulfolobus acidocaldarius*. Nature 299(5886):840–842.

Taylor KA, Dux L, Martonosi A. 1986. 3-Dimensional reconstruction of negatively stained crystals of the Ca^{2+}-Atpase from muscle sarcoplasmic reticulum. Journal of Molecular Biology 187(3):417–427.

Thon F. 1966. Zur Defokussierungsabhängigkeit des Phasenkontrastes bei der elektronenmikroskopischen Abbildung. Zeitschrift für Naturforschung A: Astrophysik Physik und Physikalische Chemie A 21(4):476–478.

Thuman-Commike PA, Chiu W. 1997. Improved common line-based icosahedral particle image orientation estimation algorithms. Ultramicroscopy 68(4):231–255.

Thuman-Commike PA, Chiu W. 2000. Reconstruction principles of icosahedral virus structure determination using electron cryomicroscopy. Micron 31(6):687–711.

Thuman-Commike PA, Tsuruta H, Greene B, Prevelige PE, King J, Chiu W. 1999. Solution X-ray scattering-based estimation of electron cryomicroscopy imaging parameters for reconstruction of virus particles. Biophysical Journal 76(4):2249–2261.

Topf M, Baker ML, John B, Chiu W, Sali A. 2005. Structural characterization of components of protein assemblies by comparative modeling and electron cryo-microscopy. Journal of Structural Biology 149(2):191–203.

Toyoshima C. 1989. On the use of holey grids in electron crystallography. Ultramicroscopy 30(3):439–443.

Trachtenberg S, DeRosier DJ. 1987. 3-Dimensional structure of the frozen hydrated flagellar filament: the left-handed filament of *Salmonella typhimurium*. Journal of Molecular Biology 195(3):581–601.

Trachtenberg S, DeRosier DJ, Macnab RM. 1987. 3-Dimensional structure of the complex flagellar filament of *Rhizobium lupini* and its relation to the structure of the plain filament. Journal of Molecular Biology 195(3):603–620.

Trachtenberg S, DeRosier DJ, Zemlin F, Beckmann E. 1998. Non-helical perturbations of the flagellar filament: *Salmonella typhimurium* SJW117 at 9.6 angstrom resolution. Journal of Molecular Biology 276(4):759–773.

Tulloch PA, Colman PM, Davis PC, Laver WG, Webster RG, Air GM. 1986. Electron and x-ray diffraction studies of Influenza neuraminidase complexed with monoclonal antibodies. Journal of Molecular Biology 190(2):215–225.

Typke D, Nordmeyer RA, Jones A, Lee JY, Avila-Sakar A, Downing KH, Glaeser RM. 2005. High-throughput film-densitometry: an efficient approach to generate large data sets. Journal of Structural Biology 149(1):17–29.

Unger VM, Kumar NM, Gilula NB, Yeager M. 1999a. Expression, two-dimensional crystallization, and electron cryo-crystallography of recombinant gap junction membrane channels. Journal of Structural Biology 128(1):98–105.

Unger VM, Kumar NM, Gilula NB, Yeager M. 1999b. Three-dimensional structure of a recombinant gap junction membrane channel. Science 283(5405):1176–1180.

Unser M, Trus BL, Steven AC. 1987. A new resolution criterion based on spectral signal-to-noise ratios. Ultramicroscopy 23(1):39–51.

Unwin N. 1993. Nicotinic acetylcholine receptor at 9 angstrom resolution. Journal of Molecular Biology 229(4):1101–1124.

Unwin N. 2005. Refined structure of the nicotinic acetylcholine receptor at 4 angstrom resolution. Journal of Molecular Biology 346(4):967–989.

Unwin PNT. 1974. Electron microscopy of stacked disk aggregate of tobacco mosaic virus protein: 2. Influence of electron irradiation on stain distribution. Journal of Molecular Biology 87(4):657.

Unwin PNT, Henderson R. 1975. Molecular structure determination by electron microscopy of unstained crystalline specimens. Journal of Molecular Biology 94(3):425–440.

Unwin PNT, Muguruma J. 1971. Transmission electron microscopy of ice. Journal of Applied Physics 42(9):3640.

Uzgiris EE, Kornberg RD. 1983. Two-dimensional crystallization technique for imaging macromolecules, with application to antigen–antibody complement complexes. Nature 301(5896):125–129.
Vainshtein BK. 1964. Structure Analysis by Electron Diffraction. Feigl EaS, JA, translator. New York: Macmillan.
Vainshtein BK. 1966. Diffraction of X-rays by Chain Molecules. Amsterdam: Elsevier.
Vainshtein BK. 1981. Modern Crystallography: I. Symmetry of Crystals, Methods of Structural Crystallography. Cardona M, Fulde P, Queisser HJ, editors. Berlin: Springer-Verlag.
Valle M, Sengupta J, Swami NK, Grassucci RA, Burkhardt N, Nierhaus KH, Agrawal RK, Frank J. 2002. Cryo-EM reveals an active role for aminoacyl-tRNA in the accommodation process. EMBO Journal 21(13):3557–3567.
Valle M, Zavialov A, Li W, Stagg SM, Sengupta J, Nielsen RC, Nissen P, Harvey SC, Ehrenberg M, Frank J. 2003. Incorporation of aminoacyl-tRNA into the ribosome as seen by cryo-electron microscopy. Nature Structural Biology 10(11): 899–906. [Corrigendum 10(12):1074.
Valpuesta JM, Carrascosa JL, Henderson R. 1994. Analysis of electron microscope images and electron diffraction patterns of thin crystals of Phi29-connectors in ice. Journal of Molecular Biology 240(4):281–287.
Valpuesta JM, Fernandez JJ, Carazo JM, Carrascosa JL. 1999. The three-dimensional structure of a DNA translocating machine at 10 angstrom resolution. Structure 7(3): 289–296.
van Bruggen EFJ, van Breemen JFL, Keegstra W, Boekema EJ, van Heel MG. 1986. Two-dimensional crystallization experiments. Journal of Microscopy 141(1):11–20.
Van Dyck D. 1975. The path integral formalism as a new description for the diffraction of high-energy electrons in crystals. Physica Status Solidi B 72:321–326.
van Heel M. 1984. Three-dimensional reconstruction with unknown angular relationships. Budapest: Electron Microscopy Foundation, pp. 1347–1348.
van Heel M. 1987. Angular reconstitution: *a posteriori* assignment of projection directions for 3-D reconstruction. Ultramicroscopy 21(2):111–123.
van Heel M. 1987. Similarity measures between images. Ultramicroscopy 21(1):95–100.
van Heel M, Frank J. 1981. Use of multivariate statistics in analyzing the images of biological macromolecules. Ultramicroscopy 6(2):187–194.
van Heel M, Keegstra W, Schutter W, van Bruggen EFJ. 1982. In: Wood EJ, editor. Structure and Function of Invertebrate Respiratory Proteins. Reading, UK: Harwood Academic.
van Heel M, Harauz G, Orlova EV, Schmidt R, Schatz M. 1996. A new generation of the IMAGIC image processing system. Journal of Structural Biology 116(1):17–24.
van Heel M, Gowen B, Matadeen R, Orlova EV, Finn R, Pape T, Cohen D, Stark H, Schmidt R, Schatz M et al. 2000. Single-particle electron cryo-microscopy: towards atomic resolution. Quarterly Reviews of Biophysics 33(4):307–369.
Venien-Bryan C, Balavoine F, Toussaint B, Mioskowski C, Hewat EA, Helme B, Vignais PM. 1997. Structural study of the response regulator HupR from *Rhodobacter capsulatus*. Electron microscopy of two-dimensional crystals on a nickel-chelating lipid. Journal of Molecular Biology 274(5):687–692.
Venien-Bryan C, Schertler GFX, Thouvenin E, Courty S. 2000. Projection structure of a transcriptional regulator, HupR, determined by electron cryo-microscopy. Journal of Molecular Biology 296(3):863–871.
Voges D, Berendes R, Burger A, Demange P, Baumeister W, Huber R. 1994. 3-Dimensional structure of membrane-bound annexin-V: a correlative electron microscopy–x-ray crystallography study. Journal of Molecular Biology 238(2):199–213.
Volkmann N, Hanein D. 1999. Quantitative fitting of atomic models into observed densities derived by electron microscopy. Journal of Structural Biology 125(2–3):176–184.
Volkmann N, Hanein D. 2003. Docking of atomic models into reconstructions from electron microscopy. Macromolecular Crystallography Pt D:204–225.
Vonck J. 1996. A three-dimensional difference map of the N intermediate in the bacteriorhodopsin photocycle: part of the F helix tilts in the M to N transition. Biochemistry 35(18): 5870–5878.

Vonck J. 2000. Structure of the bacteriorhodopsin mutant F219L N intermediate revealed by electron crystallography. EMBO Journal 19(10):2152–2160.

Vonck J, Han BG, Burkard F, Perkins GA, Glaeser RM. 1994. Two progressive substates of the M-intermediate can be identified in glucose-embedded, wild-type bacteriorhodopsin. Biophysical Journal 67(3):1173–1178.

Wade RH, Frank J. 1977. Electron microscope transfer-functions for partially coherent axial illumination and chromatic defocus spread. Optik 49(1):81–92.

Wagenknecht T, DeRosier D, Shapiro L, Weissborn A. 1981. 3-Dimensional reconstruction of the flagellar hook from *Caulobacter crescentus*. Journal of Molecular Biology 151(3): 439–465.

Wagenknecht T, Grassucci R, Frank J. 1988. Electron microscopy and computer image averaging of ice-embedded large ribosomal-subunits from Escherichia coli. Journal of Molecular Biology 199(1):137–147.

Walian PJ, Jap BK. 1990. 3-Dimensional electron diffraction of PhoE porin to 2.8 Å resolution. Journal of Molecular Biology 215(3):429–438.

Wall J, Isaacson M, Langmore JP. 1974. Collection of scattered electrons in dark field electron microscopy: 2. Inelastic scattering. Optik 39(4):359–374.

Walz T, Grigorieff N. 1998. Electron crystallography of two-dimensional crystals of membrane proteins. Journal of Structural Biology 121(2):142–161.

Walz T, Typke D, Smith BL, Agre P, Engel A. 1995. Projection map of aquaporin-1 determined by electron crystallography. Nature Structural Biology 2(9):730–732.

Walz T, Hirai T, Murata K, Heymann JB, Mitsuoka K, Fujiyoshi Y, Smith BL, Agre P, Engel A. 1997. The three-dimensional structure of aquaporin 1. Nature 387(6633):624–627.

Wang DN, Sarabia VE, Reithmeier RAF, Kuhlbrandt W. 1994. 3-Dimensional map of the dimeric membrane domain of the human erythrocyte anion-exchanger, Band-3. EMBO Journal 13(14):3230–3235.

Weber PC. 1991. Physical principles of protein crystallization. Advances in Protein Chemistry 41:1–36.

Weber PC. 1997. Overview of protein crystallization methods. Macromolecular Crystallography Pt A:13–22.

Weinkauf S, Bacher A, Baumeister W, Ladenstein R, Huber R, Bachmann L. 1991. Correlation of metal decoration and topochemistry on protein surfaces. Journal of Molecular Biology 221(2):637–645.

Wendt T, Taylor D, Trybus KM, Taylor K. 2001. Three-dimensional image reconstruction of dephosphorylated smooth muscle heavy meromyosin reveals asymmetry in the interaction between myosin heads and placement of subfragment 2. Proceedings of the National Academy of Sciences of the United States of America 98(8):4361–4366.

Williams KA. 2000. Three-dimensional structure of the ion-coupled transport protein NhaA. Nature 403(6765):112–115.

Williams KA, Geldmacher-Kaufer U, Padan E, Schuldiner S, Kuhlbrandt W. 1999. Projection structure of NhaA, a secondary transporter from *Escherichia coli*, at 4.0 angstrom resolution. EMBO Journal 18(13):3558–3563.

Williams RC, Wyckoff RWG. 1946. Applications of metallic shadow-casting to microscopy. Journal of Applied Physics 17(1):23–33.

Wilmsen HU, Leonard KR, Tichelaar W, Buckley JT, Pattus F. 1992. The aerolysin membrane channel is formed by heptamerization of the monomer. EMBO Journal 11(7): 2457–2463.

Wilson AJC. 1942. Determination of absolute from relative x-ray intensity data. Nature 150: 151–152.

Wilson-Kubalek EM. 2000. Preparation of functionalized lipid tubules for electron crystallography of macromolecules. Sphingolipid Metabolism and Cell Signaling Pt B:515–519.

Wilson-Kubalek EM, Brown RE, Celia H, Milligan RA. 1998. Lipid nanotubes as substrates for helical crystallization of macromolecules. Proceedings of the National Academy of Sciences of the United States of America 95(14):8040–8045.

Wolf SG, Mosser G, Downing KH. 1993. Tubulin conformation in zinc-induced sheets and macrotubes. Journal of Structural Biology 111(3):190–199.

Wolf SG, Nogales E, Kikkawa M, Gratzinger D, Hirokawa N, Downing KH. 1996. Interpreting a medium-resolution model of tubulin: comparison of zinc-sheet and microtubule structure. Journal of Molecular Biology 262(4):485–501.

Woodward JT, Zasadzinski JA. 1996. Thermodynamic limitations on the resolution obtainable with metal replicas. Journal of Microscopy (Oxford) 184:157–162.

Wooster WA. 1964. Microdensitometry applied to x-ray photographs. Acta Crystallographica 17(7):878–882.

Wriggers W, Birmanns S. 2001. Using Situs for flexible and rigid-body fitting of multiresolution single-molecule data. Journal of Structural Biology 133(2–3):193–202.

Wriggers W, Milligan RA, McCammon JA. 1999. Situs: a package for docking crystal structures into low-resolution maps from electron microscopy. Journal of Structural Biology 125(2–3):185–195.

Wrigley NG. 1968. Lattice spacing of crystalline catalase as an internal standard of length in electron microscopy. Journal of Ultrastructure Research 24(5–6):454–464.

Wynne SA, Crowther RA, Leslie AGW. 1999. The crystal structure of the human hepatitis B virus capsid. Molecular Cell 3(6):771–780.

Xian YJ, Hebert H. 1997. Three-dimensional structure of the porcine gastric H,K-ATPase from negatively stained crystals. Journal of Structural Biology 118(3):169–177.

Yang C, Ng EG, Penczek PA. 2005. Unified 3-D structure and projection orientation refinement using quasi-Newton algorithm. Journal of Structural Biology 149(1):53–64.

Yonekura K, Maki-Yonekura S, Namba K. 2003. Complete atomic model of the bacterial flagellar filament by electron cryomicroscopy. Nature 424(6949):643–650.

Yonekura K, Maki-Yonekura S, Namba K. 2005. Building the atomic model for the bacterial flagellar filament by electron cryomicroscopy and image analysis. Structure 13(3):407–412.

Yoshimura H, Scheybani T, Baumeister W, Nagayama K. 1994. 2-Dimensional protein array growth in thin-layers of protein solution on aqueous subphases. Langmuir 10(9):3290–3295.

Zahn R, Harris JR, Pfeifer G, Pluckthun A, Baumeister W. 1993. 2-Dimensional crystals of the molecular chaperone Groel reveal structural plasticity. Journal of Molecular Biology 229(3):579–584.

Zemlin F, Weiss K, Schiske P, Kunath W, Herrmann KH. 1978. Coma-free alignment of high-resolution electron microscopes with aid of optical diffractograms. Ultramicroscopy 3(1):49–60.

Zemlin J, Zemlin F. 2002. Diffractogram tableaux by mouse click. Ultramicroscopy 93(1):77–82.

Zernike F. 1955. How I discovered phase contrast. Science 121(3141):345–349.

Zhang PJ, Hinshaw JE. 2001. Three-dimensional reconstruction of dynamin in the constricted state. Nature Cell Biology 3(10):922–926.

Zhang PJ, Toyoshima C, Yonekura K, Green NM, Stokes DL. 1998. Structure of the calcium pump from sarcoplasmic reticulum at 8 angstrom resolution. Nature 392(6678):835–839.

Zhong SJ, Dadarlat VM, Glaeser RM, Head-Gordon T, Downing KH. 2002. Modeling chemical bonding effects for protein electron crystallography: the transferable fragmental electrostatic potential (TFESP) method. Acta Crystallographica A 58:162–170.

Zhou ZH, Chiu W. 2003. Determination of icosahedral virus structures by electron cryomicroscopy at subnanometer resolution. Advances in Protein Chemistry 64:93–124.

Zhou ZH, Prasad BVV, Jakana J, Rixon FJ, Chiu W. 1994. Protein subunit structures in the herpes-simplex virus a capsid determined from 400-kV spot-scan electron cryomicroscopy. Journal of Molecular Biology 242(4):456–469.

Zhou ZH, Hardt S, Wang B, Sherman MB, Jakana J, Chiu W. 1996. CTF determination of images of ice-embedded single particles using a graphics interface. Journal of Structural Biology 116(1):216–222.

Zhou ZH, Chiu W, Haskell K, Spears H, Jakana J, Rixon FJ, Scott LR. 1998. Refinement of herpesvirus B-capsid structure on parallel supercomputers. Biophysical Journal 74(1):576–588.

Zhou ZH, Dougherty M, Jakana J, He J, Rixon FJ, Chiu W. 2000. Seeing the herpesvirus capsid at 8.5 angstrom. Science 288(5467):877–880.

Zhou ZH, Baker ML, Jiang W, Dougherty M, Jakana J, Dong G, Lu GY, Chiu W. 2001a. Electron cryomicroscopy and bioinformatics suggest protein fold models for rice dwarf virus. Nature Structural Biology 8(10):868–873.

Zhou ZH, Liao WC, Cheng RH, Lawson JE, McCarthy DB, Reed LJ, Stoops JK. 2001b. Direct evidence for the size and conformational variability of the pyruvate dehydrogenase complex revealed by three-dimensional electron microscopy: the "breathing" core and its functional relationship to protein dynamics. Journal of Biological Chemistry 276(24):21704–21713.

Zhu J, Penczek PA, Schröder R, Frank J. 1997. Three-dimensional reconstruction with contrast transfer function correction from energy-filtered cryoelectron micrographs: procedure and application to the 70S Escherichia coli ribosome. Journal of Structural Biology 118(3): 197–219.

Zuo JM, Vartanyants I, Gao M, Zhang R, Nagahara LA. 2003. Atomic resolution imaging of a carbon nanotube from diffraction intensities. Science 300(5624):1419–1421.

Index

Abbe's diffraction theory of image formation, 50–52, 60
Aberrations, lens, 51
  chromatic, 115, 118
  field, 263
    pincushion distortion, 121
    spiral distortion, 121
  spherical, 56, 68, 115, 119
Acetyl choline receptor, nicotinic, 9, 196–197, 280
Actin, 11, 153, 201, 292
Airy pattern (Airy disc), 62, 87
Aliasing, 242–244
Alignment, 371–374
  3D projection, 374, 399–401, 404–407, 413
  conditions for, 369
  optimization of, 374
Amplitude
  of a cosine or sine function, 20–21
Amplitude object, 54
  weak, 72
Angular increment in angular refinement, 400
Angular reconstitution, 395, 397–399, 401
Angular refinement methods, 399–401, 407, 410
Angular relationships, 366
Angular space, coverage of, 369–370

Angular spacing, 366
Angular statistics, 405–406
Anomalous intensity difference, x-ray diffraction, 42–43
Anticontaminator, 125
Aperture, condenser lens, 111–115, 133–134, 136
Aperture, objective lens, 52, 61, 116
  astigmatism due to contamination of, 118
  reduction of specimen charging by, 137
Apodization, 330, 412
Aquaporin, 146, 196, 199–200
Argand diagram, 40–42, 53
Astigmatism, axial, 68, 118, 410
Atomic force microscopy, 368
Autocorrelation function, 373, 388
  rotational, 388
Average image, 380
  Fourier averaging, 102
  spatial averaging, 104
  see also Correlation averaging

B-factor, 32–33, 344, 352, 354, 359, 364
Back-projection, 340–341, 369, 389–392, 394, 407
Band limit, 241, 402
Bacteriorhodopsin, 4, 7, 15, 195, 205, 219, 224, 237, 255, 288, 295–298, 302

470 INDEX

Beam-induced movement, 159, 350
Beam tilt, corrections for, 276
    see also Coma-free imaging
Bessel function, 315
    orthogonality of, 315–316
Bilinear interpolation, 395
Bootstrapping of a reconstruction, 368
Born approximation, 78, 416
Bragg spacing, 37
Bragg's law, 36
Bright-field image, elastic, 410
Brightness, 110
Bubbling effect, see Radiation damage
Butterworth filter, 412

Ca-ATPase, 196, 201, 288
Cage effect, see Radiation damage
Capsomeres, viral, 347
Carbon film, 139, 141–142, 146, 158–160, 165–166, 402
    holey, 139–140, 141–142, 146, 158–160, 165–166
Catalase, 14, 145, 151–152, 201
CCD (charge coupled device) cameras, 10, 120, 122, 127–132, 212
Central section
    in Fourier space, 81, 369, 391
    see also Projection theorem
Charging, specimen, 137, 160
    see also Aperture, objective lens
Chirality, 168, 176–178, 395
Circumferential vector, 307
Class averages, 397
Classification, 367, 374–375
    dendrogram, 385, 386
    of the double-correlation function, 388
    hierarchical ascendant, 385–388
    by K-means, 385–388
    of orientations, 404–407
    of rotational power spectra, 388
    supervised, 374, 399, 413
    unsupervised, 374
Closure error in phasing, 42
Clustering, 380, 382, 383
Coherence
    partial, 55
    spatial (transverse), 55, 69, 109, 111, 351
    temporal (longitudinal), 55, 69, 71, 109–110
Coma-free imaging, 115

Comb function, 29, 260
Common frame of reference, 371
Common-lines method, 255–257, 356, 395, 397–399
    cross common lines, 358–360
    multiple common lines, 356–359, 401
    self common lines, 358–360
Condenser lens, 111–113
Conformational bifurcation, 370
Conformational states, 368
Conformational variability, 370, 414
    see also Heterogeneity
Conjugate product, 371
Contour maps, 279
Contrast matching, 147–150
Contrast transfer function (CTF), 116, 212, 236, 270, 351, 401
    correction of, 352–354, 366, 368, 395, 401, 410
    determination of, 212, 401–403
    effect on the cross-correlation function, 373
Convolution operation, 31
Convolution product, 27
Convolution theorem, 30–31
Correlation averaging, 103
Correspondence analysis, 377, 381–383
Coulomb potential, shielded, 5, 11, 79–80, 100
Cowley-Moodie multislice approximation, 418
Cross-correlation coefficient, 399
    locally normalized, 403, 404
Cross-correlation function, 369, 371, 372, 400
    effect of defocus on, 373
    translational, 371
Cross-correlation operation, 6
Cryo-crinkling, 166
Crystal packing, effects of, 366, 367
Crystallization, protein, 18
Crystallography, x-ray 17–48
Crystals, helical, 335
    formation on tubular lipid vesicles, 206–207, 306
Crystals, two-dimensional, 5, 167, 194
    formed on lipid monolayers, 204–210
    membrane proteins, 195–201
    S-layers, bacterial, 143, 196
    soluble proteins, 201–203
CTF, see Contrast transfer function
CTFFIND2 (program), 402

Data cloud, in multi-dimensional space, 376, 382
Data collection, strategy of, 374
Debye–Waller thermal parameter, 32–33
Deconvolution, 62
Defects, crystalline
  disclinations, 191
  dislocations, 191
  distortion, 265
  line, 191
  point, 191
  vacancies, 191
Defocus, 51, 68, 116
  determination of, 401–403
  groups, 401–404, 407, 410
  phase shift due to, 117
  Scherzer value, 68
  variation, 115, 401, 402
Delta function, Dirac, 26
  definition of, 28
Densitometer, see Microdensitometer
Density
  histogram, 238, 404
  normalization, 404
Density maps, three-dimensional, 7
  interpretation, chain-tracing, 44–46, 288–290
Depth of field, 116, 426
Detective quantum efficiency, 127–129, 217
Detector systems, 126–131
  dynamic range of, 129
Difference Fourier map, 47, 213, 300–303, 339
Differential phase residual (DPR), 408
Diffraction, electron, 12
  background subtraction, 222–223
  dynamical, 78, 416
  intensities, 212
  kinematic, 78, 416
  radial background, 219–221
  see also Scattering
Diffraction-limited images, 52, 57, 61–62
Dirac, see Delta function
Disorder
  helical, 319–325
    axial, 322
    curvature, 322
    flattening, 323
    radial, 323
    shearing, 324
    shrinkage, 324

  long-range, 34, 187, 188–189
  mosaic model of, 34
  short-range, 32, 187–188
Docking atomic models into EM maps, 12, 278, 291–293, 413
  see also Models, pseudoatomic
DPR, see Differential phase residual

EF-G, 367
Eigenimages, 380
Eigenvalue equation, 382
Eigenvalue problem, 377
Eigenvalues, 377, 378, 380, 381
  histogram of, 377, 383, 384
Eigenvector, 377, 380, 381
Electron density, 19
Electron dose, minimum, 368–369
  see also Radiation damage, critical dose
Electron exposure, 368
Electron gun,
  field emission, 108, 110
  lanthanum hexaboride, 108, 110–111, 135
  thermionic, 108, 11, 135
Electron source, see Electron gun
Electron tomography, 123, 388, 389
Electronic readout, 127, 130–131, 217, 370
EMAN (image processing system), 352
Embedding in glucose, 365
  see also Glucose-embedding
Embedding in ice, 365, 382
  see also Vitrification
Energy filter lens, 122–123, 232
  use of, 435, 438
Energy loss, 369, 432–433, 437
Energy spread, 110
Envelope function, 50, 71, 402
Equator (helical diffraction), 310
Euclidean distance, generalized, 371–372, 376
Euler identity, 25, 64
Eulerian angles, 354–357, 359, 389, 390, 392, 399, 400, 404
Ewald sphere, 34–36
  curvature of, 428–430

Factors, 381, 382
  maximum number of, 381
Factor maps, 379, 380, 382, 383
Fast Fourier transform algorithm, 43, 244–245, 371
Fermi filter, 412

Feynman path integral approximation, 417
FFT, see Fast Fourier transform algorithm
Figure of merit in phasing, 42
Flattening of specimen, 141, 145, 365
Flexibility of structural components, 366
Floating images, 330
Fluorescence spectra, 368
Fog, photographic background, 216
Fourier–Bessel transform, 317–318, 348–349
Fourier component, 23
Fourier ring correlation (FRC), 408
Fourier series, 21
Fourier shell correlation (FSC), 407–410
Fourier shift theorem, 22, 253–255
Fourier transform, 19, 25
Fraunhofer diffraction, 54
Free R-factor, 47, 291, 294
  see also R-factor
Freezing, protein crystals, 19
Fresnel diffraction, 55, 426–428
Friedel difference, 232
Friedel pairs, 224, 230
Friedel symmetry, 65–67, 437–440
Friedel's law, 66–67
Frozen-hydrated specimens, 140, 142, 150–159, 165
FSC, see Fourier shell correlation

Glass transition, 156
Glucose embedding, 145–147, 365
Grain boundaries, 193
GroEL, 413

Heavy-atom derivative, 39
Helical lattice, 307
  back side (far side), 310, 331
  front side (near side), 309, 331
  hand of, 312
  pitch of, 312
  seam in, 339
Helical reconstruction, single-particle method, 340–342
Hemocyanin, 386, 409
Heterogeneity,
  of images, 407
  of molecular species, 367
  of sample, 374
  of structure/conformation, 370, 401, 407, 413

Homogeneity, structural, 365, 366
Hybrid method of density-map interpretation, 278, 291–293

Icosahedron, description of, 344–347
Illumination, electron, 112–113
  alignment of, 114
  crossover formed by, 113
  flood-beam, 132
  pivot points for tilt and translation of, 114
  spot-scan, 132, 134–137
    dynamic focus with, 136
    effective coherence with, 135
    reduced beam-induced motion with, 135
  spot size of, 114
  tilt of, 114
Image formation, diffraction theory of, 12
  see also Abbe's diffraction theory of image formation
Image processing, 11, 62
Image quality (IQ), 257–259, 261
Image readout, see Electronic readout
Image restoration, 62
Impulse response, see Point-spread function
Indexing
  diffraction spots, 82, 220–222
  spots in a computed Fourier transform, 246–247
  see also Layer lines (helical diffraction)
Inelastic scattering, 402, 410
  energy-loss spectrum of, 433
  and its contribution to image formation, 432–437
  and no-loss electrons, 431, 435
  see also Scattering
Inner potential, 148
IQ, see Image quality
Isomorphous heavy atom replacement, 39, 212
Isoplanitic images, 59, 273
Isosurface display, 279

Kinetic studies, see Time-resolved studies

L7/L12 stalk, 367
Latent image spec, photographic, 215
Lattice, crystallographic, 168
Lattice function, 26
  reciprocal, 31
Lattice, helical, see Helical lattice

Layer lines (helical diffraction), 310–315
  height of, 325
  indexing of, 315, 325
  order of, 312, 326–330
  overlap of Bessel functions on, 318, 325
  selection rules on, 317
Lens law, 115
Linear energy transfer, 94, 96
Linear system,
  description of image formation, 59–63
  gain of, 59
  phase shift of, 59
Low-dose image technique, 131
Low-pass filtering, 410–412

Magnification, 118
Mass absorption function (coefficient), 73
Mass loss, *see* Radiation damage
Membrane proteins, integral, 195–201
  detergent solubilization of, 197
  negative purification of, 197
  reconstitution of, 197–200
Meridian (helical diffraction), 310
Metal shadowing, 142–145
  decoration during, 144
Microdensitometer, 213, 237–238
Miller indices, 23, 32, 183, 220, 228, 246
Missing cone of data, 89, 282–287, 397
Models, pseudo-atomic, 11
  see also Docking atomic models into EM maps
Modulation transfer function (MTF), 129–130
  of the digitization process, 239–244
Moliere approximation, 417
Momentum vector, 35
Motif, 21
Multivariate data analysis, 372, 374–384
Myosin, 11–12, 153, 155, 204, 206, 292

Negative staining, 140–142, 365, 369, 382
  cryo-negative staining, 142
Net (2-D lattice), 168–171, 177, 180
NMR, 413
Noise amplification, 62
Nucleation, crystal, 18
Nyquist frequency, 240, 242
Nyquist limit, 130, 240

Objective lens, 115–120
  focal length, 115
  see also Aberrations, lens

Optical density, 214, 215, 237–238
  dependence on electron exposure, 216
Optical diffraction, 235–237
Optical filtering (quasi), 259–263
Orientation determination, 374
  by 3-D projection matching, 404
  by common lines/angular reconstitution, 395
Orientational classes, 404
Orientational preferences, 382, 401, 405–407
Orientational search, 371
Overfocus, 117
Oversampling, 244

Partial occupancy, 413
Particles,
  centering of, 355, 371
  minimum number of, 368, 370
  minimum size of, 368, 369
  picking of, 403–404
  preferred orientation of, 350
  single, 5
Patterson difference map, 40
Period of a cosine or sine function, 20–21
Periodogram, 402
Phase
  of a cosine or sine function, 20–21
  crystallographic, 63
  error due to beam tilt, 114
  problem in crystallography, 26, 82
  ramp, 23
Phase Contrast Transfer Function, 50, 67–72
Phase distortion, $\gamma(s)$, 51, 68, 116
Phase extension, 295
Phase object, 52
  strong, 417
  weak, 50, 53, 63
Phase origin, 253, 255–257
  see also Fourier shift theorem
Phase residual, 295, 357
  amplitude-weighted, 331
Photographic film, 215
Point-spread function, 57, 61
  three-dimensional, 86, 287–288
Polar coordinate system, 371
Polar structures, 309
Power spectrum,
  1-D profile of, 402
  computation of, 401–403
  rotational, 388

Preferential orientation, *see* Particles, preferred orientation of
Principal component analysis, 377, 381, 385
Projection
 common-line, 397
 reference, 399, 400
 two-dimensional, 63
 weighted, 393
Projection geometry, 388–390
Projection theorem, 64, 81–82, 395
Projector lens system, 120–122
Protein biosynthesis, 366
Protein Data Bank (PDB), 48
Purple membrane, *see* Bacteriorhodopsin

R-factor
 crystallographic, 46, 290
 merging, 228
 see also Free R-factor
R^*-weighting, *see* Weighting function
Radiation damage, 6, 93–100, 368, 369
 bubbling effect in, 99
 cage effect in, 98
 critical dose for, 94, 133
 mass loss due to, 97
Radon transform, 395, 399
Random-conical data collection, 395–387, 401
Reciprocal lattice, *see* Lattice function, reciprocal
Reciprocal lattice basis vectors, 31
Reciprocal lattice lines, 83, 161, 183, 226–29
Reciprocal lattice vectors, 221
Reciprocal space, 23, 35
Reconstitution of images, 380
Reconstruction,
 algebraic reconstruction techniques (ART), 369, 393, 394
 artifacts in, directional, 407
 by back-projection, 389
 by Fourier techniques, 394, 407
 by matrix inversion, 388–389
 merging of reconstructions, 410–411
 number of projections needed for, 369, 370
 reproducibility of, 408
 by simultaneous iterative reconstruction technique (SIRT), 394
 strategy for, 374
 techniques for, 388–395
 three-dimensional, 11
 using a real-space algorithm for, 369, 394
 by weighted back-projection, 369, 391–392
Rectangle function, 27, 34, 61
 cone-shaped, 89, 283
 cubic, 92
 spherically shaped, 85
 star-shaped, 91
Refinement, crystallographic, 46–47, 212, 293–300
 including dynamical diffraction, 424–426
Refinement, single particle reconstruction
 see Alignment *and see* Angular refinement methods
Region of influence in Fourier space, 369
Reliability, standard t-test for, 338
Resolution, 43, 370, 400
 atomic, 368, 413–414
 crystallographic (Fourier), definition of, 43, 86
 criteria used for, 408–410
 estimation of, 407, 408–410
 in Fourier space, 369
 loss of, 399, 401
 Rayleigh's criterion of, 43, 86–89
Ribosome, 366, 367, 368, 383, 400, 410, 411, 413, 414
 large subunit of, 366
 projection data set of, 382
Rigid body movements, 365
Rocking of molecules, 382
Rose equation, 101
Rotation matrices, 389
Ryanodine receptor/calcium release channel, 398, 399

S-layer, *see* Crystals, two-dimensional
Scattering,
 dynamical, 78
 elastic, 368
 inelastic, 219, 368, 369
  imaginary component of the Coulomb potential, 232
  mean free path for, 430–431
  see also Inelastic scattering
 kinematic, 78
Scattering cross-section,
 atomic, 368
 differential, 436
Scattering factors, atomic, 212
 chemical bonding effect for, 296–300
Screening of data, 370

INDEX   475

Secondary structure elements, 370
Self-organized maps, 375
Shape function, 34, 248
Shape transform, 369, 395
Shannon sampling, 92
　Whittaker–Shannon sampling
　　theorem, 241
Shrinkage of specimen, 365
Signal-to-noise ratio (SNR), 370, 408
　in the cross-correlation function, 404
　requirement for alignment and
　　classification, 367
　spectral (SSNR), 408–410
Similarity of images, 385
Sinc function, 92, 184, 239, 240, 391
Single side band, 74
Sinogram, 397–399
SIRT, see Simultaneous iterative
　　reconstruction technique
SNR, see Signal-to-noise ratio
Sommerfeld, 389
Space groups, 175–182
Spatial frequency of a cosine or sine function,
　　20–21, 23
Spatial frequency vector, 23, 35
Specimen preparation
　frozen-hydrated, 13
　　see also Frozen-hydrated specimens
　glucose embedded, 13, 151, 159, 165
　　see also Glucose embedding
　see also Negative staining
Specimen stages, 123–126
　eucentric, 123
　low-temperature, 125
　side-entry, 123
　top-entry, 123
Speed, photographic, 217
SPIDER (image processing system), 402
Spot-scan imaging, see Illumination, electron
SSNR, see Signal-to-noise ratio, spectral
Statistical requirement for imaging, 369
　see also Rose equation
Structural homogeneity, see Homogeneity,
　　structural
Structure factor, 25
　average, from solution scattering, 252
　extraction of, 247–250
　modulus (amplitude) of, 26, 85, 212
　phase of, 38
Subgroups, symmetry, 182
Supergroups, symmetry, 182

Symmetry
　crystallographic, 168
　helical, 307
　icosahedral, 344–347
　layer group, 175–178
　line group, 177
　n-fold rotation, 307
　noncrystallographic, 182
　plane group, 175–178
　　two-sided, 177
　point group, 169–171
　screw-axis, 307

Temperature factor, 32, 188
　relative, 224
Thon patterns (Thon rings), 270
Tilt, beam (illumination), 75
Tilt transfer function (TTF), 273–276
Time-resolved studies, 157
Transfer function, 49, 59
　complex-valued, 60–61
　see also Contrast Transfer Function (CTF)
　see also Phase Contrast Transfer Function
Translational search, 371
Transmittance function
　lens, 55
　object, 55
Triangulation number, icosahedral, 345–347
Truncation artifacts, 412
Tubulin, 10, 45, 202, 228–229, 231, 262, 281,
　　283, 288–289, 293, 296
Twinning, 189–191, 233

Unbending, 264–270
Uncertainty principle, 369
Underfocus, 117
Unit cell, 21, 167
Unit cell vectors, 31

Variability, structural, 365
　see also Heterogeneity
Variance
　bootstrap method for estimation of, 413
　of an image, 372
　intra-class, 385
　local, 404
　three-dimensional, 374, 413
　total, of a data set, 381
Variance/covariance matrix, 381
Variational energy, 380–382
Views, minimum number of, 368

Virus particles, 402
  helical, 304–305, 339
  icosahedral, 344, 347–348, 350, 362–364
Vitrification, 153–155
Voltage, accelerating, 110

Ward's criterion, 385
Wave length
  electron, 109, 417
  x-ray, 37
Wave-front aberration, *see* Phase distortion
Weighting function, 392, 393

Wenzel–Kramers–Brillouin approximation, 417
Wiener filtering, 272, 353, 401, 410–411
Wilson plot, 34, 224
Wire-frame display, 279
Wooster effect, 214
Wrinkling (imperfect flatness), 161–166

x-ray crystallography, 393, 413

Zernike phase contrast microscope, 52–54
Zone axis, 255